Handbook of Environmental Engineering

Volume 17

Series Editors

Lawrence K. Wang
PhD, Rutgers University, New Brunswick, New Jersey, USA
MS, University of Rhode Island, Kingston, Rhode Island, USA
MSCE, Missouri University of Science and Technology, Rolla, Missouri, USA
BSCE, National Cheng Kung University, Tainan, Taiwan

Mu-Hao Sung Wang
PhD, Rutgers University, New Brunswick, New Jersey, USA
MS, University of Rhode Island, Kingston, Rhode Island, USA
BSCE, National Cheng Kung University, Tainan, Taiwan

More information about this series at http://www.springer.com/series/7645

Lawrence K. Wang • Mu-Hao Sung Wang
Yung-Tse Hung • Nazih K. Shammas
Editors

Natural Resources
and Control Processes

 Springer

Editors
Lawrence K. Wang
Lenox Institute of Water Technology
Newtonville, NY, USA

Mu-Hao Sung Wang
Lenox Institute of Water Technology
Newtonville, NY, USA

Yung-Tse Hung
Cleveland State University
Cleveland, OH, USA

Nazih K. Shammas
Lenox Institute of Water Technology
Pasadena, CA, USA

Handbook of Environmental Engineering
ISBN 978-3-319-26798-2   ISBN 978-3-319-26800-2   (eBook)
DOI 10.1007/978-3-319-26800-2

Library of Congress Control Number: 2016941950

Printed on acid-free paper

This Springer imprint is published by Springer Nature
The registered company is Springer International Publishing AG Switzerland

Preface

The past four decades have seen the emergence of a growing desire worldwide that positive actions be taken to restore and protect the environment from the degrading effects of all forms of pollution – air, water, soil, thermal, radioactive, and noise. Since pollution is a direct or indirect consequence of waste, the seemingly idealistic demand for "zero discharge" can be construed as an unrealistic demand for zero waste. However, as long as waste continues to exist, we can only attempt to abate the subsequent pollution by converting it to a less noxious or reusable form. Three major questions usually arise when a particular type of pollution has been identified: (1) How serious are the environmental pollution and natural resources crisis? (2) Is the technology to abate them available? (3) Do the costs of abatement justify the degree of abatement achieved for environmental protection and natural resources conservation? This book is one of the volumes of the *Handbook of Environmental Engineering* series. The principal intention of this series is to help readers formulate answers to the above three questions.

The traditional approach of applying tried-and-true solutions to specific environmental and natural resources problems has been a major contributing factor to the success of environmental engineering and has accounted in large measure for the establishment of a "methodology of pollution control." However, the realization of the ever-increasing complexity and interrelated nature of current environmental problems renders it imperative that intelligent planning of pollution abatement and resources recovery systems be undertaken. Prerequisite to such planning is an understanding of the performance, potential, and limitations of the various methods of environmental protection available for environmental scientists and engineers. In this series of handbooks, we will review at a tutorial level a broad spectrum of engineering systems (natural environment, processes, operations, and methods) currently being utilized, or of potential utility, for pollution abatement, natural resources conservation and ecological protection. We believe that the unified interdisciplinary approach presented in these handbooks is a logical step in the evolution of environmental engineering.

Treatment of the various engineering systems presented will show how an engineering formulation of the subject flows naturally from the fundamental principles and theories of chemistry, microbiology, physics, and mathematics. This emphasis on fundamental science recognizes that engineering practice has in recent years become more firmly based on scientific principles rather than on its earlier dependency on empirical accumulation of facts. It is not intended, though, to neglect empiricism where such data lead quickly to the most economic design; certain engineering systems are not readily amenable to fundamental scientific analysis, and in these instances we have resorted to less science in favor of more art and empiricism.

Since an environmental natural resources engineer must understand science within the context of applications, we first present the development of the scientific basis of a particular subject, followed by exposition of the pertinent design concepts and operations and detailed explanations of their applications to natural resources conservation or environmental protection. Throughout the series, methods of mathematical modeling, system analysis, practical design, and calculation are illustrated by numerical examples. These examples clearly demonstrate how organized, analytical reasoning leads to the most direct and clear solutions. Wherever possible, pertinent cost data have been provided.

Our treatment of environmental natural resources engineering is offered in the belief that the trained engineer should more firmly understand fundamental principles, be more aware of the similarities and/or differences among many of the engineering systems, and exhibit greater flexibility and originality in the definition and innovative solution of environmental system problems. In short, the environmental and natural resources engineers should by conviction and practice be more readily adaptable to change and progress.

Coverage of the unusually broad field of environmental natural resources engineering has demanded an expertise that could only be provided through multiple authorships. Each author (or group of authors) was permitted to employ, within reasonable limits, the customary personal style in organizing and presenting a particular subject area; consequently, it has been difficult to treat all subject materials in a homogeneous manner. Moreover, owing to limitations of space, some of the authors' favored topics could not be treated in great detail, and many less important topics had to be merely mentioned or commented on briefly. All authors have provided an excellent list of references at the end of each chapter for the benefit of the interested readers. As each chapter is meant to be self-contained, some mild repetition among the various texts was unavoidable. In each case, all omissions or repetitions are the responsibility of the editors and not the individual authors. With the current trend toward metrication, the question of using a consistent system of units has been a problem. Wherever possible, the authors have used the British system (fps) along with the metric equivalent (mks, cgs, or SIU) or vice versa. The editors sincerely hope that this redundancy of units' usage will prove to be useful rather than being disruptive to the readers.

The goals of the *Handbook of Environmental Engineering* series are (1) to cover entire environmental fields, including air and noise pollution control, solid waste processing and resource recovery, physicochemical treatment processes, biological treatment processes, biotechnology, biosolids management, flotation technology, membrane technology, desalination technology, natural resources, natural control processes, radioactive waste disposal, hazardous waste management, and thermal pollution control, and (2) to employ a multimedia approach to environmental conservation and protection since air, water, soil, and energy are all interrelated.

This book (Volume 17) and its future sister book (Volume 18) of the *Handbook of Environmental Engineering* series have been designed to serve as natural resources engineering reference books as well as supplemental textbooks. We hope and expect they will prove of equal high value to advanced undergraduate and graduate students, to designers of natural resources systems, and to scientists and researchers. The editors welcome comments from readers in all of these categories. It is our hope that the two natural resources engineering books will not only provide information on water resources engineering but will also serve as a basis for advanced study or specialized investigation of the theory and analysis of various natural resources systems.

This book, *Natural Resources and Control Processes* (Volume 17), covers the topics on management of agricultural livestock wastes for water resources protection, application of natural processes for environmental protection, proper deep well waste disposal, treating and managing industrial dye wastes, health effects and control of toxic lead in the environment, municipal and industrial wastewater treatment using plastic trickling filters for BOD and nutrient removal, chloride removal for recycling fly ash from municipal solid waste incinerator, recent evaluation of early radioactive disposal and management practice, recent trends in the evaluation of cementitious material in radioactive waste disposal, extensive monitoring system of sediment transport for reservoir sediment management, and land and energy resources engineering glossary.

The editors are pleased to acknowledge the encouragement and support received from Dr. Sherestha Saini, editor (Environmental Sciences) of the Springer Science + Business Media, and her colleagues, during the conceptual stages of this endeavor. We wish to thank all contributing authors for their time and effort and for having patiently borne our reviews and numerous queries and comments. We are very grateful to our respective families for their patience and understanding during some rather trying times.

Newtonville, NY, USA   Lawrence K. Wang
Newtonville, NY, USA   Mu-Hao Sung Wang
Cleveland, OH, USA   Yung-Tse Hung
Pasadena, CA, USA   Nazih K. Shammas

Contents

1 Management of Livestock Wastes for Water Resource Protection 1
Dale H. Vanderholm, Donald L. Day, Arthur J. Muehling,
Yung-Tse Hung, and Erick Butler

2 Application of Natural Processes for Environmental Protection 73
Nazih K. Shammas and Lawrence K. Wang

3 Proper Deep-Well Waste Disposal for Water Resources Protection 119
Nazih K. Shammas, Lawrence K. Wang, and Charles W. Sever

4 Treatment and Management of Industrial Dye Wastewater for Water Resources Protection 187
Erick Butler, Yung-Tse Hung, Mohammed Al Ahmad,
and Yen-Pei Fu

5 Health Effects and Control of Toxic Lead in the Environment 233
Nancy Loh, Hsue-Peng Loh, Lawrence K. Wang,
and Mu-Hao Sung Wang

6 Municipal and Industrial Wastewater Treatment Using Plastic Trickling Filters for BOD and Nutrient Removal 285
Jia Zhu, Frank M. Kulick III, Larry Li, and Lawrence K. Wang

7 Chlorides Removal for Recycling Fly Ash from Municipal Solid Waste Incinerator 349
Fenfen Zhu, Masaki Takaoka, Chein-Chi Chang,
and Lawrence K. Wang

8   Recent Evaluation of Early Radioactive Disposal Practice   371
     Rehab O. Abdel Rahman, Andrey Guskov,
     Matthew W. Kozak, and Yung-Tse Hung

**9   Recent Trends in the Evaluation of Cementitious
     Material in Radioactive Waste Disposal**   401
     Rehab O. Abdel Rahman and Michael I. Ojovan

**10   Extensive Monitoring System of Sediment Transport
     for Reservoir Sediment Management**   449
     Chih-Ping Lin, Chih-Chung Chung, I-Ling Wu,
     Po-Lin Wu, Chun-Hung Lin, and Ching-Hsien Wu

11   Glossary of Land and Energy Resources Engineering   493
     Mu-Hao Sung Wang and Lawrence K. Wang

Index   625

Editors and Contributors

Erick Butler School of Engineering and Computer Science and Mathematics, West Texas A&M University, Canyon, TX, USA

Chein-Chi Chang DC Water and Sewer Authority, Washington, DC, USA

University of Maryland, Baltimore, MD, USA

Chih-Chung Chung Disaster Prevention and Water Environment Research Center, National Chiao Tung University, Hsinchu, Taiwan, ROC

Donald L. Day Agricultural Engineering Department, University of Illinois, Urbana, Champaign, IL, USA

Andrey Guskov Division for Safety of Fuel Cycle Facilities, Scientific and Engineering Centre for Nuclear and Radiation Safety, Moscow, Russian Federation

Yung-Tse Hung Department of Civil and Environmental Engineering, Cleveland State University, Cleveland, OH, USA

Frank M. Kulick Research and Development Department, Brentwood, Reading, PA, USA

Matthew W. Kozak Intera Inc., Richland, WA, USA

Larry Li Water Group, Brentwood, Reading, PA, USA

Chun-Hung Lin Disaster Prevention and Water Environment Research Center, National Chiao Tung University, Hsinchu, Taiwan, ROC

Chih-Ping Lin Department of Civil Engineering, National Chiao Tung University, Hsinchu, Taiwan, ROC

Hsue-Peng Loh Wenko Systems Analysis, Pittsburgh, PA, USA

Nancy Loh Wenko Systems Analysis, Washington, DC, USA

Arthur J. Muehling Agricultural Engineering Department, University of Illinois, Champaign, IL, USA

Michael I. Ojovan Immobilization Science Laboratory, Department of Material Science and Engineering, The University of Sheffield, South Yorkshire, UK

Rehab O. Abdel Rahman Hot Laboratory and Waste Management Centre, Atomic Energy Authority of Egypt, Inshas, Cairo, Egypt

Charles W. Sever US Environmental Protection Agency, Dallas, TX, USA

Nazih K. Shammas Lenox Institute of Water Technology, Pasadena, CA, USA

Krofta Engineering Corporation, Lenox, MA, USA

Masaki Takaoka Environmental Engineering Department, Graduate School of Engineering, Kyoto University, Kyoto, Japan

Dale H. Vanderholm Agricultural Research Division, Institute of Agricultural and Natural Resources, University of Nebraska, Lincoln, NE, USA

Lawrence K. Wang Rutgers University, New Brunswick, NJ, USA

Lenox Institute of Water Technology, Newtonville, NY, USA

Mu-Hao Sung Wang Rutgers University, New Brunswick, NJ, USA

Lenox Institute of Water Technology, Newtonville, NY, USA

Ching-Hsien Wu Water Resources Planning Institute, Water Resources Agency, Taichung, Taiwan, ROC

I-Lin Wu Disaster Prevention and Water Environment Research Center, National Chiao Tung University, Hsinchu, Taiwan, ROC

Po-Lin Wu Disaster Prevention and Water Environment Research Center, National Chiao Tung University, Hsinchu, Taiwan, ROC

Fenfen Zhu Department of Environmental Science and Engineering, Renmin University of China, Beijing, China

Jia Zhu Water Group, Brentwood, Reading, PA, USA

Chapter 1
Management of Livestock Wastes for Water Resource Protection

Dale H. Vanderholm, Donald L. Day, Arthur J. Muehling,
Yung-Tse Hung, and Erick Butler

Contents

1	Introduction	4
	1.1 Federal Regulations	6
	1.2 State Regulations	7
2	Wastewater Characteristics	10
	2.1 General Characteristics of Wastewater	10
	2.2 Milk House Wastewater Characteristics	14
3	Waste Treatment	19
	3.1 Anaerobic Digestion	19
	3.2 Constructed Wetlands	24
	3.3 Lagoons	28
	3.4 Thermal and Biological Chemical Treatment for Biogas Production	33
	3.5 Composting	35

D.H. Vanderholm
Agricultural Research Division, Institute of Agricultural and Natural Resources,
University of Nebraska, 1665 Marsh Hawk Cir, Castle Rock, CO 80109-9593, USA
e-mail: dvanderholm@gmail.com

D.L. Day
Agricultural Engineering Department, University of Illinois, UCUI, 1796 AQero-Place,
Urbana, IL 61802-9500, USA
e-mail: DL-Day@illinois.edu

A.J. Muehling
Agricultural Engineering Department, University of Illinois, UCUI, 1501 W. John St,
Champaign, IL 61821, USA
e-mail: amuehlin@illinois.edu

Y.-T. Hung
Department of Civil and Environmental Engineering, Cleveland State University,
Cleveland, OH 44115, USA
e-mail: yungtsehung@yahoo.com; yungtsehung@gmail.com

E. Butler (✉)
School of Engineering, Computer Science, and Mathematics, West Texas A&M University,
Canyon, TX 79016, USA
e-mail: erick.ben.butler@gmail.com

© Springer International Publishing Switzerland 2016
L.K. Wang, M.-H.S. Wang, Y.-T. Hung and N.K. Shammas (eds.),
Natural Resources and Control Processes, Handbook of Environmental
Engineering, Volume 17, DOI 10.1007/978-3-319-26800-2_1

3.6 Vermicomposting 37
3.7 Summary 37
4 Land Application of Livestock Wastes 37
4.1 Description 37
4.2 Manure Handling Equipment 38
4.3 Time of Application 42
4.4 Rate of Application 42
4.5 Summary 43
5 Storage of Livestock Wastes 43
5.1 Description 43
5.2 Storage Time 43
5.3 Facilities to Store Livestock Waste 44
5.4 Storage Area Design 47
5.5 Summary 47
6 Feedlot Runoff Control Systems 48
6.1 Description 48
6.2 Runoff Control Systems 48
6.3 Summary 52
7 Odors and Gases 53
7.1 Odors: Origin and Nature 53
7.2 Sources of Odors 53
7.3 Odor Prevention 55
7.4 Greenhouse Gas Emissions 59
8 Pathogens in Livestock Industries 61
9 Livestock Waste Management Computer Software 63
10 Conclusion 64
References 65

Abstract Livestock waste management has recently become a topic of interest. Due to the increasing desire in transforming waste products into profit, it is necessary to have clear knowledge and understanding of how to handle livestock waste. This livestock waste management chapter provides insight on some treatment methods that have become popular within physical, chemical, and biological treatment methods and the design techniques to incorporate many of these methods. Finally, an introduction to some of the more modern techniques in harnessing energy from agriculture waste and its potential profits has been included. Having consulted technical papers from university extensions across such as the University of Missouri, Utah State University, North Carolina State, and materials published by the US Department of Agriculture and the US Environmental Protection Agency, this particular resource produces sections that are valuable for both the novice and also experienced within agricultural engineering.

Keywords Livestock waste management • Agriculture waste treatment • Livestock waste modeling • Biological treatment • Physical and chemical treatment • Biogas; AgSTAR

Nomenclature

AU	Number of 1000-lb animal units per animal type
BOD_5	Five-day biochemical oxygen demand
BUW	Bedding unit weight, lb/ft^3
C	Targeted rate concentration
C^*	Background rate concentration
C0	Initial concentration of conditions
Ca^{+2}	Calcium cation
CH_3COOH	Acetic acid
CO	Carbon monoxide
CO_2	Carbon dioxide
COD	Chemical oxygen demand
D	Number of days in storage period
DS	Dissolved solids
DVM	Daily volume of manure production for animal type, $ft^3/AU/day$
FR	Volumetric void ratio
FS	Fixed solids
H_2	Diatomic hydrogen
HLR	Hydraulic loading rate
k	First-order rate constant (cm/day)
Mg^{+2}	Magnesium cation
MMT of CO_{2e}	Million metric tons of CO_2 equivalent
N_2	Diatomic nitrogen
NH_3-N	Ammonia nitrogen
NH_4-N	Ammonium nitrogen
NO	Nitrous oxide
OLR	Organic loading rate
PO_4^{-2}	Phosphate ion
q	Hydraulic loading rate (cm/day)
SS	Suspended solids
TBV	Total bedding volume stored, ft^3
TKN	Total Kjeldahl nitrogen
TP	Total phosphorus
TS	Total solids
TVM	Total volume of stored manure, ft^3
TVS	Total volatile solids
TWW	Total wastewater stored, ft^3
VMD	Volume of manure production for animal type for storage period, ft^3
WB	Weight of bedding used for animal type, lb/AU/day
WV	Volume of waste stored, ft^3

1 Introduction

In recent years, livestock waste management has been a rapidly changing technology. It is subject to government regulation and sensitive to population growth patterns, community attitudes, and land use changes. It is influenced by variables such as soil type, topography, climate, crops, and livestock production practices. The evolution of larger and more concentrated livestock operations has accentuated the problems of waste management. Better management methods are necessary not only to hold down labor requirements and expense but also to minimize detrimental effects on the environment. Where animals are allowed to roam freely on pastures, such as is still done in many areas of the state, the manure from the livestock is deposited directly on the land and recycled with a minimum hazard to the environment. Even pasture production of livestock, however, requires management to prevent overgrazing, overcrowding, loss of vegetative cover, and the development of potential nonpoint sources of pollution. The facilities that cause the greatest environmental threat, however, are those in which the livestock are confined permanently or frequently on a regular basis. Figure 1.1 provides the consequences of infiltrated livestock waste.

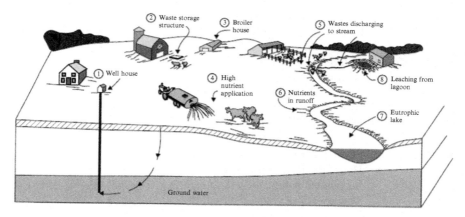

1. Contaminated well: Well water contaminated by bacteria and nitrates because of leaching through soil. (See item 4.)

2. Waste storage structure: Poisonous and explosive gages in structure.

3. Animals in poorly ventilated building: Ammonia and other gases create respiratory and eye problems in animals and corrosion of metals in building.

4. Waste applied at high rates: Nitrate toxicity and other N-related diseases in cattle grazing cool-season grasses; leaching of NO_3 and micro-organisms through soil, fractured rock, and sinkholes.

5. Discharging lagoon, runoff from open feedlot, and cattle in creek: (a) Organic matter creates low dissolved oxygen levels instream; (b) Ammonia concentration reaches toxic limits for fish; and (c) Stream is enriched with nutrients, creating eutrophic conditions in downstream lake.

6. Runoff from fields where livestock waste is spread and no conservation practices on land: P and NH_4 attached to eroded soil particles and soluble nutrients reach stream, creating eutrophic conditions in downstream lake.

7. Eutrophic conditions: Excess algae and aquatic weeds created by contributions from items 5 and 6; nitrite poisoning (brown-blood disease) in fish because of high N levels in bottom muds when spring overturn occurs.

8. Leaching of nutrients and bacteria from poorly sealed lagoon: May contaminate ground water or enter stream as interflow.

Fig. 1.1 Consequences of infiltrated livestock waste [1]

In general, the regulations do not stipulate how waste must be handled, but rather delineate the unsatisfactory practices and acceptable methods for correcting unsatisfactory situations. The decision-making process, when a farmer has to deal with correcting a problem situation, is essentially left to the farmer as to the selection of the system or combination of systems to correct the problems.

The frequent use of the term waste in this chapter is not intended to imply that we are dealing with a material of no value. The intent is to convey the understanding that the material consists of more than just the feces and urine excreted by the animals, for example, hair, soil, spilled feed, and other materials. In actuality, there is much that can be recovered and reused from this material for supplying plant fertilizers, livestock feed additives, and conversion to energy. Practical management practices to realize these and other benefits are encouraged whenever possible.

The manual has components grouped together by function and systems are composed of components with different functions. For this reason, some skipping around in the manual will be necessary when using it for planning purposes. The important thing is to insure that the components selected for the system are compatible and adequate for their purpose as well as to insure that the entire system accomplishes its management objective. English units of measurement are used in examples, although metric units are included in many tables.

Another point to consider in consistent planning is whether the failure of one component will result in failure of the entire system or if adequate flexibility is provided to permit continued operation without disastrous effects when unforeseen events happen. Often simple emergency or contingency measures can be planned into a system at various points, thereby preventing difficult situations later.

Data presented on waste production and characteristics are values generated from different parts of the United States making it nearly impossible to define consistent values. Where specific values for an individual system can be obtained, these should be used in preference to the manual values. The values found in this chapter are deemed to provide perspective on what occurs in livestock operations across the country.

Selecting a system and the individual components involved is a process that includes engineering, economics, regulatory considerations, personal preferences, and other factors. There is no single system which is best. Each component, facility, or process has advantages and disadvantages. Each of these factors mentioned in the previous sentence needs to be given consideration in order to develop the most suitable waste management system for a given situation.

The information provided in this chapter is intended to create a frame for planning and sizing waste management system components. If systems require further explanation, the reader should consult the resources for further direction on determining what constituents are necessary to create a more adequate design. It may also be necessary to obtain professional design assistance.

1.1   Federal Regulations

Federal regulations have been mandated by the US Environmental Protection Agency (US EPA) since its establishment in 1970. For the purpose of livestock waste treatment, legislation is applicable for both air and water. Air pollution research began in 1955 prior to the formation of the US EPA when the Air Pollution Act was passed to support funding and research. In 1970, the Clean Air Act required air quality standards for existing facilities and the refusal of building new infrastructure if not compliant with current legislation [2]. In addition, legislation has the US EPA control air emissions from mobile and stationary sources and establishes the National Ambient Air Quality Standards (NAAQs). NAAQs regulate hazardous air pollutants for the purpose of protecting the public health and environment and are incorporated with State Implementation Plans [3].

Nevertheless, agriculture persists with odor problems and further mandates were added later through the Clean Air Act Amendments of 1990. In the amendments, the legislation headed by the US EPA and Secrecies of Agriculture and Energy required a reduction emissions that produce acid rain and for the protection of ozone, ammonia volatilization from animal and other agricultural operations for water and soil acidification, and methane emissions from rice and livestock production for ozone depletion [2]. Figure 1.2 provides the various methods in which air pollution can be caused by the livestock industry.

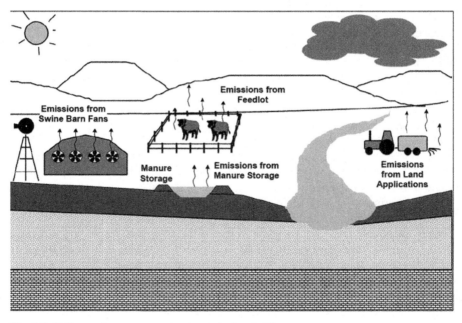

Fig. 1.2 Pathways for manure contaminants in the air [4]

Water legislation began as early as 1886 with the River and Harbors Act of 1886 and 1889. Following the induction of the US EPA, the passing of the Federal Pollution Control Act of 1972 placed federal government responsible for creating and enforcing standards for water pollution control, maintaining the integrity of the water supplies, where a goal of having 0 % discharge by 1985 was set. However, the biggest impact to water treatment in livestock wastes was the Clean Water Act of 1977. The Clean Water Act of 1977 introduces stringent legislation on feedlots and also required National Pollutant Discharge Elimination System (NPDES) permits [2].

The National Pollutant Discharge Elimination System (NDPES) regulates the quantity of waste entering navigable waters and also point sources [5]. In regard to livestock wastes, the NPDES require permits when discharging in the following conditions [2]:

1. Feeding operations consisting of 1000 animals confined for a time greater than 45 days per year and pollution less than 25 years, 24 storm event
2. Feeding operations with 300 animals discharge through a man-made device into navigable waters either from a feedlot of a man-made device
3. Hatcheries and fish farm cold water ponds that have a total of 20,000-lb animal production with 5000 lbs of food discharging 30 days per year or warm water ponds discharging 30 days per year

There have been several revisions made to NPDES permit involving concentrated animal feeding operations (CAFOs) or feedlots. The 2003 revision makes permits necessary for both open lots and CAFOs, refines the definition of CAFO requirements, and incorporates a nutrition management plan that considers faculty and land application issues where the lack of compliance can require CAFOs to point source. Proposed revisions have been suggested in 2008 and 2011 from outcomes of lawsuits submitted by both the industry and environmental interest groups. For example, in 2011 a proposal was made where it would have been required for a CAFO or its affiliated state to release information. The proposal was not mandated as the EPA decided to make additional measures to ascertain existing techniques to collect necessary information [6, 7].

1.2 State Regulations

Regulations imposed by the state will vary. There are many resources available to the user to determine which regulations are appropriate for a given state. An investigation of specific state investigation will be up to the user. A list of each state's environmental agency with associated links is in Table 1.1.

Table 1.1 List of state environmental agencies with associated links

State	State agency	Website
Alabama	Alabama Department of Environmental Management	http://www.adem.state.al.us/default.cnt
Alaska	Alaska Department of Environmental Conservation	http://dec.alaska.gov/
Arizona	Arizona Department of Environmental Quality	http://www.azdeq.gov/
Arkansas	Arkansas Department of Environmental Quality	http://www.adeq.state.ar.us/
California	California Environmental Protection Agency	http://www.calepa.ca.gov/
Colorado	Colorado Department of Public Health and Environment	https://www.colorado.gov/cdphe/
Connecticut	Connecticut Department of Energy and Environmental Protection	http://www.ct.gov/deep/site/default.asp
Delaware	Delaware Department of Natural Resources and Environmental Control	http://www.dnrec.delaware.gov/Pages/Portal.aspx
Florida	Florida Department of Environmental Protection	http://www.dep.state.fl.us/
Georgia	Georgia Environmental Protection Division	http://epd.georgia.gov/
Hawaii	Hawaii Office of Environmental Quality Control	http://health.hawaii.gov/oeqc/
Idaho	Idaho Department of Environmental Quality	http://www.deq.idaho.gov/
Illinois	Illinois Environmental Protection Agency	http://www.epa.illinois.gov/index
Indiana	Indiana Department of Environmental Management	https://secure.in.gov/idem/index.htm
Iowa	Iowa Department of Natural Resources	http://www.iowadnr.gov/Environment.aspx
Kansas	Kansas Department of Health and Environment: Division of Environment	http://www.kdheks.gov/environment/
Kentucky	Kentucky Department for Environmental Protection	http://dep.ky.gov/Pages/default.aspx
Louisiana	Louisiana Department of Environmental Quality	http://www.deq.louisiana.gov/portal/
Maine	Maine Department of Environmental Protection	http://www.maine.gov/dep/
Maryland	Maryland Department of the Environment	http://www.mde.state.md.us/Pages/Home.aspx
Massachusetts	Massachusetts Department of Environmental Protection	http://www.mass.gov/eea/agencies/massdep/
Michigan	Michigan Department of Environmental Quality	http://www.michigan.gov/deq
Montana	Montana Department of Environmental Quality	http://www.deq.mt.gov/default.mcpx
Minnesota	Minnesota Pollution Control Agency	http://www.pca.state.mn.us/

(continued)

Table 1.1 (continued)

State	State agency	Website
Mississippi	Mississippi Department of Environmental Quality	http://www.deq.state.ms.us/
Missouri	Missouri Department of Environmental Quality	http://dnr.mo.gov/env/index.html
Nebraska	Nebraska Department of Environmental Quality	http://www.deq.state.ne.us/
Nevada	Nevada Division of Environmental Protection	http://ndep.nv.gov/
New Hampshire	New Hampshire Department of Environmental Services	http://des.nh.gov/index.htm
New Mexico	New Mexico Environmental Department	http://www.nmenv.state.nm.us/
New York	New York Department of Environmental Conservation	http://www.dec.ny.gov/
North Carolina	North Carolina Department of Environment and Natural Resources	http://www.ncdenr.gov/web/guest
North Dakota	North Dakota Environmental Health	http://www.ndhealth.gov/EHS/
Ohio	Ohio Environmental Protection Agency	http://www.epa.state.oh.us/
Oklahoma	Oklahoma Department of Environmental Quality	http://www.deq.state.ok.us/
Oregon	Oregon Department of Environmental Quality	http://www.oregon.gov/deq/pages/index.aspx
Pennsylvania	Pennsylvania Department of Environmental Protection	http://www.depweb.state.pa.us/portal/server.pt/community/dep_home/5968
Rhode Island	Rhode Island Department of Environmental Management	http://www.dem.ri.gov/
South Carolina	South Carolina Department of Health and Environmental Control	http://www.scdhec.gov/HomeAndEnvironment/
South Dakota	South Dakota Department of Environment and Natural Resources	http://denr.sd.gov/
Tennessee	Tennessee Department of Environment and Conservation	http://www.state.tn.us/environment/
Texas	Texas Commission of Environmental Quality	http://www.tceq.state.tx.us/
Utah	Utah Department of Environmental Quality	http://deq.utah.gov/
Vermont	Vermont Department of Environmental Conservation	http://www.anr.state.vt.us/dec/dec.htm
Virginia	Virginia Department of Environmental Quality	http://deq.state.va.us/
Washington	Washington Department of Ecology	http://www.ecy.wa.gov/
West Virginia	West Virginia Department of Environmental Protection	http://www.dep.wv.gov/Pages/default.aspx
Wisconsin	Wisconsin Department of Natural Resources	http://dnr.wi.gov/
Wyoming	Wyoming Department of Environmental Quality	http://deq.wyoming.gov/

2 Wastewater Characteristics

2.1 General Characteristics of Wastewater

2.1.1 Terminology

Prior to evaluating the properties of wastewater, it is important to understand the general terminology associated related to quantifying the characteristics of wastewater. Overall, wastes can be evaluated based on their physical and chemical properties. Tables 1.2 and 1.3 summarize the physical and chemical properties along with characteristics from excreted beef. The most important physical properties within waste include the weight, volume, and moisture content. These properties quantify the amount of waste that must be handled and subsequently treated. Secondary physical properties evaluate categories that is found within a

Table 1.2 Physical and chemical properties of waste [2]

Physical properties	
Moisture content	Component of a waste that can be removed by evaporation and drying
Total solids	Component of a waste that is left after evaporation
Volatile solids	Component of a waste that has been removed when a waste sample is placed in a muffle furnace at 1112 °F
Fixed solids	Component of a waste that remains after a waste sample is heated in a muffle furnace at 1111 °F
Suspended solids	Component of a waste removed by means of filtration
Chemical properties	
Five-day biological oxygen demand (BOD5)	Water quality index that measures the amount of oxygen needed for microorganisms to degrade material
Chemical oxygen demand (COD)	Water quality index that determines amount of oxygen consumed by organic material

Table 1.3 Excreted beef waste characteristics [8]

Components	Units	Beef cow in confinement	Growing calf confined (450–750 lb)
Weight	lb/d-a	125	50
Volume	ft^3/d-a	2.0	0.8
Moisture	% wet basis	88	88
TS	lb/d-a	15	6.0
VS	lb/d-a	13	5.0
BOD	lb/d-a	3.0	1.1
N	lb/d-a	0.42	0.29
P	lb/d-a	0.097	0.055
K	lb/d-a	0.30	0.19

Fig. 1.3 Nitrogen
processes involved in
manure management (from
top to *bottom*:
mineralization, nitrification,
denitrification (*bottom left*),
volatilization (*bottom
right*)) (Adapted from [4])

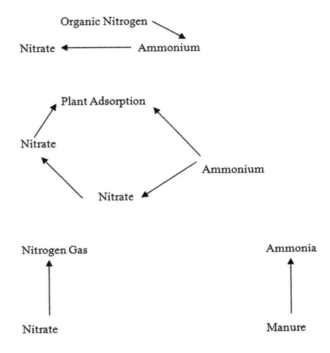

given waste. These secondary properties include total solids (TS), volatile solids
(VS), fixed solids (FS), dissolved solids (DS), and suspended solids (SS) [2].

On the other hand, chemical properties are constituted as nutrients or wastewater
quality indices. Nitrogen (N), phosphorus (P), and potassium are the elements
mainly considered as nutrients. These nutrients are further subdivided into subsequent forms that can be beneficial or detrimental to the handling of livestock.
Figures 1.3 and 1.4 summarize nitrogen and phosphorus processes that occur within
livestock waste. Five-day biochemical oxygen demand (BOD_5) and chemical
oxygen demand (COD) are two of many wastewater quality indices. These indices
are evaluated within a laboratory and are important in determining the nature of the
wastewater present. Biochemical oxygen demand (BOD_5) relates the amount of
oxygen required to degrade waste by microorganisms in 5 days at 20 °C, while
chemical oxygen demand (COD) involves the consumption of oxygen by organic
and inorganic constituents [2].

2.1.2 Wastewater Characteristics

It can be said that the type of manure in wastewater produced varies not only on
characteristics but also on the time of year. Based on the data collected between
summer and winter for cattle manure and bedding, Loehr [9] found that the ranges
for parameters are different between the summer and winter. For example, percent

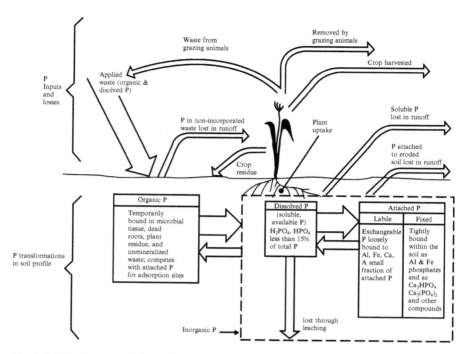

Fig. 1.4 Phosphorus cycle in relation to waste application and transformation of phosphorus in the soil profile [1]

total solids (%TS) in the winter have an average of 2.8 % versus 2.3 % in the summer. In regard to biochemical and chemical oxygen demands (BOD$_5$ and COD), the winter indicates higher values of BOD at 13,800 mg/L versus only 10,300 mg/L in the summer. Nutrient presence is higher at 2350 mg/L as N for total nitrogen in the summer, as compared with 1800-mg/L ion in the summer, and total phosphorus is 280 mg/L in the winter, while only 190 mg/L in the summer. These results can be reflected based on conditions such as precipitation and temperature [9].

In addition, having considered swine lagoon analysis in Missouri, liquid wastes are significantly higher in total solids, total nitrogen and ammonia, salts, and minerals as compared to sludge. In particular, the liquid wastes contained 3091 mg/L, as compared to only 203.843 mg/L in the solids. This trend is also noticed in terms of salts (Na 470, Ca 257, and Mg 64 mg/L versus 4.627, 6.176, and 1.514 mg/L in the liquid, respectively) [10].

Also, the waste characteristics of different industries vary. The supernatant for different animal wastes sampled from a lagoon and municipal waste treatment was compared. Poultry lagoons contained the highest concentration of wastes. The mean COD for poultry was 3700 mg/L, compared with 2050 mg/L and 1672 for the swine and dairy lagoons, respectively. This trend can be highly seen in BOD$_5$, TS, total volatile solids (TVS), suspended solids (SS), and ammonia nitrogen

(NH$_3$-N), where the poultry lagoon contained the highest amounts of all three. Nevertheless, untreated municipal wastewater has significantly lower values for every category; in some cases such as COD values, the lowest animal waste value (1672 mg/L for dairy lagoons) was four times in COD than in municipal waste and almost ten times less than the highest (poultry) [11]. Tables 1.4 and 1.5 present characteristics of manure based on various livestock types. Table 1.6 presents wastewater characteristics of swine waste.

On the other hand, while waste constituents were higher in the animal waste, the untreated municipal wastewater contained higher amounts of trace metals, specifically cadmium, chromium, copper, and lead. In fact, in examining copper,

Table 1.4 Total manure, nitrogen, phosphorus, and potassium excreted by different livestock species [12]

Livestock type	Fresh manure (gal/day)	N (lb/day)	P$_2$O$_5$ (lb/day)	P (lb/day)	K$_2$O (lb/day)	K (lb/day)
Beef cattle (1000-lb body weight)	7.5	0.34	0.25	0.11	0.29	0.24
Dairy cow (1000-lb body weight)	11	0.41	0.17	0.074	0.32	0.27
Swine (100-lb body weight)	1	0.045	0.034	0.015	0.036	0.030
Poultry (4-lb body weight)	0.028	0.0029	0.0026	0.0011	0.0015	0.0012

Note: Livestock type is based on 1000-lb body weight

Table 1.5 Manure characteristics per animal [13]

Animal type	Average weight (lb)	Total manure production (cu.ft/day)	Total solid production (lb/day)	Volatile solid production (lb/day)
Swine				
Nursery pig	35	0.04	0.39	0.30
Growing pig	65	0.07	0.72	0.55
Finishing pig	150	0.16	1.65	1.28
Gestation sow	275	0.15	0.82	0.66
Sow and litter	375	0.36	2.05	1.64
Boar	350	0.19	1.04	0.84
Cattle				
Dairy	1000	1.39	12.00	10.00
Beef	1000	0.95	8.50	7.20
Poultry				
Layers	4	0.0035	0.064	0.048
Broilers	2	0.0022	0.044	0.034

Table 1.6 Swine waste characteristics [2]

Component	Units	Grower 40–200 lb	Replacement gift	Sow gestation	Lactation	Boar	Nursing/ nursery pig 0–40 lb
Weight	ln/d/1000[a]	6.40	32.80	27.20	60.00	20.50	106.00
Volume	ft^3/d/1000[a]	1.00	0.53	0.44	0.96	0.33	1.70
Moisture	%	90.00	90.00	90.80	90.00	90.70	90.00
TS	% w.b.	10.00	10.00	9.20	10.00	9.30	10.00
	lb/d/1000[a]	6.34	3.28	2.50	6.00	1.90	10.60
VS	lb/d/1000[a]	5.40	2.92	2.13	5.40	1.70	8.80
FS	lb/d/1000[a]	0.94	0.36	0.37	0.60	0.30	1.80
COD	lb/d/1000[a]	6.06	3.12	2.37	5.73	1.37	9.80
BOD$_5$	lb/d/1000[a]	2.08	1.08	0.83	2.00	0.65	3.40
N	lb/d/1000[a]	0.42	0.24	0.19	0.47	0.15	0.60
P	lb/d/1000[a]	0.16	0.08	0.06	0.15	0.05	0.25
K	lb/d/1000[a]	0.22	0.13	0.12	0.30	0.10	0.35
TDS		1.29					
C:N ratio		7	7	6	6	6	8

[a]Average daily production for weight range noted. Increase solids and nutrients by 4 % for each 1 % feed waste more than 5 %

the range for copper was between 190 and 440 mg/L for poultry lagoons; however, in untreated municipal wastewater, it was found that the range of copper was between 20 and 3360 mg/L, almost four times as much for the averages of these ranges. With the exception of arsenic and cadmium, poultry lagoons consistently had higher amounts of trace elements [11].

2.2 Milk House Wastewater Characteristics

Milk house wastewater is generated from various sources within the dairy industry. These sources include but are not limited to [14]:

- Wash water from cleaning bulk tanks
- Cleaning of milk pipelines
- Cleaning of milking units
- Cleaning equipment
- Cleaning of milk house floor
- Remnant within the milk pipelines, receiver, and bulk tanks
- Chemicals
- Water softener recharge
- Manure
- Bedding

- Floor dirt and grit
- Washing the udders of the cows

Typical milk house and diary wastewater characteristics are listed in Tables 1.7 and 1.8.

The Wisconsin National Resource Conservation Service (NRCS) describes three constituents within milk house wastewater—solids, phosphorus, and ammonia nitrogen and chlorides. Solids contain manure, primarily made of lignin and cellulose. These are a major producer of milk house wastewater. Solids usually have a concentration range between 1600 and 7000 mg/L. Depending on the source, some solids can be comprised of high-concentration biochemical oxygen demand (BOD). For example, it has been determined that raw waste milk can have a BOD concentration of 100,000 mg/L [15].

The presence of phosphorus has been attributed to daily cleaning operations such as pipeline washing or the presence of cleaning chemicals such as detergents and acid rinses, many of which can have 3.1–10.6 % phosphorus by weight. Phosphorus in milking house centers is usually soluble and can cause eutrophication [15].

Ammonia is found in manure, urine, and decomposed milk. The discharge of milk house wastewater with substantial concentrations of ammonia can be toxic to fish. On the other hand, chlorides are also found in urine, milking system cleaners

Table 1.7 Characteristics of milk house wastewater [14]

Parameter	Final effluent tank (mg/L)	Design (mg/L)
BOD$_5$	500–2600	1200
Total solids (TS)	200–1000	450
Fats, oils, grease	90–500	225
	30–100	65
Total phosphorus	21–100	55
pH	6.2–8.0	7.5
Temperature	53–70	–

Table 1.8 Dairy waste characterization—milking center [15, 16]

Component	Units	Milk house only	Milk house and parlor	Milk house, parlor, and holding area	Milk house, parlor, and holding area
Volume	ft^3/day/1000 head	0.22	0.60	1.40	1.60
Water volume	gal/day/1400-lb cow	2.3	6.3	14.7	16.8
Moisture	%	99.72	99.40	99.70	98.50
COD	lb/1000 gal	25.30	41.70	–	–
BOD$_5$	lb/1000 gal	–	8.37	–	–
N	lb/1000 gal				
P	lb/1000 gal	0.58	0.83	0.23	0.83
K	lb/1000 gal	1.50	2.50	0.57	3.33

and sanitation, and water softening generation. The presence of chlorides can have an impact on the salinity of the wastewater being treated [15].

The daily operations within a milk house require daily cleaning of equipment and pipelines. The University of Minnesota Extension describes a four-stage cleaning process. Cleaning begins with rinsing the transfer lines to remove any raw milk that may remain. Next, organic material is removed by a detergent with an active chlorine concentration of 100 mg/L. This detergent raises the pH above 11. Then, an acid rise is completed to reduce inorganic material. The pH is lowered to around 3.5 to prevent bacteria formation and neutralize any detergent residue that may remain. Finally, chlorine with a concentration of 200 mg/L is added to kill microorganisms in the line. The process of cleaning equipment and pipelines accounts for an additional source of wastewater that needs to be treated prior to any discharge [14].

2.2.1 Treatment of Milk House Wastewater

There are several treatment methods for milk house wastewater. Table 1.9 lists several treatment methods that are being used in the state of Minnesota. For example, a viable option of treating milk house wastewater is through a two-stage septic system. It is important to note that wastewater entering into the tank does not include waste milk from cows. Waste milk will be disposed with manure.

Table 1.9 Treatment methods for milk house wastewater treatment [17, 18]

Treatment method	Description	Requirements
Chemical batch reactor	Coagulation and flocculation	Effluent BOD ≤ 205 mg/L Discharge into infiltration/filtration system
Bark bed	Soil infiltration with 18–24 inches of bark wood Pressure distribution system disperses effluent	Requirement of soil texture to a minimum of 3 ft bedrock. Treatment consists of three processes: 1. Primary treatment is completed by two septic tanks. Tanks are designated based on an HRT of 3 days or the volume whichever is greater 2. Infiltration area 3. Distribution system: the system consists of a pump, transferring pipe. Effluent traveling to the pipe must have a minimum velocity of 2 ft/s. The transferring pipe must have a diameter of 2 inches with a drainage slope of 1 %. Distribution is done through gravel bed or a chamber system

(continued)

Table 1.9 (continued)

Treatment method	Description	Requirements
Aeration and media filtration	Aerobic treatment or recirculating media filter	Treatment will consist of three processes: 1. Primary treatment will use two septic tanks. Design requirements similar to bark bed primary treatment 2. Aerobic treatment follows primary treatment where the goal must be less than 200 mg/L effluent BOD 3. Following aerobic treatment the discharge will enter an infiltration/filtration system
Irrigation	Treatment consists of water filled within the tank that will be dispersed onto crops	1. A proper site for irrigation consists of a location where 20 % of materials from 2 ft below the buffer zone pass through a #200 sieve 2. The irrigation area must have a minimum of 3 % slope, where the down gradient should be 50 ft away from karst, surface water, or any private wells 3. Treatment consists of using a septic tank. Design requirements are similar to bark bed primary treatment 4. Wastewater moves to a 3-day holding dosing tank with piping for distribution and pumping
Vegetated treatment dosing system	Wastewater from a septic tank is distributed onto vegetation by a sloping elevated pipe where the upslope side of the pipe is enclosed	1. Both siting and primary treatment use similar design criteria as previously mentioned 2. Treated waste from a septic system will travel through a distribution system to a dosing tank by a perforated pipe with perforations between 1/2 and 1 inch diameter. The pipe is elevated 1–1.5 ft above the ground 3. Determination of vegetated area is based on either a flow depth no greater than 0.5 ft using a treatment time of 15 min and a Manning constant of 0.24 or the smallest area that can handle a design loading rate no greater than 0.9 inches/week

Treatment by the septic system is contingent on the strength of the wastewater leaving the parlor and also time spent in the septic tanks [17, 18].

Wastewater is pretreated using two septic tanks consisting of inlet and outlet baffles. The tanks remove settable solids, fats, and grease and inhibit contamination throughout the remaining sections of the treatment plant. In the state of the Minnesota, tank sizing is based on either a hydraulic retention time of 3 days or a volume of 1000 gallons whichever is greater. In addition, Minnesota requires 4 ft of soil cover. Prior to exiting the septic tank, the wastewater passes through an effluent filter. The effluent filter prevents suspended solids from leaving the septic tank [17, 18].

Next wastewater moves through a bark bed. The bark bed combines soil with bark and shredded wood. The depth of the bark bed is between 18 and 24 inches. The purpose of the mixture is to prevent the soil in colder climates and allows for more oxygen transfer which in turn increases the rate of degradation at the soil-effluent interface. The sizing and application within the bark bed are determined by the soil type. Typical bark beds consist of a depth of 2 ft of soil to the bedrock or groundwater. Sizing of the bed is computed by taking the loading rate of the soil (contingent on soil type) and dividing it by the total wastewater volume. The loading rate is read from a table based on soil type. Presented values consider a BOD_5 concentration of 750 mg/L, flow rate of 5 gallons per day, and a BOD_5 loading rate of 0.0062 lbs/gallon. Bark beds can also be sized using hydraulic loading as well [19].

Another treatment method that can be employed is the use of constructed wetlands. Because constructed wetlands are not unique to milk house waste treatment, they will be discussed in Sect. 4: Wastewater Treatment.

Nevertheless, literature has discussed the efficiency of constructed wetlands for treating dairy wastewater. A three-celled surface wetland was used to treat dairy wastewater. The study compared the performance of the summer and winter seasons. The results found that total suspended solids (TSS), total phosphorus (TP), and total Kjeldahl nitrogen (TKN) were reduced in the summer as compared to the winter. In addition, BOD_5 removal was lower than 30 mg/L during the summer months as compared to the winter months. Finally, fecal coliform removal was approximately 31 % [20].

To avoid eutrophication in a local surface waterbody, a three-celled parallel free-water surface wetland was used to treat dairy wastewater. The treatment process began with the concrete settling pad for the purpose of eliminating solids prior to entry into the wetland. Following treatment into the constructed wetland, a three-sump pump transfers the wastewater into a holding pond. The authors concluded that BOD_5, conductivity, total dissolved solids (TDS), TSS, TKN, TP, phosphate, ammonia, nitrate, nitrite, and fecal coliform bacteria were generally reduced by the wetland. In addition, all parameters with the exception of nitrate and nitrate were diminished from the settling pad to the holding pond. Fecal coliform was reduced provided that cows were kept from grazing in the constructed wetlands [21].

2.2.2 Conservation

Along with dairy wastewater treatment, water conservation is another important facet to properly handle wastewater. Water conservation is important because it provides the dairy plant owners an opportunity to reduce the cost for treatment. In general, wastewaters with high BOD_5 concentration discharged into a municipal wastewater treatment system incur high costs. It can also become expensive for on-site treatment as well; therefore, water conservation efforts provide owners an opportunity to save on funds. In addition, methods have a positive impact on areas where water resources are currently being depleted and can also reduce the potential of stringent legislation. In the dairy industry, water reuse can reduce fresh water demand to 1 gal of water/1 gal of milk produced provided proper management of goals along with regularly scheduled maintenance [22].

3 Waste Treatment

3.1 Anaerobic Digestion

Anaerobic digestion is the fermentation of organic waste by hydrolytic microorganisms into fatty acid chains, carbon dioxide (CO_2), and hydrogen (H_2). Short fatty acids are then converted into acetic acid (CH_3COOH), H_2, CO_2, and microorganisms. Acetic acid forms biogas, a combination of methane (CH_4), CO_2, and trace elements by means of methanogenic bacteria. Occasionally biogas can form hydrogen sulfide by sulfate-reducing bacteria. In general CH_4 in biogas produces between 55 % and 80 %, while approximately 65 % is found in animal manure [23].

The processes in anaerobic digestion are driven by temperature, moisture, and solid content. There are three major temperature ranges defined—psychrophilic (<20 °C), mesophilic (35–40 °C), and thermophilic (51–57 °C). Ideally an anaerobic digester should operate at temperatures greater than 35 °C. A moisture content of 60–99 % is ideal, while solid content in the digester should be less than 15 % [24].

Recently, there has been a big interest in anaerobic digestion for the purpose of energy conversion [25]. Since 1996, the Environmental Protection Agency has partnered with the US Department of Agriculture, National Resource Conservation Service (NCRS), and the US Department of Energy to develop a program known as AgSTAR, an opportunity for monetary support in projects related to anaerobic digestive systems. In 1998, the program began by promoting seven farm digesters across the country [26].

There have been reports of profit being made on the energy that has been captured through the use of livestock manure. These values have greatly depended upon the monetary cost of electricity. For example, if one were to sell electricity in Wisconsin and California, a 1000-head dairy farm with manure production would

Net present value ($) of digester per head

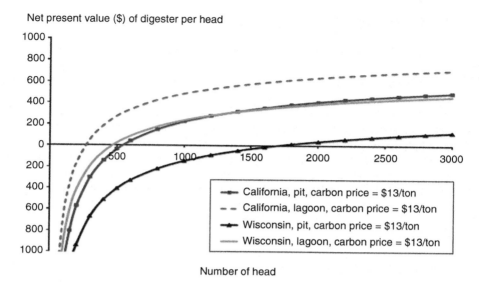

Number of head

Fig. 1.5 Net value in dollars of digesters per head versus number of head [25]

be worth about $56,000 and $77,500 in Wisconsin and California, respectively [25]. Statistically speaking, it was found that in 2009 the approximately 151 biogas systems that have been installed within the state of Wisconsin produced about 11.6 megawatts of electricity, enough for use by 10,000 homes. Within January 2007 and June 2008 alone, 150,000 kilowatt hours (kWh) of electricity was produced by farms that had 2000 head of animals and 440,000 kWh of electricity for those between 2000 and 4500 [27]. Figure 1.5 indicates the net value of dollars based on the digester per number of head of cattle. Figure 1.6 indicates the number of dairies operating at a given carbon price per operation size.

There are a plethora of reasons why AgSTAR has become a popular consideration for the development of biogas. Consider that the state of Wisconsin has spent between $16 and $18 billion each year for coal energy imports, whereas about $853 million for transportation [27]. If the state of Wisconsin, rich in manure and crop remains, waste components from dairy processing, and fats and greases can convert this material into fuel, it would create an infrastructure that would be safer and easier to be controlled as compared to the current energy options on the market today.

A recent 2013 study conducted by the US Environmental Protection Agency (US EPA) evaluating the AgSTAR program found that anaerobic digesters reduce greenhouse gases by 1.73 million metric tons of CO_2 equivalent (MMCTCO$_{2e}$). This is because methane is captured and burned before entering into the atmosphere. On the other hand, anaerobic digesters produced 840 million kWh in 2013. These benefits were contingent on the type of anaerobic digester applied. For example, the most commonly used digesters in the United States were complete mixed and mixed plug flow [28]. Biogas production is also dependent upon the type

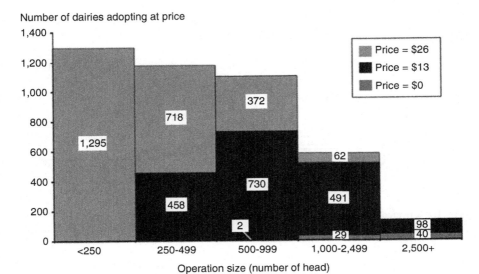

Fig. 1.6 Number of dairies operating at a given carbon price versus operation size [25]

Notes: Numbers at higher prices are additive to those for lower prices: for example, at a price of $13/ton, an additonal 491 opeations of size 1,000-2,499 head are predicted to adopt, for a total of 520 operations of this size. At a carbon price of $13/ton, no operation smaller than 250 head is predicted to adopt. At a carbon price of $0, no operation with fewer than 500 head and 2 operations 500-999 head are predicted to adopt.

Animal type	Average weight (kg)	Biogas/animal/day (m^3)
Dairy	625	1.3
Beef	447	0.32
Swine	70	0.14
Poultry	1.2	0.0092

Table 1.10 Biogas production by animal [23]

of livestock. Table 1.10 provides information concerning the daily production of biogas per animal type.

3.1.1 Types of Anaerobic Digesters

There are six types of anaerobic digesters—covered anaerobic lagoons, plug flow, continually stirred tank reactor, fixed film, induced blanket reactor, and anaerobic sequencing batch reactors. Table 1.11 reports the characteristics of three of the six anaerobic digesters (covered anaerobic lagoons, plug-flow digester, and mixed). The selection of the appropriate anaerobic digester is determined by appropriate parameters such as the geographic location. Covered anaerobic lagoons form biogas from manure stored in structures, are low cost, are simplistic in design, and are

Table 1.11 Characteristics of various anaerobic digester types [23]

Anaerobic digestion system	OLR COD/m^3/kg	HRT (day)
Covered anaerobic lagoon	0.05–0.2	60–360
Plug-flow digester	1–6	18–20
Mixed	1–10	5–20

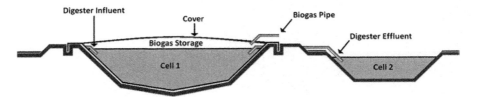

Fig. 1.7 Covered lagoon digester [29]

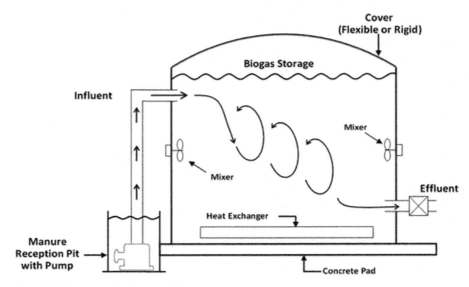

Fig. 1.8 Complete mix digester [29]

manageable. There are two types of covers—full and partial. Production of biogas by a covered anaerobic lagoon depends on the temperature. Therefore, covered lagoons are more appropriate in areas of warmer climate. Biogas production in a covered lagoon is collected in pipes at the top of the digester and then transported by using a low vacuum. From there, the remaining biogas is then flared. Additional characteristics of a covered anaerobic lagoon include high total solid (TS) concentration, organic loading rate (OLR) of 0.2–0.5-kg chemical oxygen demand (COD)/m^{3}day, and a hydraulic retention time (HRT) of 60–360 days [23]. Figures 1.7 and 1.8 are diagrams of a covered lagoon digester and a completely mixed digester.

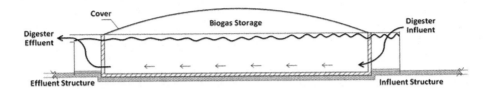

Fig. 1.9 Plug-flow digester [29]

On the other hand, manure in a plug-flow digester enters undigested and leaves digested. A typical plug-flow digester includes concrete and geosynthetic material for gas collection. Manure enters into the digester and is limited to 11–14 % total solid concentration, $1 - 6$-kg COD/m^3d OLR, and an HRT between 20 and 30 days. In a continually stirred tank reactor, manure enters into a tank and is mixed to maintain a consistent concentration throughout the reactor. Unlike a plug-flow digester which is limited to 6-kg COD/m OLR, the maximum allowable organic loading rate for total solids entering a continually stirred tank reactor is 10-kg COD/m^3day. In addition, the hydraulic retention time is shorter than a plug-flow reactor ranging between 5 and 20 days [23]. Figure 1.9 is a diagram of a plug-flow digester.

A fixed-film digester is an attached growth reactor with fixed-film media. When waste enters into the fixed-film digester, anaerobic biomass attaches to the fixed-film media. Typical fixed-film digesters have a low HRT between 0.5 and 4 days. Influent manure in a fixed-film digester has an OLR between 5 and 10-kg COD/m^3 with a solid concentration less than 1 % [23].

Finally, an induced blanket reactor forms a sludge blanket by digesting the waste. Manure moves upward from the bottom of the reactor to the top. Inside the blanket, manure moves upward contacting with anaerobic biomass to become digested. At the top of the tank, the biogas is created while the sludge blanket moves back to the bottom of the reactor. There are two types of blanket reactors—upflow anaerobic sludge blanket digester (UASB) and induced blanket reactor (IBR). UASB involve low concentration of solids, while IBR usually handle high solid concentrations [30].

The cost of an anaerobic digester application has been contingent on the type. In the design and construction of a system, the price involves the initial cost of the system and its operation and maintenance (O&M). The US Department of Agriculture (USDA) reported values on 38 different digesters. The overall cost of anaerobic digesters has been estimated to be between $114,000 and $326,000. Operation and maintenance (O&M) was found to be contingent on the type of waste. The O&M for swine waste was 2.3 % of the initial cost for the system, while dairy was 7 % [23].

Within the last 5 years, other anaerobic digestion processes have been tested. A specific type of anaerobic digestion design is known as a temperature-phased anaerobic digestion reactor. Temperature-phased anaerobic digestion (TPAD) is a system that completes treatment in two stages at two temperatures—during the first stage, the digester operates at a temperature at the highest thermophilic temperatures, approximately 55 °C, while the second stage at the lower ended mesophilic conditions or approximately 35 °C. When using a TAPD for livestock waste, the advantages are significant as the digester is capable of increasing a higher probability of bioconversion and methane production, with lower hydraulic retention times (HRT) and also size reduction [31]. Harikishan and Sung used a temperature-phased anaerobic digestion process (TPAD) to treat livestock wastewater for the purpose of analyzing dairy cattle manure. Having organic loadings of 1.87–5.82-g VS/l/day, 36–41 % of volatile solids were removed, converting 0.52–0.62-l methane/g VS. In addition, fecal coliform and *Salmonella* counts meet USEPA Class A standards [31].

Other authors have researched and found results under different conditions. King et al. [32] used a 3-year pilot In-Storage Psychrophilic Anaerobic Digester (ISPAD) to consider swine manure and if it is able to handle psychrophilic conditions and be able to complete anaerobic digestion and successfully produce methane. Results based on the microbial community analysis were able to produce methane, provided that volatile solids (VS) had a rate of 44.6 dm^3/kg day at 35°, 9.8 dm^3/kg day at 18°, and 8.5 dm^3/kg day at 8°. In addition, the ISAD reduces organic matter content by 24 % [32]. Rao et al. [33] used a self mixed anaerobic digester (SMAD) combined with a multistage high-rate biomethanation process where the authors were capable of reducing volatile solids (VS) by 58 % and producing a methane yield of 0.16 m^3/ kg, with a loading rate of 3.5-kg VS/m^3 day and a hydraulic retention time (HRT) of 13 days. The authors considered using the opportunity to reduce the loading rate and reduce the hydraulic retention time and percent treatment [33].

3.2   Constructed Wetlands

3.2.1   Description

The purpose of a constructed wetland is to provide a low maintenance treatment system that creates a quality effluent for areas that have a high volume of wastewater. Constructed wetlands house wastewater within wide channels. These channels also support plant life that grows by using the nutrients from the wastewater. There are four major processes employed in constructed wetlands—sedimentation, filtration, plant uptake (oxygen is provided at the plant root for waste decomposition), and biological decomposition (plants provide adequate binding sites for microorganisms) [15].

The basic idea of a wetland is to maintain moist conditions for pollutants to be trapped and broken down by the plant that are contained within them. In addition, constructed wetlands take advantage of combining anaerobic and aerobic conditions that persist through the wetland. The majority of constructed wetland design consists of using either subsurface flow or surface flow. Surface flow wetlands consist of having a "free-water zone" about 30-cm deep on top of a soil layer where the majority of the plant growth would occur. The advantage of designing a wetland by this manner is that it would place microbial growth in the best advantage to occur in the areas where the water and its contaminants would be. Subsurface flow wetlands, also known as "root zone method," remove the "free-water zone" for the purpose of allowing direct contact between plant material and contaminants present [34]. There are several design parameters that are necessary for treatment—hydraulic loading rate, length-to-width ratio, bottom slope, water depth, and vegetation [35].

The water depth of a constructed wetland is usually between 20- and 40-cm deep. The advantage of using surface constructed wetlands is the biological, physical methods that are employed within the system. Microbial activity (biological) degrades much of the organic materials, while colloids are either settled within the wetland or can become filtered out (physical). Nitrogen is capable of being removed by means of nitrification (the formation of nitrate from ammonium nitrogen [2]) and denitrification (formation of atmospheric nitrogen from nitrates [2]), while ammonia is volatilized by the use of algal photosynthesis. If any phosphorus is removed, it is by means of wetland plants eventually by either absorption or precipitation [36].

3.2.2 Constructed Wetland Types

Literature recognizes three major types of constructed wetlands—free-water surface (FWS), vegetated submerged or subsurface system, and floating aquatic plant systems (FAP) [37]. Figures 1.10, 1.11, and 1.12 are drawings of each type of constructed wetland.

In a free-water surface system, the wastewater depth is usually shallow, anywhere between 6 and 18 inches with a flat-bottom slope. Because of their shallow depths, FWS wetlands usually degrade wastewater under aerobic conditions. When wastewater enters an FWS, it moves above the sediment having direct contact with the plants at the surface. However, the efficiency of FWS treatment is contingent upon the presence of microorganism located throughout the surface. Nevertheless, microorganisms attach themselves to plant stems and/or litter below the water surface or at the soil/plant-root matrix creating the proper environment for wastewater treatment. Prior to entry of an FWS, a pretreatment system to remove settling and floating solids or ammonia is recommended [37]. FWS constructed wetlands have been proven to reduce BOD_5 and TSS to 30 mg/L and ammonia and ammonium-nitrogen to 10 mg/L [39]. In addition, to the effluent quality, FWS wetlands are very common in livestock operations because they are inexpensive and can be in operation year round [37].

Under the National Resource Conservation Service guidelines, an FWS is to be designed based on a 25-year storm event depending on the state. Sometimes a detention pond downstream may be necessary to meet this requirement. The sizing

Fig. 1.10 Free-water
surface (FWS) constructed
wetland [38]

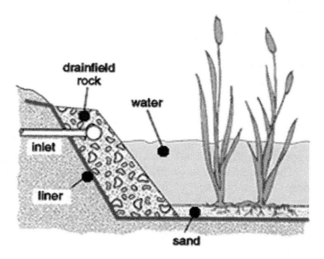

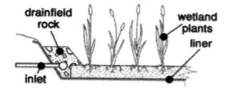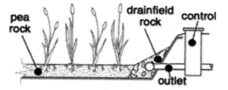

Fig. 1.11 Subsurface constructed wetland [38]

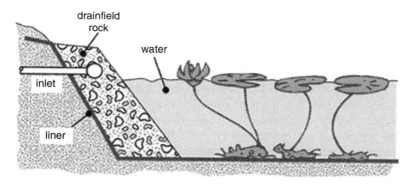

Fig. 1.12 Floating aquatic plant (FAP) constructed wetland [38]

of an FWS is done by using one of two methods—presumptive method or the field test method. The presumptive method assumes a BOD_5 concentration, while the field test method is based on an actual daily measurement of BOD_5 from the given livestock operation [39]. The presumptive method approximates a pollutant entering into a wetland by reviewing the BOD_5 or nitrogen concentration and then applies the value to an areal loading rate (typically 65-lb BOD_5/acre/day). The presumptive method has been well known since the Tennessee Valley Authority (TVA) introduced it in 1989 [37].

The field test method requires a collection of samples and analysis based on BOD_5 and total nitrogen (TN). Some of the important factors examined include average daily flow, temperature, and decay rate constant. The data collection from the field test is used to determine the size of the wetland. The purpose of the field test method is to ensure that the design of the wetland does not exceed discharge limits [37].

On the other hand, in vegetated submerged systems, wastewater flows within the sediment bed having more contact with the plant roots. The sediment bed is usually made of rock, gravel, and soils. Vegetation is usually planted at the top of the wetland [37]. Because wastewater flows at lower depths, wastewater is usually degraded at anaerobic conditions. The slope of this wetland ranges between 2 % and 6 %. Sizing of submerged systems is contingent on flow rate, influent, and desired outflow BOD_5 [39]. Vegetated submerged systems are not as prolific as surface flow wetlands. This is because the sediment beds can easily accumulate solids. Also, the beds can be very expensive to construct. Nevertheless, vegetated submerged systems can be used to treat wastewater with low flows and solids [37].

Finally, floating aquatic plant systems comprise of one or more ponds. The ponds are designed for plants to grow and float at the top of the ponds. Each pond is designed for a depth between 3 and 5 ft for the purpose of avoiding non-desired plant species to grow and become prominent within the system and gives the plant access to nutrients within the wastewater. There are several factors for appropriately harvesting. These include the number, size, and arrangement of ponds and the technique for harvesting. There are two major plant species in FAP systems—water hyacinths and duckweed [37].

3.2.3 Constructed Wetland Design

Constructed wetland design is usually consisting of first-order models under plug-flow conditions, alternating between looking for values of BOD, TSS, ammonium, and fecal coliforms [34]:

$$\ln\left[(C - C^*) \div (C_0 - C^*)\right] = -\frac{k}{q} \qquad (1.1)$$

where

$C_0 = $ initial concentration of conditions
$C = $ targeted rate concentration
$C^* = $ background rate concentration
$k = $ first-order rate constant (cm/day)
$q = $ hydraulic loading rate (cm/day)

An alternative method to designing a constructed wetland would be the use of regression equations for one had the desire to consider looking at multiple components at one time.

Stone et al. [40] used constructed wetlands, particularly marsh-pond-marsh wetland system at North Carolina A&T University. Six wetland systems with the dimensions of 11×40 m treat ammonia nitrogen and phosphorus. First-order kinetics were 3.7–4.5 m/day for total-N and 4.2–4.5 m/day for P, much lower than the typical model rate constant [40].

In addition, the US Environmental Protection Agency has tracked several constructed wetlands that have been used for the purpose of waste treatment. Seven locations to treat three different waste types—swine, dairy, and poultry—were constructed. A project in Duplin County, North Carolina uses construction wetlands to remove total Kjeldahl nitrogen (TKN) from swine wastewater, as it was observed a major factor effecting treatment was loading rates of TKN (3-kg/ha/day TKN) was able to remove between 91 % and 96 % TKN, while 10 kg/ha/d only removed approximately 73 %. A wetland in Essex, Ontario, reduced TSS (97 %), BOD_5 (97 %), and 99 % fecal coliforms and 95 % *E. coli* from dairy farm milk house wastewater. Auburn University used a constructed wetland for poultry lagoon considered in a series of five wetlands at 3.1 cm/day, loading rate of 145 kg/ha/day for chemical oxygen demand (COD), and 30-kg/ha/day total TKN at a maximum 49.8 % BOD_6, 60.7 % COD, and 36.8 % PO_4 [41].

3.3 *Lagoons*

A lagoon is an earthen basin that treats wastewater and stores both liquids and solids [2]. Lagoons can store wastewater, manure, or rainfall runoff [42]. Lagoons are capable of reducing BOD and chemical oxygen demand (COD), nitrogen, and odors [2]. Lagoons can take a round, square, or rectangular shape with a typical length-to-width ratio of 3:1 [43]. In addition, lagoons can be situated as a single- or multiple-stage lagoon system. A single lagoon is divided into three major volumes—sludge storage, treatment, and effluent storage. Above the effluent storage is a freeboard for the purpose of protecting the lagoon from storm situations [44]. Figure 1.13 provides a cross-sectional area of a lagoon.

In the sludge storage, sludge settles at the bottom of the lagoon and is digested at the top of the layer. Over time sludge will accumulate within this layer until it becomes equal to the liquid present. The treatment volume is located above sludge storage consisting of manure at the bottom. Biological degradation converts sludge into organic acids and other compounds. The products of organic acids include methane and carbon dioxide, hydrogen sulfide, ammonia, and volatile organics. Treated wastewater not leaving the lagoon is stored in the effluent storage section. Effluent is stored for the purpose of watering crops [44].

Lagoons are designed based on a 25-year, 24-h storm event. This value is contingent on the location of the lagoon as the 25-year, 24-h storm event varies

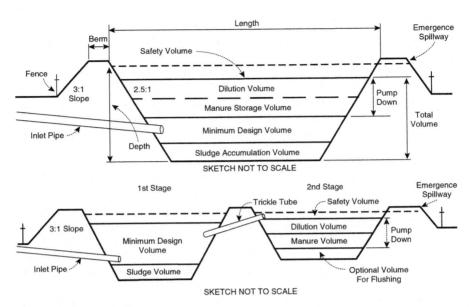

Fig. 1.13 Cross-sectional area of lagoons [13]

across the country. The design loading into the lagoon is determined by the number, size, and the species of animal, along with the geographical location of the lagoon. Prior to land application, dewatering the lagoon is very important. Frequency of dewatering is contingent on the salt concentration and the soil type [45].

The sizing of a lagoon is based on the volume, depth, and the pH. The volume of a lagoon is contingent on the loading rate of volatile solids per 1000 ft^3. This is a function of temperature. The depth of a lagoon is predicated on the precipitation and evaporation rates where the lagoon is located. A typical minimum depth is 6 ft but can be 10 ft for colder climates. However, these values are general and are contingent on the type of lagoon constructed. The optimum pH should be maintained at 6.5 to avoid inhibiting methane bacteria. Anytime when the pH is below 6.5, lagoons will experience a high organic loading [2].

Before construction of a lagoon, it is imperative that a soil and groundwater study is done. This is to ensure that sensitive areas are protected from any discharged from the lagoon. These areas would be any region that leads to surface runoff. Avoid areas that are geologically unstable [42]. Pretreatment of wastewater may be beneficial to reduce odor if the BOD$_5$ loading rate is 50-lb BOD$_5$/ac/day and the depth of the pond is between 6 and 20 ft [43]. In addition, lagoons should be in close proximity if manure is scraped into the lagoon or below the manure source [42].

Lagoon maintenance is important for controlling odors. Lagoons should be analyzed for the presence of algal blooms. Algal blooms occur in basins that have high loading of nutrients (nitrogen and phosphorus). If a lagoon is void of algal blooms, ensure that aerobic lagoons do not become anaerobic. Anaerobic conditions can produce products that can cause odors. The operator should also check and if necessary provide adequate dilution of waste prior to entry into the lagoon and avoid overloading [46]. This can be accomplished by using a combination of runoff and wash water [45]. If odors still persist, lime addition to the lagoon can reduce the presence of odors [46].

Lagoon operators should also evaluate the species of algae and check for the presence of weeds and grasses and protect from erosion and unauthorized access. A healthy lagoon should have green algae. Blue-green and filamentous algae can clump within a lagoon blocking the sun. Gray, black, or purple algae are very unhealthy for a lagoon. The presence of weeds can cause a lagoon to short circuit, thereby affecting the flow of wastewater within the unit. Grass covers on the slopes, and level surfaces of the lagoon can be beneficial but should be mowed and properly fertilized and should be checked for food, trash, or scum on or near the premise. These items should be discarded. Trees or any bushes should not be present near the berm of a lagoon and should be removed [46]. This will also protect the embankments [44]. In the event of erosion, operators should determine the source and make necessary adjustments to the lagoon if necessary. Unauthorized activity can be avoided by placing fences and warning signs adjacent to the lagoon [46].

Finally, operators should also monitor the sludge storage and sludge depth. Remove excess sludge that has accumulated within the lagoon [44].

3.3.1 Anaerobic Lagoons

Anaerobic lagoons are the most common lagoon used for treatment of livestock wastewater. One of the biggest reasons is because anaerobic bacteria have a higher rate of organic decomposition as compared with aerobic bacteria [42]. This is because anaerobic bacteria operate in environments without molecular oxygen a condition that does not require constant maintenance. Generally, anaerobic lagoons are usually very deep. Ranges for depth can vary on the region [46]. For example, the University of Missouri Extension and the State of Mississippi state that lagoons can have depths between 8 and 20 ft [42, 43]. Based on treatment desired, lagoons can be designed to be completed as single stage with no secondary treatment or in multiple stages where further treatment is completed by additional lagoons [45]. Figure 1.14 is a diagram of a two-stage anaerobic lagoon system.

Anaerobic lagoon can be circular, square, or rectangular. A length-to-width ratio of 3:1 for rectangular anaerobic is desired, with earthen dike and bank slopes between 2:1 and 3:1 [42, 43, 45]. Anaerobic lagoons should have a 1-foot spillway below the top of the berm where inlets should be located on the longest side of the lagoon [42].

During the wastewater treatment process, anaerobic lagoons separate into top and bottom layers. At the top of the lagoon, less dense materials such as oils float to the top of the lagoon, while sludge settles the bottom. The presence of oils and other materials prevents oxygen entry maintaining anaerobic conditions within the system [46].

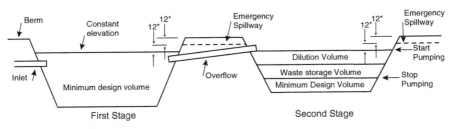

Fig. 1.14 Two-stage anaerobic lagoons for livestock manure treatment [47]

Fig. 1.15 Floating aerator [47]

Anaerobic lagoons are sized based on the volatile solid (VS) loading rate. These values can be expressed in 1000 ft³/day or lb VS/1000 ft³/day. These numbers are affected by the climate. For example, in South Carolina the volatile solid loading rate is 5-lb VS/1000 ft³/day, while Iowa has a VS loading rate of 3.5-lb VS/1000³/ day [48].

Nevertheless, anaerobic lagoons are problematic because of odors. These odors are a product of hydrogen sulfide, ammonia, organic acids [49], and methane. Odors can also be caused by winter to fall and summer to fall turnover within the lagoon or during land application [42]. There are many solutions that can resolve persisting odor problems in a pond. Anaerobic lagoons can be covered to prevent the release of methane gas exiting the system. Anaerobic lagoons can also have induced aerobic layers at the top of the lagoon. This can be done by including a floating cover or aerating the top of the lagoon at very low rates [44]. Figure 1.15 is a floating aerator.

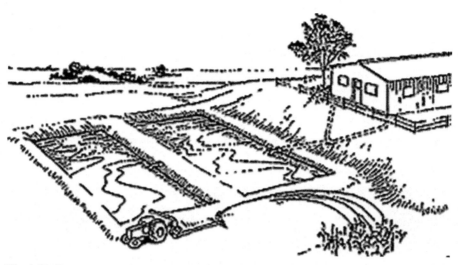

Fig. 1.16 Two-cell aerobic lagoon to treat swine waste [50]

3.3.2 Aerobic Lagoons

Aerobic lagoons degrade organic matter by the application of dissolved oxygen throughout the lagoon. Because dissolved oxygen persists throughout the lagoon, odors are not present within the system. In order to maintain aerobic conditions, aerobic ponds are shallow but require a large land requirement. These ponds are more commonly found in warm and sunny climates. There are two subcategories of aerobic lagoons—naturally and mechanically aerated [46]. Figure 1.16 is a diagram of an aerobic lagoon.

Naturally aerobic oxidation ponds reduce organic materials within wastewater by using either oxygen from the atmosphere or algae by means of photosynthesis [46]. Wind on the pond surface also mixes with the water within the oxidation pond [44]. These ponds are very shallow with a minimum depth between 1 and 5 ft and with a maximum of 5 and 6 ft [46]. The main design parameter is the organic loading rate which is typically 50-lb BOD_5/acre of surface area [49]. Nevertheless, naturally aerobic lagoons are not often used for the treatment of livestock wastewater.

Mechanically aerated lagoons mix oxygen throughout the lagoon by mechanical means. The need for supplying energy can make these lagoons expensive. In many cases solar or wind power supplies the power to operate aeration equipment. Also the lagoons can be designed to have anaerobic segments to reduce energy requirement [46]. Compared with naturally aerobic, mechanically aerated lagoons do not have a large area requirement, but usually have a depth of 10 ft. However, in addition to being more expensive, mechanically aerated lagoons tend to generate more sludge, have a high tendency for foaming, and may require additional treatment such as a septic tank to collect and remove solids [49].

3.3.3 Facultative Lagoons

Facultative lagoons are basins that operate in both aerobic and anaerobic conditions. These lagoons can be arranged as a two-staged pond system where each pond has a depth of 4 ft or as a single pond with a depth of 6 ft [43]. Facultative lagoons usually have three layers. At the top is an aerobic layer. This layer receives sunlight and wind promoting the process of photosynthesis and provides oxygen. The middle layer is a facultative layer. In this layer of the lagoon, anaerobic and aerobic conditions exist. The extent as to which condition is prominent is contingent on the location of the geographical location of the lagoon [46]. Bacteria that can thrive in anaerobic or aerobic conditions (facultative bacteria) are present in this region [51] are commonly found in this layer. The bottom layer is anaerobic. This layer contains an accumulation of sludge from lagoon activities [46]. Because of the layering of the lagoon, odors can be minimized [51].

3.4 Thermal and Biological Chemical Treatment for Biogas Production

3.4.1 Description

Recent developments have occurred where there has been a call for the conversion of livestock wastes that can be used for energy, specifically biofuels. To summarize, biochemical processes are transforming organic materials to fuels by means of various processes such as anaerobic photosynthesis. Following a biochemical process, the remaining solid and slurry within the reactor become viable as a reusable resource such as a fertilizer [52].

Thermochemical processes transition organic matter into gas, fuels, or other carbon residuals by the use of high temperatures to physically convert the bonds of organic matter. Some of the major chemical conversion procedures include combustion, pyrolysis, gasification, and liquefaction [52].

3.4.2 Pyrolysis

Pyrolysis ultimately transfers a given biomass into either char or a volatile gas that can form bio-oil or combustible pyrolytic oils. Slow pyrolysis methods have been used to form char, an entity that has the benefit of producing energy for coal combustion plants, or as an addendum to soil. Some authors have found that chars from various pyrolytic processes are capable of having better absorption than those made from granular activation carbon [52].

There are two major types of pyrolysis—fast and slow/moderate. Fast pyrolysis is a pyrolytic process that consists of using high heat rate and residence time. The resultant products include low molecular weight or an insoluble organic compound such as tar. Reactor examples include bubble fluidized bed, circulating fluidized bed, and vacuum reactor. The requirements within fast pyrolysis include a particle size less than 1 mm. Slow/moderate pyrolysis is the antithesis of fast as it requires a long vapor residence time and low heat rate. The resultant products are charcoal, depending on the concentration of lignin and hemicelluloses. Examples include rotary kiln and moving bed reactor [53].

Pyrolysis applications have been experimented with considering various manure types. It has been determined that the effectiveness of char production was based upon manure type and the conditions, as it was observed that organic materials differ between two different waste types [54].

3.4.3 Direct Liquefaction

Direct liquefaction is another thermochemical process that converts organic material and specifically hydrolyzes lignin components into various organic oils. Ideal conditions for liquefaction would be having very high pressures (5–20 MPa) and low temperatures (250–350 °C). Following the process, the remainders of direct liquefaction are nonreactive and stable and are then converted into oil-based compounds with high molecular weights [52].

The process of liquefaction begins by the bonds of organic material that is broken into simpler compounds resulting into the forms of chars, instead of the process of oils. To prevent the formation of chars, solvents are typically added to slow down higher order solid-state reactions, reducing condensation and the subsequent char formation. Examples of the solvents that are used include dioxane, MDSO, DMF, acetone, and methyl alcohol [53].

3.4.4 Gasification

Gasification operates at high temperatures and atmospheric pressure within the range of 800–1300 °C for the purpose of producing chars and a low-energy fuel. The gasification process has three components:

First, pyrolysis, or the conversion of organic materials into both tars and hydrogen-based combustible fuels.

Second, exothermic reactions with the presence of oxygen can occur to remove the bonds within the organic material at high temperatures.

Third, methanation or the formation of methane from hydrogen and carbon monoxide proceeds where the conditions consist of lower temperatures [52].

A fixed-bed 10 kW power, countercurrent atmospheric pressure gasifier was capable of achieving a gas product made from either high ash feedlot manure (HFB) or poultry litter biomass (HLB) that consisted of the following product: H_2: $5.8 \pm 1.7\%$, CO: $27.6 \pm 3.6\%$, CH_4: $1.0 \pm 0.5\%$, CO_2: $6.7 \pm 4.3\%$, and N_2: $59.0 \pm 7.1\%$. Ideal processes included air-blown gasification for the purpose of having a higher energy fuel [55]. Also, nickel or aluminum can be used as a catalyst to prevent cracking tar [56].

Priyadarsan et al. [57] completed gasification studies for the production of both cattle manure and chicken litter biomass under batch mode where it was determined that the molar composition of gas was 27–30 % CO, 7–10 % H_2, 1–3 % CH_4, 2–6 % CO_2, and 51–63 % N_2 based on the use of air mass flow rate of 1.48 and 1.97 kg/g, where particle sizes are 9.4 and 5.15 mm, respectively [57].

3.5 Composting

There are many reasons to compost. Composting is done to reduce organic material, degrade dead livestock, and reduce disease transmission at a low cost. There are several factors that affect the quality of composting—carbon-to-nitrogen ratio, moisture content, temperature, and the type of composting materials [58]. A proper carbon-to-nitrogen ratio reduces the odors while the temperature affects the microbial degradation [59]. The temperature affects degradation processes. During the winter season, degradation can be reduced in some places by 20 % [60]. Composting materials include sawdust, wood chips, and litter. Composting consists of microorganisms (bacteria and fungi) degrading organic materials within the compost pile to simple products [58].

The general composting values are shown in Table 1.12 below. These values are based on manure composting. Composting consists of primary and secondary processes. In primary composting, the temperature is raised and the organic material is degraded. As composting progresses, degradation begins slowly and the temperature is reduced. Eventually degradation ends and the material left idle [59] in a process known as curing. Curing maintains the conditions within the pile. It also allows items such as bones to be degraded [60].

There are two types of composting facilities—bins and piles or windrows. These are contingent on the type of livestock industry. Bins are used in poultry and swine. Beef and dairy cattle use piles or windrows [58]. Windrows or piles place materials into rows at triangular cross-sections. They are usually combined with bulking agents [61]. Aeration occurs by turning the piles by using front-end loader or compost turners [62]. Piles constructed in arid regions will need to receive outside moisture. This can be done by using a high-pressure nozzle from holding ponds or lagoon wastewater. On the other hand, piles in areas with precipitation may need to

Factor	Value
C:N ratio	25–40:1 (*optimum*: 30:1)
Moisture	40–65 % (*optimum*: 50 %)
Temperature	43–66 (*optimum*: 54–60 C), >71 not ideal
Site selection	1–5 % (2–3 % account for runoff and erosion)

Table 1.12 Factors that affect composting [58]

be covered to protect from odor production [62]. Bins can be designed to have dimensions of 6×8 ft with a wall height between 5 and 6 ft. Bins can be constructed of 2×6 or 2×8 lumber or using plywood with a 2×6 to provide support behind the plywood [63]. The foundation of bins can be made up of pallets, gravel, concrete, and bare soil [64].

There are two entities that can be composted—manure and dead livestock. Dead animal composting is an option to remove livestock carcasses without having detrimental effects on the environment [58]. Dead animal composting maintains aerobic conditions provided gases and liquids are taken away from the system [59]. Livestock operators should consider state requirements to decide what the state requirement of handling dead animals is. For example, in the state of Kansas, composting facilities of dead livestock require a roof and floor to sustain moisture and avoid groundwater contamination with a fence surrounding the facility [58]. The process of composting is contingent on the size of the carcass materials [59].

A dead animal composting pile begins with a layer of sawdust 1–2 ft in depth. The dead livestock are then spread evenly across the sawdust layer [59]. Animals are laid on the side in an attempt to maximize the space for livestock [60]. Another layer of sawdust 2 ft in depth covers the dead animals. This second layer of sawdust maintains heat, prevents odors from escaping, and collects liquids and air to encourage microbial activity within the pile [59]. The amount of sawdust needed is contingent on the type of livestock to be composted. A rule of thumb for sawdust application is in every 1000 lb of carcass, apply 7.4 yd^3 of sawdust in the dimensions of 9×10 ft [60]. When livestock need to be added to the composting pile, the top sawdust layer is removed exposing the dead animal layer. At this point, the new animals are added and then covered up with a new sawdust layer. To maintain the quality of the pile, it is advised that the pile is turned every 90 days. Once composting is complete, the products can be land applied or reused in other capacities [59]. This will usually happen anywhere between 4 and 12 months of composting time [60].

While composting dead livestock is advantageous, there are several concerns involved. These can include leachate of fluids from the carcasses entering into surface and groundwater and disease-spreading pathogens [64]. Therefore, it is necessary to consider the best place to site the place for composting dead livestock. Changes can include placing the facilities away from the water table, away from low permeable soils, and downwind from neighbors. Facilities should also be constructed away from livestock to suppress disease potential. Livestock operators should also have an emergency plan in case of outbreak [60]. For additional protection the livestock operator can create a barrier wall to prevent access to the composting pile. The barrier can be 4 ft high using four steel t-posts with concrete floors, wooden walls, and a metal roof [64].

3.6 Vermicomposting

An alternative method of treating wastes that has been used related to composting is vermicomposting. Vermicomposting is a method where earthworms digest a small portion of organic matter where the majority becomes waste in a form known as worm casts. The processes involved in earthworm digestion are typically physical mechanical, grinding and mixing, and biological or microbial decomposition in nature. In vermicomposting, waste is added to the system. It must be added into the system in thin layers for the purpose of increasing degradation. There is great competition between earthworms and microorganisms for the carbon sources. Application of waste can change—it will either increase or decrease productivity [65].

Vermicomposting treatment technology has been used extensively in animal excretion, sewage, and agroindustrial wastes but not animal manures. Therefore, Loh et al. [65] treated cattle and goat manures using the earthworm, *Eisenia fetida*. The experiments found that total C, P, and K were high in goat manure worm casts as compare to cattle, whereas cattle worm casts were richer in N content. In addition, cattle manure had a higher biomass and reproductive performance as well along with a higher cocoon production per worm [65]. Other studies have been compiled on cow, buffalo, horse, donkey, goat, animal [66], dairy [67], and pig [68, 69] to name a few. Aira and Domingues use a continuous feeding reactors to compare two different types of pig slurry composted by 500 earthworms (*Eisenia fetida*); microbial biomass specifically with 3 kg of pig slurry; loss of C not related by the pig slurry rate; rate of manure-earthworm relationships [70].

3.7 Summary

There are many treatment methods that can be considered for the handling of wastes that persist within the livestock industry. An operator must consider what is available in regard to space and the desired treatment needed in order to make an appropriate decision on selecting the proper treatment method.

4 Land Application of Livestock Wastes

4.1 Description

Land application is a waste management technique that involves recovery of nutrients from manure by plants for the purpose of producing a crop [2]. The classification of manure depends on the percent of dry matter present and the type of livestock waste industry. Manure can be in liquid (less than 5 % dry matter), semiliquid (5–10 % dry matter), or solid (greater than 15 % dry matter) form.

Generally, beef and poultry industry handle solid manure, while dairy and swine manure are usually in liquid form [71].

Regardless of the industry, the nutrient content is a primary focus for application. Nutrient content within the manure is affected by the type of animal species, the process for handling of manure, livestock housing, bedding system, diet, temperature, and the nutrients present. The primary nutrients of concern are nitrogen, phosphorus, and potassium. The nitrogen presence affects the type of plants and quality of the produce. There are two important forms of nitrogen that must be considered—organic nitrogen and ammonium nitrogen. When organic nitrogen enters into soils, it is mineralized into inorganic nitrogen. Mineralization is contingent on the temperature and time of year. Warm and moist soils are better for the degradation of organic nitrogen as compared with cool and dry soils. Ammonium nitrogen is converted to organic nitrogen by plants in a process known as nitrification. Twenty-five to fifty percent of organic nitrogen is converted to ammonium nitrogen. However, improper application of manure can lead to volatilization or the conversion of ammonium nitrogen to ammonia nitrogen. This becomes problematic because ammonia nitrogen dissipates into the atmosphere On the other hand, potassium and phosphorus must be converted to inorganic forms in order for it to be of use by plants [72]. Manure can also be problematic because it can produce various gases. These gases can have grave effects depending on the concentration. Table 1.13 summarizes the major gases found in manure. Previous treatment methods can effect land application. Table 1.14 discusses the various treatment processes and their effects on land application. Therefore, the type of handling equipment and time and rate of application should be considered if an operator is to consider land application.

4.2 Manure Handling Equipment

The equipment necessary for handling manure depends on the type of manure. Each operator must make a decision of handling manure that best distributes the nutrients to the crops being planted. Depending on the type of manure handled, there are unique pieces of equipment that are used in order to safely move the manure onto the field.

Table 1.13 Manure gases [73]

Gas	Effects (percent indicates percent or concentration in ppm)
Ammonia (NH_3)	Eye irritation (<1 %)
	Coughing, irritation of the throat, eyes, lungs (3–5 %)
Carbon dioxide (CO_2)	Difficulty breathing, drowsiness, headaches (3–6 %)
	Death (>30 %)
Methane (CH_4)	Asphyxiation (5–15 %)
Hydrogen sulfide (H_2S)	Dizziness irritation, headache (50 ppm)
	Death

Table 1.14 Various wastewater and biosolid treatment processes and methods and their effects on land application processes [74]

Process/ method	Process definition	Effects on biosolids	Effect on land application process
Wastewater treatment process			
Thickening	Low force separation of water and solids by gravity, flotation, or centrifugation	Increase solid content by removing water	Lowers transportation costs
Stabilization method			
Digestion (anaerobic and/or aerobic)	Biological stabilization through conversion of organic matter to carbon dioxide, water, and methane	Reduces biological oxygen demand, pathogen density, and attractiveness of the material to vectors (disease-spreading organisms)	Reduces the quality of biosolids
Alkaline stabilization	Stabilization through the addition of alkaline materials (e.g., lime, kiln dust)	Raises pH. Temporarily decreases biological activity. Reduces pathogen density and attractiveness of the material to vectors	High pH immobilizes metals as long as pH levels are maintained
Heat drying	Drying of biosolids by increasing temperature of solids during wastewater treatment	Destroys pathogens, eliminates most of water	Greatly reduces sludge volume
Chemical and physical processes that enhance the handling of stabilized biosolids			
Conditioning	Processes that cause biosolids to coagulate to aid in the separation of water	Improves sludge dewatering characteristics. May increase dry solid mass and improve stabilization	The ease of spreading may be reduced by treating biosolids with polymers
Dewatering	High force separation of water and solids. Methods include vacuum filters, centrifuges, filter and belt presses, etc.	Increase solid concentration to 15–45 %. Lower nitrogen (N) and potassium (K) concentrations. Improve ease of handling	Reduces land requirements and lowers transportation costs
Advanced stabilization method			
Composting	Aerobic, thermophilic, biological stabilization in a windrow, aerated static pile, or vessel	Lowers biological activity, destroys most pathogens, and degrades sludge to humus-like material	Excellent soil conditioning properties. Contains less plant available N than other biosolids

4.2.1 Solid Manure

Solid manure is incorporated at the surface by using spreaders that are trucked mounted or trailer towed. Regardless of the type of spreader, manure can be spread at the side or the rear. Nevertheless, rear manure spreaders are more likely used today [71]. For example, livestock operators in the state of Missouri primarily use rear-end box-type spreaders with beaters. These spreaders can consist of a conveyor with chains to move manure from the front of the spreader to the beaters or a front end gate that moves the manure to the beaters [75]. Once it is moved to the rear, the beaters scatter the manure onto the ground [71].

Rear-end box-type spreaders can have single, horizontal, or double vertical beaters. However, each beater type is limited in its ability to properly distribute nutrients. Single beaters cannot spread manure homogenously onto the land. Horizontal beaters only spread manure in areas of close proximity to the trailer. Double vertical beaters spread manure very wide and thin. Overall, rear-end box-type spreaders have a problem with spreading manure homogenously onto the land. They are also very heavy and have the potential to compact soils if land application is done in the fall and spring. Similar to box-type spreaders, truck-mounted spreaders apply manure using double beaters in various horizontal or vertical configurations. Regardless of application, solid manure handling should be applied within 24 h. This is to ensure the minimization of nutrient loss, the presence of odors, nutrient runoff, and compaction [71].

Since the application of solid manure can generate odors, there are methods that can be done to suppress odors in manure land application. These include placing a cover over solid manure not being applied, using a chemical treatment such as alum (also advantageous for preventing ammonia volatilization), and considering the wind direction when applying onto the surface. There are also mechanisms that can be employed that can better spread the manure upon entry on the field. These include a tandem disk or a field cultivator. Solid manure can also be pretreated by drying or composting [76].

4.2.2 Semisolid Manure

Semisolid manure is handled by using spreaders with an end gate. The configuration can range from side discharge or a V-shaped hopper. Each of these can be handled by power take-off (PTO) or ground wheel tractor spreaders or a truck-mounted spreader. The process of application consists of moving the manure by augers to be flung at the point of emission on the spreader. Manure is flung either by using a rotating or flail-type expeller. A rotating expeller directly flings manure, while in a flail-type expeller, manure travels from a hopper onto a rotating shaft with chain-suspended hammers. Once the manure is on the hammers, it is tossed onto land [75].

4.2.3 Liquid Manure

Liquid manure can be applied at or below the surface. Surface application of liquid manure is completed by fixed sprinklers, hand-carried sprinkler, traveling guns, or central pivot irrigators [75]. Factors that control application by irrigation equipment are nozzle size and pressure. These affect the size of the drops applied to the surface. Larger size drops are greatly preferred to control the loss of nitrogen and decrease odors [76]. A recommended size is greater than 150 microns. Other ways include adding dilution water or drop nozzles [77]. Surface application of manure is preferred in areas where odors and nutrient loss are minimal [75]. Figures 1.17 and 1.18 provide diagrams of irrigation systems.

Subsurface application injects liquid manure below the surface where it is then applied below the soil surface by a self-propelled application. Manure can also be transferred by a drag hose or a tractor-drawn applicator. The method chosen is determined by the size of the operation. Usually larger operations opt to use a drag hose or a tractor-drawn applicator. When liquid waste is applied below the surface,

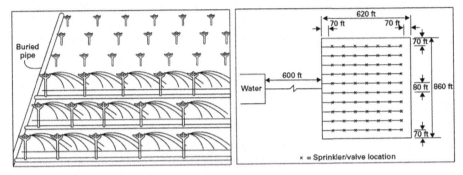

Fig. 1.17 Irrigation system to apply liquid manure [75]

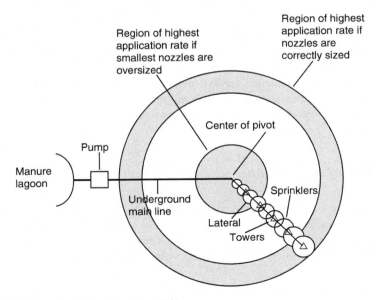

Fig. 1.18 Center pivot irrigation system [78]

injectors have chisels that break up material or sweeps which uniformly apply the liquid manure below the root surface to avoid leaching [75]. Chisel-type knives also prohibit odors and volatilization, while sweep-knife injection reduces volatilization, denitrification, and material degradation [72].

Subsurface is preferred to surface application for several reasons. First, subsurface reduces the potential of ammonia nitrogen emissions [75], greenhouse gases, and odors [71]. For example, research has shown odors to be reduced by 90 % when incorporating a subsurface method [77]. Second, subsurface application reduces runoff potential availing more nitrogen to plants. Third, subsurface injection spreads the manure so it does not have an impact on the surface of the soil. Despite its many advantages, subsurface application is energy intensive; requires more maintenance, time, and management; has higher equipment costs; and is incapable of being used on rocky soils. Therefore, assessment should be made to determine whether or not subsurface injection is a more viable option than any surface application method [79].

4.3 Time of Application

The time when manure is applied determines nutrient availability to plants. Spring is the best season for manure application because nutrients are broken down into the soils during the growing season. Organics are quickly broken down in the soils increasing nitrogen availability. Summer applications are appropriate if growing hay, pasture, and warm-season grasses and if application is completed by travel guns or the central pivot system. Applying manure during the fall is only appropriate if temperature stays below 50 °F [79]. This is because manure is immobilized and remains in the soil [72, 79] leading to more time for degradation. But when the temperature is above 50°, nitrification, leaching, and denitrification occur [72]. Winter application of manure is never recommended as manure hardly enters the soil and has a higher potential for runoff [79] into surface waters. If manure application is a necessity in the winter, apply at low concentrations or during periods of snow melt [72].

4.4 Rate of Application

The amount of nitrogen, type of manure, how manure is applied and used, and additional economic or environmental are the factors that determine how frequently manure will be applied to a given crop. The University of Minnesota Extension provides four steps to determine the process by properly determining the rate of application [72]:

- Determine the nutrient needs of the crop.
- Analyze the nutrient content within the manure.

- Uncover the nutrient available to the crop.
- Compute the rate of application.

4.5 *Summary*

In summary, the purpose of land application attempts to resolve the issue of losing nutrients that are vital to the growth of crops. Manure should be land applied uniformly to avoid the volatilization of nitrogen into the atmosphere. It should also maintain the potassium (K) and phosphorus (P) on the field. The time of application should be considered in order to have nutrients maintained within the soils and avoid any subsequent losses that occur during improper times of application [80]. Manure application should be done to avoid the presence of odors [75] and other potential issues. The rate of application depends on the crop's needs.

5 Storage of Livestock Wastes

5.1 *Description*

Most often the treatment of livestock waste is done for the purpose of recycling products back within the system. This can include land application for growing plants. However, there may be times when the conditions are not conducive for reusing treated wastes. Therefore, livestock wastes must be stored until the appropriate conditions take place. There are several factors that should be considered when deciding whether or not to store manure: first, if the soil is saturated, wet, frozen, or snow covered or if the soil will compact under the weight of manure handling equipment; second, if the temperature and/or humidity creates a proper environment for the generation of odors; third, if a livestock operation may not have the proper equipment or personnel available to apply manure at the present moment; fourth, if the cropping schedule may require temporary storage; and finally, if there is a higher volume of manure and wastewater than what can be handled [81]. There are several methods for storing wastes. These methods are employed usually based on the time of storage and type of waste treated—i.e., solid, semisolid, and liquid wastes.

5.2 *Storage Time*

Livestock wastes can be stored either on a short-term or long-term basis. When wastes are stored for 60–90 days or up to 180 days, it is termed as short-term storage. Short-term storage is a viable option when poor weather conditions persist or when setup is not appropriate to properly handle manure. Short-term storage is also used in mild climates or when growing crops [82]. However, it is very seldom for operators to store liquid manure on a short-term basis. Dairy wastes are the most appropriate to be stored short term.

There are many methods for storing manure on a short-term basis. These can include stacking within a field, covered with a plastic sheet, or storage in a detention pond. Manure can also be scraped into open lots in mounds or inside pole sheds. Regardless of the method, the operator should choose to avoid any contamination of water supplies or exposure to bacteria from the manure [83].

Long-term storage can last for approximately 180 days. Facilities are available to hold solid, semisolid, or liquid wastes. For example, walls and slabs can stack solid manure, while semisolid pumps or scrapers help transport waste into areas designated for storage. Liquid waste is usually transported by pumps or pipes [83]. Sometimes manure can be held for longer than 180 days. For example, waste is stored for 6 months for the purpose of application on annual row or small grain crops. In the center and upper midwest, storage can happen for a full year if fall applications are unsuccessful because of wet conditions [81].

5.3 Facilities to Store Livestock Waste

There are many facilities that can be used to store manure. However, the practicality of each facility depends if the operator is storing solid, semisolid, or liquid manure. Table 1.15 provides an estimated cost for manure storage facilities.

5.3.1 Solid Manure Storage

The objective in storing solid manure is to reduce the volume, odor, and the potential for runoff. Solid manure is stored based on climate and industry. Because the evaporation rate is greater than precipitation, arid regions can store solid manure

Table 1.15 Estimated costs for manure storage facilities. Numbers based on 500,000 gallon capacity [84]

Storage type	Cost ($/1000 gallon)
Naturally lined earthen basin	25–36
Clay-lined earth basin using clay on-site	50–70
Clay-lined earth basin using clay from off-farm borrow site (depending on hauling distance)	80–100
Earthen basin with plastic liner	100–140
Earthen basin with concrete	120–280
Aboveground precast concrete tank	200–250
Aboveground concrete tank poured in place	230–270

in a different fashion as compared to regions that retain precipitation. Arid regions simply store manure in stacks or piles. In the beef and dairy industry, manure is composted using windrows or piles, while in the poultry industry, the manure is contained inside stack houses. On the other hand, non-arid regions require the solid manure to be walled with a concrete bottom and covered with a roof. If solid manure is not housed in this manner, it could also be composted [82]. However, there are alternatives for non-arid region storage of solid manure. Purdue University Extension states that if manure is dried and bedding is added to form a solid, it can be stored on concrete pads. Concrete pad storage of manure reduces the potential of groundwater leaching and runoff provided the operator constructs a roof [85].

5.3.2 Semisolid Manure Storage

Pits are a main way to store semisolid manure. Pits in general are a viable option for waste storage because they can reduce waste volume and reduce the production of odors provided they are properly maintained. Pits can be fabricated from concrete or a coated metal or can be completely made of earth. Manure is transferred into them by means of slated floors. Fabricated pits can be constructed for a location completely above, partially above, or below the surface of the ground. The process of transferring semisolid manure is by scraping or flushing the manure from its source. Equipment used for transferring can include collection sump pumps or by gravity depending where the pit is located. Semisolid manure should be agitated before transfer to ensure all suspended solids are relocated into the pit [82].

Pits made from earthen structures are capable of housing large quantities of semisolid wastes. Therefore, operators will need to ensure ample space is available if a pit from earth is to be used [82]. The incorporation of manure at the bottom of the pit protects the pit from leaching nutrients. This is especially advantageous for very clayey soils. Pits are also lined to protect leaching from the walls. The change in fluid levels can alter the stability of the pit, leading to the formation of cracks [85]. In addition, earthen structures require vegetative cover. Maintenance is then necessary for its upkeep. As with fabricated pits, manure entering into an earth structured pit also requires agitation. Transporting semisolid wastes into the pit is easily done with the use of a built-in access ramp. This can make hauling and transporting waste very time consuming. Nevertheless, earthen pits can be a culprit for odor production so proper maintenance is necessary. Despite the time-consuming hauling and the high potential for odors, earthen pits are less expensive as compared to fabricated pits [82].

5.3.3 Liquid Manure Storage

Facilities that can store liquid manure can include lagoons, runoff holding ponds, and storage tanks. Table 1.16 provides a detailed description of the solid content within liquid manure systems. Lagoons are a beneficial option for storing liquid

Table 1.16 Solid content
for liquid manure
systems [75]

System	Solid content
Manure pit	
Swine	4–8 %
Cattle	10–15 %
Holding pond	
Pit overflow	1–3 %
Feedlot runoff	<1 %
Dairy bard wastewater	<1 %
Lagoon, single or first stage	
Swine	½–1 %
Cattle	1–2 %
Lagoon, second stage	<1/2 %

manure because they can house liquid manure for 6–24 months [85], can be cost effective per animal, and reduce odors [82]. Lagoons provide a mechanism for liquid waste to be treated prior to land application [85]. Lagoons require a higher volume than treatment of semisolid manure and must consider the temperature, climate, and the volume of wastewater to be housed. Biological activities in the lagoon are maintained by replenishing the lagoon with dilution water and prevent salt buildup. This should be monitored during high rates of evaporation [85]. Lagoons should also be monitored to avoid a buildup of settled solids [86]. More information on lagoons can be found in Sect. 3.

Runoff holding ponds are typically used for storage during rainfall events. This means that any liquid manure housed must be pumped out following the event [82]. Holding ponds are designed to be smaller than lagoons. This reduces the rate of degradation within the pond. Erosion and overflow are controlled by installing a 12-inch spillway. To maintain liquids within the holding ponds, a settling basin is set up to collect 60–75 % of the solid manure. This allows waste removal to be completed by irrigation systems [85].

Storage tanks for liquid waste can be made from glass, concrete, or earth. Similar to pits storage tanks can be placed above, partially above, or underground. A storage tank is divided into five major sections—residual volume, manure storage, wash water, rainfall and evaporation, and safety volume depth. The residual volume comprises of volume 6–12 inches from the bottom of the tank. Above the residual volume houses the manure. The manure is pumped into this section of the tank and can be stored for 3–6 months. The wash water stores wash or fresh water. If the tank is open, the net rainfall and evaporation section collects any rainfall that may occur. Finally, the safety volume depth provides adequate space to handle a 25-year, 24-h storm event. Depending on the type of material, storage tanks will have a different depth [87].

5.4 Storage Area Design

The storage of manure has been published by the US Department of Agriculture (USDA) and follows the following calculation based on storage volume [2]:

$$VMD = AU \times DVM \times D \qquad (1.2)$$

where

VMD = volume of manure production for animal type for storage period, ft^3
AU = number of 1000-lb animal units per animal type
DVM = daily volume of manure production for animal type, $ft^3/AU/day$
D = number of days in storage period

The second equation calculates the bedding storing volume (BV):

$$BV = FR \times WB \times AU \times D \times BUW \qquad (1.3)$$

where

FR = volumetric void ratio (values range from 0.3 to 0.5)
WB = weight of bedding used for animal type, lb/AU/day
BUW = bedding unit weight, lb/ft^3

The bedding storage value can be multiplied by 0.5 to calculate the volumetric void ratio.

Sizing for a liquid and slurry waste storage can be calculated from the following equation:

$$WV = TVM = TWW = TBV \qquad (1.4)$$

where

WV = volume of waste stored, ft^3
TVM = total volume of stored manure, ft3
TWW = total wastewater stored, ft^3
TBV = total bedding volume stored, ft^3

5.5 Summary

The type of manure affects the manure facility chosen. Within the types of manure, there are various facilities that can house manure. Each facility should be analyzed carefully before installation. This ensure that the proper facility is constructed based on the needs of the operation.

6 Feedlot Runoff Control Systems

6.1 Description

Section 2 of this chapter indicates that feedlots are required to have NPDES permits as defined in the Clean Water Act of 1977 [88]. This limits the amount of discharge that can occur at a particular location. A major source of discharge from feedlots is runoff. There are several different systems that properly contain runoff. Many of systems have been discussed in prior sections, and therefore information concerning the significance for runoff control will only be presented. Runoff control protects a feedlot from the presence of weeds, odors, and insects. The collected water provides an alternative source for fertilizers and irrigation water [89].

6.2 Runoff Control Systems

6.2.1 Description

The processes of a runoff control systems are multifaceted. A runoff control system captures and reroutes rain or snowmelt. It can also provide a method to treat runoff before it is to be discharged. There are two major subsets of runoff control systems—full containment and discharge runoff control systems. Full containment systems (also known as clean water diversion systems) include the use of terraces, channels, and roof gutters [88].

6.2.2 Clean Water Diversion

The purpose of diversion is to control runoff entry into holding ponds and settling basins [90]. In addition, precipitation is prevented from invading manure storage systems preventing the potential for creating polluted runoff [89].

6.2.3 Discharge Runoff Control

Discharge runoff control systems include settling basins and runoff holding ponds. Settling basins are a runoff control system that separates liquid from solids. The separation of liquids from solids allows liquids to be further treated by methods such as storage ponds. Solids settle to the bottom while the liquids remain at the top. There are several processes that will cause solids to separate from liquids. These include risers, slotted board, or porous dams. Settling basins consist of channels or boxes made of concrete or earth. Cleaning the basin is necessary to allow for solid placement. The cleaning of the basin should be done if 50 % of the basin is filled with solids. Solids are taken from the basin and led away from the feedlot.

If cleaning is not permissible, an alternative method is to increase the size of the basin by 25–50 %. Scrapers, high-pressure water systems, and metal screens prevent the system from being clogged. Figure 1.19 is a diagram of a solid-liquid separator [89]. Figure 1.20 depicts a system to handle runoff. Figure 1.21 provides a diagram of a settling basin.

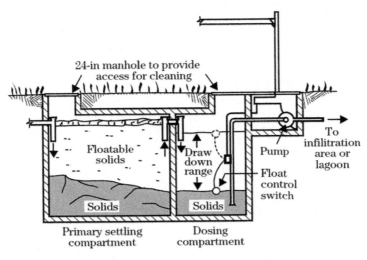

Fig. 1.19 Solid-liquid separator [8]

Fig. 1.20 Lot runoff handling system for milking wastewater [91]

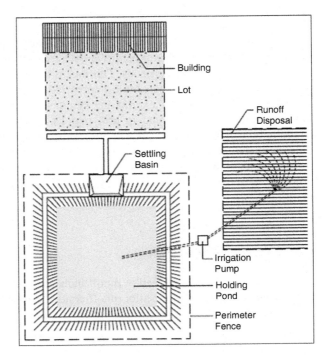

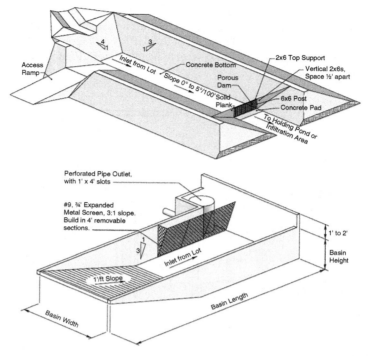

Fig. 1.21 Settling basins for manure management—earthen sidewall settling basins (top) and concrete settling basins (bottom) [91]

Runoff holding ponds receive and store liquid runoff from settling basins. This process can happen 15–30 min before entry into a settling basin [92]. In general, they are smaller than holding ponds. This means that when wastewater is collected, it will only remain in the ponds for a short period of time. They must be dewatered by using equipment such as a sprinkling systems or perforated pipes. However, if holding ponds are constructed in arid regions, dewatering is not necessary as evaporation will be sufficient. Water removed from the holding pond can be applied onto crops [89]. Figure 1.22 is a diagram of a holding pond.

In general, holding ponds are designed based on a 25-year, 24-h storm [89]. The volume chosen for the pond is also contingent upon the time of storage permitted [92].

6.2.4 Vegetative Filter Strips

Another method for control feedlot runoff includes vegetative filter strips. Vegetative filter strips (VFS) are a feedlot runoff control system consisting of vegetation. This vegetation is grown in close proximity to the feedlot reducing constituents

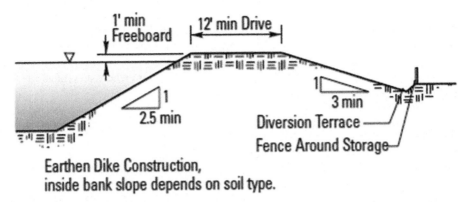

1' min Freeboard

12' min Drive

1
2.5 min

1
3 min

Diversion Terrace

Fence Around Storage

Earthen Dike Construction, inside bank slope depends on soil type.

Fig. 1.22 Holding pond for storing milk house wastewater [91]

such as sediments, nutrients, pesticides [93], and COD [90]. In a VFS system, vegetation uptakes pollutants from runoff prohibiting transport beyond the feedlot. The removal of these particulates from the runoff results in clean water [94]. Associated processes include settling, filtration, dilution, pollution absorption, and infiltration [95]. VFS systems are capable of removing 60–70 % suspended solids, 70–80 % nitrogen [90], 7–100 % phosphorus, and 64–87 % pathogen removal [96]. VFS systems create a mechanism that can reduce nonpoint pollution runoff. Several factors affect the efficiency of a VFS system. These include the type of pollutant, soil type, vegetation, state of flow, and current plant status [93].

The nature of the pollutant is important in determining its ability to be treated by vegetative filter strips. Vegetative filter strips are capable of reducing particulate-bound pollutants in comparison with soluble particulates. Various processes incorporated within VFS are able to be removed by the system as compared with soluble particulates which can only be removed by sedimentation. The type of soil is important because of the various processes that occur within soil. Sandy-loam soils with a depth between 3 and13 ft or clay soil 26–145 ft are ideal for VFS. Vegetation should be dense and rough and must be able to reduce the surface velocity so that collected solids are kept within the system. Flow entering into the VFS system should be overland sheet flow as compared to concentrated flow. Overland sheet flow prevents sediments from leaving the VFS system lowering the velocity of the wastewater within the system [93]. Sheet flow is also uniform throughout the system and is shallow [96]. Channelized or channel flow differs from overland flow because runoff flows through a narrow channel such as a gated terrace or a waterway. This presents a problem because water flows a velocity that is higher than one in channelized flow. Channelized flow also requires more land because the strip will need to be longer to accommodate the channel [95]. Loading into the VFS system is also inconsistent. As a result, channelized flow includes a reduction in treatment and an increase in erosion [96].

There are two types of VFS systems—vegetated infiltration basin (VIB) and vegetative treatment area (VTA). A VTA system plants vegetation downslope from crops or livestock housing. On the other hand, VIB is similar to a VTA with the exception of a berm for runoff collection. Included within the treatment system is the presence of aerobic bacteria to treat nitrogen by means of nitrification. When wastewater enters into the VIB system, nutrients are absorbed into the soil and are used by plants. Runoff is collected through tiles in the system where it is transferred to other wastewater treatment systems [96].

A VFS system is most effective if it has a depth less than 1.5 ft. In this scenario, uptake of pollutants by plants is more feasible. Pollutant removal efficiency is also affected by the length of the VFS—the longer the VFS, the more efficient the treatment [93]. A recommended length for a VFS system is 100 ft or 1 ft/animal unit whichever value is greater. However, the ground slope will affect the length of the system chosen. A 0–2 % slope can have a minimum length of 100 ft, while a 6 % slope a minimum of 300 ft [90]. Other recommendations for design include 200-ft length for a 1-year, 2-h storm and 300 ft for a 0.5 % slope to 860 ft for a 4 % slope. VFS treatment system should include a pretreatment step to settle solids from the runoff [96].

There are many types of vegetation that can be used with a VFS system. The University of Kentucky Extension states the type of vegetation planted within the system is contingent on the season. Five plants are suggested—tall fescue, orchard grass, timothy, Bermuda grass, and gama grass. Tall fescue is an option because it is capable of using nutrients when planted. However, it cannot be used for grazing. Orchard grass not only removes nutrients but unlike tall fescue is capable of being used for grazing but only up to 4 inches. Timothy grass is a viable option for horses and cattle to graze provided grazing is limited to 4 inches. Bermuda grass is a quality choice because it is capable of reducing nutrients and also drought resistant. Bermuda grass can grow up to 8 inches, while grazing is limited to 3–4 inches. Planting gamma grass will absorb nutrients deep from within the treatment system [94].

6.3   Summary

In summary, this section presents several feedlot runoff control systems that are available to divert runoff coming from a feedlot. Feedlot operators must consider the characteristics of each control system and consult the state legislation in order to understand what are the design requirements and limitations to use the treatment method chosen by the feedlot.

7 Odors and Gases

7.1 Odors: Origin and Nature

Dispersed odors form in the air travel and can cause great discomfort for those that live in close proximity to livestock operations. There are three major causes for odor compounds in livestock operations—"the livestock themselves, animal housing facilities, feedlots, and feed storage facilities; manure storage structures; and application of livestock manure to agricultural land" [97]. Particular examples include anaerobic degradation of organics in manure, feed, and silage. Odors caused by anaerobic digestion increase in intensity when temperatures are warm. Also if manure becomes wet, it can also be a major cause for odors [98]. In feedlot operations, incomplete fermentation of nutrients by bacteria in manure produces odors [99].

Odors can spread in the air as a gas. Dust particles can also be agents to carry odors. When particles that cause odors come into contact with dust particles, they are absorbed and carried along. The effectiveness of odors spreading is contingent on the weather. Very humid days maintain the odors in the area, while dry and windy days will disperse them [98]. Rainfall can also increase the emission of odors. If rain water remains on the ground surface, anaerobic conditions can occur on the manure [99].

7.2 Sources of Odors

The major sources of odors are gases, anaerobic decomposition of manure, and other various compounds. The compounds that can provide the biggest issue include volatile fatty acids, mercaptans, esters, carbonyls, aldehydes, alcohols, ammonia, and amines [100]. A major proponent of odors is the formation of volatile fatty acids.

The reason why volatile fatty acids cause so many odors is because of the volatile organic compounds that are present within the manure. Volatile fatty acid presence within manure varies between animal types. For example, the majority of compounds found in pig manure include acetic, propionic, n-butyric, iso-butyric, n-valeric, isovaleric, n-caproic, and iso-caproic acids. These organic compounds vary with the amount of carbon atoms present within the system, where butyric, valeric, and caproic being the highest amount of odor. Other potential dangers for volatile fatty acids increase toxic pathogens within soil base [101].

One can state that the majority of VFAs have carbon numbers between 2 and 9. In addition, the presence of bacteria *Eubacteria, Peptostreptococcus, Bacteroides, Streptococcus, Escherichia, Megasphaera, Propionibacterium, Lactobacilli*, and *Clostridium* is also noted for contributing to the major problems associated with volatile fatty acids [102]. Volatile fatty acids are generated during the process of fermentation, when carbohydrates are broken down from sugars into pyruvate which is then fermented into volatile fatty acids in anaerobic conditions. Therefore, the lack of aerobic conditions such as incomplete microbial decomposition or other anaerobic treatment methods is the major cause of this potential issue [97].

Aromatic compounds are a major concern within animal manure due to the presence of indole, skatole, p-cresol, phenols, and 4-ethylphenol. Under anaerobic conditions, bacteria such as *Bifidobacterium*, *Clostridium*, *Escherichia*, *Eubacteria*, and *Propionibacterium* use aromatic amino acids such as tyrosine, phenylalanine, and tryptophan [97].

Sulfate-reducing bacteria typically cause the presence of hydrogen sulfide due to the reduction of amino acids cysteine and methionine. Sulfur-reducing bacteria typically use sulfate as a terminal electron acceptor transforming sulfate compounds into hydrogen sulfate. The most common bacteria heavily involved in this process are *Desulfovibrio desulfuricans*, *Veillonella*, *Megasphaera*, and the enterobacteria [97].

Ammonia emissions causing odor are commonly attributed to ammonia volatilization. The reason behind such a problem can be contributed to the animal species, diet, and age. For example, urea, the nitrogen compound within urine, typically forms ammonium and bicarbonate ions by means of urease enzymes. Nitrogen found in feces is broken down by bacteria, where it transfers from proteins to amino acids and eventually into ammonium. The time in which this occurs depends on the time, temperature, concentration, and pH [103].

One of the more common entities that are emitted through livestock waste is the presence of hydrogen sulfide. Hydrogen sulfate odor emissions commonly occur from the anaerobic decomposition of sulfur [104]. One of the most common methods of forming hydrogen sulfate is due to the efforts of sulfate-reducing bacteria [105].

Table 1.17 Odor emission strategies for livestock housing [47]

Method	Description
Filtration and biofiltration	1. Filtration traps 45 % 5–10 µm particles; 40–70 % particles greater than 10 µm
Biofilters	1. Biofilters trap and biologically degrade particles; remove odorous emissions 2. Biofilters can remove 90 % odors, including 90 % hydrogen sulfide and 74 % ammonia
Impermeable barriers	1. Dust particles retain odors preventing movement 2. Impermeable barriers such as windbreak walls or dams are very effective
Oil sprinkling	1. Application of vegetable oil can control dust movement 2. Study applying oil reduced hydrogen sulfide concentrations by 40–60 %
Landscaping	1. Application of trees and shrubs 2. Landscaping reduces particulate movement and dilutes the concentration of emissions

Table 1.18 Examples of odor emission strategies for manure storage [47]

Method	Description
Solid separation	1. Removal of large materials, typically the size of a screen opening 2. Removal of large material reduces the loading rates and thereby producing less odors during decomposition of remaining material 3. Solid separation uses processes such as sedimentation, screening, filtration, or centrifugation
Anaerobic digestion	1. Under anaerobic conditions, odors are biologically reduced from manure 2. Anaerobic digestion encapsulates manure maintaining odors
Additives	1. Application of additional enzymes or chemicals to dilute manure under anaerobic conditions
Impermeable cover	1. Coverage of a manure storage area will control odors from gases 2. Impermeable covers can control wind and radiation
Permeable covers	1. Coverage of a manure storage area to control the contact between manure and radiation and wind velocity 2. Emission rates are reduced 3. Permeable covers create an aerobic zone encouraging aerobic microorganism growth
Aeration	1. Application of oxygen by mechanical means to maintain aerobic conditions 2. Aeration can cause an increase in ammonia emissions
Composting	1. Composting provides an aerobic environment reducing the creation of odors 2. A more viable option for those that handle solid manure because of high maintenance required to maintain suitable decomposition conditions

7.3   Odor Prevention

There are various methods to prevent the spreading of odors. These can include animal nutrition management, manure treatment and handling, waste treatment methods, and better livestock operation management. Tables 1.17 and 1.18 provide various methods to mitigate odors.

7.3.1   Animal Nutrition Management

One of the best ways to reduce odors is to alter animal nutrition. If livestock feed contains more crude protein concentration or blood meal, it will lead to the production of odors. Studies have shown that feeding livestock crystalline amino acids or peppermint as compared to a diet heavy with crude protein can reduce odorous manure. Barley-based diets can also reduce odors by 25 % as compared with a diet dominated by sorghum [47, 106]. Fecal starches, proteins, and lipids should be eliminated as much as possible. This will prevent incomplete fermentation which is a main cause of odors [99].

In addition to changing the diet of the animals, the operator should consider a change in feeding schedule. An appropriate feeding schedule could be feeding the animals at sunrise, at noon, and at sunset. This cannot only eliminate the presence of odors but also control the emission of dust in the atmosphere from cattle that move their hooves on the ground. As a reminder, dust can be used as an agent to transfer odors [99].

7.3.2 Manure Treatment and Handling

Another method for reducing odors is to consider the treatment and handling of manure. First, operators can incorporate additives to manure. Additives can be chemical or biological. Additives can be applied to overpower the presence of an odor and reduce the ability for odors to be smelt, can absorb constituents in manure that cause odors, or can slow microbial degradation to reduce odors [100]. Choices for additives are based on the product and the rate and frequency of application [47]. Manure can also be chemically treated. The University of Arkansas Extension recommends several options for chemical treatment. These include sodium bisulfate (PLT), ferric sulfate granular (ferric-3), alum, and zeolite [106].

Next, solid separation can be used to better hand manure. Solid separation processes include sedimentation, screening, filtration, and centrifugation. This process attempts to remove constituents that cannot pass through a specified screen size. The removal of these materials decreases biological degradation and thereby reduces odors [47]. Solid separation also reduces odors by reducing the organic loading. Usually solids are separated before entering a treatment basin such as a lagoon. Some of the materials removed include cattle waste fiber and grit. There are several machines employed for solid separation. These include vibrating screens, sloping stationary screens, or pressure-rolling mechanical separator. Solid separation can occur within a gravity settling basin, earthen settling basin, rectangular metallic, or a concrete settling tank [49].

Finally, operators can make strategic choices in how they land apply manure for the sole purpose of preventing the spread of odors. Spreading manure can be done in the morning or when the sun is present and on days when the direction of the wind is away from the neighbors [100]. Manure can also be applied during the early evening for better wind dispersion [47]. It is best for the livestock operators to choose the weekdays as opposed to weekends when neighbors will most likely not be at home [106]. When manure is applied, it should be applied quickly, in large quantities, and based solely on the needs of the crop [47]. Operations should employ a liquid waste management schedule [106].

If liquid manure is applied by irrigation equipment, operators can make choices on nozzle size of the sprayers. An alternative would be using a low-rise, low-pressure, trickling system. Application of liquid manure should be done in close range to avoid the spread of odors [47]. Instead of the land application of manure by irrigation, operators can also make the decision to inject manure directly into soils as compared to choosing surface application [106].

When solid manure is not directly applied, operators can select to cover the manure before use. There are two types of covers—impermeable and permeable. Impermeable covers prevent manure storage facilities from the emission of odors into the atmosphere. The covers can also reduce the effects of wind and radiation.

Impermeable covers can reduce odors by 90 %. Cover efficiency is contingent on the presence of wind and snow [47].

On the other hand, permeable covers (biocovers) are used to cover places for anaerobic digesters or manure storage facilities [47, 106]. Biocovers can consist of straw, cornstalks, peat moss, foam geotextile fabric, or Leka rock [47]. Biocovers can also include used closed-cell polyurethane foam with or without zeolite. Biocovers remove radiation from the surface of the manure storage facility and also reduce the impact of the wind blowing [106]. Biocovers contain an aerobic zone where aerobic microorganisms thrive on the presence of chemical constituents within the manure. These constituents reduced by the microorganisms reduced the odors. The reduction of odors is contingent upon the material used. Covers that are primarily made of straw reduce odors by 50 %, while 85 % of odors are reduced when the cover consists of a floating mat or corrugated materials [47].

As an alternative to biocovers, manure storage facilities can be aerated to supply molecular oxygen. This will assist in reducing odors. Nevertheless, aeration can be dangerous because nitrogen is volatilized into the atmosphere as ammonia. Therefore, great care should be taken to prevent this from occurring [47].

7.3.3 Waste Treatment Methods

There are many waste treatment methods that can reduce the potential of creating odors. First, operators can install filters to separate odor causing particles within the air. There are two potential filters available—mechanical and biofilters. Mechanical filtration devices are capable of removing odors from particles. There are indications that 45 % of odors are caused by particles with a size between 5 and 10 microns, while 80 % are caused by particles greater than 10 microns. Mechanical filtration has been proven to reduce odors between 40 % and 70 % [47].

Biofilters capture particles where aerobic bacteria degrade them to create products that do not cause odors [47]. Biofilters are supplied air by natural ventilation. The presence of air and adequate environmental conditions allows for the bacteria to grow within the system [98]. Bacteria grow on media consisting of wood chips or compost [106]. For these reasons, biofilters are inexpensive as compared to mechanical filtration. Efficiency of a biofilter is contingent on oxygen concentration, temperature, residence time, and moisture content [47]. The design of biofilters is contingent on the volume of air needed to be treated [106]. Biofilters have been successful in removing 40 % of hydrogen sulfide [47]. It has also been reported that biofilters remove 90 % of odors [106]. Biofilters are also capable of filter odor causing liquids from manure storage [98].

By means of rock wool packing material, Yasuda et al. were able to produce 8.2–12.2-mg N/100 g sample of nitrification and 1.42–4.69 mg N/100 g of denitrification [107]. Ro et al. found that a polyvinyl alcohol (PVA)-powered activated carbon biofilter removed 80 % ammonia nitrogen with hydrogen sulfide removal at 97 % [108]. Kastner et al. ranged between 25 % and 95 % ammonia nitrogen concentration removed in waste from swine production, where the major factors that depended on the treatment efficiency were residence inlet time and ammonia concentration [109].

Second, anaerobic digestion is a feasible treatment method to reduce odors. The biological degradation of constituents under anaerobic conditions can reduce the odors significantly in organic material. The products from anaerobic digestion can be safely placed in a liquid storage facility [98]. A study using anaerobic digestion for degradation of dairy waste reported a 50 % reduction in odors provided the waste remained in the digester for 20 days. While anaerobic digestion is an expensive method, it can be viable for some operators [47]. Anaerobic digestion can be profitable as it produces biogas [98]. More information about anaerobic digestion is presented in Sect. 3 of this chapter.

Various enzymes such as peroxidase, specifically horseradish peroxidase (HRP) and tyrosinate [110], are used to control odors. Horseradish peroxidase (HRP) has become a new method in research for deodorization because the quantity of peroxidase containing within the plant is capable of transforming aromatic compounds into free radicals or quinones which ultimately form non-odor compounds [111].

Govere et al. [111] experimented with pilot-scale reactors with volumes between 20 and 120 L using minced horseradish comparing effectiveness between the addition of either calcium peroxide or hydrogen peroxide to deodorize swine wastewater. From the results, it was determined that the addition of horseradish was capable of completing removing odors [111].

The management of lagoons serves a way of reducing odors. A healthy lagoon will degrade organic materials into constituents that do not produce odors. Odors can be reduced in a lagoon if manure contains a dilution of 1–2 %. Lagoons should also be refrained from having a high solid concentration. When high solids are present, a lagoon is overloaded. Overloaded lagoons change the conditions from aerobic to anaerobic thereby creating odors [98].

7.3.4 Livestock Operations Management

Livestock operators can mitigate the spread of odors by providing better management of the buildings and facilities. This can include disposing unused or even moldy feed, fix leaks, if necessary replace or repair pipes, and designate a location to dispose dead animals. Another alternative is to increase ventilation within these areas. Ventilation can be supplied by mechanical or natural means. Mechanical methods of ventilation include fans and fresh air inlets. If cost is a barrier, an alternative is to use natural methods. Openings, change in roof slope, and rearranging the orientation of the building are ways that a livestock operator can generate natural ventilation within a building or facility. Despite the fact it saves on energy, natural ventilation may be inhibited by environmental circumstances so the operator should make a wise decision on which method should be chosen [100].

In addition to ventilation, livestock operators can introduce landscape onto the premises to contain odors. Landscaping provides an opportunity to prevent the constituents that cause odors from further leaving the operation. These constituents are either dispersed or diluted. Landscaping also gives an aesthetic appeal to the area. Trees and shrubs are the two most commonly entities planted [47].

The design and maintenance of feedlot pens should be reviewed to better prevent odor mitigation. Feedlot pens should maintain a dry surface to prevent the formation of anaerobic conditions on the surface. This means that each pen should be designed to have proper drainage. Having a pen maintain a slope between 4 % and 6 % will provide adequate drainage and prevent pens from accumulating standing water. Also, pen scraping should occur once every 3–4 months [99].

7.3.5   Summary

With many people leaving municipalities and inner-ring suburbs for rural and farmland communities, the discussion on odor mitigation will continue to increase. Therefore, it is important for livestock operators to develop good relationships with the residents living in close proximity to livestock operation facilities. Regardless of the method(s) chosen, the ultimate goal should be to provide neighbors the ability to feel as liberated as possible from the presence of odors.

7.4   Greenhouse Gas Emissions

Recent developments have discussed the relationship between greenhouse gas emissions and livestock. This chapter will discuss some of the current issues related to the relationship between greenhouse gas and livestock waste. The purpose of discussion is not to take sides but rather present what is currently found in literature.

Greenhouse gases consist of carbon dioxide (CO_2), methane (CH_4), and nitrous oxide (N_2O). Carbon dioxide is considered a primary greenhouse gas because in general only 9 % of greenhouse gas emissions are caused by CH_4 and N_2O [114]. However, in the livestock sector, CH_4 emissions is 21 times the carbon dioxide emission, while N_2O is 310 times the CO_2 emissions. This is because animals produce methane during the process of enteric fermentation, while nitrous oxides are formed during the degradation of manure when nitrification and denitrification occur. In general, greenhouse gases maintain the temperature of the earth to 15 °C. The current debate with greenhouse gases involves global warming and climate change. This debate has been whether or not greenhouse gases cause a change in climate [112]. It was reported that from 2001 to 2010 greenhouse gas emissions from crop and livestock, operations increased by 14 % [113]. In 2012, it was estimated that the agriculture industry released 526 million metric tons of carbon dioxide equivalent (MMT of CO_{2e}) plus 62 MMT of CO_{2e} related to operating electric products [112].

According to the US EPA, greenhouse gases have caused 9 % of the total greenhouse gas emissions in the United States, while the United Nations

(UN) have stated 18 % of global emissions have been caused by greenhouse gases. There are many sources of greenhouse gases reported. The United Nations mentions that greenhouse gas emissions are caused by livestock feeding, manure management, livestock processing, and transportation of livestock products. On the contrary, the US EPA states that greenhouse gases have been caused by crop and livestock production. Other sources have stated that deforestation (34 %) and ruminant digestion (25 %) are additional factors that must be considered [112].

According to the University of Missouri Extension's paper titled "Agriculture and Greenhouse Gas emissions," there are four major areas that have been major contributors to greenhouse gases in the agriculture sector—crop and soil management, livestock manure management, enteric fermentation, and agricultural carbon sequestration. These values are contingent on the US production of greenhouse gases in 2012, data produced by the US EPA [112]:

1. *Crop and soil management.* Agricultural crop and soil management produced 307 MMT of CO_{2e} or 48 % of the total greenhouse gas emissions within the agricultural sector. Ninety-eight percent of all emissions from greenhouse gas were because of N_2O. This has been attributed to the fact that cropland has produced more N_2O than lands that are grasslands. In addition, fertilization, manure application, crop residue collection, planting nitrogen-fixed crops and forage, and using soils with organic materials are major practices that lead to N_2O emissions. N_2O emissions occur in the Corn Belt, cropped land in California and the Mississippi Valley, rice production, and burnt fields.

2. *Livestock manure management.* Manure management accounted for 71 MMT of CO_{2e} in greenhouse gas emissions. Most of the greenhouse gases produced in livestock manure are CH_4. The major causes of greenhouse gases include anaerobic decomposition of liquids and slurry. N_2O in manure management is caused by manure, urine, and aerobic and anaerobic degradation. The dairy cattle industry produced 47 % of CO_2 emissions, while the beef cattle industry was responsible for 71 % of CH_4.

3. *Enteric fermentation.* As previously stated, enteric fermentation causes the majority of CH_4 emissions. Enteric fermentation produced a greenhouse gas total of 141 MMT of CO_{2e}. Varying factors determine the production of enteric fermentation. These include the number of livestock and the type of feed.

4. *Agricultural carbon sequestration.* Land use and forestry were responsible for 979 MMT of CO_{2e} or 15 % of overall green house gas emissions. A relationship between land use and carbon sequestration was made. This relationship analyzed the carbon sequestration of land in 2012 and its state 20 years before. Land that remained grassland was capable of sequestering carbon where loses only occurred because of drought. This has also been the case when land was converted into grasslands. On the contrary, land that remained cropland or converted into cropland carbon was not sequestered. However, land that remained cropland was able to sequester carbon provided the organic content remained between 1 % and 6 %.

A more recent study was completed by Caro et al. to assess the global green-house gas emissions between 1961 and 2010. Analysis compared the livestock greenhouse gas emissions between developing and developed countries. The results from the study concluded that global greenhouse gas emissions increased by 51 %, where the primary source of greenhouse gas emissions was caused by enteric fermentation. In general, the generation of greenhouse gases decreased overall. However, there was a difference in the trends for developing and developed countries. Greenhouse gas emissions in developed countries increased in the 1970s and then gradually decreased by 23 %. On the contrary, greenhouse gas emissions increased in developing countries by 117 %. The authors attributed increase to changes in economic and ideological changes. The signature year for these changes occurred in 1989. These countries transitioned from being focusing heavy on importing to exporting. With regard to the various livestock industries, the beef cattle industry was accountable for 54 % of greenhouse gases, while only 17 % was due to the dairy industry [113].

The development of numbers has created an interesting stir within the scientific community. Various authors have published papers that attempt to support the values generated by entities recognizing global climate change (e.g., US EPA, UN, and the International Panel on Climate Change (IPCC)). However, authors such as Herrero et al. request for a reduction in ambiguity and more consistency in methodologies used to quantify greenhouse gas emissions within livestock. The areas of concern include the exclusion of CO_2 production by livestock, quantifying emissions due to land use and land change, global warming potential of methane, and the overall allocation of processes to livestock. With a more accurate picture, the authors state that the discussion of greenhouse gas emissions in livestock can improve [114]. Regardless of an individual's stance on greenhouse gas emissions and global warning, the discussion of the livestock industry's role in greenhouse gas emission will continue.

8 Pathogens in Livestock Industries

Pathogens are an issue within the livestock industry. The impact from pathogenic outbreak causes a loss in productivity for the livestock operation by becoming detrimental to the animals, the business, and employees. Pathogens can also be harmful to the public and the environment. Survival of pathogens is predicated on the temperatures, the pH, the amount of microbial activity, the routes of transfer, and the applicable host. The routes of transfer for pathogens include fecal-to-oral, foodborne, aerosol, or human-to-animal contact. The applicable hosts can range from humans, farm animals, and other carriers such as flies. There are four major categories of pathogens—viruses, bacteria, mycotic agents, and parasites [61].

For example, contact with viruses for a period of time can lead to illness and death and can limit the product from livestock. Viruses are classified as enveloped and non-enveloped viruses. Enveloped viruses persist within animal manure and

Table 1.19 Examples of each type of pathogen (123)

Pathogen	Example
Viruses	Animal enteroviruses, *Rotaviruses*, *Hepatitis E*
Bacteria	*Aeromonas hydrophila* *Aerobacter* *Bacillus anthracis* *Chlamydia* *E. coli* *Salmonella*
Mycotic agent	*Histoplasmosis capsulatum* *Pneumocystis carinii*
Parasites	Protozoa *Ascaris and Ascariasis* *Cryptosporidium parvum* *Giardia* *Toxoplasmosis*

Table 1.20 Pathogen treatment methods (123)

Pathogen	Method
Dry techniques	Composting
Physical treatment	Sand filtration or dry beds Sedimentation and screening
Biological treatment	Lagoon Anaerobic digestion Sequencing batch reactor Constructed wetlands Overland flow
Disinfection	Chlorine Ozone Chlorine dioxide Lime stabilization Pasteurization

can stay for a long period of time without treatment and storage, while non-enveloped viruses are incapable of being destroyed with any treatment method. On the other hand, mycotic agents are not a major concern within the livestock industry and are usually dangerous in soils or self-contained with the body of an animal or human [61]. Examples of each pathogen category are listed in Table 1.19.

Livestock operators can know the quantity of pathogens within its waste by using organisms known as fecal indicator organisms. Fecal indicator organisms are surrogate organisms used in the laboratory as a method for quantifying pathogenic presence. Typically *E. coli* has been used as a fecal indicator organism but recent studies have used other organisms such as coliphages and *C. perfringens* spores. An adequate choice for a fecal indicator organism must fulfill a series of criteria. Fecal indicator organisms must:

1. Exist in the same conditions as pathogen.
2. Have a life span similar to pathogens.

3. Withstand disinfectants and unfavorable conditions.
4. Easily detectable.
5. Distributed randomly.
6. Portray similar risks in humans as pathogens.

As an alternative, testing for microorganisms can include culture-specific microorganisms, antibiotic resistance patterns, molecular fingerprinting, genotype, and chemical indicators [61].

There are various treatment methods that can be used to reduce the pathogens within livestock waste. The treatment of livestock waste can use dry techniques, physical treatment, biological treatment, and chemical treatment. Examples of treatment techniques found within each category are shown in Table 1.20. Many of these methods have been discussed in grave detail in previous sections [61].

The presence of pathogens can have a major impact on livestock operations. While this section is not extensive, it does attempt to provide a summary of major pathogen categories, their associated impacts, and the potential treatment methods.

9 Livestock Waste Management Computer Software

Within the recent century transition, there has been the presence of computer modeling tools that are capable of being used to predict livestock wastes. For example, the animal waste management software tool (AWM) is a computer program designed to determine parameters such as waste storage facilities, waste treatment lagoons, and utilization [2]. Other options include the collaboration between the University of South Carolina's Earth Science and Resource Institute and the Natural Resources Conservation Service (NCRS) in South Carolina to develop a suite of products that include the geospatial tools [ArcGIS] and a nutrient management planning software AFOPro© [115].

Ideas of the use of software for livestock waste management have not been limited to just the United States. A program known as Integrated Swine Manure Management (ISMM) is an integrated decision support system (DSSs) used by Canadian province decision-makers to control manure, considering various criteria such as environmental, agronomic, social and health, greenhouse gas emission, and economic factors [116]. The introduction of computer software for livestock management can be very significant for those that are planning to provide a consistent method of managing livestock. Nevertheless, it is still important to remember that computer software is a "tool" but does not replace proper education and understanding of what is needed for proper livestock waste within the given area.

10   Conclusion

This chapter provides a plethora of information concerning livestock waste management from treatment, handling, and storage. While this not an all-encompassing manual for all given conditions, it can be used as a catalyst for research and exploration in how to properly maintain and manage livestock waste for a given industry.

Glossary of Terms Related to Livestock Waste

Anaerobic digestion is the fermentation of organic waste by hydrolytic microorganisms into fatty acid chains, carbon dioxide (CO_2), and hydrogen (H_2). Short fatty acids are then converted into acetic acid (CH_3COOH), H_2, CO_2, and microorganisms.

Biogas is a product from anaerobic digestion containing gases such as methane (CH_4), CO_2, and trace elements. Biogas can be used as a source of energy.

Chemical oxygen demand (COD) is a wastewater quality index that determines the amount of oxygen consumed by wastes.

Concentrated animal feeding operations (CAFO) raises livestock within a restricted space. It is also known as feedlot.

Constructed wetland is a treatment method that uses plants (most commonly water hyacinth and duckweed) to degrade organic material.

Denitrification converts nitrate into atmospheric nitrogen using microorganisms known as denitrifiers.

Eutrophication is the condition of a waterbody (particularly a lake) where molecular oxygen levels have been depleted. Eutrophication most commonly occurs when nutrient levels are high within the waterbody forming the presence of algal blooms. When eutrophication occurs, all organisms that rely on molecular oxygen to survive will die.

Five-day biochemical oxygen demand (BOD_5) is a wastewater quality index that determines the amount of oxygen required for microorganisms to degrade a given substance within a 5-day period.

Lagoon is a basin that treats wastewater and stores waste. There are three major types of lagoons—anaerobic, aerobic, and facultative.

Liquid manure contains dry matter less than 5 %.

Mesophilic is a state in an anaerobic digester or composting when the temperature remains between 35 °C and 40 °C.

National Pollutant Discharge Elimination System (NDPES) regulates the quantity of waste entering navigable waters and point sources. It was first introduced by the US EPA in the Clean Water Act of 1977. Livestock waste operations are required to have NPDES permits to discharge. State legislation defines the operations that require NPDES permit.

Nitrification is the process of converting ammonium nitrogen (NH_4^+) into nitrate (NO_3^{2-}) with an intermediate step of producing nitrite (NO_2^-). Nitrification is converted by nitrogen-fixing bacteria (nitrifiers).

Psychrophilic is a state in an anaerobic digester or composting when the temperature remains below 20 °C.

Semisolid manure contains 5–10 % dry matter.

Solid manure contains dry matter greater than 15 %.

Thermophilic is a state in an anaerobic digester or composting when the temperature remains between 51 °C and 57 °C.

Volatilization is a phase change process that converts constituents into gaseous form. The most common volatilization experienced is ammonia volatilization or the conversion of ammonium nitrogen to ammonia nitrogen. This is problematic for livestock operations because plant nitrogen is lost for plant uptake.

References

1. US Department of Agriculture and National Conservation Resource Service (1999) Agriculture wastes and water, air, and animal resources part 651—agriculture waste management field handbook
2. US Department of Agricultural and National Conservation Resource Service (1996) National Engineering Handbook (NEH) part 651—agricultural waste management field handbook
3. US Environmental Protection Agency (2015) Summary of the clean air act. Available at: http://www2.epa.gov/laws-regulations/summary-clean-air-act. Last accessed 6 Mar 2015
4. University of Nebraska-Lincoln Extension (2009) Managing livestock manure to protect environmental quality, EC174. Available at: http://www.ianrpubs.unl.edu/epublic/live/ec179/build/ec179.pdf. Last accessed 6 Mar 2015
5. US Environmental Protection Agency (2015) Summary of the clean water act. Available at: http://www2.epa.gov/laws-regulations/summary-clean-water-act. Last accessed 6 Mar 2015
6. US Department of Agriculture and National Conservation Service (2009) Chapter 1: laws, regulations, policy, and water quality criteria. Agriculture wastes and water, air, and animal resources part 651—agriculture waste management field handbook. Available at: http://directives.sc.egov.usda.gov/OpenNonWebContent.aspx?content=25878.wba. Last accessed 6 Mar 2015
7. US Environmental Protection Agency (2011) United States Protection Agency Office of Water, Office of Wastewater Management Water Permits Division October 2011 Proposed NPDES CAFO reporting rule. Available at: http://water.epa.gov/polwaste/npdes/afo/upload/2011_npdes_cafo_factsheet.pdf. Last accessed 6 Mar 2015
8. US Department of Agriculture and National Conservation Resource Service (2012) Agriculture wastes and water, air, and animal resources part 651—agriculture waste management field handbook. http://directives.sc.egov.usda.gov/viewerFS.aspx?id=3851. Accessed 6 Mar 2015
9. Loehr RC (1974) Agricultural waste management. Academic, New York
10. Fullhage CD (1981) Performance of anaerobic lagoons as swine waste storage and treatment facilities in Missouri. Am Soc Agric Eng:225–227

11. Payne VWE, Shipp JW, Miller FA (1981) Supernatant characteristics of three animal waste lagoons in North Alabama. Am Soc Agric Eng:240–243
12. Illinois Environmental Protection Agency (2015) Part 560: design criteria for field application of livestock waste. Available at: http://web.extension.illinois.edu/clmt/Workbook/WK_FILES/IEPA_FLD.PDF. Last accessed 6 Mar 2015
13. Iowa State University Extension (1995) Design and management of anaerobic lagoons in Iowa for animal manure storage and treatment. Pm-1590
14. Janni KA, Schmidt DR, Christopherson SH (2007) Milk house wastewater characteristics. University of Minnesota Extension. Publication 1206. Available at: http://www.extension.umn.edu/agriculture/manure-management-and-air-quality/wastewater-systems/milkhouse-wastewater-characteristics/docs/milkhouse-wastewater-characteristics.pdf. Last accessed 6 Mar 2015
15. Holmes BJ, Struss S (2015) Milking center wastewater guidelines: a companion document to wisconsin NRCS standard 629. Available at: http://clean-water.uwex.edu/pubs/pdf/milking.pdf. Last accessed 6 Mar 2015
16. Pennsylvania Nutrient Management Program (2015) Section 2: milk house wastewater characteristics. Pennsylvania State Extension. http://extension.psu.edu/plants/nutrient-management/planning-resources/other-planning-resources/milkhouse-wastewater-characterisitics. Last accessed 6 Mar 2015
17. Schmit D, Janni K (2015) Milk house wastewater treatment system design workshop. University of Minnesota Extension. Available at: http://www.extension.umn.edu/agriculture/manure-management-and-air-quality/wastewater-systems/docs/intro-milkhouse-wastewater-treatment.pdf. Last accessed 6 Mar 2015
18. National Resources Conservation Service (2015) Conservation practice standard: waste treatment. NCRS Minnesota. No. 629–1. Available at: http://efotg.sc.egov.usda.gov/references/public/MN/629mn.pdf. Last accessed 6 Mar 2015
19. Schmidt DA, Janni JA, Christopherson SH (2008) Milk house wastewater guide. University of Minnesota Extension. Available at: http://www.extension.umn.edu/agriculture/manure-management-and-air-quality/wastewater-systems/milkhouse-wastewater-design-guide. Last accessed 6 Mar 2015
20. Newman JM, Cluasen JC (1997) Seasonal effectiveness of a constructed wetland for processing milk house wastewater. Wetlands 17(3):375–382
21. Reaves PP, DuBowy PJ, Miller BK (1994) Performance of a constructed wetland for dairy waste treatment in lagrange county, Indiana. In: Proceedings of a workshop on constructed wetlands for animal waste management. Available at: http://www.lagrangecountyhealth.com/Documents/CWDairyFarm.pdf. Last accessed 6 Mar 2015
22. Rausch KD, Powell GM (1997) Diary processing methods to reduce water use and liquid waste load. MF-2071. March 1997. Available at: http://www.fpeac.org/dairy/DairyWastewater.pdf. Accessed 6 Mar 2015
23. US Department of Agriculture and National Conservation Resource Service (2007) An analysis of energy production costs from anaerobic digestion systems on US livestock production facilities. Washington, DC. Available at: http://directives.sc.egov.usda.gov/OpenNonWebContent.aspx?content=22533.wba. Last accessed 6 Mar 2015
24. Sharvelles S, Loetscher L (2015) Anaerobic digestion of animal wastes in Colorado. Colorado State University Extension. Fact Sheet No. 1.2271. Available at: http://www.ext.colostate.edu/pubs/livestk/01227.pdf. Last accessed 6 Mar 2015
25. Key N, Sneeringer S (2011) Carbon prices and the adoption of methane digesters on dairy and hog farms. United States Department of Agriculture Economic Research Service Economic Brief Number 16
26. Moser MA, Mattocks RP, Gettier S, Roos K (1998) Benefits, costs, and operating experience at seven new agricultural anaerobic digester. Available at: http://www.epa.gov/agstar/documents/lib-ben.pdf. Last accessed 6 Mar 2015

27. Wisconsin Bioenergy Initiative (2011) The biogas opportunity in Wisconsin: 2011 strategic plan. University of Wisconsin Extension. Available at: http://energy.wisc.edu/sites/default/files/pdf/Biogas%20Opportunity%20in%20Wisconsin_WEB.pdf. Last accessed 6 Mar 2015

28. US Environmental Protection Agency (2014) 2103 use and AD in the livestock sector. Available at: http://www.epa.gov/agstar/documents/2013usebenefits.pdf. Last accessed 6 Mar 2015

29. US Environmental Protection Agency (2014) Anaerobic digesters. http://www.epa.gov/agstar/anaerobic/ad101/anaerobic-digesters.html. Last accessed 6 Mar 2015

30. Hamilton DW (2015) Anaerobic digestion of animal manures: types of digesters. Oklahoma Cooperative Extension Service. BAE-1750. Available at: http://pods.dasnr.okstate.edu/docushare/dsweb/Get/Document-7056/BAE-1750web2014.pdf. Last accessed 6 Mar 2015

31. Harikishan S, Sung S (2003) Cattle waste treatment and Class A biosolid production using temperature-phased anaerobic digester. Adv Environ Res 7(3):701–706

32. King SM, Barrington S, Guiot SR (2011) In-storage psychrophilic digestion of swine manure: accumulation of the microbial community. Biomass Bioenergy 35(8):3719–3726

33. Rao AG, Prakash SS, Jospeh J, Reddy AR, Sarma PN (2011) Multi-stage high rate biomethanation of poultry litter with self mixed anaerobic digester. Bioresour Technol 102 (2):729–735

34. Hill V (2003) Prospects for pathogen reductions in livestock wastewaters: a review. Crit Rev Environ Sci Technol 33(2):187–235

35. Knight RL, Payne VWE, Borer RE, Clarke RA, Pries JH (2000) Constructed wetlands for livestock wastewater management. Ecol Eng 15(1):41–55 ['34a']

36. Vymazal J (2006) Constructed wetlands for wastewater treatment. Ecol Stud 190:69–96

37. National Resource Conservation Service (2000) Chapter 3: constructed wetlands. In Part 637 Environmental Engineering Handbook. Available at: http://directives.sc.egov.usda.gov/OpenNonWebContent.aspx?content=25905.wba. Last accessed 6 Mar 2015

38. Gustafon D, Anderson J, Christopherson SH, Axler R (2002) Constructed wetlands. Available at: http://www.extension.umn.edu/environment/water/onsite-sewage-treatment/innovative-sewage-treatment-systems-series/constructed-wetlands/index.html. Last accessed 6 Mar 2015

39. Cronk JK (1996) Constructed wetlands to treat wastewater from dairy and swine waste: a review. Agric Ecosyst Environ 58(2):97–114

40. Stone KC, Poach ME, Hunt PG, Reddy GB (2004) Marsh-pond-marsh constructed wetland design analysis for swine lagoon treatment. Ecol Eng 23(2):127–133

41. Payne Engineering and CH2M Hill (1997) Constructed wetlands for animal waste treatment: a manual on performance design and operation with case histories. Document no. 855B97001. US Environmental Protection Agency

42. Pfost D, Fulhage C, Rastorfer D (2000) Anaerobic lagoons for storage/treatment of livestock manure. University of Missouri-Columbia Research Extension. Available at: http://extension.missouri.edu/p/EQ387. Last accessed 6 Mar 2015

43. Mississippi Department of Environmental Quality (2008) Chapter 100: wastewater treatment ponds (lagoons). Available at: http://www.deq.state.ms.us/MDEQ.nsf/pdf/SRF_NPELF40100/$File/NPELF40-100.doc?OpenElement. Last accessed 6 Mar 2015

44. Hamilton D (2015) Lagoons for livestock waste treatment. Oklahoma Cooperative Extension Service. BAE-1736. Available at: http://pods.dasnr.okstate.edu/docushare/dsweb/Get/Document-7615/BAE-1736web2011.pdf. Last accessed 6 Mar 2015

45. Funk T, Bartzis G, Treagust J (2015) Designing and managing livestock waste lagoons in Illinois. University of Illinois at Urbana-Champaign College of Agriculture Cooperative Extension Service. Circular 1326. Available at: http://www.aces.uiuc.edu/vista/html_pubs/LAGOON/lagoon.html. Accessed 6 Mar 2015

46. Miller R (2011) How a lagoon works for livestock wastewater treatment. Utah State University Cooperative Extension. Available at: http://extension.usu.edu/files/publications/publication/AG_WasteManagement_2011-01pr.pdf. Accessed 6 Mar 2015

47. Powers W (2015) Practices to reduce odor from livestock operations. Iowa State University. Available at: https://store.extension.iastate.edu/Product/pm1970a-pdf. Last accessed 6 Mar 2015

48. Chastain JP, Henry S (2015) Chapter 4: management of lagoons and storage structures for swine manure. Available at: http://www.clemson.edu/extension/livestock/camm/camm_files/swine/sch4_03.pdf. Accessed 6 Mar 2015

49. Barker J (1996) Lagoon design and management for livestock waste treatment and storage. North Carolina State University Cooperative Extension. EBAE 103–83. Available at: http://www.bae.ncsu.edu/extension/ext-publications/waste/animal/ebae-103-83-lagoon-design-barker.pdf. Last accessed 6 Mar 2015

50. Dickey EC, Brumm M, Shelton DP (2009) Swine manure management systems. University of Nebraska-Lincoln. G80-531-A. Available at http://infohouse.p2ric.org/ref/32/31081.htm. Last accessed 6 Mar 2015

51. Alabama A&M, Auburn Universities (2015) Sizing swine lagoons for odor control. Alabama Cooperative Extension System. Circular ANR-1900. Available at: http://www.aces.edu/pubs/docs/A/ANR-1090/ANR-1090-low.pdf. Last accessed 6 Mar 2015

52. Cantrell KB, Ducey T, Ro KS, Hunt PG (2008) Livestock waste-to-bioenergy generation opportunities. Bioresour Tecnol 99(17):7941–7953

53. Zhang L, Xu C, Champagne P (2010) Overview of recent advances in thermo-chemical conversion of biomass. Energy Convers Manag 51(15):969–982

54. Zhang SY, Hong RY, Cao JP, Takarada T (2009) Influence of manure types and pyrolysis conditions on the oxidation behavior of manure char. Bioresour Technol 100(18):4278–4283

55. Priyadarsan S, Annamalai K, Sweeten JM, Mukhtar S, Holtzapple MT (2004) Fixed-bed gasification of feedlot manure and poultry litter biomass. Trans ASABE 47(5):1689–1696

56. Zhang SY, Huang FB, Morishita K, Takarada T (2009) Hydrogen production from manure by low temperature gasification. In: Power and Energy Engineering Conference, APPECC. Asia-Pacific. IEEE, pp 1–4

57. Priyadarsan S, Annamalai K, Sweeten JM, Mukhtar S, Holtzapple MT, Mukhtard S (2005) Co-gasification of blended coal with feedlot and chicken litter biomass. Proc Combust Inst 30(2):2973–2980

58. Kansas Department of Health and Environmental Bureau of Waste Management (2015) Composting at livestock facilities. Available at: http://www.kdheks.gov/waste/compost/compostingatlivestockfacilitiesinfosheet.pdf. Last accessed 6 Mar 2015

59. Keener H, Elwell D, Mescher T (2015) Composting swine morality principles and operation. The Ohio State University Extension. AEX-711-97. Available at: http://ohioline.osu.edu/aex-fact/0711.html. Last accessed 6 Mar 2015

60. Bass T (2015) Livestock mortality composting: for large and small operations in the semi-arid west. Montana State University Extension. Available at: http://www.ext.colostate.edu/pubs/ag/compostmanual.pdf. Last accessed 6 Mar 2015

61. Sosbey MD, Khatib LA, Hill VR, Alocija E, Pillai S (2015) Pathogens in animal waste and the impact of waste management practices on their survival, transport, and fate. Available at: http://munster.tamu.edu/Web_page/Research/Ecoli/pathogens-animalagriculture.pdf. Last accessed 6 Mar 2015

62. Avermann B, Mukhtar S, Heflin K (2006) Composting large animal carcasses. Texas Cooperative Extension. E-422. Available at: http://tammi.tamu.edu/largecarcassE-422.pdf. Last accessed 6 Mar 2015

63. Iowa State University Extension (2015) Composting dead livestock: a new solution to an old problem. Available at: https://store.extension.iastate.edu/Product/sa8-pdf. Last accessed 6 Mar 2015

64. Payne J, Pugh B (2015) On-farm mortality composting of livestock carcasses. Oklahoma Cooperative Extensive Surface. BAE-1749. Available at: http://poultrywaste.okstate.edu/files/BAE1749%20On-Farm%20Mortality.pdf. Accessed 6 Mar 2015

65. Loh TC, Lee YC, Liang JB, Tan D (2006) Vermicomposting of cattle and goat manures by *Eisenia fetida* and their growth and reproduction performance. Bioresour Technol 96 (1):111–114

66. Garg VK, Yadav YK, Sheoran A, Chand S, Kaushik P (2006) Livestock excreta management through vermicomposting using an epigenetic earthworm *Eisenia fetida*. Environmentalist 26 (4):269–276

67. Mupondi LT, Mnkeni PNS, Muchaonyerwa P (2011) Effects of a precomposting step on the vermicomposting of dairy manure-waste paper mixtures. Waste Manag Res 29(2):219–228

68. Mupondi LT, Mnkeni PNS, Muchaonyerwa P (2010) Effectiveness of combined thermophilic composting and vermicomposting on biodegradation and sanitization of mixtures of dairy manure and waste paper. Afr J Biotechnol 9(30):4754–4763

69. Lee JS, Choi DC (2009) A study on organic resources for pig manure treatment by vermicomposting. J Livest Hous Environ 15(3):289–296

70. Aira M, Domingues J (2008) Optimizing vermicomposting of animal waste: effects of rate of manure application on carbon loss and microbial stabilization. J Environ Manag 88 (4):1525–1529

71. Rahman S, Widerholt R (2012) Options for land application of solid manure. North Dakota State University Extension. NM1613. Available at: http://www.ag.ndsu.edu/manure/docu ments/nm1613.pdf. Last accessed 6 Mar 2015

72. Hernandez JA, Schmitt MA (2012) Manure management in Minnesota. University of Minnesota Extension. WW-03353. Available at: http://www.extension.umn.edu/agriculture/ manure-management-and-air-quality/manure-application/manure-management-in-minne sota/docs/manure-management-in-minnesota.pdf. Accessed 6 Mar 2015

73. Field B (1980) Beware of on-farm manure. Purdue University Cooperation Extension Service. Available at: https://www.extension.purdue.edu/extmedia/S/S-82.html. Accessed 6 Mar 2015

74. Evanylo GK (2015) Chapter 10: land application of biosolids. In: Haering KC, Evanylo GK (eds) The mid-Atlantic nutrient management handbook. MAWP 06–02. February 2006. Available at: http://www.mawaterquality.org/capacity_building/mid-atlantic%20nutrient% 20management%20handbook/chapter10.pdf. Last accessed 6 Mar 2015

75. Pfost DL, Fulhage CD, Alber O (2001) Land application equipment for livestock and poultry management. University of Missouri-Columbia Research Extension. Available at: http:// extension.missouri.edu/explorepdf/envqual/eq0383.pdf. Accessed 6 Mar 2015

76. Jacobson L, Lorimor L, Bicudo J, Schmidt J (2001) Lesson 44: emission control strategies for land application. MidWest Plan Service. Available at: http://www.extension.org/mediawiki/ files/2/26/LES_44.pdf. Last accessed 6 Mar 2015

77. Zhao L, Rausch JN, Combs TL (2015) Odor control for land application of manure. The Ohio State University Extension. Available at: http://ohioline.osu.edu/aex-fact/pdf/odor_control. pdf. Last accessed 6 Mar 2015

78. Jarrett AR, Graves RE (2002) Irrigation of liquid manure with center-pivot irrigation systems. Penn State Extension. F-256. http://pubs.cas.psu.edu/FreePubs/pdfs/F256.pdf. Last accessed 6 Mar 2015

79. Fulhage C (2015) Land application considerations for animal manure. University of Missouri-Columbia University Extension. Available at: http://extension.missouri.edu/ explorepdf/envqual/eq0202.pdf. Last accessed 6 Mar 2015

80. Rise M (2012) Livestock application of livestock and poultry manure. The University of Georgia Cooperative Extension. Circular 826. Available at: http://extension.uga.edu/publica tions/files/pdf/C%20826_3.PDF. Last accessed 6 Mar 2015

81. Harrison JD, Smith DR (2004) Manure storage: process improvement for animal feeding operations. Utah State University Cooperative Extension. AG/AWM-01-1. Available at: http://extension.usu.edu/files/publications/factsheet/AG_AWM-01-1.pdf. Last accessed 6 Mar 2015

82. Harrison JD, Smith DR (2004) Types of manure storage: process improvement for animal feeding operations. Utah State University Cooperative Extension. AG/AWM-01-2. Available at: http://extension.usu.edu/files/publications/factsheet/AG-AWM-01-2.pdf. Last accessed 6 Mar 2015

83. Virginia Department of Environmental Quality (2015) Livestock manure and storage facilities. Virginia Cooperative Extension. Publication 442–909. Available at: http://pubs.ext.vt.edu/442/442-909/442-909_pdf.pdf. Last accessed 6 Mar 2015

84. Harrison JD, Smith D (2004) Manure storage selection: process improvement for animal feeding operations. Utah State University Cooperative Extension. AG/AWM-01-3. http://extension.usu.edu/files/publications/factsheet/AG_AWM-01-3.pdf. Last accessed 6 Mar 2015

85. Sutton AL (1990) Animal agriculture's effect on water quality waste storage. Purdue University Cooperative Extension Service. WQ-8. Available at: https://www.extension.purdue.edu/extmedia/WQ/WQ-8.html. Last accessed 6 Mar 2015

86. Harrison JD, Smith D (2004) Animal manure removal methods for manure storage facilities. Utah State University Extension. AG/AWM-05. Available at: http://extension.usu.edu/files/publications/factsheet/AG-AWM-05.pdf. Last accessed 6 Mar 2015

87. Fulhage CD, Pfost DL (1993) Storage tanks for liquid dairy waste. University of Missouri-Columbia Extension. WQ306. Available at: http://extension.missouri.edu/publications/DisplayPrinterFriendlyPub.aspx?P=WQ306. Last accessed 6 Mar 2015

88. Minnesota Department of Agriculture (2015) Conservative practices Minnesota conservation funding guide. Available at: https://www.mda.state.mn.us/en/protecting/conservation/practices/feedlotrunoff.aspx. Last accessed 6 Mar 2015

89. Dickey EC, Bodman GR (1992) Management of feedlot runoff control system. Cooperative Extension Service—Great Plains States. GPE-7523. Available at: http://digitalcommons.unl.edu/biosysengfacpub/262?utm_source=digitalcommons.unl.edu%2Fbiosysengfacpub%2F262&utm_medium=PDF&utm_campaign=PDFCoverPages. Last accessed 6 Mar 2015

90. Lorimor JC, Shouse S, Miller W (2002) Vegetative filter strips for open feedlot runoff treatment. Iowa State University Extension. PM1919. Available at: https://store.extension.iastate.edu/Product/pm1919-pdf. Last accessed 6 Mar 2015

91. Stowell R, Zulovich J (2008) Chapter 8: manure and effluent management. In: Dairy freestall housing and equipment, 7th edn. Midwest Plan Service. MWPS-7. http://www.public.iastate.edu/~mwps_dis/mwps_web/87zgGwEKj.QDg.pdf. Last accessed 6 Mar 2015

92. Nye JC, Jones DD, Sutton AL (1976) Runoff control systems for open livestock feedlots. Purdue University Cooperative Extension Service. ID-114-W. Available at: https://www.extension.purdue.edu/extmedia/ID/ID-114-W.html. Last accessed 6 Mar 2015

93. Rahman S, Rahman A, Wiederholt R (2011) Vegetative filter strips: reduce feedlot runoff pollutants. North Dakota State University Extension Service. NM1591. Available at: http://www.ag.ndsu.edu/manure/documents/nm1591.pdf. Last accessed 6 Mar 2015

94. Higgins S, Wightman S, Smith R (2012) Enhanced vegetative strips for livestock facilities. University of Kentucky Cooperative Extension Service. Available at: http://www2.ca.uky.edu/agc/pubs/id/id189/id189.pdf. Last accessed 6 Mar 2015

95. Dickey EC, Vanderholm DH (1981) Vegetative filter treatment of livestock feedlot runoff. J Environ Qual 10(3):279–284. Available at:http://www.pcwp.tamu.edu/docs/lshs/endnotes/vegetative%20filter%20treatment%20of%20livestock%20feedlot%20runoff-27479 26786/vegetative%20filter%20treatment%20of%20livestock%20feedlot%20runoff.pdf. Last accessed 6 Mar 2015

96. Koelsch RK, Lorimor JC, Mankin KR (2006) Vegetative treatment systems for management of open lot runoff: review of literature. Appl Eng Agric 22(1):141–153. Available at: http://digitalcommons.unl.edu/biosysengfacpub/5?utm_source=digitalcommons.unl.edu%2Fbiosysengfacpub%2F5&utm_medium=PDF&utm_campaign=PDFCoverPages. Last accessed 6 Mar 2015.

97. Rappert S, Muller R (2005) Odor compounds in waste gas emissions from agricultural operations and food industries. Waste Manag 25(9):887–907
98. Leggett J, Graves RE (1995) Odor control for animal production operations. Penn State Extension. G79. Available at: http://pubs.cas.psu.edu/freepubs/pdfs/G79.pdf. Last accessed 6 Mar 2015
99. Rahman S, Mukhtar S, Wiederholt R (2008) Managing odor nuisance and dust from cattle feedlots. North Dakota State University. NM-1391. Available at: http://www.ag.ndsu.edu/manure/documents/nm1391.pdf. Last accessed 6 Mar 2015
100. Chastain JP (2015) Chapter 9: odor control from poultry facilities. Available at: http://www.clemson.edu/extension/livestock/camm/camm_files/poultry/pch9_03.pdf. Last accessed 6 Mar 2015
101. Conn KL, Topp E, Lazarovits G (2006) Factors influencing the concentration of volatile fatty acids, ammonia, and other nutrients in stored liquid pig manure. J Environ Qual 36 (2):440–447
102. Chi F-H, Lin PH-P, Leu M-H (2005) Quick determination of malodor-causing fatty acids in manure by capillary electrophoresis. Chemosphere 60(9):1262–1269
103. McCroy DF, Hobbs PJ (2000) Additives to reduce ammonia and odor emissions from livestock wastes: a review. J Environ Qual 30(2):345–355
104. Clark OG, Morin B, Zhang YC, Sauer WC, Feddes JJR (2005) Preliminary investigation of air bubbling and dietary sulfur reduction to mitigate hydrogen sulfide and odor from swine waste. J Environ Qual 34(6):2018–2023
105. Cook KL, Whitehead TR, Spensce C, Cotta MA (2008) Evaluation of the sulfate-reducing bacterial population associated with stored swine slurry. Anaerobe 14(3):172–180
106. Liang Y, VanDevender K (2015) Managing livestock operation to reduce odor. University of Arkansas Research Service Cooperation Extension Service. FSA3007. Available at: http://www.uaex.edu/publications/pdf/FSA-3007.pdf. Last accessed 6 Mar 2015
107. Yasuda T, Kuroda K, Fukumoto Y, Hanajima D, Suzuki K (2009) Evaluation of full-scale biofilter with rock wool mixture treating ammonia gas from livestock manure composting. Bioresour Technol 100(4):1568–1572
108. Ro KS, McConnell LL, Johnson MH, Hunt PG, Parker DL (2008) Livestock air treatment using PVA-coated powered activated carbon biofilter. Appl Eng Agric 24(6):791–798
109. Kastner JR, Das KC, Crompton B (2004) Kinetics of ammonia removal in a pilot-scale biofilter. Trans ASABE 47(5):1867–1878
110. Ye FX, Zhu RF, Ying LI (2009) Deodorization of swine manure slurry using horseradish peroxidase and peroxides. J Hazard Mater 167(1):148–153
111. Govere EM, Tonegawa M, Bruns MA, Wheeler EF, Kephart KB, Voigt JW, Dec J (2007) Using minced horseradish roots and peroxides for the deodorization of swine manure: a pilot scale study. Bioresour Technol 98(6):1191–1198
112. Massey R, McClure H (2014) Agriculture and greenhouse gas emissions. University of Missouri-Columbia Extension. Available at: http://extension.missouri.edu/explorepdf/agguides/agecon/g00310.pdf. Accessed 6 Mar 2015
113. Caro D, Davis SJ, Bastianoni S, Caldeira K (2014) Global and regional trends in greenhouse gas emissions from livestock. Clim Chang 126(1–2):203–216
114. Herrero M, Gerber P, Vellinga T, Garnett T, Leip A, Opio C, Westhoek HJ (2011) Livestock and greenhouse gas emissions: the important of getting it right. Anim Feed Sci Technol 126:779–792
115. Henry ST, Kloot RW, Evans M, Hardee G (2003) Comprehensive nutrient management plans and the tools used to develop them in South Carolina. In: Proceedings of the 9th International Symposium Agricultural and Food Processing Wastes proceedings. Research Triangle Park, North Carolina, October 2003
116. Karmakar SN, Ketia M, Lague C, Agnew J (2010) Development of expert system modeling based decision support system for swine manure management. Comput Electron Agric 71(1):88–95

Chapter 2
Application of Natural Processes
for Environmental Protection

Nazih K. Shammas and Lawrence K. Wang

Contents

1 Aquaculture Treatment: Water Hyacinth System 75
  1.1 Description 75
  1.2 Applications 75
  1.3 Limitations 76
  1.4 Design Criteria 76
  1.5 Performance 77
2 Aquaculture Treatment: Wetland System 78
  2.1 Description 78
  2.2 Constructed Wetlands 79
  2.3 Applications 80
  2.4 Limitations 80
  2.5 Design Criteria 81
  2.6 Performance 81
3 Evapotranspiration System 82
  3.1 Description 82
  3.2 Applications 84
  3.3 Limitations 84
  3.4 Design Criteria 84
  3.5 Performance 85
  3.6 Costs 85

N.K. Shammas (✉)
Lenox Institute of Water Technology, Pasadena, CA, USA

Krofta Engineering Corporation, Lenox, MA, USA
e-mail: n.k.shammas@gmail.com

L.K. Wang
Lenox Institute of Water Technology, Newtonville, NY, USA

Rutgers University, New Brunswick, NJ, USA
e-mail: LawrenceKWang@gmail.com

© Springer International Publishing Switzerland 2016
L.K. Wang, M.-H.S. Wang, Y.-T. Hung and N.K. Shammas (eds.),
Natural Resources and Control Processes, Handbook of Environmental
Engineering, Volume 17, DOI 10.1007/978-3-319-26800-2_2

4 Land Treatment: Rapid Rate System 86
   4.1 Description 86
   4.2 Applications 88
   4.3 Limitations 88
   4.4 Design Criteria 88
   4.5 Performance 89
   4.6 Costs 90
5 Land Treatment: Slow Rate System 92
   5.1 Description 92
   5.2 Applications 93
   5.3 Limitations 94
   5.4 Design Criteria 94
   5.5 Performance 95
   5.6 Costs 95
6 Land Treatment: Overland Flow System 97
   6.1 Description 97
   6.2 Application 99
   6.3 Limitations 99
   6.4 Design Criteria 99
   6.5 Performance 100
   6.6 Costs 101
7 Subsurface Infiltration 103
   7.1 Description 103
   7.2 Applications 105
   7.3 Limitations 105
   7.4 Design Criteria 106
   7.5 Performance 106
   7.6 Design Example 1: Elevated Sand Mound at a Crested Site 108
   7.7 Design Example 2: Elevated Sand Mound at a Sloping Site 110
References 113

Abstract Aquaculture or the production of aquatic organisms (both flora and fauna) under controlled conditions has been practiced for centuries, primarily for the generation of food, fiber, and fertilizer. The water hyacinth and a host of other organisms like duckweed, seaweed, midge larvae, and alligator weeds are used for wastewater treatment. Water hyacinth system, wetland system, evapotranspiration system, rapid rate land treatment system, slow rate land treatment system, overland flow land treatment system, and subsurface infiltration have also been applied. This chapter describes the above applications and explains their practice, limitations, design criteria, performance, and costs.

Keywords Natural processes • Aquatic organisms • Aquaculture • Water hyacinth system • Wetland • Subsurface Infiltration • Overland flow land treatment system • Evapotranspiration • Elevated sand mound design • Performance • Costs

1   Aquaculture Treatment: Water Hyacinth System

1.1   Description

Aquaculture or the production of aquatic organisms (both flora and fauna) under controlled conditions has been practiced for centuries, primarily for the generation of food, fiber, and fertilizer. The water hyacinth (*Eichhornia crassipes*) appears to be the most promising organism for wastewater treatment and has received the most attention [1]. However, other organisms are being studied. Among them are duckweed, seaweed, midge larvae, alligator weeds, and a host of other organisms. Water hyacinths are large fast-growing floating aquatic plants with broad, glossy green leaves and light lavender flowers. A native of South America, water hyacinths are found naturally in waterways, bayous, and other backwaters throughout the South. Insects and disease have little effect on the hyacinth and they thrive in raw, as well as partially treated, wastewater. Wastewater treatment by water hyacinths is accomplished by passing the wastewater through a hyacinth-covered basin (Fig. 2.1), where the plants remove nutrients, BOD_5, suspended solids, metals, etc. Batch treatment and flow-through systems, using single and multiple cell units, are possible. Hyacinths harvested from these systems have been investigated as a fertilizer/soil conditioner after composting, animal feed, and a source of methane when anaerobically digested [2].

1.2   Applications

Water hyacinths are generally used in combination with (following) lagoons, with or without chemical phosphorus removal. A number of full-scale systems are in operation, most often considered for nutrient removal and additional treatment of secondary effluent [1–3]. Also, research is being conducted on the use of water hyacinths for raw and primary treated wastewater or industrial wastes, but present data favor combination systems. Very good heavy metal uptake by the hyacinth has been reported. Hyacinth treatment may be suitable for seasonal use in treating wastewaters from recreational facilities and those generated from processing of

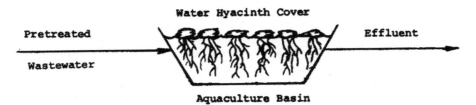

Fig. 2.1   Aquaculture treatment: water hyacinth system (Source: US EPA [2])

agricultural products. Other organisms and methods with wider climatological applicability are being studied. The ability of hyacinths to remove nitrogen during active growth periods and some phosphorus and retard algae growth provides potential applications in [2, 3]:

(a) The upgrading of lagoons
(b) Renovation of small lakes and reservoirs
(c) Pretreatment of surface waters used for domestic supply
(d) Storm water treatment
(e) Demineralization of water
(f) Recycling fish culture water
(g) For biomonitoring purposes

1.3   Limitations

Climate or climate control is the major limitation. Active growth begins when the water temperature rises above 10 °C and flourishes when the water temperature is approximately 21 °C. Plants die rapidly when the water temperature approaches the freezing point; therefore, greenhouse structures are necessary in northern locations. Water hyacinths are sensitive to high salinity. Removal of phosphorus and potassium is restricted to the active growth period of the plants.

Metals such as arsenic, chromium, copper, mercury, lead, nickel, and zinc can accumulate in hyacinths and limit their suitability as a fertilizer or feed material. The hyacinths may also create small pools of stagnant surface water which can breed mosquitoes. Mosquito problems can generally be avoided by maintaining mosquito fish in the system. The spread of the hyacinth plant itself must be controlled by barriers since the plant can spread and grow rapidly and clog affected waterways. Hyacinth treatment may prove impractical for large treatment plants due to land requirements. Removal must be at regular intervals to avoid heavy intertwined growth conditions. Evapotranspiration can be increased by two to seven times greater than evaporation alone.

1.4   Design Criteria

Ponds, channels, or basins are in use. In northern climates covers and heat would be required. Harvesting and processing equipment are needed. Operation is by gravity flow and requires no energy. Hyacinth growth energy is supplied by sunlight. All experimental data is from southern climates where no auxiliary heat was needed. Data is not available on heating requirements for northern climates, but it can be assumed proportional to northern latitude of location and to the desired growth rate of hyacinths.

Design data vary widely. Table 2.1 shows the design criteria for water hyacinth systems [4]. The following ranges refer to hyacinth treatment as a tertiary process on secondary effluent [2]:

Table 2.1 Design criteria for water hyacinth systems

Factor	Aerobic non-aerated	Aerobic non-aerated	Aerobic aerated
Influent wastewater	Screened or settled	Secondary	Screened or settled
Influent BOD_5, mg/L	130–180	30	130–180
BOD_5 loading, kg/ha-d	40–80	10–40	150–300
Expected effluent, mg/L			
BOD_5	<30	<10	<15
SS	<30	<10	<15
TN	<15	<5	<15
Water depth, m	0.5–0.8	0.6–0.9	0.9–1.4
Detention time, days	10–36	6–18	4–8
Hydraulic loading, m^3/ha-d	>200	<800	500–1000
Harvest schedule	Annually	Twice per month	Monthly

Source: US EPA [4]

(a) Depth should be sufficient to maximize plant rooting and plant absorption.
(b) Detention time depends on effluent requirements and flow, range 4–15 days.
(c) Phosphorus reduction 10–75 %.
(d) Nitrogen reduction 40–75 %.
(e) Land requirement is usually high 2–15 acres/MG/day.

1.5 Performance

Process appears to be reliable from mechanical and process standpoints, subject to temperature constraints. In tests on five different wastewater streams including raw wastewater and secondary effluents, the following removals were reported [2]:

(a) BOD_5: 35–97 %
(b) TSS: 71–83 %
(c) Nitrogen: 44–92 %
(d) Total P: 11–74 %

Takeda and coworkers [3] reported using aquaculture wastewater effluent for strawberry production in a hydroponic system which reduced the final effluent phosphorus concentration to as low as 0.1 mg/L which meets the stringent phosphorus discharge regulations. There is also evidence that in aquaculture system coliform, heavy metals and organics are also reduced, as well as pH neutralization.

Hyacinth harvesting may be continuous or intermittent. Studies indicate that average hyacinth production (including 95 % water) is on the order of 1000–10,000 lb/day/acre. Basin cleaning at least once per year results in harvested hyacinths. For further detailed information on water hyacinth systems, the reader is referred to references [5–13].

2 Aquaculture Treatment: Wetland System

2.1 Description

Aquaculture-wetland systems for wastewater treatment include natural and artificial wetlands as well as other aquatic systems involving the production of algae and higher plants (both submerged and emergent), invertebrates, and fish. Natural wetlands, both marine and freshwater, have inadvertently served as natural waste treatment systems for centuries; however, in recent years, marshes, swamps, bogs, and other wetland areas have been successfully utilized as managed natural "nutrient sinks" for polishing partially treated effluents under relatively controlled conditions. Constructed wetlands can be designed to meet specific project conditions while providing new wetland areas that also improve available wildlife wetland habitats and the other numerous benefits of wetland areas. Managed plantings of reeds (e.g., *Phragmites* spp.) and rushes (e.g., *Scirpus* spp. and *Schoenoplectus* spp.) as well as managed natural and constructed marshes, swamps, and bogs have been demonstrated to reliably provide pH neutralization and reduction of nutrients, heavy metals, organics, BOD_5, COD, SS, fecal coliforms, and pathogenic bacteria [2, 4].

Wastewater treatment by natural and constructed wetland systems is generally accomplished by sprinkling or flood irrigating the wastewater into the wetland area or bypassing the wastewater through a system of shallow ponds, channels, basins, or other constructed areas where the emergent aquatic vegetation has been planted or naturally occurs and is actively growing (see Fig. 2.2). The vegetation produced as a result of the system's operation may or may not be removed and can be utilized for various purposes [2]:

(a) Composted for use as a source of fertilizer/soil conditioner
(b) Dried or otherwise processed for use as animal feed supplements
(c) Digested to produce methane

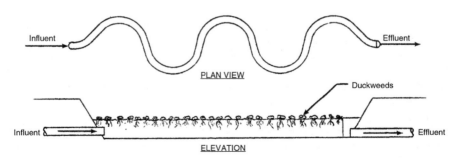

Fig. 2.2 Aquaculture treatment: wetland system (Source: US EPA [2])

2.2 Constructed Wetlands

Constructed wetlands are classified as a function of water flow [2, 4]: surface and subsurface which are known as free water surface (FWS) and subsurface flow system (SFS) [also termed vegetated submerged bed, VSB]. When simply expressed, constructed wetland treatment technology makes artificial receiving water and its vegetation part of the treatment process. In comparison to algae, the higher forms of plant life floating (duckweed, water hyacinths), submerged, and emergent (cattails, rushes, and reeds) perform less efficiently per unit weight of biomass.

FWS constructed wetland treatment conceptually relies on attached growth bacterial performance, receiving oxygen from the evapotranspiration response of the aquatic vegetation. Practically, the dominant bacterial action is anaerobic. The ammonium and nitrogen removal mechanisms [14–17] are a combination of aerobic oxidation, particulate removal, and synthesis of new plant protoplasm.

An FWS wetland is nothing more than a lagoon, except that a far greater expanse is needed to maximize the productivity per unit area. In practice, very large systems may achieve significant, if not complete, nitrogen oxidation, with surface reaeration contributing to the oxygen supply. Some nitrification and denitrification undoubtedly occur in all systems.

If it is assumed that the wetland vegetation will not be harvested, as is the case with natural wetland systems, its capacity for nitrogen control is finite, reflecting the site-specific vegetation and the ability to expand in the available space. Thus, the bigger the natural wetland that is called part of the process, the better, since there is dilution of the wastewater to the point that it is no longer significant in comparison to the naturally occurring background flow and water quality.

Constructed FWS wetlands yield a managed vegetative habitat that becomes an aquaculture system. Examination of the evolution of this technology shows the emergence of concepts that include organic load distribution or artificial aeration to avoid aesthetic nuisances and emphasis on plants that grow the fastest. Duckweed and water hyacinth systems (classified as aquaculture) have been reported to achieve long-term total nitrogen residuals of less than 10 mg/L and may be manageable, with harvesting and sensitive operation, to values of less than 3 mg/L on a seasonal, if not sustained, basis.

Submerged-flow constructed wetlands are simply horizontal-flow gravel filters with the added component of emergent plants within the media. They have been classically used for BOD removal following sedimentation and/or additional BOD and SS removal from lagoon effluents as with FWS approaches. This technology has the potential for high-level denitrification when a nitrified wastewater is applied; the naturally occurring environment promotes anoxic (denitrification) pathways for oxidized nitrogen elimination.

Ultimately, the success or failure of the wetland approach for nitrogen control may rest with the harvest of the vegetation, the need for backup (so that areas under harvest have the backup of areas in active growth), and often natural seasonal growth and decay cycles. If biomass production is an unacceptable goal, the designer should think of a more tolerant mixed vegetation system that minimizes the need to harvest the accumulated vegetation and maximizes the promotion of concurrent or staged nitrification and denitrification in some fashion. Conceptually, the optimization has to begin with promotion of nitrogen oxidation systems that may be shallow (better aeration for attached and suspended bacterial growth) with vegetation that minimizes light penetration and avoids as much algal growth as possible. Cyclic staging, recycle, forced aeration, and mixing represent some of the enhancements that naturally follow [17].

2.3 Applications

Several full-scale systems are in operation or under construction [18]. Wetlands are useful for polishing treated effluents. They have potential as a low-cost, low energy-consuming alternative or addition to conventional treatment systems, especially for smaller flows. Wetlands have been successfully used in combination with chemical addition and overland flow land treatment systems. Wetland systems may also be suitable for seasonal use in treating wastewaters from recreational facilities, some agricultural operations, or other waste-producing units where the necessary land area is available [18]. Potential application as an alternative to lengthy outfalls extended into rivers, lakes etc. and as a method of pretreatment of surface waters for domestic supply, storm water treatment, recycling fish culture water, and biomonitoring purposes.

2.4 Limitations

Temperature (climate) is a major limitation since effective treatment is linked to the active growth phase of the emergent vegetation. Tie-ins with cooling water from power plants to recover waste heat have potential for extending growing seasons in colder climates. Enclosed and covered systems are possible for very small flows.

Herbicides and other materials toxic to the plants can affect their health and lead to poor treatment. Duckweeds are prized as food for waterfowl and fish and can be seriously depleted by these species. Winds may blow duckweeds to the shore if windscreens or deep trenches are not employed. Small pools of stagnant surface water which can allow mosquitoes to breed can develop, but problems can generally be avoided by maintaining mosquito fish or a healthy mix of aquatic flora and fauna in the system. Wetland systems may prove impractical for large treatment plants due to the large land requirements. They also may cause loss of water due to increases in evapotranspiration.

2.5   Design Criteria

Natural or artificial marshes, swamps, bogs, shallow ponds, channels, or basins could be used. Irrigation, harvesting, and processing equipment are optional. Aquatic vegetation is usually locally acquired.

Design criteria are very site and project specific. Available data vary widely. Values below refer to one type of constructed wetland system used as a tertiary process on secondary effluent [2]:

(a)   Detention time = 13 days.
(b)   Land requirement = 8 acres/MG/day.
(c)   Depth may vary with type of system, generally 1–5 ft.

2.6   Performance

Process appears reliable from mechanical and performance standpoints, subject to seasonality of vegetation growth. Low operator attention is required if properly designed.

Tables 2.2 and 2.3 illustrate the capacities of both natural and constructed wetlands for nutrient removal [4]. In test units and operating artificial marsh facilities using various wastewater streams, the following removals have been reported for secondary effluent treatment (10-day detention) [2]:

(a) BOD_5, 80–95 %
(b)   TSS, 29–87 %
(c)   COD, 43–87 %

Table 2.2 Nutrient removal from natural wetlands

Project	Flow, m^3/day	Wetland type	Percent reduction			
			TDP^a	NH_3-N	NO_3-N	TN^b
Brillion Marsh, WI	757	Marsh	13	–	51	–
Houghton Lake, MI	379	Peatland	95	71	99^c	–
Wildwood, FL	946	Swamp/ marsh	98	–	–	90
Concord, MA	2309	Marsh	47	58	20	–
Bellaire, MI	$1,136^d$	Peatland	88	–	–	84
Cootes Paradise, Town of Dundas, Ontario, Canada	–	Marsh	80	–	–	60–70
Whitney Mobile Park, Home Park, FL	227	Cypress dome	91	–	–	89

Source: US EPA [4]
[a]Total dissolved phosphorus
[b]Total nitrogen
[c]Nitrate and nitrite
[d]May to November only

Table 2.3 Nutrient removal from constructed wetlands

Project	Flow, m³/day	Wetland type	BOD₅, mg/L Influent	BOD₅, mg/L Effluent	SS, mg/L Influent	SS, mg/L Effluent	Percent reduction BOD₅	Percent reduction SS	Hydraulic surface loading rate, m³/ha-d
Listowel, Ontario [12]	17	FWSª	56	10	111	8	82	93	–
Santee, CA [10]	–	SFSᵇ	118	30	57	5.5	75	90	–
Sydney, Australia [13]	240	SFS	33	4.6	57	4.5	86	92	–
Arcata, CA	11,350	FWS	36	13	43	31	64	28	907
Emmitsburg, MD	132	SFS	62	18	30	8.3	71	73	1,543
Gustine, CA	3,785	FWS	150	24	140	19	84	86	412

Source: US EPA [4]
ªFree water surface system
ᵇSubsurface flow system

(d) Nitrogen, 42–94 % depending upon vegetative uptake and frequency of harvesting
(e) Total P, 0–94 % (high levels possible with warm climates and harvesting)
(f) Coliforms, 86–99%
(g) Heavy metals, highly variable depending on species

There is also evidence of reductions in wastewater concentrations of chlorinated organics and pathogens, as well as pH neutralization without causing detectable harm to the wetland ecosystem.

Residuals are dependent upon type of system and whether or not harvesting is employed. Duckweed, for example, yields 50–60 lb/acre/day (dry weight) during peak growing period to about half of this figure during colder months. For further detailed information on wetland systems, the reader is referred to references [19–23].

3 Evapotranspiration System

3.1 Description

Evapotranspiration (ET) system is a means of on-site wastewater disposal that may be utilized in some localities where site conditions preclude soil absorption. Evaporation of moisture from the soil surface and/or transpiration by plants is the mechanism of ultimate disposal. Thus, in areas where the annual evaporation rate equals or exceeds the rate of annual added moisture from rainfall and wastewater application, ET systems can provide a means of liquid disposal without danger of surface or groundwater contamination.

If evaporation is to be continuous, at least three conditions must be met [2]:

(a) There must be a continuous supply of heat to meet the latent heat requirement, approximately 590 cal/g of water evaporated at 15 °C.

(b) A vapor pressure gradient must exist between the evaporative surface and the atmosphere to remove vapor by diffusion, convection, or both. Meteorological factors, such as air temperature, humidity, wind velocity, and radiation, influence both energy supply and vapor removal.

(c) There must be a continuous supply of water to the evaporative surface. The soil material must be fine textured enough to draw up the water from the saturated zone to the surface by capillary action but not so fine as to restrict the rate of flow to the surface.

Evapotranspiration is also influenced by vegetation on the disposal field and can theoretically remove significant volumes of effluent in late spring, summer, and early fall, particularly if large silhouette, good transpiring bushes and trees are present.

A typical ET bed system (Fig. 2.3) consists of a 1½–3-ft depth of selected sand over an impermeable plastic liner. A perforated plastic piping system with rock cover is often used to distribute pretreated effluent in the bed. The bed may be square shaped on relatively flatland or a series of trenches on slopes. The surface area of the bed must be large enough for sufficient ET to occur to prevent the water level in the bed from rising to the surface.

Beds are usually preceded by septic tanks or aerobic units to provide the necessary pretreatment. Given the proper subsurface conditions, systems can be constructed to perform as both evapotranspiration and absorption beds. Nearly three-fourths of all the ET beds in operation were designed to use both disposal methods. Mechanical evaporators have been developed, but are not used at full scale.

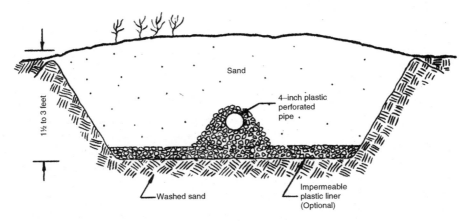

Fig. 2.3 Section through an evapotranspiration bed (Source: US EPA [2])

3.2 Applications

There are estimated to be 4,000–5,000 year-round evapotranspiration beds in operation in the USA, particularly in the semiarid regions of the southwest.

ET beds are used as an alternative to subsurface disposal in areas where these methods are either undesirable due to groundwater pollution potential or not feasible due to certain geological or physical constraints of land. The ET system can also be designed to supplement soil absorption for sites with slowly permeable soils. The use of ET systems for summer homes extends the range of application, which is otherwise limited by annual ET rates. Since summer evaporation rates are generally higher and plants with high transpiration rates are in an active growing state, many areas of the country can utilize ET beds for this seasonal application.

3.3 Limitations

The use of an evapotranspiration system is limited by climate and its effect on the local ET rate. In practice, lined ET bed systems are generally limited to areas of the country where pan evaporation exceeds annual rainfall by at least 24 in. The decrease of ET in winter at middle and high latitudes greatly limits its use. Snow cover reflects solar radiation, which reduces EF. In addition, when temperatures are below freezing, more heat is required to change frozen water to vapor. When vegetation is dormant, both transpiration and evaporation are reduced. An ET system requires a large amount of land in most regions. Salt accumulation may eventually eliminate vegetation and, thus, transpiration. Bed liner (where needed) must be kept watertight to prevent the possibility of groundwater contamination. Therefore, proper construction methods should be employed to keep the liner from being punctured during installation.

3.4 Design Criteria

Design of an evapotranspiration bed is based on the local annual weather cycle. The total expected inflow based on household wastewater generation and rainfall rates is compared with an average design evaporation value established from the annual pattern. It is recommended to use a 10-year frequency rainfall rate to provide sufficient bed surface area [2]. A mass balance is used to establish the storage requirements of the bed. Vegetative cover can substantially increase the ET rate during the summer growing season, but may reduce evaporation during the non-growing season. Uniform sand in the size range of D_{50} of approximately 0.10 mm is capable of raising water about 3 ft to the top of the bed. The polyethylene liner thickness is typically greater than or equal to 10 mil. Special attention should be paid to storm water drainage to make sure that surface runoff is drained away from the bed proximity by proper lot grading.

3.5 Performance

Performance is a function of climate conditions, volume of wastewater, and physical design of the system. Evapotranspiration is an effective and reliable means of domestic wastewater disposal. An ET system that has been properly designed and constructed is an efficient method for the disposal of pretreated wastewater and requires a minimum of maintenance. Healthy vegetative covers are aesthetically pleasing, and the large land requirement, although it limits the land use, does conserve the open space. Neither energy is required, nor is head loss of any value incurred.

3.6 Costs

The following site-specific costs serve to illustrate the major components of an evapotranspiration bed in Boulder, Colorado, with an annual net ET rate in the range of 0.04 gpd/ft^2 [2]. A 200-gpd household discharge would require a 2-ft deep bed with an area of approximately 5000 ft^2. All costs have been adjusted to 2016 US dollars (USD) using the Cost Index for Utilities [24].

Construction cost in 2016 USD:

Building sewer with 1,000-gal septic tank, design, and permit	2,020
Excavation and hauling (375 yd^3)	2,900
Liner (5,200 ft^2)	1,900
Distribution piping (625 ft)	850
Sand (340 yd^3) and gravel (38 yd^3)	5.060
Supervision and labor	1,390
Total	14,120 USD

Annual operation and maintenance cost:

Pumping septage from septic tank (every 3–5 years)	13.5–57
Total	13.5–57 USD

The construction cost for this particular system would be approximately USD 2.83/ft^2, which is consistent with a reported national range of USD 2.07 to USD 4.52/ft^2. The cost of an evapotranspiration bed is highly dependent upon local material and labor costs. As shown, the cost of sand is a significant portion of the cost of the bed. The restrictive sand size requirement makes availability and cost sensitive to location.

If an aerobic pretreatment unit is used instead of the septic tank, add USD 756–7,560 to the construction cost and an amount of USD 166–580/year to the annual operation and maintenance cost.

4 Land Treatment: Rapid Rate System

The land-based technologies have been in use since the beginning of civilization. Their greater value may be the use of the wastewater for beneficial return (agricultural and recharge) in water-poor areas, as well as nitrogen control benefits. If nitrogen control benefits are desired, some key issues arise concerning the type of plant crop with its growing and harvesting needs and/or the cycling of the water application and restorative oxygenation resting periods. Native soils and climate add the remaining variables.

Generally, the wastewater applications are cyclic in land-based technologies, making some form of storage or land rotation mandatory to ensure the restorative oxygenation derived from the resting period. Surface wastewater applications allow additional beneficial soil aeration (plowing, tilling, and raking), which can become mandatory for the heavily loaded systems after an elapsed season, or number of loading cycles. Actual surface cleaning programs, to remove the plastic, rubber, and other debris found in pretreated municipal wastewaters, also may be necessary, although not at the frequency used for beneficial soil aeration.

In this and the following sections, detailed information on the four most common land-based technologies will be provided. Subsurface, slow, and rapid infiltration systems do not discharge to surface waters and conceptually may allow a more relaxed nitrogen control standard in comparison to the overland flow system, depending on local groundwater regulations.

4.1 Description

Rapid rate infiltration was developed approximately 100 years ago and has remained unaltered since then. It has been widely used for municipal and certain industrial wastewaters throughout the world. Wastewater is applied to deep and permeable deposits, such as sand or sandy loam usually by distributing in basins (Fig. 2.4) or infrequently by sprinkling, and is treated as it travels through the soil matrix by filtration, adsorption, ion exchange precipitation, and microbial action [25]. Most metals are retained on the soil; many toxic organics are degraded or adsorbed. An underdrainage system consisting of a network of drainage pipe buried below the surface serves to recover the effluent, to control groundwater mounding, or to minimize trespass of wastewater onto adjoining property by horizontal subsurface flow. To recover renovated water for reuse or discharge, underdrains are usually intercepted at one end of the field by a ditch. If groundwater is shallow,

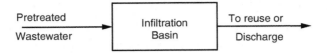

Fig. 2.4 Flow diagram of land treatment using rapid rate system (Source: US EPA [2])

underdrains are placed at or in the groundwater to remove the appropriate volume of water [2]. Thus, the designed soil depth, soil detention time, and underground travel distance to achieve the desired water quality can be controlled. Effluent can also be recovered by pumped wells.

Basins or beds are constructed by removing the fine-textured topsoil from which shallow banks are constructed. The underlying sandy soil serves as the filtration media. Underdrainage is provided by using plastic, concrete (sulfate resistant if necessary), or clay tile lines. The distribution system applies wastewater at a rate which constantly floods the basin throughout the application period of several hours to a couple of weeks. The waste floods the bed and then drains uniformly away, driving air downward through the soil and drawing fresh air from above. A cycle of flooding and drying maintains the infiltration capacity of the soil material. Infiltration diminishes slowly with time due to clogging. Full infiltration is readily restored by occasional tillage of the surface layer and, when appropriate, removal of several inches from the surface of the basin. Preapplication treatment to remove solids improves distribution system reliability, reduces nuisance conditions, and may reduce clogging rates. Common preapplication treatment practices include the following:

(a) Primary treatment for isolated locations with restricted public access [26].
(b) Biological treatment for urban locations with controlled public access.
(c) Storage is sometimes provided for flow equalization and for nonoperating periods.

Nitrogen removals are improved by [17, 27]:

(a) Establishing specific operating procedures to maximize denitrification
(b) Adjusting application cycles
(c) Supplying an additional carbon source
(d) Using vegetated basins (at low rates)
(e) Recycling portions of wastewater containing high nitrate concentrations
(f) Reducing application rates

Rapid rate infiltration systems require relatively permeable, sandy to loamy soils. Vegetation is typically not used for nitrogen control purposes but may have value for stabilization and maintenance of percolation rates. The application of algae-laden wastewater to rapid infiltration systems is not recommended because of clogging considerations but could be considered with attendant additional tolerance for surface maintenance, drying and soil aeration needs.

4.2   Applications

Rapid infiltration is a simple wastewater treatment system that is [2]:

(a) It is less land intensive than other land application systems and provides a means of controlling groundwater levels and lateral subsurface flow.
(b) It provides a means of recovering renovated water for reuse or for discharge to a particular surface water body.
(c) It is suitable for small plants where operator expertise is limited.
(d) It is applicable for primary and secondary effluent and for many types of industrial wastes, including those from breweries, distilleries, paper mills, and wool-scouring plants [26, 28, 29].

In very cold weather, the ice layer floats atop the effluent and also protects the soil surface from freezing. Generated residuals may require occasional removals of top layer of soil. The collected material is disposed of on-site.

4.3   Limitations

The rapid infiltration process is limited by [2]:

(a) Soil type
(b) Soil depth
(c) The hydraulic capacity of the soil
(d) The underlying geology
(e) The slope of the land

Nitrate and nitrite removals are low unless special management practices are used.

4.4   Design Criteria

The design criteria for rapid rate system can be summarized as follows [2]:

(a) Field area 3–56 acres/MG/day
(b) Application rate 20–400 ft/year, 4–92 in./week
(c) BOD_5 loading rate 20–100 lb/acre/day
(d) Soil depth 10–15 ft or more
(e) Soil permeability 0.6 in./h or more
(f) Hydraulic loading cycle 9 h to 2 weeks' application period, 15 h to 2 weeks' resting period
(g) Soil texture sands, sandy barns
(h) Basin size 1–10 acres, at least two basins/site

Table 2.4 Loading cycles for high-rate infiltration systems

Loading cycle objective	Applied wastewater	Season	Application period, d[a]	Drying period, d
Maximize infiltration rates	Primary	Summer	1–2	5–7
		Winter	1–2	7–12
	Secondary	Summer	1–3	4–5
		Winter	1–3	5–10
Maximize nitrogen removal	Primary	Summer	1–2	10–14
		Winter	1–2	12–16
	Secondary	Summer	7–9	10–15
		Winter	9–12	12–16
Maximize nitrification	Primary	Summer	1–2	5–7
		Winter	1–2	7–12
	Secondary	Summer	1–3	4–5
		Winter	1–3	5–10

Source: US EPA [25]
[a]Regardless of season or cycle objective, application periods for primary effluent should be limited to 1–2 days to prevent excessive soil clogging

(i) Height of dikes 4 ft, underdrains 6 or more ft deep
(j) Application techniques: flooding or sprinkling
(k) Preapplication treatment: primary or secondary

Designs can be developed that foster only nitrification or nitrification and denitrification [17, 27]. Nitrification is promoted by low hydraulic loadings and short application periods (1–2 days) followed by long drying periods (10–16 days). Denitrification can vary from 0 % to 80 %. For significant denitrification, the application period must be long enough to ensure depletion of the soil (and nitrate nitrogen) oxygen. Higher denitrification values predictably track higher BOD: nitrogen ratios. Enhancement may be promoted by recycling or by adding an external driving substrate (methanol). Nitrogen elimination strategies also may reduce the drying period by about half to yield lower overall nitrogen residuals with higher ammonium-nitrogen concentrations. Suggested loading cycles [25] to maximize infiltration rates, nitrogen removal, and nitrification rates are given in Table 2.4.

4.5 Performance

The effluent quality is generally excellent where sufficient soil depth exists and is not normally dependent on the quality of wastewater applied within limits. Well-designed systems provide for high-quality effluent that may meet or exceed primary drinking water standards. Percent removals for typical pollution parameters are [2]:

(a) BOD_5, 95–99%
(b) TSS, 95–99%

(c) Total N, 25–90%
(d) Total P, 0–90% until flooding exceeds adsorptive capacity [30]
(e) Fecal coliform, 99.9–99.99 + % [31]

The process is extremely reliable, as long as sufficient resting periods are provided. However, it has a potential for contamination of groundwater by nitrates. Heavy metals could be eliminated by pretreatment techniques as necessary. Monitoring for metals and toxic organics is needed where they are not removed by pretreatment. The process requires long-term commitment of relatively large land areas, although small by comparison to other land treatment systems [32, 33].

4.6   Costs

The construction and operation and maintenance costs are shown in Figs. 2.5 and 2.6, respectively [2]. The costs are based on 1973 (Utilities Index = 149.36, EPA Index 194.2, ENR Index = 1850) figures. To obtain the values in terms of the present 2016 USD, using the Cost Index for Utilities [24], multiply the costs by a factor of 5.50.

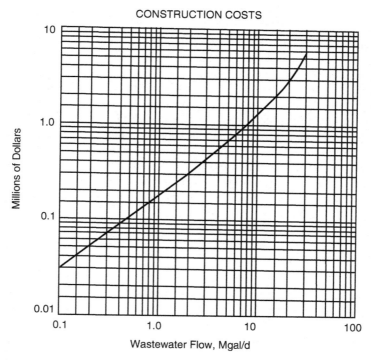

Fig. 2.5 Construction costs for rapid rate system (Source: US EPA [2]). (To elevate costs to 2016 multiply by a factor of 5.50)

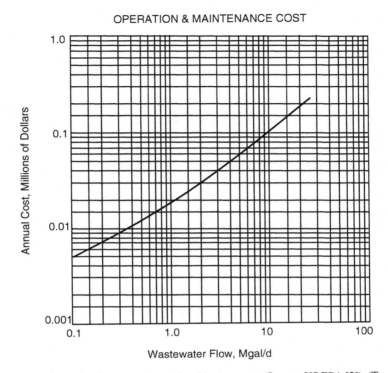

OPERATION & MAINTENANCE COST

Fig. 2.6 Operation and maintenance costs of rapid rate system (Source: US EPA [2]). (To elevate costs to 2016 multiply by a factor of 5.50)

Assumptions applied in preparing the costs given in Figs. 2.5 and 2.6:

(a)   Application rate 182 ft/year.

(b)   Construction costs include field preparations (removal of brush and trees) for multiple unit infiltration basins with 4-ft dike formed from native excavated material, and storage is not assumed necessary.

(c)   Drain pipes buried 6–8 ft with 400-ft spacing, interception ditch along length of field, and weir for control of discharge; gravel service roads and 4-ft stock fence around perimeter.

(d)   O & M cost includes inspection and unclogging of drain pipes at outlets, annual tilling of infiltration surface and major repair of dikes after 10 years, high-pressure jet cleaning of drain pipes every 5 years, annual cleaning of interceptor ditch, and major repair of ditches, fences, and roads after 10 years.

(e)   Costs of pretreatment monitoring wells, land, and transmission to and from pretreatment facility not included.

5   Land Treatment: Slow Rate System

5.1   Description

Slow-rate land treatment represents the predominant municipal land treatment practice in the USA. In this process, wastewater is applied by sprinkling to vegetated soils that are slow to moderate in permeability (clay barns to sandy barns) and is treated as it travels through the soil matrix by filtration, adsorption, ion exchange, precipitation, microbial action, and plant uptake (Fig. 2.7). An underdrainage system consisting of a network of drainage pipe buried below the surface serves to recover the effluent, to control groundwater, or to minimize trespass of leachate onto adjoining property by horizontal subsurface flow. To recover renovated water for reuse or discharge, underdrains are usually intercepted at one end of the field by a ditch. Underdrainage for groundwater control is installed as needed to prevent waterlogging of the application site or to recover the renovated water for reuse. Proper crop management also depends on the drainage conditions. Sprinklers can be categorized as hand moved, mechanically moved, and permanent set, the selection of which includes the following considerations [2]:

(a)   Field conditions (shape, slope, vegetation, and soil type)
(b)   Climate
(c)   Operating conditions
(d)   Economics

Vegetation is a vital part of the process and serves to extract nutrients, reduce erosion, and maintain soil permeability. Considerations for crop selection include:

(a)   Suitability to local climate and soil conditions
(b)   Consumptive water use and water tolerance
(c)   Nutrient uptake and sensitivity to wastewater constituents
(d)   Economic value and marketability
(e)   Length of growing season
(f)   Ease of management
(g)   Public health regulations

Common preapplication treatment practices include the following:

(a)   Primary treatment for isolated locations with restricted public access and when limited to crops not for direct human consumption
(b)   Biological treatment plus control of coliform to 1,000 MPN/100 mL for agricultural irrigation, except for human food crops to be eaten raw

Fig. 2.7 Flow diagram of land treatment using slow rate system (Source: US EPA [2])

Table 2.5 Potential adverse effects of wastewater constituents on crops

Constituent level				
Problem and related constituent	No problem	Increasing problems	Severe problems	Crops affected
Salinity (EC_W), mmho/cm	<0.75	0.75–3.0	>3.0	Crops in arid climates only
Specific ion toxicity from root absorption Boron, mg/L	<0.5	0.5–2	2.0–10.0	Fruit and citrus trees 0.5–1.0 mg/L; Field crops 1.0–2.0 mg/L; Grasses 2.0–10.0 mg/L
Sodium, adj-SAR[a]	<3	3.0–9.0	>9.0	Tree crops
Chloride, mg/L	<142	142–355	>355	Tree crops
Specification toxicity from foliar absorption Sodium, mg/L	<69	>69	–	Field and vegetable crops under sprinkler
Chloride, mg/L	<106	>106	–	Application
Miscellaneous NH_4-N + NO_3-N, mg/L	<5	5–30	30	Sugar beets, potatoes, cotton, grains
HCO_3, mg/L	<90	90–520	>520	Fruit
pH, units	6.5–8.4	4.2–5.5	<4.2 and >8.5	Most crops

Source: US EPA [25]
[a]Adjusted sodium adsorption ratio

(c)   Secondary treatment plus disinfection to 200 MPN/100-mL fecal coliform for public access areas (parks)

   Wastewaters high in metal content should be pretreated to avoid plant and soil contamination. Table 2.5 shows the wastewater constituents that have potential adverse effects on crops [25]. Forestland irrigation is more suited to cold weather operation, since soil temperatures are generally higher, but nutrient removal capabilities are less than for most field crops.

5.2   Applications

Slow rate systems produce the best results of all the land treatment systems. Advantages of sprinkler application over gravity methods include [34]:

(a)   More uniform distribution of water and greater flexibility in range of application rates
(b)   Applicability to most crops
(c)   Less susceptibility to topographic constraints
(d)   Reduced operator skill and experience requirements

   Underdrainage provides a means of recovering renovated water for reuse or for discharge to a particular surface water body when dictated by senior water rights and a means of controlling groundwater. The system also provides the following benefits:

(a) An economic return from the use of water and nutrients to produce marketable crops for forage
(b) Water and nutrient conservation when utilized for irrigating landscaped areas

5.3 Limitations

The slow rate process is limited by [2]:

(a) Soil type and depth
(b) Topography
(c) Underlying geology
(d) Climate
(e) Surface and groundwater hydrology and quality
(f) Crop selection
(g) Land availability

Crop water tolerances, nutrient requirements, and the nitrogen removal capacity of the soil-vegetation complex limit hydraulic loading rate [35]. Climate affects growing season and will dictate the period of application and the storage requirements. Application ceases during period of frozen soil conditions. Once in operation, infiltration rates can be reduced by sealing of the soil. Limitations to sprinkling include adverse wind conditions and clogging of nozzles. Slopes should be less than 15 % to minimize runoff and erosion. Pretreatment for removal of solids and oil and grease serves to maintain reliability of sprinklers and to reduce clogging. Many states have regulations regarding preapplication disinfection, minimum buffer areas, and control of public access for sprinkler systems.

The process requires long-term commitment of large land area, i.e., largest land requirement of all land treatment processes [36]. Concerns with aerosol carriage of pathogens, potential vector problems, and crop contamination have been identified, but are generally controllable by proper design and management.

5.4 Design Criteria

The design criteria for slow rate system can be summarized as follows [2]:

(a) Field area 56–560 acres/MG/day
(b) Application rate 2–20 ft/year, 0.5–4 in./week
(c) BOD_5 loading rate 0.2–5 lb/acre/day
(d) Soil depth 2–5 ft or more
(e) Soil permeability 0.06–2.0 in./h
(f) Minimum preapplication treatment primary
(g) Lower temperature limit 25 °F
(h) Particle size of solids less than one-third of the sprinkler nozzle diameter
(i) Underdrains 4–8 in. diameter, 4–10 ft deep, and 50–500 ft apart and pipe material plastic, concrete (sulfate resistant, if necessary), or clay

5.5   *Performance*

Effluent quality is generally excellent and consistent regardless of the quality of wastewater applied [37]. Percent removals for typical pollution parameters when wastewater is applied through more than 5 ft of unsaturated soil are:

(a) BOD_5, 90–99 + %
(b) TSS, 90–99 + %
(c) Total N, 50–95 % depending on N uptake of vegetation
(d) Total P, 80–99 %, until adsorptive capacity is exceeded [38]
(e) Fecal coliform, 99.99 + % when applied levels are more than 10 MPN/100 mL

This treatment is capable of achieving the highest degree of nitrogen removal. Typically, nitrogen losses due to denitrification (15–25 %), ammonia volatilization (0–10 %), and soil immobilization (0–25 %) supplement the primary nitrogen removal mechanism by the crop [17]. The balance of the nitrogen passes to the percolate. Typical design standards require preservation of controlling depths to groundwater and establishing nitrogen limits in either the percolate or groundwater as it leaves the property site. Nitrogen loading to the groundwater is often the controlling consideration in the design. For further detailed information on slow rate infiltration systems, the reader is referred to references [39–44].

5.6   *Costs*

The construction and operation and maintenance costs are shown in Figs. 2.8 and 2.9, respectively [2]. The costs are based on 1973 (Utilities Index = 149.36, EPA Index 194.2, ENR Index = 1850) figures. To obtain the values in terms of the present 2016 USD, using the Cost Index for Utilities [24], multiply the costs by a factor of 5.50.

Assumptions applied in preparing the costs given in Figs. 2.8 and 2.9:

(a) Yearly average application rate: 0.33 in./day.
(b) Energy requirements: solid set spray distribution requires 2100 kwh/year/ft of TDH/MG/d capacity. Center-pivot spraying requires an additional 0.84×10^6 kwh/year/acre (based on 3.5 days/week operation) for 1 MG/d or larger facilities (below 1 MG/day, additional power = 0.84 to 1.35×110^6 kwh/year/acre).
(c) Clearing costs are for brush with few trees using bulldozer-type equipment.
(d) Solid set spraying construction costs include lateral spacing, 100 ft; sprinkler spacing, 80 ft along laterals; 5.4 sprinklers/acre; application rate, 0.20 in./h; 16.5-gpm flow to sprinklers at 70 psi; flow to laterals controlled by hydraulically operated automatic valves; laterals buried 18 in.; mainlines buried 36 in.; all pipe 4-in. diameter and smaller that is PVC; and all larger pipe that is asbestos cement (total dynamic head = 150 ft).

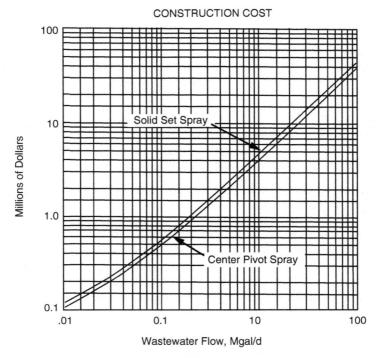

Fig. 2.8 Construction cost of slow rate system (Source: US EPA [2]). (To elevate costs to 2016 multiply by a factor of 5.5)

(e) Center-pivot spraying construction costs include heavy-duty center-pivot rig with electric drive; multiple units for field areas over 40 acres; maximum area per unit, 132 acres; and distribution pipe that is buried 3-ft deep

(f) Underdrains are spaced 250 ft between drain pipes. Drain pipes are buried 6–8-ft deep with interception ditch along length of field and weir for control of discharge.

(g) Distribution pumping construction costs include structure built into dike of storage reservoir, continuously cleaned water screens, pumping equipment with normal standby facilities, piping and valves within structure, and controls and electrical work.

(h) Labor costs include inspection and unclogging of drain pipes at outlets and dike maintenance.

(i) Materials costs include for solid set spraying, replacement of sprinklers and air compressors for valve controls after 10 years; for center-pivot spraying, minor repair parts and major overhaul of center-pivot rigs after 10 years; high-pressure jet cleaning of drain pipes every 5 years, annual cleaning of interceptor ditch, and major repair of ditches after 10 years; distribution pumping repair work performed by outside contractor and replacement parts; and scraping and patching of storage receiver liner every 10 years.

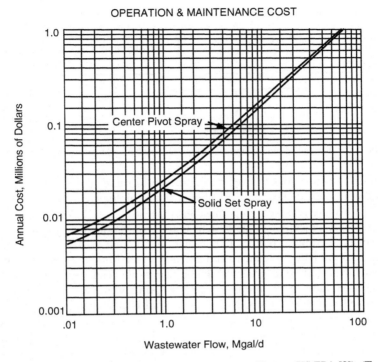

Fig. 2.9 Operation and maintenance cost of slow rate system (Source: US EPA [2]). (To elevate costs to 2016 multiply by a factor of 5.50)

(j)   Storage for 75 days is included; 15-ft dikes (12-ft wide at crest) are formed from native materials (inside slope 3:1, outside 2:1); rectangular shape on level ground; 12-ft water depth; multiple cells for more than 50-acre size; asphaltic lining; 9-in. riprap on inside slope of dikes.

(k)   Cost of pretreatment, monitoring wells, land, and transmission to and from land treatment facility not included.

6   Land Treatment: Overland Flow System

6.1   Description

Wastewater treatment using the overland flow system is relatively new. It is now extensively used in the food processing industry. Very few municipal plants are in operation and most are in warm, dry areas. A flow diagram of the system is shown in Fig. 2.10. Wastewater is applied over the upper reaches of sloped terraces and is

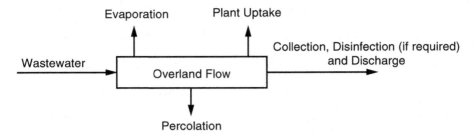

Fig. 2.10 Flow diagram of land treatment using overland flow system (Source: US EPA [2])

treated as it flows across the vegetated surface to runoff collection ditches. The wastewater is renovated by physical, chemical, and biological means as it flows in a thin film down the relatively impermeable slope.

A secondary objective of the system is for crop production. Perennial grasses (reed canary, bermuda, redtop, tall fescue, and Italian rye) with long growing seasons, high moisture tolerance, and extensive root formation are best suited to overland flow. Harvested grass is suitable for cattle feed. Biological oxidation, sedimentation, and grass filtration are the primary removal mechanisms for organics and suspended solids. Nitrogen removal is attributed primarily to nitrification/denitrification and plant uptake. Loading rates and cycles are designed to maintain active microorganism growth on the soil surface. The operating principles are similar to a conventional trickling filter with intermittent dosing. The rate and length of application are controlled to minimize severe anaerobic conditions that result from overstressing the system. The resting period should be long enough to prevent surface ponding, yet short enough to keep the microorganisms in an active state. Surface methods of distribution include the use of gated pipe or bubbling orifice. Gated surface pipe, which is attached to aluminum hydrants, is aluminum pipe with multiple outlets. Control of flow is accomplished with slide gates or screw adjustable orifices at each outlet. Bubbling orifices are small diameter outlets from laterals used to introduce flow. Gravel may be necessary to dissipate energy and ensure uniform distribution of water from these surface methods. Slopes must be steep enough to prevent ponding of the runoff, yet mild enough to prevent erosion and provide sufficient detention time for the wastewater on the slopes. Slopes must have a uniform cross slope and be free from gullies to prevent channeling and allow uniform distribution over the surface. The network of slopes and terraces that make up an overland system may be adapted to natural rolling terrain. The use of this type of terrain will minimize land preparation costs. Storage must be provided for nonoperating periods. Runoff is collected in open ditches. When unstable soil conditions are encountered or flow velocities are erosive, gravity pipe collection systems may be required. Common preapplication practices include the following: screening or comminution for isolated sites with no public access and screening or comminution plus aeration to control odors during storage or application for urban locations with no public access [45, 46]. Wastewaters high in metal content should be pretreated to avoid soil and plant contamination.

A common method of distribution is with sprinklers. Recirculation of collected effluent is sometimes provided and/or required. Secondary treatment prior to overland flow permits reduced (as much as two-third reduction) land requirements. Effluent disinfection is required where stringent fecal coliform criteria exist.

6.2   Application

Because overland flow is basically a surface phenomenon, soil clogging is not a problem. High BOD_5 and suspended solids removals have been achieved with the application of raw comminuted municipal wastewater. Thus, preapplication treatment is not a prerequisite where other limitations are not operative. Depth to groundwater is less critical than with other land systems. It also provides the following benefits: an economic return from the reuse of water and nutrients to produce marketable crops or forage and a means of recovering renovated water for reuse or discharge. This type of applications is preferred for gently sloping terrain with impermeable soils.

6.3   Limitations

The process is limited by soil type, crop water tolerances, climate, and slope of the land. Steep slopes reduce travel time over the treatment area and, thus, treatment efficiency. Flatland may require extensive earthwork to create slopes. Ideally, slope should be 2–8 %. High-flotation tires are required for equipment. Cost and impact of the earthwork required to obtain terraced slopes can be major constraints. Application is restricted during rainy periods and stopped during very cold weather [47]. Many states have regulations regarding preapplication disinfection, minimum buffer zones, and control of public access.

6.4   Design Criteria

The design criteria for overland flow system can be summarized as follows [2]:

(a) Field area required, 35–100 acres/MG/day
(b) Terraced slopes 2–8 %
(c) Application rate, 11–32 ft/year, 2.5–16 in./week
(d) BOD_5 loading rate, 5–50 lb/acre/day
(e) Soil depth, sufficient to form slopes that are uniform and to maintain a vegetative cover
(f) Soil permeability, 0.2 in./h or less

Table 2.6 Design loadings for overland flow systems

Preapplication treatment	Application rate m^3/h m	Hydraulic loading rate cm/day
Screening/primary	0.07–0.12[a]	2.0–7.0[b]
Aerated cell (1-day detention)	0.08–0.14	2.0–8.5
Wastewater treatment pond[c]	0.09–0.15	2.5–9.0
Secondary[d]	0.11–0.17	3.0–10.0

Source: US EPA [48]
[a]m^3/h m × 80.5 = gal/h ft
[b]cm/d × 0.394 = in./day
[c]Does not include removal of algae
[d]Recommended only for upgrading existing secondary treatment

(g) Hydraulic loading cycle, 6–8-h application period, 16–181week resting period
(h) Operating period, 5–6 days/week
(i) Soil texture clay and clay loams

Below are representative application rates for 2–8 % sloped terraces:

In./week	Pretreatment	Terrace length, ft
2.5–8	Untreated or primary	150
6–16	Lagoon or secondary	120

Generally, 40–80% of applied wastewater reaches collection structures, lower percent in summer and higher in winter (southwest data). Table 2.6 shows the required pretreatment and allowed application and hydraulic rates [48].

6.5 Performance

Percent removals for comminuted or screened municipal wastewater over about 150 ft of 2–6 % slope:

(a) BOD_5, 80–95 %
(b) Suspended solids, 80–95 %
(c) Total N, 75–90 %
(d) Total P, 30–60 %,
(e) Fecal coliform, 90–99.9 %

The addition of alum $[Al_2(SO_4)_3]$, ferric chloride $[FeCl_3]$, or calcium carbonate $[CaCO_3]$ prior to application will increase phosphorus removals.

Little attempt has been made to design optimized overland flow systems with a specific objective of nitrogen control. Their performance depends on the same fundamental issues: nitrification-denitrification, ammonia volatilization, and harvesting of crops. When measured, overland flow systems designed for secondary treatment often reveal less than 10-mg/L total nitrogen [49]. For further detailed information on overland flow systems, the reader is referred to references [12, 50–52].

6.6   *Costs*

The construction and operation and maintenance costs are shown in Figs. 2.11 and 2.12, respectively [2]. The costs are based on 1973 (Utilities Index = 149.36, EPA Index 194.2, ENR Index = 1850) figures. To obtain the values in terms of the present 2016 USD, using the Cost Index for Utilities [24], multiply the costs by a factor of 5.50.

Assumptions applied in preparing the costs given in Figs. 2.11 and 2.12:

(a) Storage for 75 days included.
(b) Site cleared of brush and trees using bulldozer-type equipment; terrace construction: 175–250-ft wide with 2.5 % slope (1400 yd/acre of cut). Costs include surveying, earthmoving, finish grading, ripping two ways, disking, land planning, and equipment mobilization.
(c) Distribution system: application rate, 0.064 in./h; yearly average rate of 3 in./week (8 h/day, 6 days/week); flow to sprinklers, 13 gpm at 50 psi; laterals 70 ft from top of terrace, buried 18 in.; flow to laterals controlled by hydraulically operated automatic valves; mainlines buried 36 in.; all pipe 4 in. diameter and smaller is PVC; and all larger pipe is asbestos cement.

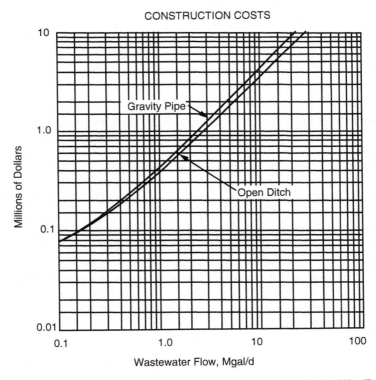

Fig. 2.11 Construction cost of overland flow treatment system (Source: US EPA [2]). (To elevate costs to 2016 multiply by a factor of 5.50)

OPERATION & MAINTENANCE COSTS

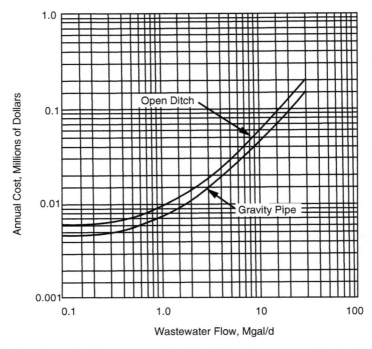

Fig. 2.12 Operation and maintenance cost of overland flow treatment system (Source: US EPA [2]) (To elevate costs to 2016 multiply by a factor of 5.50)

(d) Open ditch collection: network of unlined interception ditches sized for a 2 in./h storm; culverts under service roads; and concrete drop structures at 1,000-ft intervals.

(e) Gravity pipe collection: network of gravity pipe interceptors with inlet/manholes every 250 ft along submains; storm runoff that is allowed to pond at inlets; each inlet/manhole that serves 1,000 ft of collection ditch; and manholes every 500 ft along interceptor mains.

(f) O & M cost includes replacement of sprinklers and air compressors for valve controls after 10 years and either biannual cleaning of open ditches with major repair after 10 years or the periodic cleaning of inlets and normal maintenance of gravity pipe and also includes dike maintenance and scraping and patching of storage basin liner every 10 years.

(g) Costs for pretreatment, land, transmission to site, disinfection, service roads, and fencing not included.

7 Subsurface Infiltration

Subsurface infiltration systems are capable of producing a high degree of treatment; with proper design, they can provide a nitrified effluent, and denitrification can be achieved under certain circumstances. Keys to their success are the adequacy of the initial gravel infiltration zone for solid capture and the following unsaturated zone of native or foreign soils. Failure to provide an oxygenated environment by either resting or conservative loadings can lead to failure. Denitrification under gravity loading is likely to be small, but may be improved through pressure/gravity dosing concepts of liquid application to the trenches [53].

Subsurface infiltration wastewater management practices are embodied in the horizontal leach fields that routinely serve almost one-third of the United States population that use more than 20 million septic tanks in their individual non-sewered establishments and homes [2]. In recent years, they have also been advanced for collective service in small isolated communities.

7.1 Description

A septic tank followed by a soil absorption field is the traditional on-site system for the treatment and disposal of domestic wastewater from individual households or establishments. The system consists of a buried tank where wastewater is collected and scum, grease, and settleable solids are removed by gravity separation and a subsurface drainage system where clarified effluent percolates into the soil. Precast concrete tanks with a capacity of 1,000 gallons are commonly used for house systems. Solids are collected and stored in the tank, forming sludge and scum layers. Anaerobic digestion occurs in these layers, reducing the overall volume. Effluent is discharged from the tank to one of three basic types of subsurface systems, absorption field [53], seepage bed [53, 54], or seepage pits [55]. Sizes are usually determined by percolation rates, soil characteristics, and site size and location. Distribution pipes are laid in a field of absorption trenches to leach tank effluent over a large area (Fig. 2.13). Required absorption areas are dictated by state

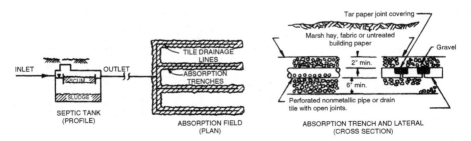

Fig. 2.13 Septic tank absorption field (Source: US EPA [2])

and local codes. Trench depth is commonly about 24 in. to provide minimum gravel depth and earth cover. Clean, graded gravel or similar aggregate, varying in size from one-half to 2½ inches, should surround the distribution pipe and extend at least two inches above and six inches below the pipe. The maintenance of at least a 2-ft separation between the bottom of the trench and the high water table is required to minimize groundwater contamination. Piping typically consists of agricultural drain tile, vitrified clay sewer pipe, or perforated, nonmetallic pipe. Absorption systems having trenches wider than 3 ft are referred to as seepage beds. Given the appropriate soil conditions (sandy soils), a wide bed makes more efficient use of available land than a series of long, narrow trenches.

Many different designs may be used in laying out a subsurface disposal field. In sloping areas, serial distribution can be employed with absorption trenches by arranging the system so that each trench is utilized to its capacity before liquid flows into the succeeding trench. A dosing tank can be used to obtain proper wastewater distribution throughout the disposal area and give the absorption field a chance to rest or dry out between dosings. Providing two separate alternating beds is another method used to restore the infiltrative capacity of a system. Aerobic units may be substituted for septic tanks with no changes in soil absorption system requirements.

In areas where problem soil conditions preclude the use of subsurface trenches or seepage beds, mounds can be installed (Fig. 2.14) to raise the absorption field above ground, provide treatment, and distribute the wastewater to the underlying soil over

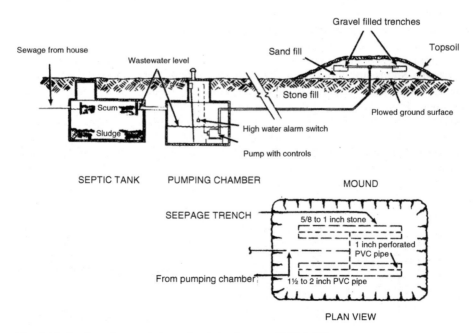

Fig. 2.14 Septic tank mound absorption field (Source: US EPA [2])

a wide area in a uniform manner [2, 56, 57]. A pressure distribution network should be used for uniform application of clarified tank effluent to the mound. A subsurface chamber can be installed with a pump and high water alarm to dose the mound through a series of perforated pipes. Where sufficient head is available, a dosing siphon may be used. The mound must provide an adequate amount of unsaturated soil and spread septic tank effluent over a wide enough area so that distribution and purification can be effected before the water table is reached.

The mound system requires more space and periodic maintenance than conventional subsurface disposal system, along with higher construction costs. System cannot be installed on steep slopes nor over highly (120 min/in.) impermeable subsurface. Seasonal high groundwater must be deeper than two feet to prevent surfacing at the edge of the mound [2].

An alternative to the mound system is a new combined distribution and pretreatment unit to precede the wastewater application to the subsurface infiltration systems [58]. The new system is based on pumping of septic tank effluent to one or more units filled with lightweight clay aggregates. The wastewater is distributed evenly over the 2.3-m^2 surface of the pretreatment filter. The filter effluent is then applied to the subsurface infiltration system.

7.2 Applications

Subsurface infiltration systems for the disposal of septic tanks effluents are used primarily in rural and suburban areas where economics are favorable. Properly designed and installed systems require a minimum of maintenance and can operate in all climates.

7.3 Limitations

The use of subsurface effluent disposal fields is dependent on the following factors and conditions [2]:

(a) Soil and site conditions
(b) The ability of the soil to absorb liquid
(c) Depth to groundwater
(d) Nature of and depth to bedrock
(e) Seasonal flooding
(f) Distance to well or surface water

A percolation rate of 60 mm/in. is often used as the lower limit of permeability. The limiting value for seasonal high groundwater should be 2 ft below the bottom of the absorption field. When a soil system loses its capacity to absorb septic tank effluent, there is a potential for effluent surfacing, which often results in odors and, possibly, health hazards.

Table 2.7 Required areas
of subsurface infiltration
absorption fields

Percolation rate, min/in.	Required area per bedroom, ft^2
1 or less	70
3	100
5	125
10	165
15	190
30	250
45	300
60	330

Source: US EPA [2]

7.4   Design Criteria

Absorption area requirements for individual residences are given in Table 2.7. The area required per bedroom is a function of the percolation rate; the higher the rate, the smaller is the required area [2].

Design criteria for the mound system are as follows [2, 56, 57, 59]: design flow 75 gal/person/day, 150 gal/bedroom/day, basal area based on percolation rates up to 120 min/in., mound height at center approximately 3.5–5 ft, and pump (centrifugal) that must accommodate approximately 30 gpm at required TDH. The design standards for a mound are shown in Table 2.8. It includes four steps as illustrated in the following two examples. The steps are [59]:

1. Flow estimation
2. Design of the absorption trenches
3. Dimensioning the mound
4. Checking for limiting conditions

Properly designed, constructed, and operated septic tank systems have demonstrated an efficient and economical alternative to public sewer systems, particularly in rural and sparsely developed areas. System life for properly sited, designed, installed, and maintained systems may equal or exceed 20 years.

7.5   Performance

Performance is a function of the following factors [2]:

(a) Design of the system components
(b) Construction techniques employed
(c) Rate of hydraulic loading
(d) Area geology and topography
(e) Physical and chemical composition of the soil mantle
(f) Care given to periodic maintenance

Table 2.8 Standards for elevated sand mounds

1. Elevated sand mounds shall utilize absorption trench distribution design, and shall not be installed on land with a slope greater than twelve (12) percent. The trenches shall be installed with the long dimension of the trench parallel to the land contour. The minimum spacing between trenches shall be four feet (1.2 m), a maximum trench width of four feet (1.2 m) shall be permitted. No more than three (3) parallel trenches may be installed within a mound.

2. A minimum total trench length of forty (40) feet (12 m) shall be provided in each trench in mounds constructed in soils with percolation rates of 60 to 120 minutes per inch when two or more trenches are used.

3. The required bottom area of the trench or trenches shall be based upon a flow of 150 gallons (570 L) per bedroom per day with a application rate of 1.2 gallons/day/square foot (49 L/d/m²).

4. The effective basal area of the mound for soils with a percolation rate of 61 to 120 minutes per inch is to be calculated on a maximum application rate of 0.24 gallons/day/square foot (10 L/d/m²).

5. The area of sand fill shall be sufficient to extend three feet (0.9 m) beyond the edge of the required absorption (or stone) area before the sides are shaped to a 4:1 slope.

6. The effective basal area of the mound for soils with a percolation rate of 61 to 120 minutes per inch is to be calculated on a maximum application rate of 0.24 gallons/day/square foot (10 L/d/m²).

7. The effective basal area of the mound for soils with a percolation rate of 5 to 60 minutes per inch is to be calculated on an application rate of 0.74 gallons/day/square foot (29 L/d/m²).

Source: [59]

Pollutants are removed from the effluent by natural adsorption and biological processes in the soil zone adjacent to the field. BOD, SS, bacteria, and viruses, along with heavy metals and complex organic compounds, are adsorbed by soil under proper conditions. However, chlorides and nitrates may readily penetrate coarser, aerated soils to groundwater.

Leachate can contaminate groundwater when pollutants are not effectively removed by the soil system. In many well-aerated soils, significant densities of homes with septic tank-soil absorption systems have resulted in increasing nitrate content of the groundwater. Soil clogging may result in surface ponding with potential esthetic and public health problems. The sludge and scum layers accumulated in a septic tank must be removed every 3–5 years. For further detailed information on subsurface infiltration systems, the reader is referred to references [60–65].

Additional technical information on natural biological treatment processes and terminologies can be found from the literature [66–79].

7.6 Design Example 1: Elevated Sand Mound at a Crested Site

Given
A three-bedroom home, a percolation rate of 80 min/in. at a depth of 24 in. below ground surface, a crested site with a land slope of 2 %, high groundwater is 36 in. below ground surface, and all other site factors that are satisfactory [59].

Design
1. Estimate daily flow in gpd = Q
   Q = number of bedrooms × 150 gpd/bedroom (see Sect. 7.4 Design Criteria)
   = 3 × 150 = 450 gpd.
2. Absorption trench system
   Application rate in sand fill in $gpd/ft^2 = 1.2\ gpd/ft^2$ (from Table 2.8)

   $$\text{Trench bottom area} = \text{daily flow in gpd/application rate in sand fill in } gpd/ft^2$$
   $$= 450\,gpd/1.2\,gpd/ft^2$$
   $$= 375\,ft^2$$

   L_1 = total length of trench = (trench bottom area)/trench width
   Using a trench width = a = 3 ft (see Fig. 2.15)
   $$= 375\,ft^2/3\,ft$$
   $$= 125\,ft$$
   Using two trenches, length per trench = 125 ft/2 = 62.5 ft
   Use two trenches, 3-ft wide and 65-ft long
   At least 4 ft must be provided between trenches (b). For crested sites it is desirable to provide at least 10 ft between trenches. This design will use a spacing, b = 10 ft.
3. Dimensioning the sand fill portion of the mound
   The length of the mound is computed by adding the length of the top (horizontal extent) of the sand fill (l + 2c) and the horizontal distances on each end of the top needed to provide a side slope of one vertical to four

horizontal. Note that 12 in. of fill must exist below the trench and a trench is 9-in. deep. Thus:

L = mound length = $L_1 + 2c + [(2$ sides $\times 4$ vertical $\times$ vertical thickness of fill, in.$)/$
(1 horizontal $\times$ 12 in./ft)]

$L = 65 + (2 \times 3) + [(2 \times 4 (9$ in. $+ 12$ in.$) / (12$ in./ft)]$

$L = 65 + 6 + 14$

L = 85 ft the length of the mound at the crest of this site

The width of the mound (W) is computed in a similar manner using the width of the trenches (a) and number of trenches and trench spacing (b). Thus:

$W = 2a + b + 2c + [(2$ sides $\times 4$ vertical $\times$ vertical thickness of fill, in.$)/$
(1 horizontal $\times$ 12 in./ft)]

$W = (2 \times 3) + 10 + 6 + 14$

$W = 36$ ft, 0% slope

For a 2 % slope, the vertical fall is about 2 % $\times$ 36 ft $= 0.36$ ft in a horizontal distance of 18 ft. The approximate additional width of the mound on a crested site with 2 % slopes is $0.36 \times 4 \times 2 = 2.9$ ft. ≈ 3 ft. Thus:

W = total mound width = $36 + 3 = 39$ ft.

The mound length, L, at the downslope base of the sand fill will be 88 ft. Note that additional land area is needed for the mound for placement of the topsoil over the sand fill portion of the mound.

4. Checking for limiting conditions

The effective basal area of the sand fill (shaded area in plan view of Fig. 2.15) in the mound below and downslope of the trenches must be large enough to absorb the estimated daily waste flow. In calculating this basal area, exclude the portions of the mound on each side of the end the trench or trenches and their extension downslope.

In this calculation:

A = effective basal area of mound, ft^2 = shaded area in plan view,

Q = estimated daily flow, gpd,

R = natural soil infiltration rate, gpd/ft^2.

Thus, $R = Q/A$, expressed in gpd/ft^2.

In this case, a crested site is used and the trench length, L_1 by the sum of the trench width, a, the horizontal extent of fill beyond the trench, c, and the horizontal distance needed to provide the side slope (one-half of the correction used for total mound width).

$$A = L_1(a + c + 7 + 1.5)$$
$$A = 65 (3 + 3 + 7 + 1.5)$$
$$A = 942 \text{ ft}^2$$

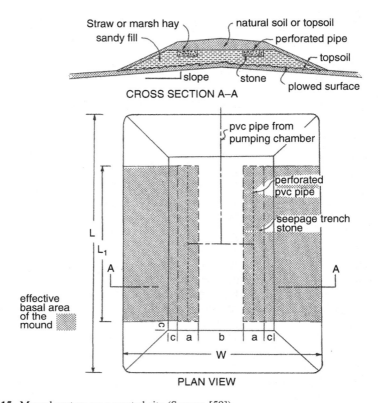

Fig. 2.15 Mound system on a crested site (Source: [59])

$$Q = 450/2 = 225 \text{ gpd per trench}$$
$$R = Q/A = 225/942 = 0.24 \text{ gpd/ft}^2$$

From Table 2.8, R must not exceed 0.24 gpd/ft^2

This design is satisfactory. If calculated R exceeds the maximum, increase the size of the mound downslope or trench length to provide a satisfactory value of R.

7.7 Design Example 2: Elevated Sand Mound at a Sloping Site

Given

A four-bedroom home, a percolation rate of 45 min/in. at a depth of 24 in. below ground surface, land slope is 4 %, high groundwater is 36 in. below ground surface, and all other site factors that are satisfactory [59].

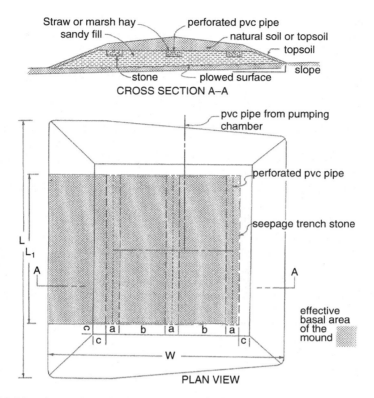

Fig. 2.16 Mound system on a level or sloping site (Source: [59])

Design

1. Estimate daily flow in gpd = Q

    Q = number of bedrooms × 150 gpd/bedroom (see Sect. 7.4 Design Criteria)
       = 4 × 150 = 600 gpd

2. Absorption trench system

    Application rate in sand fill in gpd/ft² = 1.2 gpd/ft² (from Table 2.8)

    Trench bottom area = daily flow in gpd/application rate in sand fill in gpd/ft²
                       = 600 gpd/1.2 gpd/ft²
                       = 500 ft²

    L_1 = total length of trench = (trench bottom area)/trench width
    Using a trench width = a = 4 ft (see Fig. 2.16)
                       = 500 ft²/4 ft
                       = 125 ft

    Where the natural soil percolation rate is faster than 60 min/in., it is desirable to limit trench length to 50 ft for ease in designing the pressure distribution system. Generally it is best to design a mound with long trenches.

Number of trenches = 125/50 = 2.5; use three trenches.

Using three trenches, length per trench = 125 ft/3 = 41.67 ft; use a trench length of 42 ft

Use three trenches, 4-ft wide and 42-ft long.

At least 4 ft must be provided between trenches (b). Use a spacing, b = 4 ft.

3. Dimensioning the sand fill portion of the mound

The length of the mound is computed by adding the length of the top (horizontal extent) of the sand fill (1 + 2c) and the horizontal distances on each end of the top needed to provide a side slope of one vertical to four horizontal. Note that 12 in. of fill must exist below the trench and a trench is 9-in. deep. Thus:

L = mound length = L_1 + 2c + [(2 sides × 4 vertical × vertical thickness of fill, in.)/
    (1 horizontal × 12 in./ft)]
L = mound length at the trench = 42 + (2 × 3) + [(2 × 4(9 in. + 12 in.)/(12 in./ft)]
L = 42 + 6 + 14 = 62 ft
L = 62 ft the length of the mound at the upslope base of the fill.

The width of the mound (W) is computed in a similar manner using the width of the trenches (a) and number of trenches and trench spacing (b). Thus:

W = 3a + 2b + 2c + [(2 sides × 4 vertical × vertical thickness of fill, in.)]/
    (1 horizontal × 12 in./ft)]
W = (3 × 4) + 2 × 4 + 6 + 14
W = 40 ft, 0 % slope

The approximate additional downslope width of the mound at a sloping site with a 4 % slope is:

(4 vertical 1 horizontal)(3a + 2b + c + 14/2) (4 %/100)
4 [(3 × 4) + (2 × 4) + 3 + 7] (0.04) = 4.8 ft; use 5 ft.
W = total mound width = 40 + 5 = 45 ft

The mound width correction upslope is negligible. Thus the mound width of the sand fill portion measured from the center of the mound will be 20-ft upslope and 25-ft downslope. The mound length, L, at the downslope base of the sand fill will be 72 ft. Note that additional land area is needed for the mound for placement of the topsoil over the sand fill portion of the mound.

4. Checking for limiting conditions

The effective basal area of the sand fill (shaded area in plan view of Fig. 2.16) in the mound below and downslope of the trenches must be large enough to absorb the estimated daily waste flow. In calculating this basal area, exclude the portions of the mound upslope from the trenches and the portion of the mound on each side of the ends of the trenches and their extension downslope.

In this calculation:

A = effective basal area of mound, ft^2 = shaded area in plan view
Q = estimated daily flow, gpd
R = natural soil infiltration rate, gpd/ft^2.
Thus, R = Q/A, expressed in gpd/ft^2.

A is computed by multiplying the trench length, L_1, by the sum of the trench widths, the spaces between trenches, the horizontal extent of fill beyond the last trench, and the horizontal distance needed to provide the side slope (including the correction for total mound width).

$$A = L_1(3a + 2b + c + 7 + 5)$$
$$A = 42[(4 \times 3) + (2 \times 4) + 3 + 7 + 5]$$
$$A = 1470 \ ft^2$$

$$Q = 600 \ gpd$$
$$R = Q/A = 600/1470 = 0.41 \ gpd/ft^2$$

From Table 2.8, R must not exceed 0.74 gpd/ft^2

This design is satisfactory. If calculated R exceeds the maximum, increase the size of the mound by increasing spacing between trenches or by increasing trench length to provide a satisfactory value of R.

References

1. FINS Information Service (2015) Constructed wetland for aquaculture wastewater. Aquatic Technology. http://fins.actwin.com/aquatic-plants/month.9612/msg00372.htlm
2. U.S. EPA (1980) Innovative and alternative technology assessment manual, EPA/430/9-78-009. U. S Environmental Protection Agency, Washington, DC
3. Takeda F, Adler PR, Glen DM (2015) Strawberry production linked to aquaculture wastewater treatment, International Society for Horticultural Science. ISHS III International Strawberry Symposium, Veldhoven, 1 Sept 1997. Abstract can be located at www.actahort.org/books/439/439_113.htm
4. U. S. EPA (1988) Design manual: constructed wetlands and aquatic plant systems for municipal wastewater treatment, EPA/625/1-88/022. U. S. Environmental Protection Agency, Office of Research and Development, Washington, DC, Sept 1988
5. Metcalf and Eddy (2003) Wastewater engineering treatment and reuse, 4th edn. McGraw Hil, New York
6. Vesilind A (2003) Wastewater treatment plant design. Water Environment Federation and IWA Publishing, Alexandria
7. Wang LK, Pereira NC (eds) (1986) Handbook of environmental engineering, vol. 3, Biological treatment processes. The Humana Press, Totowa, NJ, p 520
8. Microtack (2015) Organic aquaculture and wastewater treatment supplies. TechOzone, Bangkok. www.microtack.com
9. DeBusk WF, Reedy KR (1987) Wastewater treatment using floating aquatic macrophytes: contaminant removal processes and management strategies. In: Reedy KR, Smith WH (eds)

Aquatic plants for water treatment and resource recovery. Magnolia Publishing, Macon, GA, pp 27–48

10. Reedy KR, Sutton DL (1984) Water hyacinth for water quality improvement and biomass production. J Environ Qual 14:459–462

11. Tchobanoglous G, Maitski F, Thomson K, Chadwick TH (1989) Evolution and performance of city of San Diego pilot scale aquatic wastewater treatment system using water hyacinth. J Water Pollut Control Fed 61(11/12):1625–1635

12. Reed SC, Crites RW (1984) Handbook of land treatment systems for industrial and municipal wastes. Noyes, Park Ridge

13. Crites RW, Middlebrooks EJ, Bastian RK, Reed SC (2014) Natural wastewater treatment processes, 2nd edn. CRC Press/Taylor and Francis Group, Boca Raton

14. Shammas NK (1971) Optimization of biological nitrification. PhD dissertation, Microfilm Publication, University of Michigan, Ann Arbor

15. Wang LK (1978) Chemistry of nitrification-denitrification process. J Environ Sci 21:23–28

16. Wang LK, Aulenbach DB (1986) BOD and nutrient removal by biological A/O process systems. US Department of Commerce, National Technical Information Service Springfield, VA, PB88-168430/AS, Dec 1986, pp 12

17. U. S. EPA (1993) Manual nitrogen control. EPA/625/R-93/010. U. S. Environmental Protection Agency, Office of Research and Development, Washington, DC, Sept 1993

18. Hung YT, Gubba S, Lo H, Wang LK, Yapijakis C, Shammas NK (2003) Application of wetland for wastewater treatment. OCEESA J 20(1):41–46

19. Crites RW (1996) Constructed wetlands for wastewater treatment and reuse. Presented at the Engineering Foundation conference, environmental and engineering food processing industries XXVI, Santa Fe

20. WPCF (1990) Natural systems for wastewater treatment, Manual of practice # FD-16, Water Pollution Control Federation, Alexandria, Feb 1990

21. Hammer DA (ed) (1989) Constructed wetlands for wastewater treatment; municipal, industrial and agricultural. Lewis Publishers, Chelsea

22. Wang JC, Aulenbach DB, Wang LK (1996) Energy models and cost models for water pollution controls. In: Misra KB (ed) Clean production. Springer, Berlin, pp 685–720

23. Wang LK, Krougzek JV, Kounitson U (1995) Case studies of cleaner production and site remediation. UNIDO-Registry DTT-5-4-95. United Nations Industrial Development Organization (UNIDO), Vienna, Apr 1995, 136 pages

24. US ACE (2015) Yearly average cost index for utilities. In: Civil works construction cost index system manual, 1110-2-1304. U.S. Army Corps of Engineers, Washington, DC. PDF file is available on the Internet at http://www.nww.usace.army.mil/Missions/CostEngineering.aspx

25. U. S. EPA (1981) Process design manual, land treatment of municipal wastewater, EPA 625/1-81-013. U. S. Environmental Protection Agency, Center for Environmental Research Information, Cincinnati

26. Satterwhite MB, Condike BJ, Stewart GL (1976) Treatment of primary sewage effluent by rapid infiltration. Army Corps of Engineers, Cold Region Research and Engineering Laboratory, Hanover, NH, Dec 1976

27. Crites RW (1985) Nitrogen removal in rapid infiltration system. J Environ Eng, ASCE 111:865

28. Smith DG, Linstedt KD, Bennett ER (1979) Treatment of secondary effluent by infiltration-percolation. EPA-600/2-79-174, U. S. Environmental Protection Agency, Ada, Aug 1979

29. Bouwer H (1974) Renovating secondary effluent by groundwater recharge with infiltration basins. Conference on recycling treated municipal wastewater through forest and cropland, U. S. Environmental Protection Agency, EPA-660/2-74-003

30. Kioussis DR, Wheaton FW, Kofinas P (1999) Phosphate binding polymeric hydrogels for aquaculture wastewater remediation. Aquac Eng 19(3):163–178

31. Gerba CP, Lance JC (1979) Pathogen removal from wastewater during ground water recharge. Proceedings of symposium on wastewater reuse for groundwater recharge, Pomona, 6–7 Sept 1979

32. Aulenbach DB (1979) Long term recharge of trickling filter effluent into sand. EPA-600/2-79-068. U. S. Environmental Protection Agency, Ada, Mar 1979
33. Leach E, Enfield CG, Harlin C Jr (1980) Summary of long-term rapid infiltration system studies. EPA-600/2-80-165. U. S. Environmental Protection Agency, Ada, July 1980
34. U. S. EPA (1975) Evaluation of land application systems. EPA-430/9-75-001. U. S. Environmental Protection Agency, Washington, Mar 1975
35. Shammas NK (1991) Investigation of irrigation water application rates to landscaped areas in Ar-Riyadh. J Eng Sci 3(2):147–165
36. Stone R, Rowlands J (1980) Long-term effects of land application of domestic wastewater: Mesa, Arizona irrigation site. EPA-600/2-80-061. U. S. Environmental Protection Agency, Ada, Apr 1980
37. Uiga A, Crites RW (1980) Relative health risks of activated sludge treatment and slow rate land treatment. J Water Pollut Control Fed 52(12):2865–2874
38. Tofflemire TJ, Chen M (1977) Phosphate removal by sands and soils. In: Loehr RC (ed) Land as a waste management alternative. Ann Arbor Science, Ann Arbor
39. U. S Department of Agriculture, Soil Conservation Service (1986) Trickling irrigation, Chapter 7. In: Irrigation, SCS national engineering handbook. U. S. Government Printing Office, Washington, DC
40. Jenkins TF, Palazzo EJ (1981) Wastewater treatment by a slow rate land treatment system. U. S. Army Corps of Engineers, Cold Region Research and Engineering Laboratory, CRREL Report 81–14, Hanover, Aug 1981
41. Loehr RC (ed) (1977) Land as a waste management alternative. Ann Arbor Science, Ann Arbor
42. Overman AR (1979) Wastewater irrigation at Tallahassee, Florida. U. S. Environmental Protection Agency, EPA-600/2-79-151. U. S. Environmental Protection Agency, Ada, Aug 1979
43. Duscha LA (1981) Dual cropping procedure for slow infiltration of land treatment of municipal wastewater. Department of the Army, Engineering Technical Letter # 1110-2-260, Mar 1981
44. Shammas NK, El-Rehaili A (1986) Tertiary filtration of wastewater for use in irrigation. Symposium on the effect of water quality on the human health and agriculture, Al-Khobar, Oct 1986
45. Smith RG, Schroeder ED (1985) Field studies of the overland flow process for the treatment of raw and primary treated municipal wastewater. J Water Pollut Control Fed 57:7
46. Perry LE, Reap EJ, Gilliand M (1982) Evaluation of the overland flow process for the treatment of high-strength food processing wastewaters. In: Proceedings of the 14th Mid-Atlantic industrial waste conference, University of Maryland, June 1982
47. de Figueredo RF, Smith RG, Schroeder ED (1984) Rainfall and overland flow performance. J Environ Eng, ASCE 110:678
48. U. S. EPA (1981) Process design manual, land treatment of municipal wastewater; supplement on rapid infiltration and overland flow, EPA 625-1-81-13a. U. S. Environmental Protection Agency, Center
49. Johnston J, Smith RF, Schroeder ED (1988) Operating schedule effects on nitrogen removal in overland flow wastewater treatment systems. Presented at the 61st annual water pollution control federation conference, Dallas
50. Witherow JL, Bledsoe BE (1983) Algae removal by the overland flow process. J Water Pollut Control Fed 55:1256
51. Smith RG, Schroeder ED (1983) Physical design of overland flow systems. J Water Pollut Control Fed 55:3
52. Wang LK (1987) Wastewater treatment by biological physicochemical two-stage process system. In: Proceedings of the 41st annual purdue industrial waste conference, Lafayette, IN, pp 67

53. Otis RJ, Plews GD, Patterson DH (1977) Design of conventional soil absorption trenches and beds, ASAE, Proceedings of the 2nd national home sewage treatment symposium, Chicago, Dec 1977
54. Bendixen TW, Coulter JB, Edwards GM (1960) Study of seepage beds. Robert A. Taft Sanitary Engineering Center, Cincinnati
55. Bendixen TW, Thomas RE, Coulter JB (1963) Report of a study to develop practical design criteria for seepage pits as a method for disposal of septic tank effluents. NTIS Report # PB 216 931, Cincinnati, pp 252
56. Converse JC, Carlile BL, Peterson GB (1977) Mounds for the treatment and disposal of septic tank effluent, ASAE, Proceedings of the 2nd national home sewage treatment symposium, Chicago, Dec 1977
57. Converse JC (1978) Design and construction manual for Wisconsin mounds, small scale waste management project. University of Wisconsin, Madison, pp 80
58. ASAE (2015) A new combined distribution and pretreatment unit for wastewater soil infiltration systems. American Society of Agricultural Engineer, Technical Library, http://asae. frymulti.com
59. Ten-State Standards (1996) Recommended standards for individual sewage systems. Great Lakes-Upper Mississippi River Board of State Sanitary Engineers. NY State Dept. of Health, Albany, NY
60. Krof FW, Laak R, Healey KA (1977) Equilibrium operation of subsurface absorption systems. J Water Pollut Control Fed 49:2007–2016
61. Mellen WL (1976) Identification of soils as a tool for the design of individual sewage disposal systems. Lake County Health Department, Waukegan, IL, pp 67
62. Bernhardt AP (1978) Treatment and disposal of wastewater from homes by soil infiltration and evapotranspiration. University of Toronto Press, Toronto, pp 173
63. Bernhart AP (1974) Return of effluent nutrients to the natural cycle through evapotranspiration and subsoil-infiltration of domestic wastewater, ASCE, Proceedings of the national home sewage disposal symposium, Chicago, Dec 1974
64. NEHA (1979) On-site wastewater management. National Environmental Health Association, Denver, pp 108
65. U. S. EPA (1980) Design manual: onsite wastewater treatment and disposal systems, EPA 625/1-80-012. U. S. Environmental Protection Agency, Office of Research and Development, Municipal Environmental Research Laboratory, Cincinnati, Oct 1980
66. Krofta M, Wang LK (1999) Treatment of household wastes, septage and septic tank effluent. Technical Report PB-2000-101750INZ. U. S. Department of Commerce, National Technical Information Service, Springfield, VA
67. White KD, Wang LK (2000) Natural treatment and on-site processes. Water Environ Res 72 (5):1–12
68. Worrell Water Technologies (2015) Tidal wetland living machine. Charlottesville. www. worrellwater.com
69. Aulenbach DB, Clesceri NL (2005) Treatment by application onto land. In: Wang LK, Pereira NC, Hung YT (eds) Biological treatment processes. Humana Press, Totowa, NJ
70. Clesceri NL, Aulenbach DB, Roetzer JF (2005) Treatment by surface application. In: Wang LK, Pereira NC, Hung YT (eds) Biological treatment processes. Humana Press, Totowa, NJ
71. Wang LK, Shammas NK, Hung YT (eds) (2005) Biosolids treatment processes. Humana Press, Totowa, NJ
72. Fabrizi L (2015) Natural treatment of wastewater. Lenntech, Netherlands. http://www. lenntech.com/natural-wastewater-treatment.htm
73. WEF (2010) Natural systems for wastewater treatment. Water Environment Federation, MOP FD-16, 3rd Edn. Water Environment Federation, Alexandria
74. Sharma S, Rousseau D (2011) Natural systems for water and wastewater treatment and reuse. SWITCH scientific conference, 24–26 Jan 2011, Paris, UNESCO, IHE, Delft

75. Winans K, Speas-Frost S, Jerauld M, Clark M, Toor G (2012) Small-scale natural wastewater treatment systems: principles and regulatory framework. University of Florida, Institute of Food and Agricultural Sciences (IFAS), SL365
76. Roth TM (2015) Natural wastewater treatment. Onsite Water Treatment J, Santa Barbara, Sept/Oct 2005, http://forester.net/ow_0509_land.html. Accessed 2015
77. Vymazal J (2011) Constructed wetlands for wastewater treatment: five decades of experience. Environ Sci Technol 45(1):61–69
78. Wang LK, Shammas NK, Evanylo GK, Wang MHS (2014) Engineering management of agricultural land applications for watershed protection. In: Wang LK, Yang CT (eds) Modern resources engineering. Springer, New York, pp 571–642
79. Wang MHS, Wang LK (2015) Environmental water engineering glossary. In: Yang CT, Wang LK (eds) Advances in water resources engineering. Springer, New York, pp 471–556

Chapter 3
Proper Deep-Well Waste Disposal for Water Resources Protection

Nazih K. Shammas, Lawrence K. Wang, and Charles W. Sever

Contents

1   Introduction 121
2   Regulations for Managing Injection Wells 123
3   Basic Well Designs 126
4   Evaluation of a Proposed Injection Well Site 132
    4.1   Confinement Conditions 133
    4.2   Potential Receptor Zones 134
    4.3   Subsurface Hydrodynamics 136
5   Five Potential Hazards: Ways to Prevent, Detect, and Correct them 137
    5.1   Fluid Movement During Construction, Testing, and Operation of the System . 138
    5.2   Failure of the Aquifer to Receive and Transmit the Injected Fluids 139
    5.3   Failure of the Confining Layer 139
    5.4   Failure of an Individual Well 141
    5.5   Failures Because of Human Error 141
6   Economic Evaluation of a Proposed Injection Well System 142
7   Use of Injection Wells in Wastewater Management 143
    7.1   Reuse for Engineering Purposes 143
    7.2   Injection Wells as a Part of the Treatment System 144
    7.3   Storage of Municipal Wastewaters for Reuse 144
    7.4   Storage of Industrial Wastewaters 145
    7.5   Disposal of Municipal and Industrial Sludges 145
8   Use of Injection Wells for Hazardous Wastes Management 146
    8.1   Identification of Hazardous Wastes 147

N.K. Shammas (✉)
Lenox Institute of Water Technology, Pasadena, CA, USA

Krofta Engineering Corporation, Lenox, MA, USA
e-mail: n.k.shammas@gmail.com

L.K. Wang
Lenox Institute of Water Technology, Newtonville, NY 12128-0405, USA

Rutgers University, New Brunswick, NJ, USA
e-mail: lawrenceKWang@gmail.com

C.W. Sever
US Environmental Protection Agency, Dallas, TX, USA

© Springer International Publishing Switzerland 2016
L.K. Wang, M.-H.S. Wang, Y.-T. Hung and N.K. Shammas (eds.),
Natural Resources and Control Processes, Handbook of Environmental
Engineering, Volume 17, DOI 10.1007/978-3-319-26800-2_3

8.2   Sources, Amounts, and Composition of Injected Wastes 148
8.3   Geographic Distribution of Wells 151
8.4   Design and Construction of Wells 151
8.5   Disposal of Radioactive Wastes 154
9   Protection of Usable Aquifers 156
9.1   Pathway 1: Migration of Fluids Through a Faulty Injection Well Casing 156
9.2   Pathway 2: Migration of Fluids Upward Through the Annulus Between
      the Casing and the Well Bore 157
9.3   Pathway 3: Migration of Fluids from an Injection Zone Through
      the Confining Strata 158
9.4   Pathway 4: Vertical Migration of Fluids Through Improperly
      Abandoned or Improperly Completed Wells 160
9.5   Pathway 5: Lateral Migration of Fluids from Within an Injection
      Zone into a Protected Portion of Those Strata 163
9.6   Pathway 6: Direct Injection of Fluids into or Above an Underground
      Source of Drinking Water 165
10   Case Studies of Deep-Well Injection 166
10.1 Case Study 1: Pensacola, FL (Monsanto) 167
10.2 Case Study 2: Belle Glade, FL 170
10.3 Case Study 3: Wilmington, NC 173
11   Practical Examples 176
11.1 Example 1 176
11.2 Example 2 177
11.3 Example 3 177
11.4 Example 4 178
11.5 Example 5 179
11.6 Example 6 179
Appendix: US Yearly Average Cost Index for Utilities [39] 180
References 180

Abstract Deep-well injection is one of feasible technologies for water conservation and disposal of hydrofracturing (hydraulic fracturing) process wastes and similar hazardous wastes. This chapter introduces the regulations for managing injection wells, basic well designs, well evaluation, economic analysis, and the methods to prevent, detect and correct potential hazards. Practical application examples and design examples for water storage, waste water disposal, sludge disposal and well analyses are included with emphasis on water resources protection.

Keywords Deep-well injection • Regulations • Well design • Well evaluation • Economic analysis • Hazard prevention, detection and correction • Application examples • Design examples • Water storage and reuse • Waste disposal • Sludge disposal • Water resources protection

Nomenclature

A   Area, (ft^2)
B   Leakage factor
C   Compressibility $(psi)^{-1}$

D   Dispersion coefficient
H   Reservoir thickness (ft)
h_c Thickness of confining layer (ft)
h_c' Thickness of second confining layer (ft)
I   Hydraulic gradient (ft/ft)
k   Average permeability (millidarcy)
k_c Vertical permeability of confining layer (millidarcy)
k_c' Vertical permeability of second confining layer (millidarcy)
L   Leakage factor for semiconfined aquifer $= \sqrt{khhc/kc}$
P   Coefficient of permeability (gal/d/ft^2)
P_{DL} Dimensionless pressure for semiconfined reservoirs
P_i Initial formation pressure (ft of water or psi)
P_1 Hydrostatic pressure in the base of freshwater (ft of water or psi)
P_2 Hydrostatic pressure in the injection zone (ft of water or psi)
P_r Reservoir pressure at radius r (ft of water or psi)
P_u Upward pressure
P_d Downward pressure
Q   Flow or injection rate (ft^3/d, gpm or barrels/d)
r   Radial distance from well bore (ft)
s   Change in pressure (ft of water or psi)
S   Coefficient of storage
t   Time (d)
t_D Dimensionless time
T   Transmissibility (gal/d/ft)
u   1.87r^{2}S/Tt (centipoises $=$ cp)
V   Q t $=$ cumulative volume of waste injected (ft^3)
v   Fluid velocity (ft/d)
W(u)   Well function of u given in Table 3.1
β   Formation volume factor $= \dfrac{\text{Volume of liquid at reservoir temperature and pressure}}{\text{Volume of liquid at standard temperature and pressure}}$
Φ   Porosity expressed as a decimal
γ   Radial distance from well bore with dispersion (ft)
μ   Viscosity
π   3.14

1   Introduction

Underground injection is the technology of placing fluids underground, in porous formations of rocks, through wells or other similar conveyance systems [1, 2]. While rocks such as sandstone, shale, and limestone appear to be solid, they can contain significant voids or pores that allow water and other fluids to fill and move through them. Man-made or produced fluids (liquids, gases, or slurries) can move into the pores of rocks by the use of pumps or by gravity. The fluids may be water, wastewater, or water mixed with chemicals. Injection well technology can predict the capacity of rocks to contain fluids and the technical details to do so safely.

Underground wastewater disposal and storage by well injection are being used by both industries [3–9] and municipalities [3, 4, 6, 10–13] to help solve environmental problems. Facilities across the USA discharge a variety of hazardous and nonhazardous fluids into more than 400,000 injection wells [2]. A national goal established by Congress in the Clean Water Act of 1972 was to cease all discharges of pollutants to navigable waters by 1985 [14]. While treatment technologies exist, it would be very costly to treat and release to surface waters the billions and trillions of gallons of wastes that industries produce each year. Agribusiness and the chemical and petroleum industries all make use of underground injection for waste disposal. When wells are properly sited, constructed, and operated, underground injection is an effective and environmentally safe method to dispose of wastes.

Subsurface disposal of liquid wastes into aquifers is based on the concept that such wastes can be injected through wells into confined geologic strata not having other uses, thereby providing long-term isolation of the waste material. Subsurface storage and artificial recharge involve the concept that highly treated municipal and other wastewaters are valuable and should be reused. In a nation where water deficiencies or management problems are forecast for the foreseeable future, storage of treated wastewaters for reuse is destined to become a major element for consideration in water resource management [15].

Injection for the extraction of salt started in France in the ninth century and later in China. The first documented project for the disposal of oil field brine (salt water produced along with oil and gas), in the same formation where it originated, started in Texas in 1938. Enhanced recovery of oil, which is the injection of water or other fluids into a formation to extract additional oil and gas, probably started early in the 1930s. Industrial-waste injection started in 1950 with Dow Chemical injecting industrial fluids. In the 1950s, DuPont Chemical Corporation started to inject some of its industrial waste into deep wells [16].

The earliest use of an injection well for municipal wastewater disposal in the USA was in 1959 at the Collier Manor Sewage Treatment Plant in Pompano Beach, FL [3, 13]. During 13 years of operation, the City of Pompano Beach injected about 3 billion gallons (11.4 Mm^3) of secondary-treated wastewater into a cavernous "boulder zone" through two wells 1,000–1,400 ft (305–427 m) deep [17].

Underground space is recognized [18] as a natural resource of considerable value. A small percentage of this space, like the "boulder zone," consists of large caverns capable of receiving and transmitting extremely large volumes of wastewater for a single injection well. But most space underground consists of the area available between sand grains in the rock strata. The percentage of this space available for fluid storage and movement depends upon how much clay and silt are present and the amount and type of cementing material present [19].

The porosity of a rock is essentially its interstitial pore space. It can be expressed quantitatively as the ratio of the volume of the pore space to the total volume of the rock and generally is stated as a percentage. Gravel or sand that is clean, uniform, and free of clay and silt will have about 30–40 % of its volume available for storage space. Gravel or sand containing abundant clay and silt, cementing material, or a precipitant from injected wastes may have as little as 5–15 % of its volume available. Fractures, joints, and solution channels in cemented rock formations, such as limestone, are additional types of pore space that contribute to porosity, but are difficult to measure.

Virtually all of this subsurface pore space is already occupied by natural water, either fresh or mineralized to some extent. Thus, injection does not usually involve the filling of unoccupied space, but rather consists of the compression or displacement of existing fluids. Since the compressibility of water is small, creation of significant volumes of storage space through this mechanism requires disposal strata that underlie a large geographic area.

2   Regulations for Managing Injection Wells

The most accessible freshwater is stored in geologic formations called aquifers. These aquifers feed our lakes; provide recharge to our streams and rivers, particularly during dry periods; and serve as resources for 92 % of public water systems in the USA [2]. Many people in the country also rely on groundwater for their private drinking water wells. Injection of fluids can potentially contaminate aquifers that supply drinking water to households and public water systems. Direct injection into a drinking water aquifer, injection into a zone that is not isolated from a drinking water aquifer, or poor performance of an injection well can contaminate groundwater. Because contamination of groundwater can be very persistent and difficult to remediate, it is important to ensure that contaminants do not enter groundwater [16].

The US EPA defines an injection well as any bored, drilled, or driven shaft or dug hole, where the depth is greater than the largest surface dimension that is used to discharge fluids underground. This definition covers a wide variety of injection practices that range from more than 100,000 technically sophisticated and highly monitored wells which pump fluids into isolated formations up to two miles below the Earth's surface to the far more numerous on-site drainage systems, such as septic systems, cesspools, and stormwater wells, that discharge fluids a few feet underground [2, 19].

Congress, in passing the Safe Drinking Water Act (SDWA) in 1974, gave the US EPA the authority to control underground injection to protect underground drinking water sources [14]. In 1979 and then again in 1980, the US EPA developed the Statement of Basis and Purpose for the UIC (Underground Injection Control) Program, to support the regulations that were proposed and then finalized in those years. The US EPA published final technical regulations for the UIC Program in 1980. The regulations set minimum standards state programs must meet to receive primary enforcement responsibility (primacy) of the UIC Program, In 1981, Congress passed amendments to the SDWA which allowed for the delegation of the UIC Program for injection wells to states if the program was effective in protecting underground sources of drinking water (USDW) and included traditional program components such as oversight, reporting, and enforcement [16].

The US EPA groups underground injection into five classes for regulatory control purposes [2, 20]. Each class includes wells with similar functions and construction and operating features so that technical requirements can be applied consistently to the class. Class I includes the emplacement of hazardous and nonhazardous fluids (industrial and municipal wastes) into isolated formations beneath the lowermost USDW [3]. Because they may inject hazardous waste, Class I wells are the most strictly regulated and are further regulated under the Resource Conservation and Recovery Act. Class II includes injection of brines and other fluids associated with oil and gas production; Class III encompasses injection of fluids associated with solution mining of minerals; Class IV addresses injection of hazardous or radioactive wastes into or above a USDW and is banned unless authorized under other statutes for groundwater remediation. Class V includes all underground injections not included in Classes I–IV. Class V wells inject nonhazardous fluids into or above an USDW and are typically shallow, on-site disposal systems. Injection practices or wells which are not covered by the UIC Program include other individual residential waste disposal systems that inject only sanitary waste and commercial waste disposal systems that serve fewer than 20 persons that inject only sanitary waste.

Class I injection wells are sited such that they inject below the lowermost USDW and a confining zone above an injection zone [21]. Injection zone reservoirs typically range in depth from 1,700 to over 10,000 ft below the surface [3]. Class I wells are mainly used in the following industries: petroleum refining, metal production, chemical production [22], pharmaceutical production, commercial disposal, municipal disposal, and food production. There are 272 active Class I injection facilities nationwide. Of these, 51 are hazardous and 221 are nonhazardous. These 272 facilities maintain approximately 529 Class I injection wells that are scattered throughout the USA in 19 states [3]. The greatest concentrations are located in the Gulf Coast, Great Lakes, and the Floridan peninsular geographical regions.

The oil and gas injection wells, Class II, account for a large proportion of the fluids injected in the subsurface [23, 24]. Typically, when oil and gas are extracted, large amounts of salt water (brine) are also brought to the surface. This salt water can be very damaging if it is discharged in surface water. Instead, all states require that this brine be injected into formations similar to those from which it was extracted [25]. Over two billion gallons of brine are injected daily into injection wells in the USA [26].

Mining wells, Class III, are used in the mining of a number of minerals. In general, the technology [27] involves the injection of a fluid, usually called lixiviant, which contacts an ore which contains minerals that dissolve in the fluid. The pregnant fluid (lixiviant nearly saturated with components of the ore) is pumped to the surface where the mineral is removed from the fluid. The US EPA protects drinking water from contamination from mining wells by implementing regulations that set minimum standards. These regulations require mining well operators to case and cement their wells to prevent the migration of fluids into an underground drinking water source; never inject fluid between the outermost casing and the well bore; and test the well casing for leaks at least once every 5 years.

Some of the practices using mining wells are [27]: (a) Salt solution mining started in France in the ninth century. The Chinese used boring techniques to develop wells and extract brines at the beginning of the eighteenth century. In the USA, the salt industry started recovering brine at the end of the eighteenth century. The process consists of pumping water into the salt formation and extracting the salt from the resulting fluid after retrieval. More than 50 % of the salt used in the USA is obtained this way. (b) In situ leaching of uranium is the practice of injecting a fluid to leach out the uranium salts and pumping it back to the surface where the uranium is extracted. Eighty percent of the uranium extracted in the USA is produced this way. (c) The production of sulfur using the Frasch process is one of the earliest uses of the technology. Traditionally, superheated steam is injected in order to recover a sulfur solution.

Shallow hazardous and radioactive injection wells, Class IV, are prohibited unless the injection wells are used to inject contaminated groundwater that has been treated and is being injected into the same formation from which it was drawn. These wells are authorized by rule [28] for the life of the well if such subsurface emplacement of fluids is approved by the US EPA, or a state, pursuant to provisions for the cleanup of releases under the Comprehensive Environmental Response, Compensation, and Liability Act of 1980 (CERCLA) or pursuant to requirements and provisions under the Resource Conservation and Recovery Act (RCRA).

The wells in the shallow injection wells, Class V, are as diverse as they are similar. This category came about after all the easy definable wells were put into classes I through IV. In September 1999, the US EPA completed The Class V Underground Injection Control Study, a comprehensive study proposed of most types of Class V wells [29]. In May 2001, the US EPA determined that existing federal Underground Injection Control (UIC) regulations at the time are adequate to prevent Class V injection wells from endangering underground sources of drinking water [30]. The final determination would be based on the agency's evaluation of existing data collected for The Class V Underground Injection Control Study (EPA/816-R-99-014). In June 2003, septic system and stormwater discharges were regulated as Class V wells (816-F-03-002 and 816-F-03-001, respectively).

Typically, Class V injection wells are shallow "wells," such as septic systems and drywells, used to place nonhazardous fluids directly below the land surface. However, Class V wells can be deep, highly sophisticated wells. The US EPA estimates there are more than 650,000 Class V wells in the USA [29]. Class V wells are located in every state, especially in unsewered areas where the population is likely to depend on groundwater for its drinking water source.

Class V wells are convenient and inexpensive means to dispose of a variety of nonhazardous fluids. Some Class V wells are agricultural drainage wells, stormwater drainage wells, large-capacity septic systems, wastewater treatment effluent wells, spent brine return flow wells, mine backfill wells, aquaculture waste disposal wells, solution mining wells, in situ fossil fuel recovery wells, special drainage wells, experimental wells, aquifer remediation wells, geothermal electric power wells, geothermal direct heat wells, heat pump/air conditioning return flow wells, saltwater intrusion barrier wells, aquifer recharge and aquifer storage and recovery wells, subsidence control wells, and industrial wells [29].

Injection wells have the potential to inject contaminants that may cause our underground sources of drinking water to become contaminated. The UIC Program prevents this contamination by setting minimum requirements. The goals of the US EPA's UIC Program are to prevent contamination by keeping injected fluids within the well and the intended injection zone, or in the case of injection of fluids directly or indirectly into a USDW, to require that injected fluids not cause a public water system to violate drinking water standards or otherwise adversely affect public health. These minimum requirements affect the siting of an injection well and the construction, operation, maintenance, monitoring, testing, and, finally, the closure of the well [2]. All injection wells require authorization under general rules or specific permits. Finally, states may apply to have primary enforcement responsibility (primacy) for the UIC Program. The US EPA provides grant funds to all delegated programs to help pay for program costs. States must provide a 25 % match on the US EPA funds [31]. To date (August 2004), 33 states, Guam, the Commonwealth of the Northern Mariana Islands, and Puerto Rico have obtained primacy for all classes of injection wells. Seven states share primacy with the US EPA. The US EPA administers UIC programs for the remaining ten states, the Virgin Islands, American Samoa, and Indian country [102].

3   Basic Well Designs

Design of a casing program depends primarily on well depth, character of the rock sequence, fluid pressures, type of well completion, and the corrosiveness of the fluids that will contact the casing. Where fresh groundwater supplies are present, a casing string (surface casing) is usually installed to below the depth of the deepest groundwater aquifer immediately after drilling through the aquifer (Fig. 3.1). One or more smaller-diameter casing strings are then set, with the bottom of the last string just above or through the injection horizon, the latter determination depending on whether the hole is to be completed as an open hole or gravel packed or is to be cased and perforated.

The annulus between the rock strata and the casing is filled with a cement grout. This is done to protect the casing from external corrosion, to increase casing strength, to prevent mixing of the waters contained in the aquifers behind the casing, and to forestall travel of the injected waste into aquifers other than the disposal horizon.

Cement should be placed behind the complete length of the surface casing and behind the entire length of the smaller-diameter casing strings also or at least for a sufficient length to provide the desired protection. It is suggested that at least 1½ in. of annular space be allowed for proper cementing. Casing centralizers, other

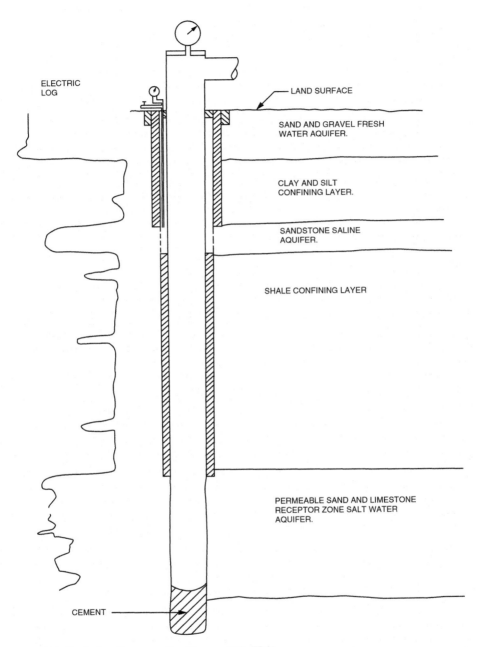

ELECTRIC
LOG

LAND SURFACE

SAND AND GRAVEL FRESH
WATER AQUIFER.

CLAY AND SILT
CONFINING LAYER.

SANDSTONE SALINE
AQUIFER.

SHALE CONFINING LAYER

PERMEABLE SAND AND LIMESTONE
RECEPTOR ZONE SALT WATER
AQUIFER.

CEMENT

Fig. 3.1 Typical well construction (Source: U.S. EPA)

equipment, and techniques such as stage cementing can give added assurance of a good seal between the strata and the casing and should be encouraged where applicable.

The majority of injection wells constructed to date have been for one of three basic purposes: injection plus external monitoring for leakage through confining layers near the well (Fig. 3.2), injection plus internal monitoring for leaks in the injection tubing and casing (Fig. 3.3), and injection only (Fig. 3.4). Numerous variations in design have been used to accomplish these three basic purposes.

Wells designed for injection plus monitoring of the confining layer generally are constructed like the well in Fig. 3.2. In these types of wells, the fluid chemistry and pressures outside the well in an aquifer overlying the receptor are monitored at the wellhead for changes that would indicate leakage. This monitoring is accomplished either by leaving the annulus exposed to the aquifer to be monitored as in Fig. 3.2 or by some method of accessing the aquifer inside the injection well casing. In designing a well for this purpose, the engineer should remember that to obtain a sample of fluids from the monitored aquifer, fluids in the pipes must be pumped out. If these fluids are saline, then their disposal may be a problem. One variation is to fasten a small 0.25–0.5 in. pipe (usually stainless steel or neoprene) to the outside of the inner casing (in the annulus) extending from land surface to the top of the aquifer being monitored. By sampling through this "drop" pipe, the volume of fluids to be disposed is minimized. Another variation is to drill a 4–6 in. larger hole and then attach a 2–3 in. monitoring pipe outside the inner casing (in the annulus) from land surface to the aquifer being monitored.

Wells designed for self-inspection of internal leaks are constructed similar to the well in Fig. 3.3. A sampling tube is installed inside the casing to a depth near the bottom of the injection well. Either a seal is placed outside of this tubing near its bottom to prevent fluid circulation in the annulus and the annulus is pressurized or low-density fluid capable of causing back pressure, such as kerosene, is placed in the annulus. Pressure is measured at the land surface to detect fluid movement into or out of the annulus. Pressure changes in the annulus must be correlated with changes in injection rate, temperature, specific gravity of the fluids, and other factors to avoid false interpretation. In wells using this type of design, leaks in both the injection tubing and the well casing are quickly detected. Industrial waste should be injected through separate interior tubing rather than the well casing itself. This is particularly important when corrosive wastes are being injected. A packer can be set near the bottom of the tubing to prevent corrosive wastes from contacting the casing. Additional corrosion protection can be provided by filling the annular space between the casing and the tubing with oil or water containing a corrosion inhibitor.

Wells designed for injection purposes only are less expensive to construct, but lack the ability for self-monitoring described above. The well shown in Fig. 3.4 is typical of this type design. Separate wells must be drilled for monitoring purposes where required.

Temperature logs, cement bond logs, and other well-logging techniques can be required as a verification of the adequacy of the cementing. Cement should be pressure tested to assure the adequacy of a seal.

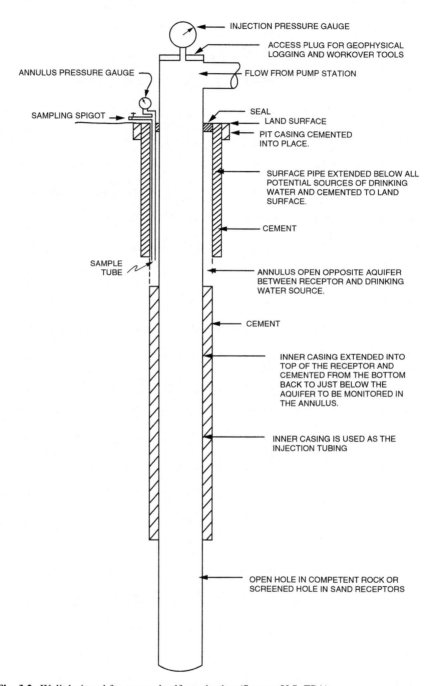

INJECTION PRESSURE GAUGE

ACCESS PLUG FOR GEOPHYSICAL
LOGGING AND WORKOVER TOOLS

ANNULUS PRESSURE GAUGE

FLOW FROM PUMP STATION

SAMPLING SPIGOT

SEAL

LAND SURFACE

PIT CASING CEMENTED
INTO PLACE.

SURFACE PIPE EXTENDED BELOW ALL
POTENTIAL SOURCES OF DRINKING
WATER AND CEMENTED TO LAND
SURFACE.

CEMENT

SAMPLE
TUBE

ANNULUS OPEN OPPOSITE AQUIFER
BETWEEN RECEPTOR AND DRINKING
WATER SOURCE.

CEMENT

INNER CASING EXTENDED INTO
TOP OF THE RECEPTOR AND
CEMENTED FROM THE BOTTOM
BACK TO JUST BELOW THE
AQUIFER TO BE MONITORED IN
THE ANNULUS.

INNER CASING IS USED AS THE
INJECTION TUBING

OPEN HOLE IN COMPETENT ROCK OR
SCREENED HOLE IN SAND RECEPTORS

Fig. 3.2 Well designed for external self-monitoring (Source: U.S. EPA)

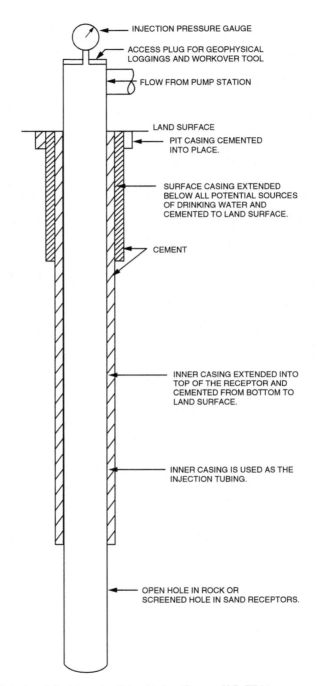

Fig. 3.3 Well designed for internal self-monitoring (Source: U.S. EPA)

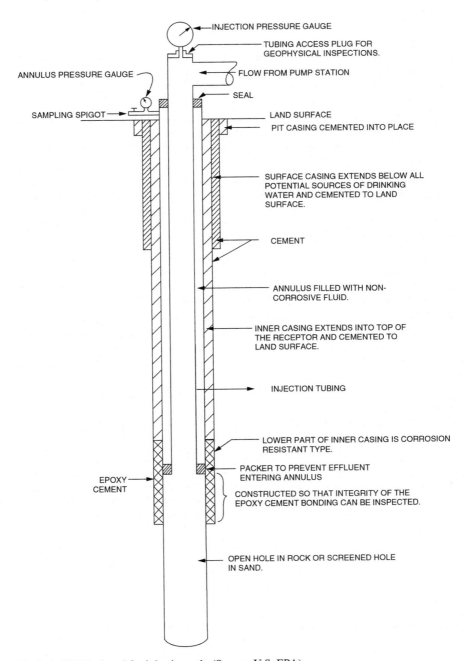

Fig. 3.4 Well designed for injection only (Source: U.S. EPA)

Neat Portland cement (no sand or gravel) is the basic material for cementing. Many additives have been developed to impart some particular quality to the cement. Additives can, for example, be selected to give increased resistance to acid, sulfates, pressure, temperature, and so forth. Other additives reduce the viscosity of the cement until it flows like water.

It is frequently desired to increase the acceptance rate of injection wells by chemical or mechanical treatment of the injection zone. Careful attention should be given to stimulation techniques, such as hydraulic fracturing, perforating, and acidizing, to ensure that only the desired intervals are treated and that no damage to the casing or cement occurs.

The type of wellhead equipment can be a consideration in cases where the buildup of high back pressure is a possibility. In such cases, the wellhead should be designed to "bleed off" backflows into holding tanks or pits before pressures reach a hazardous level. High back pressures can be developed by chemical reactions in the formation. For example, at Louisville, KY, where ferric chloride solutions had been injected into dolomite and limestone, for several years an excessive buildup of carbon dioxide gas pressure caused a blowout during routine maintenance in 1980.

Surface equipment often includes holding tanks and flow lines, filters, other treatment equipment, pumps, monitoring devices, and standby facilities.

Surface equipment associated with an injection well should be compatible with the waste volume and physical and chemical properties of the waste to ensure that the system will operate as efficiently and continuously as possible. Experience with injection systems has revealed the difficulties that may be encountered because of improperly selected filtration equipment and corrosion of injection pumps.

Surface equipment should include wellhead pressure and volume monitoring equipment, preferably of the continuous recording type. Where injection tubing is used, it is advantageous to monitor the pressure of both the fluid in the tubing and in the annulus between the tubing and the casing. Pressure monitoring of the annulus is a means of detecting tubing or packer leaks. An automatic alarm system should signal the failure of any important component of the injection system. Filters should be equipped to indicate immediately the production of an effluent with too great an amount of suspended solids.

4   Evaluation of a Proposed Injection Well Site

It would be impossible to cover all the potential problems that could develop during the construction and operation of a disposal well system. But a safe economical injection well system that will function properly can be built at most sites in geographic areas suitable for this method of disposal.

Before proceeding with an evaluation, answers to the following two questions should be obtained from the regulatory authorities.

1. What criteria will determine the degree of pretreatment that will be required before fluids can be injected?
2. What restrictions have been placed upon water quality in receptor aquifers?

Pretreatment requirements vary from state to state. One state authority may prohibit emplacement of toxic wastes underground, while another will prohibit its surface discharge and encourage injection. At least one state agency pushes for pretreatment of all wastes to drinking water standards before injection. Many state agencies and most environmental groups push for "nondegradation" of the environment. But for successful operation of the system, all wastewaters must be pretreated until the fluid to be injected is compatible with the environment in the receptor aquifer.

A common mistake made evaluating deep-well injection systems is to underestimate the degree of pretreatment required. After a well system is completed is not the time to find that after the extensive pretreatment required, the wastes are in fact suitable for surface disposal. But before pretreatment can be determined, certain basic relationships between disposal operations and the physical environment must he examined. The important areas include the receptor zones, the confinement conditions, and the subsurface hydrodynamics. Data on these subjects are available from state and federal agencies, as well as local consultants, and should be assembled and evaluated. If the results are favorable, then additional data should be collected by constructing test wells. Various potential confining layers and receptors then need to be tested. Final evaluation of data may show the disposal of wastewaters into underground formations to be an unwise solution to a disposal problem not only from the standpoint of damage to the environment but from the standpoint of overall costs.

4.1 Confinement Conditions

Confining layers, although generally required by regulatory agencies, are not always essential. For example, in one of the southeastern states, wastes are being injected into the Knox Dolomite where it is about 5,000 ft (1,500 m) thick and intensely fractured vertically. Porosity is low, averaging about 10 %. Wastes being injected are heavy, having a specific gravity of about 1.2. The heavy wastes after injection move in response to gravity downward to the base of the Knox where they are permanently stored. The saltwater-freshwater boundary that now lies at about 4,000 ft (1,200 m) is being displaced upward at a rate of about 1 ft/year.

Confining layers are rarely impermeable to waste movements. They just retard the movement. Most wastes are capable of slowly moving through even the denser clays. However, as movement takes place, the wastes are subjected to ion exchange, osmosis, filtration, absorption, and other forms of treatment. The rate of leakage through a confining bed (Q) and the velocity of fluid movement (v) can be determined from Eqs. (3.1) and (3.2).

$$Q = PIA \tag{3.1}$$

$$v = PI/\phi \tag{3.2}$$

where

Q = Rate of leakage, ft^3/d
v = Velocity of fluid, ft/d
A = Leakage area, ft^2
P = Permeability, $ft^3/d/ft^2$
I = Hydraulic gradient, ft/ft
ϕ = Porosity expressed as a decimal

Possibly the most important laboratory tests to be conducted on samples collected during drilling of the test wells are vertical permeability and ion-exchange capacity on cores collected from the confining layer overlying the receptor. The vertical permeability, together with thickness of the confining bed and the anticipated differential pressure across the confining bed, should be used to predict the velocity at which wastewater will travel through the confining bed. The general permeabilities of rocks are shown in Table 3.1. Data generated from ion exchange and other tests may show that toxic wastes can meet drinking water use or other applicable standards after being subjected to subsurface treatment. These tests also indicate the physical and chemical changes that may take place with time during movement through the confining bed.

4.2 Potential Receptor Zones

A basic requirement of a receptor zone or a combination of zones is that it be capable of receiving and transmitting the volume of wastewater planned for injection. State regulatory agencies frequently place additional requirements on receptors based upon depth and water chemistry. For example, some state agencies

Table 3.1 Relationship of the coefficient of permeability to potential of a stratum for use as a receptor or as a confining layer

Rock type	Flow potential	Permeability range, md
Cavernous limestone	Excellent receptor	3×10^6 to 1×10^9
Gravel	Good receptor	1×10^4 to 3×10^6
Sands and sandy silts	Poor receptor or confining layer	1×10^{-2} to 1×10^4
Clay, shale	Good confining layer	1×10^{-6} to 1×10^{-2}

prohibit injection into receptors at a depth less than 2,000 ft (610 m) or where the native fluids contain less than 10.000 mg/L total dissolved solids.

Another requirement is that changes in the physical and chemical properties of the wastewater and in the receptor after injection be compatible with the goals of injection. These changes usually can be grouped as follows:

1. Changes in the wastewater induced by the environment in the receptor
2. Changes in the wastewater caused by chemical reactions with the receptor rocks
3. Changes in the wastewaters caused by chemical reactions with fluids in the receptor
4. Physical and chemical changes in the receptor resulting from reactions with the wastewater

Precipitants formed as a result of these types of changes can plug the receptor and cause the system to fail.

Knowledge of the complete chemical character of the wastewater after the pretreatment is of the foremost importance in evaluating a potential receptor. This knowledge, plus data about the physical subsurface environment available from drilling oil wells in the area or from a test well drilled at the proposed site, should enable a company to forecast the chemical stability of its waste. Knowledge of the mineralogy of the aquifer and the chemistry of interstitial fluids collected from a test well should indicate the reactions to be anticipated during injection. Laboratory tests can be performed with rock cores, formation fluids, and wastewater samples to confirm anticipated reactions.

Selm and Hulse [32] list reactions between injected and interstitial fluids that can cause the formation of plugging precipitates. These include precipitation of alkaline earth metals such as calcium, barium, strontium, and magnesium as relatively insoluble carbonates, sulfates, orthophosphates, fluorides, and hydroxides. Precipitation of other metals such as iron, aluminum, cadmium, zinc, manganese, and chromium as insoluble carbonates, bicarbonates, hydroxides, orthophosphates, and sulfides can also occur. Also, the precipitation of oxidation-reduction reaction products can occur.

Carbonate rocks generally are excellent receptors for acid wastewaters. The soluble carbonate rocks neutralize the acid and cause precipitation of many of the above metals. Where the volume of precipitant is significantly less than the volume of carbonate rock dissolved, the system will work safely. If not, then the receptor pores will generally plug and the system will fail. Undesirable effects of the reaction of acid wastes with carbonate receptors could be the evolution of carbon dioxide gas that might retard fluid movement if present in excess of its solubility.

Marine sand receptors containing clays such as montmorillonite will pass saline wastewater without change, but the clays may swell to many times their original volume when in contact with freshwater. Such swelling effectively reduces permeability and may cause well failures.

4.3   Subsurface Hydrodynamics

The dynamics of subsurface fluids in the receptor and overlying aquifers must be understood to the extent that the direction and rate of movement of any wastewater injected and any native fluids displaced by this waste can be estimated. Using data collected from a test well and other means, the pressure buildup in affected aquifers with time should be estimated.

A well injecting at a constant rate into an extensive confined receptor aquifer produces an area of influence that expands with time. As the formation pressure is increased, flow is radially away from the injection well, but not a steady-state flow. Theis [33] hypothesized a close analogy between groundwater flow and heat conduction and developed the following nonequilibrium equation (Fig. 3.5) for determination of the coefficients of transmissibility (T) and storage (S).

$$T = \frac{114.60QW(u)}{s} \tag{3.3}$$

$$S = uT/[1.87(r^2/t)] \tag{3.4}$$

where

T = Coefficient of transmissibility
S = Coefficient of storage
Q = Flow or injection rate (gpm or barrels/d)

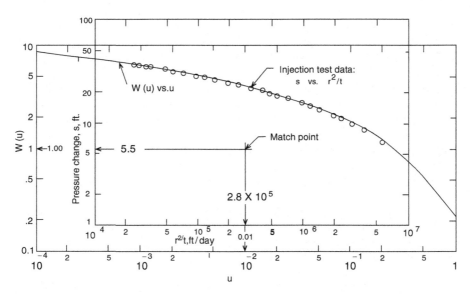

Fig. 3.5 Theis method of superposition for solution of the nonequilibrium equation (Source: U.S. EPA)

Table 3.2 Values for W(u) for values of u

u	1.0	2.0	3.0	4.0	5.0	6.0	7.0	8.0	9.0
$\times 10^{-1}$	1.82	1.22	0.91	0.70	0.56	0.45	0.37	0.31	0.26
$\times 10^{-2}$	4.04	3.35	2.96	2.68	2.47	2.30	2.15	2.03	1.92
$\times 10^{-3}$	6.33	5.64	5.23	4.95	4.73	4.54	4.39	4.26	4.14
$\times 10^{-4}$	8.03	7.94	7.53	7.25	7.02	6.84	6.69	6.55	6.44
$\times 10^{-5}$	10.94	10.24	9.84	9.55	9.33	9.14	8.99	8.86	8.74
$\times 10^{-6}$	13.24	12.55	12.14	11.85	11.63	11.45	11.29	11.16	11.04
$\times 10^{-7}$	15.54	14.85	14.44	14.15	13.93	13.75	13.60	13.46	13.34
$\times 10^{-8}$	17.84	17.15	16.74	16.46	16.23	16.05	15.90	15.76	15.65
$\times 10^{-9}$	20.15	19.415	19.05	18.76	18.54	18.35	18.20	18.07	17.95
$\times 10^{-10}$	22.45	21.76	21.35	21.06	20.84	20.66	20.50	20.37	20.25
$\times 10^{-11}$	24.75	24.06	23.65	23.36	23.14	22.96	22.81	22.67	22.55
$\times 10^{-12}$	27.05	26.36	25.96	25.67	25.44	25.26	25.11	24.97	24.86
$\times 10^{-13}$	29.36	28.66	28.26	27.97	27.75	27.56	27.41	27.28	27.16
$\times 10^{-14}$	31.66	30.97	30.56	30.27	30.05	29.87	29.71	29.58	29.46
$\times 10^{-15}$	33.96	33.27	32.86	32.58	32.35	32.17	32.02	31.88	31.76

Source: Wenzel [43]

$W(u)$ = The well function of u as shown in Table 3.2
s = The pressure change at r, ft of water
r = The radius from the point of injection to the point of observation, ft
t = The time since injection started, d

The Theis method is a graphical one based on superposition of curves. $W(u)$ is plotted against u on logarithmic paper then s is plotted against r^2/t using paper of the same scale. The two plots are superimposed with the coordinate axes parallel and shifted until the position with most of the two curves matched is found as shown in Fig. 3.5. The coincident values of $W(u)$, s, and r^2/t are noted. By substituting these values into the above formulas, S and T for the receptor aquifer are determined (see Example 1 in Sect. 11).

5 Five Potential Hazards: Ways to Prevent, Detect, and Correct them

Problems with injection wells generally can be related to failures in one or more of the five areas listed below [34, 35]:

1. Lack of consideration of all fluid movements
2. Failure of the receptor to receive and transmit the wastes
3. Failure of the confining layer
4. Failure of an individual well either in design, construction, or operation
5. Failure because of human error

Experience and the use of good common sense will avoid most of these problems. When injection wells' arc are properly designed, installed, operated, and maintained, they are no different from any other good piece of equipment. If properly used, they can play an important role in removing toxic and hazardous substances from the immediate human environment.

5.1   Fluid Movement During Construction, Testing, and Operation of the System

Because of lack of knowledge about fluid movements, one drilling company placed salt water generated during pump testing of a receptor zone into an unlined storage pit on their site. The salt water, because of its heavy specific gravity, seeped downward into the local drinking water aquifer. It then moved laterally without mixing and some 8 months later contaminated a nearby city's drinking water supply.

Each well should be designed so that local freshwater supplies are protected from contamination by a separate casing set into the top of their underlying confining layer and cemented back to land surface before the confining layer is breached during construction. Mud pits should be lined with impervious material to prevent seepage into the shallow freshwater strata. All materials used during construction, such as salt, mud, acid, and the like, should be stored in such a way as to assure that materials spilled from damaged containers and the like will not contaminate the freshwater supply.

Another example of a problem is the company that sought to cut costs by eliminating consulting fees. They hired a local, very capable, water well contractor to design and construct their injection and monitoring wells. He did an excellent job of designing the injection well, four monitoring wells into the receptor aquifer, plus one monitoring well into an overlying aquifer. The only problem was that he did not understand subsurface hydrodynamics and located these entire monitor wells within 150 ft of the injection well. The waste front passed the deep monitor wells within 3 days after injection started. Four replacement monitor wells had to be drilled some distance away. Monies wasted on the first four wells (about $140,000) far exceeded the cost of hiring an experienced groundwater geologist.

Making sure that the flow capacity in the test equipment is representative of the proposed permanent facilities under design is also important. An abandoned gas well being tested for conversion into a disposal well showed a capacity of only 50 gpm on gravity flow [36]. The project was almost abandoned when an engineer realized that the friction loss in the temporary piping was higher than in the permanent facilities. After the piping was changed, the injection tests showed the well would handle 450 gpm on gravity flow.

5.2   Failure of the Aquifer to Receive and Transmit the Injected Fluids

Failure of an aquifer generally is caused by lack of naturally developed permeability in the receiving zone or by filling of the pore space with either suspended solids from the effluent or precipitants formed by chemical or biological reactions in the receptor. Wellhead injection pressure should be continuously monitored so that these problems can be detected early and failure avoided. A reduction of the ability of the receiving zone to accept and transmit wastes from any cause will increase the pressure at the wellhead.

Plugging of the receptor zone is by far the most common operational problem where the receptor is a sand aquifer. Most plugging problems can be avoided by one or more of the following:

1. Detailed coring to study the size and shape of pore spaces in the receptor
2. Detailed chemical analyses of fluids and rocks in the receptor
3. Biological cultures of both receptor fluids and wastes
4. Analysis of pressures and temperatures in the receptor
5. Changes expected in the wastes after injection, and
6. Proper cleanup during completion of the well

One injector ran extensive compatibility tests in his own labs at room temperatures and pressures with no indication that a problem might exist. However, a few weeks after injection began, the injection pressures started rising above the predicted pressures. New tests conducted under environmental conditions similar to those in the receptor showed that minor changes in fluid pH caused precipitants to form that partially plugged the receptor.

Another injector's effluent was found to be incompatible with the native fluids in the receptor. If the pH was raised slightly, a precipitant formed; when the pH was lowered, gases were released. The solution: keep the two fluids separated by injecting a compatible buffer ahead of the effluent. The importance of compatibility tests cannot be overemphasized. These tests should be run as close to actual well conditions as possible.

5.3   Failure of the Confining Layer

The effectiveness of the confining layer at each site should be thoroughly investigated by monitoring changes in formation pressures and chemical water qualities during testing of the disposal wells. The most frequent cause for such failures is the creation or the opening up of vertical fractures in a receptor (Fig. 3.6), then propagation of these fractures by continued injection until they radiate through the confining layer. For example, a company was permitted to inject its wastes at a maximum surface pressure of 150 psi, a rather conservative

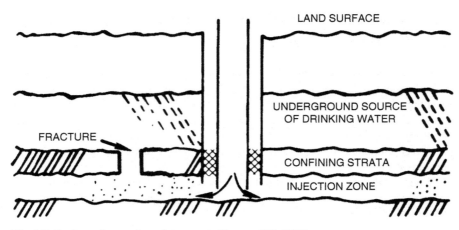

Fig. 3.6 Faulty or fractured confining strata (Source: U.S. EPA)

pressure for most operations. But the specific gravity of the fluids was not taken into account by either the permitting agency or the company. During the operative history of this system, each time the injection pressure approached 149 psi, it would miraculously (so the company officials thought) decline. Soon the pressure in nearby shallow observation wells began to rise. Then wastes were detected in these observation wells and the company had to abandon its injection well system. They built surface treatment facilities at great expense. The cause of the problem was uncontrolled vertical hydraulic fracturing of the confining layers at a pressure of 149 psi. The bottom-hole pressure was about 650 psi, well within the vertical fracturing gradient for the shallow unconsolidated silts and clays that made up the confining layer at this site.

Hydraulic fracturing can be avoided by adequately testing the injection system prior to injecting any effluent. Testing and operations should be planned so that operating pressures never exceed the bottom-hole pressures reached during testing.

It is almost inconceivable that an entire system would fail considering the operational history of existing wells and provided that all the precautions and testing described herein are incorporated into the evaluation of a site. However, should this occur, the disposal of effluent into the subsurface is not a strictly one-way process. If pressures begin to build up or saline water begins to increase in an aquifer above the injection zone, which would indicate leakage of a saline front moving out ahead of the waste front, the entire disposal system could be reversed by installing pimps into the disposal wells and pumping the injected fluids back up to land surface. The effluent can then be disposed via an alternate method. Much of the injected effluent can be recovered by this means. The remaining effluent would probably stop moving as pressures are reduced.

5.4 Failure of an Individual Well

An example of extreme well failure was the injector who went to great expense to install and cement into place fiberglass casing through which they could inject acids without installing injection tubing. However, they overlooked the fact that a cement plug had to be drilled out of the bottom of the fiberglass casing, the hole deepened and screens installed. Of course the driller did not use centralizers and his drill stem fractured the fiberglass. With time, acids in the effluent ate through the cement and then entered a shallow aquifer above the monitoring point. This company now has abandoned its injection well system and has drilled more than 20 monitoring wells to keep track of the movement of wastes lost through the fiberglass casing. The local regulatory agency has declared that injection wells will never be used in their state again. Inspection of this well with a caliper upon completion would have detected the fractures, which would then have been repaired and the failure avoided.

After completion of all construction, each well should be thoroughly inspected and tested using all practical geophysical and hydrogeologic methods available [37]. Constructing each well using the best available technology followed by extensive testing and inspection prior to its use for disposal of effluents is the best practical way to avoid well failure.

Should effluents leak through a break in the casing, packers can then be installed to isolate the break. The zone between the packers can be pumped to recover the leaked fluids. Leaks of this type can usually be repaired. It is recommended that an emergency standby well be constructed at each site so that in the event of mechanical failure of some of the pumping equipment or a need develops for servicing or inspecting an injection well, the well that needs servicing can be shut down, and the emergency standby well can be put into operation. Also, this emergency standby well can be utilized for observational purposes to gain additional information on changes in water quality in the aquifer during injection.

5.5 Failures Because of Human Error

An example of human error is an injector who noticed a sharp increase in the injection pressure. The pressure continued to rise until the maximum head pressure of the pump was reached and then the flow began to decrease. Investigation of the problem revealed that one of the cartridges in the polishing filter had been left out.

Davis and Funk [36] cited another excellent example of the importance of human error at a site where a new disposal well system was installed and shut in pending the startup of a new plant. Compatibility tests had shown the presence of freshwater-sensitive clays in the receptor. During plant construction, the transfer lines to the disposal well systems were pressure tested with freshwater by the contractor and left full. When the well was put into service by the company, the freshwater was displaced into and plugged the receptor aquifer by causing hydration and swelling of the freshwater-sensitive clays. The result was a 6,000 ft-deep $250,000 posthole.

Experience is the best solution to human error. However, installing automatic shutoff and alarm systems at key locations in the system will minimize the effects of human error.

6   Economic Evaluation of a Proposed Injection Well System

Costs for constructing and operating an injection well system vary tremendously from one area of the country to another [38]. They are lower in areas with active petroleum exploration because of competitive bidding, equipment availability, and the availability of energy sources. Costs shown herein are adjusted using US Army Corps of Engineers Cost Index for Utilities of 302.25 in 1981 to 819.11 in 2016 (See Appendix). Costs are broken down into general costs and indirect costs.

Table 3.3 shows the range in capital costs experienced in the southeastern USA by industries and municipalities for the construction of injection well systems. An engineer should be able to take these cost data, adjust them for inflation and other differences in the local area, add pretreatment capital costs, and make a reasonable estimate of the total costs for disposal of a company's wastes by well injection. These costs are based upon an average well depth of 3,500 ft (1,067 m).

Operation and maintenance costs range from 80 cents to about $2.00/1,000 gal wastewater injected (2004 energy costs). This variation is caused primarily by differences in wellhead injection pressures monitoring requirements. The average cost is about $1.30/1,000 gal injected.

Table 3.3 Capital costs for injection well systems

Average flow Q MGD	Cost 2016 million US $
0.5	0.84
1	0.90
2	0.97
3	1.22
5	1.70
10	2.42
20	4.12
30	6.29
50	10.16
100	24.22

Conversion factor: 1 MGD = 3.785 MLD

Legal fees, permitting fees, changes in insurance rates, and other indirect costs can amount to as much as 10 % of construction costs and should be recognized during cost estimating.

7 Use of Injection Wells in Wastewater Management

Numerous pro and con considerations must be investigated before deciding upon a method of discharge for a waste fluid. For industries in certain locations, the disposal or storage of wastes in the subsurface by use of well injection may be the most environmentally acceptable practice available. The difference between storage and disposal is that storage implies the existence of a plan for recovery of the material within a reasonable time, whereas disposal implies that no recovery of the material is planned at a given site. Either operation will require essentially the same type of information prior to injection. However, the attitude of the regulatory agencies toward evaluation of a proposal to use the well injection method could be quite different for each [1, 3, 7, 16, 21, 29].

7.1 Reuse for Engineering Purposes

The reinjection of fluids produced with hydrocarbons into oil-producing horizons to maintain oil field pressures has been practiced for years. Land subsidence in the vicinity of many oil and gas fields and in areas of substantial groundwater production is generally considered to be caused by the withdrawal of fluids. In California, for example, treated wastewaters are being injected to retard the rates of subsidence by repressurizing subsurface formations.

Along many coastal areas [40], the heavy withdrawal of potable groundwater for municipal, industrial, and other uses from freshwater aquifers has caused saltwater encroachment laterally within the aquifer systems. In such areas, treated wastewaters may be injected into the aquifer system to create a hydraulic barrier and hold back the encroaching salt water [33, 41–43]. For example, since 1965, tertiary-treated wastewater has been injected at Bay Park, NY, into a shallow artesian sand aquifer used for public water supply [42] to create a hydraulic barrier against saltwater encroachment.

In areas with water shortages, or where for environmental reasons other methods of discharge are not practical, it may be desirable to inject highly treated wastewaters into the subsurface for purposes of groundwater storage. Wells designed for recharge purposes will see increased use as a tool for capturing and storing highly treated wastewaters such as that would come from drain tiles underlying industrial or municipal land spreading disposal sites. This method should be analyzed considering engineering feasibility and health effects.

7.2 Injection Wells as a Part of the Treatment System

The use of injection wells for subsurface treatment of the wastewaters deserves greater consideration. The emplacement of acid wastewaters in limestone, marble, dolomite, or other carbonate formations where the acid will be neutralized through chemical reaction with the rock formation is generally more environmentally sound than surface treatment with a combination of strip mining of carbonate rocks, plus surface discharge of neutralized wastewater, plus land tilling of the solids precipitated during neutralization [44, 45]. A key requirement for successful well injection is the determination that the volume of solids that may be precipitated is substantially less than the volume of rock dissolved during neutralization. An example of this operation is Kaiser's disposal well system at Mulberry, FL, where about 150 gal/min (9.4 L/s) of fluosilicic acid (pH less than 1.0) and sodium chloride (6.5 %) is being injected into a porous dolomite (Lawson Formation of the Upper Cretaceous age) at a depth of about 4,500 ft (1,372 m). The ratio of volume of dolomite dissolved to volume of precipitant formed is about 11:7. The permeability of the receiving aquifer is increasing with time, indicating that the system is operating as planned and that the precipitant is not plugging the formation pore space.

Where odors may be a problem, where the rate of treatment is slow or where economics are favorable, the subsurface can be used for chemical treatment and for biological treatment by anaerobic and facultative microorganisms. One of the world's largest nylon plants, Monsanto plant located near Pensacola, FL, since 1963, has been disposing of its wastewaters containing nitric acid (pH 2.3) and other nitrogen compounds into the Ocala Limestone of the Eocene age. Backflushing experiments carried out by the US Geological Survey [45] in 1968 showed that the pH of the injected fluid increases rapidly in the aquifer accompanied by rapid denitrification and generation of carbon dioxide, nitrogen, methane, and other gases. Nitrate concentrations decreased from 3,000 to near 0 mg/L in less than 75 min.

Still another use of the subsurface for treatment is the storage of wastewaters containing radioactive minerals that have relatively short half-lives.

7.3 Storage of Municipal Wastewaters for Reuse

Under certain conditions, a double benefit can be realized by injecting a good quality wastewater effluent into a saline aquifer: potentially harmful viruses and bacteria that might survive the treatment process are removed from the human environment, and the effluent displaces a poorer quality (salty) groundwater, thus creating a reserve of potentially usable water in underground storage.

Several deep-injection wells have been constructed in Florida, Hawaii, Louisiana, Illinois, and Texas for storage of secondary-treated wastewater effluent into salt-water aquifers. Secondary- and tertiary-treated municipal wastewaters are of such good quality and in such large volumes (5–50 MGD) (220–2,200 L/s) that it is much too valuable to waste in areas where water shortages are forecast for the foreseeable future. The storage of treated wastewater for future reuse is receiving increased attention in long-range management planning [6, 11–13, 15, 40, 42, 46]. Expansion of this method of reuse as a tool of long-range water quality and water resource management is being encouraged by the US Environmental Protection Agency (US EPA) [47] and many state regulatory agencies as long as measures are taken to protect the public health. The method is particularly adaptable and acceptable when the planned reuse is for agricultural or other nonpotable demands.

Esmail and Kimbler [48] in their investigation of the technical feasibility of storing freshwater in saline aquifers concluded that the rate of injection and the permeability of the receiving aquifer were the two principal factors that control the recovery of the stored freshwater. Recovery is inversely proportional to the aquifer's permeability and directly proportional to the rate of injection.

7.4 Storage of Industrial Wastewaters

In this era of rapidly changing economics, developing technology, increasing energy costs, and demands for reuse and recycling, what is today a waste product may tomorrow become a valuable byproduct [7]. At each plant, serious consideration should be given to separating streams of wastewater that contain chemicals with a potential for future reuse; these reusable chemicals could then be injected in a manner whereby they can be recovered at a future date,

7.5 Disposal of Municipal and Industrial Sludges

One of the developments is a method [49] of dewatering, compacting, and disposing of sludges and other solid wastes from municipal- and industrial-waste treatment plants using the elastic rebound properties of subsurface strata to compress and dewater the waste. Hydraulic pressure is used to compress the rock and create a large opening into which the sludge is placed.

Water separated from the sludge migrates through the receptor stratum radially away from the injection well, which is the point of greatest pressure. Immediately following injection, the volume of the opening in the receptor stratum is slightly less than, but directly proportional to, the volume of waste injected. But as dewatering takes place, the volume of the opening is reduced, and sludge compaction begins. During compaction, the volume reduction is proportional to the amount of its suspended solid content, for example, if the sludge contained 95 % water and 5 % solids, then the compacted volume will be about one-twentieth of the injected volume. Deformation properties such as elasticity and plasticity allow the overburden to absorb the increased thickness, except that if large volumes are emplaced at shallow depth, a small but measurable rise in land surface would be expected.

The end product of this method is a series of very thin (0.001–0.01 in.) pancakes of sludge.

8 Use of Injection Wells for Hazardous Wastes Management

Injection of hazardous wastes into deep wells began in the USA in the 1960s [3]. At that time, the chemical industry was looking for a safe, relatively inexpensive method for disposing of high volumes of waste that could be considered toxic. Technology was borrowed from the oil and gas industry to develop this new form of disposal.

Currently (2002), there are 163 hazardous waste deep-injection wells in the USA located at 51 facilities. Most are found in Texas [78] and Louisiana [18]. Eleven of the facilities are commercial hazardous waste injection facilities. These are the only facilities that can accept hazardous waste generated off-site for injection. Ten of them are located in the Gulf Coast region while one is located in the Great Lakes region [3]

A few regulatory agencies prohibit the disposal of toxic waste underground. Others require the best available pretreatment before injection on the premise that the safety of the method is maximized by this approach. The US Public Health Service [50] considers that only concentrated toxic wastes that cannot otherwise be satisfactorily disposed of should be considered for deep-well injection. The US EPA [18] considered such disposal to be temporary until new technology becomes available, enabling more assured environmental protection. However, as the result of the Hazardous and Solid Waste Amendments of 1984, the US EPA published special regulations for deep wells injecting hazardous waste. In addition to making the requirements for these wells more stringent, the regulations require that each well operator provide a demonstration that the hazardous waste will not be released from the injection zone for at least 10,000 years or will be rendered nonhazardous by natural processes [16].

The injection of troublesome industrial wastewaters into subsurface formations via deep wells is a relatively simple low-cost disposal procedure that has attracted the attention of many manufacturing companies, particularly of the refining and chemical industries [51–55].

Deep-well disposal of hazardous wastes has been demonstrated to be technically feasible in many areas of the country [5–9, 50]. However, ill-sited and improperly designed or constructed wells can result in serious pollution problems. The effects of subsurface injection and the fate of injected wastewaters should be adequately researched to ensure protection of the integrity of the subsurface environment.

8.1 Identification of Hazardous Wastes

Wastes are defined as hazardous for purposes of regulatory control in 40 CFR Part 261. In this regulation, wastes are classified as hazardous either by being listed in tables within the regulation or by meeting certain specified characteristics. Thus, under 40 CFR Part 261, hazardous wastes are known either as *listed* or *characteristic wastes*. Some listed wastestreams, such as spent halogenated solvents, come from many industries and processes. Other listed wastestreams, such as API separator sludges from the petroleum refining industry, come from one particular industry and one process. Radioactive wastes are not covered by 40 CFR Part 261. A characteristic waste is not listed, but is classified as hazardous because it exhibits one or more of the following four characteristics [56]:

(a) *Toxicity*
   A waste is toxic if the extract from a representative sample of the waste exceeds specified limits for eight elements and four pesticides using extraction procedure (EP) toxicity test methods. The elements are arsenic, barium, cadmium, chromium, lead, mercury, selenium, and silver. The pesticides include endrin, methoxychlor, toxaphene, 2,4-D, and 2,4,5-TP Silvex. Note that this narrow definition of toxicity relates to whether a waste is defined as hazardous for regulatory purposes; in the context of this chapter, toxicity has a broader meaning because many deep-well-injected wastes have properties that can be toxic to living organisms.

(b) *Reactivity*
   Reactivity describes a waste's tendency to interact chemically with other substances. Many wastes are reactive, but it is the degree of reactivity that defines a waste as hazardous. Hazardous reactive wastes are those which are normally unstable and readily undergo violent change without detonating, react violently with water, form potentially explosive mixtures with water, generate toxic gases or fumes when combined with water, contain sulfide or cyanide and are exposed to extreme pH conditions, or are explosive. Because deep-well-injected wastestreams are usually dilute (typically less than 1 % waste in water), hazardous reactivity is not a significant consideration in deep-well injection, although individual compounds may exhibit this property at higher concentrations than those which exist in the wastestream. Nonhazardous reactivity is, however, an important property in deep-well injection, since when a reactive waste is injected, precipitation reactions that can lead to well plugging may occur.

(c) *Corrosivity*
   Corrosive wastes are defined as those wastes with a pH $\leq 2 \geq 12.5$ (i.e., the waste is very acidic or very basic). Beyond its importance in defining a waste as hazardous, the corrosivity of wastes is also a property of concern to deep-well injection systems and operations. Corrosive wastes may damage the injection system, typically by electrochemical or microbiological means. Corrosion of injection well pumps, tubing, and other equipment can lead to

hazardous waste leaking into strata not intended for injection. For information on various types of electrochemical corrosion relevant to the injection well system, the reader is referred to [57]. Other recommended sources include [58–62], which discuss saturation and stability indexes for predicting the potential for corrosion or scaling (accumulation of carbonate and sulfate precipitates) in injection wells. The Stiff and Davis Index is recommended by Warner and Lehr [57] as most applicable to deep-well injection of hazardous wastes, because it is intended for use with highly saline groundwaters. Additionally, Ostroff [63] provides examples of how to use the index, Watkins [64] describes procedures that test for corrosion, and Davis [65] thoroughly discusses microbiological corrosion of metals.

(d) *Ignitability*

As noted, deep-well-injected wastes are relatively dilute. Therefore, ignitability is not a significant consideration in deep-well injection, although in a concentrated form, individual compounds may exhibit this property. Ignitability has no further implications for the fate of deep-well-injected waste.

8.2   *Sources, Amounts, and Composition of Injected Wastes*

The sources, amounts, and composition of injected hazardous wastes are a matter of record, since the Resource Conservation and Recovery Act (RCRA) requires hazardous waste to be manifested (i.e., a record noting the generator of waste, its composition or characteristics, and its volume must follow the waste load from its source to its ultimate disposal site). The sources and amounts of injected hazardous waste can be determined, therefore, based on these records. Table 3.4 shows the estimated volume of deep-well-injected wastes by industrial category in 1983 [56]. More than 11 billion gallons of total hazardous waste were injected in 1983. Organic chemicals (51 %) and petroleum refining and petrochemical products (25 %) accounted for three-quarters of the volume of injected wastes. The remaining 24 % was divided among six other industrial categories: miscellaneous

Table 3.4 Estimated volumes of deep-well-injected wastes by industrial category, 1983

Industrial category	Volume MG/year	Percent of total
Organic chemical	5,868	50.9
Petroleum refining and petrochemical products	2,888	25.0
Miscellaneous chemical products	687	6.0
Agricultural chemical products	525	4.6
Inorganic chemical products	254	2.2
Commercial disposal	475	4.1
Metals and minerals	672	5.8
Aerospace and related industry	169	1.5
Total	11,539	100.0

chemical products, agricultural chemical products, inorganic chemical products, commercial disposal, metals and minerals, and aerospace and related industry.

Although the general composition of each shipment of wastes to an injection well may be known, a number of factors make it difficult to characterize fully the overall composition of industrial wastewaters at any one well. These factors include (1) variations in flow, in concentrations, and in the nature of organic constituents over time; (2) biological activity that may transform constituents over time; and (3) physical inhomogeneity (soluble and insoluble compounds) [66]. Further, the exact composition of the shipment may not be known because of chemical complexity [66]. An example of the complexity of organic wastes is illustrated in Roy et al. [67], which presents an analysis of an alkaline pesticide-manufacturing waste. This waste contained more than 50 organic compounds, two-fifths of which could not be precisely identified.

Although no systematic database exists on the exact composition of deep-well-injected wastes in a survey of 209 operating waste injection wells, Reeder et al. [68] found that 53 % injected one or more chemicals identified in that study as hazardous. The US EPA gathered data for 108 wells (55 % of total active wells) that were operated in 1983. Table 3.5 summarizes the total quantity of undiluted waste in six major categories, provides a breakdown of average concentrations of constituents for which data were available, and indicates the number of wells involved. A little more than half the undiluted waste volume was composed of nonhazardous inorganics (52 %). Acids were the most important constituent by volume (20 %), followed by organics (17 %). Heavy metals and other hazardous inorganics made up less than 1 % of the total volume in the 108 wells. About a third of the wells injected acidic wastes and about two-thirds injected organic wastes. Although the percentage of heavy metals by volume was low, almost one-fifth of the wells injected wastes containing heavy metals.

An injected wastestream typically is composed of the waste material and a large volume of water. Because the data in Table 3.5 include only 55 % of the injection wells that were active in 1983, it is not possible to estimate precisely the percentage of waste to the total volume of injected fluid shown in Table 3.4. However, if the same total proportions apply to all wells, wastes made up 3.6 % of the total volume of injected fluid (36,000 mg/L). This percentage agrees well with an independent estimate for a typical injection ratio of 96 % water and 4 % waste [69].

Table 3.6 also shows that the average concentration of all the acidic wastes exceeded 40,000 mg/L. Concentrations of metals ranged from 1.4 mg/L (chromium) to 5,500 mg/L (unspecified metals, probably containing multiple species). Five of the 18 organic constituents exceeded 10,000 mg/L (total organic carbon, organic acids, formaldehyde, chlorinated organics, and formic acid); four exceeded 1,000 mg/L (oil, isopropyl alcohol, urea nitrogen, and organic peroxides).

Table 3.5 Waste characteristics of 108 hazardous waste wells active in 1983 in the USA

Waste type/components	Gallons[a]	Average concentration (mg/L)	No of wells
Acids	44,140,900(20.3)[b]		35 (32.4)[b]
Hydrochloric acid		78,573	15
Sulfuric acid		43,000	6
Nitric acid		75,000	2
Formic acid		75,000	2
Acid, unspecified		44,900	12
Heavy metals	1,517,600(0.7)		19(17.6)
Chromium		1,4	11
Nickel		600	5
Metals, unspecified		5,500	2
Metal hydroxides		1,000	1
Hazardous inorganics	89,600(<0.1)		4(3.7)
Selenium		0.3	2
Cyanide		391	2
Organics	39,674,500(17.4)		71(65.7)
Total organic carbon (TOC)		11,413	24
Phenol		805	22
Oil		3,062	6
Organic acids		10,000	3
Organic cyanide		400	3
Isopropyl alcohol		1,775	3
Formaldehyde		15,000	2
Acetophenone		650	2
Urea "N"		1250	2
Chlorinated organics		35,000	2
Formic acid		75,000	2
Organic peroxides		4,950	2
Pentachlorophenol		7.6	2
Acetone		650	2
Nitrile		700	1
Methacrylonitrile		22	1
Ethylene chloride		264	1
Carbon tetrachloride		970	1
Nonhazardous inorganics	118,679,700 (52.0)	–	50 (46.3)
Others	22,964,600 (9.9)	–	33 (30.5)
Total	228,021,800[c]		108

[a]Gallons of nonaqueous wastes before dilution and injection
[b]Number in parentheses is the percentage of total
[c]Excludes overlaps between organics and acids

Table 3.6 Density of brines

Specific gravity of water at 60 °F	Approximate total solids in parts per million, mg/L	Weight in psi of pressure per ft^a
1.000	none	0.433
1.010	13,500	0.437
1.020	27,500	0.441
1.030	41,400	0.445
1.040	55,400	0.450
1.050	69,400	0.454
1.060	83,700	0.459
1.070	98,400	0.463
1.080	113,200	0.467
1.090	128,300	0.471
1.100	143,500	0.476
1.110	159,500	0.480
1.120	175,800	0.485
1.130	192,400	0.489
1.140	210,000	0.493

^{a}Averages approximately 0.0043 psi/ft for each increase of 0.01 in density

8.3 Geographic Distribution of Wells

The use of wells for disposal of industrial wastes dates back to the 1960s, when it was implemented primarily in response to more stringent water pollution control regulations [70]. The number of industrial-waste injection wells more than doubled between 1967 and 1986.

The state distribution shows some interesting patterns. Class I injection wells are concentrated in two states, Texas (112 wells) and Louisiana (70 wells), which have a total of 69 % of all wells in 1986. The growth from 1984 to 1986 has been concentrated in Texas, with a 38% increase, from 81 to 112 wells. The only other states to show a significant increase from 1984 to 1986 are Indiana (13 proposed wells) and California (7 proposed wells). Nine states have had industrial-waste injection wells in the past but did not have any permitted Class I wells in 1986 (Alabama, Colorado, Iowa, Mississippi, Nevada, North Carolina, Pennsylvania, Tennessee, and Wyoming). One state (Washington) had a Class I well in 1986, but no record of industrial wastewater injection before that year.

The 1989 US EPA data show that the heavy concentration of hazardous waste injection wells is in three geologic basins: Gulf Coast, Illinois Basin, and the Michigan Basin.

8.4 Design and Construction of Wells

The following description of the design and construction of deep-injection wells is adapted from Donaldson [71], Donaldson et al. [72], and the US EPA [73].

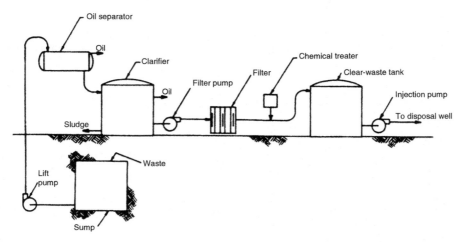

Fig. 3.7 Aboveground components of a subsurface waste disposal system (Source: U.S. EPA)

8.4.1 Surface Equipment Used in Waste Disposal

Figure 3.7 shows the surface equipment used in a typical subsurface waste disposal system. Detailed discussion of surface treatment methods can be found in Warner and Lehr [57]. The individual elements are:

(a) A *sump tank* or an open 30,000–50,000-gal steel tank is commonly used to collect and mix wastestreams. An oil layer or, in a closed tank, an inert gas blanket is often used to prevent air contact with the waste. Alternatively, large, shallow, open ponds may provide sufficient detention time to permit sedimentation of particulate matter. Such ponds often are equipped with cascade, spray, or forced-draft aerators to oxidize iron and manganese salts to insoluble forms that precipitate in the aeration ponds.

(b) An *oil separator* is used when the waste contains oil because oil tends to plug the disposal formation. The waste is passed through a settling tank equipped with internal baffles to separate the oil from the waste.

(c) A *clarifier* removes such particulate matter as polymeric flocs, dirt, oil, and grease. It is often a tank or a pond in which detention time is long enough to allow suspended particles to settle gradually. The process also may be accelerated by adding a flocculating agent such as aluminum sulfate, ferric sulfate, or sodium aluminate. Tank clarifiers are often equipped with a mechanical stirrer, sludge rake, and surface skimmer that continuously remove sludge and oil.

(d) A *filter* is used in some cases when coagulation and sedimentation do not completely separate solids from the liquid waste in areas where sand and sandstone formations are susceptible to plugging. Filters with a series of metal screens coated with diatomaceous earth or cartridge filters typically are used. Where limestone formations with high solution porosity are used for injection, filtration is usually not required.

(e) A *chemical treater* is used to inject a bactericide if microorganisms could cause fouling of injection equipment and plugging of the injection reservoir.

(f) An unlined steel *clear-waste tank* typically is used to hold clarified waste before injection. The tank is equipped with a float switch designed to start and stop the injection pump at predetermined levels.

(g) An *injection pump* is used to force the waste into the injection zone, although in very porous formations, such as cavernous limestone, the hydrostatic pressure of the waste column in the well is sufficient. The type of pump is determined primarily by wellhead pressures required, the volume of liquid to be injected, and the corrosiveness of the waste. Single-stage centrifugal pumps are used in systems that require wellhead pressures up to about 150 psi, and multiplex piston pumps are used to achieve higher injection pressures.

8.4.2   Injection Well Construction

Most injection wells are drilled using the rotary method, although depending on the availability of equipment and other site-specific factors, reverse-rotary or cable-tool drilling may be used. The construction of an injection well incorporates several important elements: (1) bottom-hole and injection-interval completion (2), casing and tubing (3), packing and cementing (4), corrosion control, and (5) mechanical integrity testing. A detailed discussion of the technical aspects of industrial-waste injection well construction can be found in Warner and Lehr [57]. The US EPA [73] also presents a survey of well construction methods and materials used for 229 hazardous waste injection wells.

1. Two types of *injection well completions* are used with hazardous waste injection wells:

   (a) *Open-hole* completion typically is used in competent formations such as limestone, dolomite, and consolidated sandstone that will stand unsupported in a borehole. In 1985, 27 % of Class I wells were of this type, with most located in the Illinois Basin.

   (b) *Gravel-pack* and *perforated* completions are used where unconsolidated sands in the injection zone must be supported. In gravel-pack completions, the cavity in the injection zone is filled with gravel, or, more typically, a screen or liner is placed in the injection zone cavity before the cavity is filled with gravel. In perforated completions, the casing and cement extend into the injection zone and are then perforated in the most permeable sections. In 1985, 53 % of Class I wells were perforated and 17 % were screened [73].

2. *Casing* and *tubing* are used to prevent the hole from caving in and to prevent aquifer contamination by confining wastes within the well until they reach the

injection zone. Lengths of casing of the same diameter are connected together to form casing strings. Usually two or three casing strings are used. The outer casing seals the near-surface portion of the well (preferably to below the point where aquifers containing less than 10,000 mg/L total dissolved solids, potential underground sources of drinking water, are located). The inner casing extends to the injection zone. Tubing is placed inside the inner casing to serve as the conduit for injected wastes, and the space between the tubing and casing is usually filled with kerosene or diesel oil after packing and cementing are completed.

3. *Packers* are used at or near the end of the injection tubing to plug the space, called the annulus, between the injection tubing and the inner casing. Cement is applied to the space between the outer walls of the casing and the borehole or other casings. Portland cement is used most commonly for this purpose, although when acidic wastes are injected, special acid-resistant cements are sometimes used in the portion of the well that passes through the confining layers.

4. *Corrosion control* can be handled several ways: (1) by using corrosion-resistant material in constructing the well, (2) by treating the wastestream through neutralization or other measures, and (3) by cathodic protection.

5. *Mechanical integrity testing* is required by the US EPA regulations to ensure that an injection well has been constructed or is operating without (1) significant leakage from the casing, tubing, or packer or (2) upward movement of fluid through vertical channels adjacent to the well bore. Types of mechanical integrity tests include the following:

(1) Pressure test
(2) Monitor annual pressure
(3) Temperature log
(4) Noise log
(5) Radioactive tracer log
(6) Cement bond log
(7) Caliper log
(8) Casing condition log

A detailed discussion of mechanical integrity can be found in the US EPA [74].

8.5 Disposal of Radioactive Wastes

Radioactive contaminants created by nuclear fission and other means differ from the usual industrial plant waste in their ability to emit radiation. There is no known method for neutralizing radioactivity, but radioactive isotopes decay and thus lose their activity with time.

The Halliburton Company Inc. has developed (British Patent L 054740) an improved method for the disposal of a radioactive waste by mixing the waste with cement to form slurry then injecting it through a well. A horizontal fracture is developed hydraulically, and the slurry is injected into the opening and caused to harden in place. This method, which has been used successfully since about 1967 at Oak Ridge, TN, is available for disposal of low- and intermediate-level radioactive wastes. In using this method, a conventional injection well is drilled, generally to a depth less than 1,000 ft, but until an impermeable formation is transversed. The nonpermeable formation is perforated by, for example, gun perforations. A fracturing fluid, which may be the waste-cement slurry, is pumped into the well under sufficient pressure (greater than 1 psi/ft of depth) to exceed the formation breakdown (fracture) pressure. The formation will fracture or part generally in a horizontal direction at this depth. The waste-cement slurry is injected into the fracture, and the fracture is sealed by allowing the cement to harden.

The same procedure then may be repeated at other depths within the well. To cause the cement to harden, the slurry may need an absorption-type clay such us attapulgite. At other times, an agent such as calcium lignosulfonate that reacts chemically with the cement and retards its setting time may need to be added.

Another little-used but technically sound method for the disposal of radioactive wastes is the injection through well radioactive or toxic materials into strata of low permeability at depth of 1,000–20,000 ft or at such a depth that the wastes are removed from the biosphere. This method involves displacement of formation connate water with liquid waste. Rocks of low permeability that are impregnated with waste become permanent storage receptors if the differential pressures across the receptor strata are maintained at a minimum. They will be retained almost indefinitely as a film held by molecular adhesion on the wall of the interstices [75–77].

Other factors also play a role in the retention of fluids in formations of low permeability. For example, the greater the amount of total interstitial surface in a rock or unconsolidated material, the greater is its specific retention. As would be expected, it is found that as the effective diameter of the material's grain decreases, the specific retention generally increases because the total exposed surface area increases with decreasing grain size.

Capillarity is important in the retention of fluids in any granular material such as sedimentary rocks. The openings between the granules are interconnected in all directions, with the result that capillary forces act out in all directions within such materials. The moisture film around particles is held so tightly that it strongly resists any forces tending to displace it. The degree of its resistance to movement is expressed by its capillary potential, which is a measure of the force required to move this moisture from the soil.

Ion exchange is an important geochemical process. Many ions in hazardous waste products may be removed from wastewaters by means of this process. Clay minerals exhibit a marked capacity for the exchange of cations; in fact, all clay formations possess some ion-exchange capacity. Clay minerals exhibiting good ion exchange are kaolinite, halloysite, montmorillonite, illite, vermiculite, chlorite, sepiolite, attapulgite, and polygorskite. Of these, the montmorillonites are noted for the highest cation-exchange capacity, and the kaolins are noted for the most rapid rate of exchange.

The cohesive property of fluids plays an important part in their retention or movement in porous rocks. Cohesion is the ability of a fluid, or other substances, to stick together and resist separation.

The adhesive properties of a rock also play an important part in the movement or retention of a fluid in a porous rock. Adhesion is a measure of a fluid's ability to stick to the surfaces of other materials, such as to rock material in a formation.

The salty water found in deeply buried sedimentary formation generally is ancient seawater called connate water. Understanding the openings or pore spaces of the rock materials that build up on the ocean floor during geologic time and contain the connate water is the key to understanding the value of these natural spaces as permanent receptors for storage of hazardous and radioactive wastes.

9   Protection of Usable Aquifers

Any aquifer that contains water that is both economically practical and technologically feasible to use for domestic, agricultural, industrial, or other purposes should be protected. Such aquifers are vulnerable to contamination by either the injected fluids or fluids being displaced by the injected fluids migrating into the aquifer to be protected. A variety of measures described below can be used to assure that injection well systems will not contaminate protected aquifers [78].

To protect groundwater, the well must be constructed so as to prevent contamination by (a) keeping injected fluids within the injection well casing and within the intended injection zone and (b) keeping formation fluids displaced by the injected fluids from migrating into a protected aquifer. There are six major pathways [79] by which fluids may move and contaminate aquifers. The following discussion describes each pathway and summarizes a way to prevent migration through that pathway.

9.1   Pathway 1: Migration of Fluids Through a Faulty Injection Well Casing

The casing of a well can serve a variety of purposes. It supports the well bore to prevent collapse of the geologic formations into the hole and consequent loss of the well; it serves as the conductor of injected fluids from the land surface to the intended injection zone and supports other components of the well. If a well casing is defective, injected fluids may escape through the defect and enter the protected aquifer (Fig. 3.8). Such migration can contaminate an underground source of drinking water.

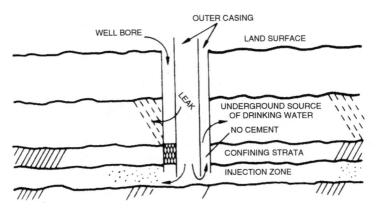

Fig. 3.8 Faulty well construction (Source: U.S. EPA)

To detect migration of fluids through leaks in the casing, periodic tests of the casing's integrity should be made. Several types of casing inspection tests are commercially available. For example, the downhole TV camera [80] can "see" what is wrong on the inside of the casing. The downhole casing inspection log called Vertilog [81] is a system for making a quantitative measurement of corrosive damage, indicating if the metal loss is internal or external and if it is isolated or circumferential. Holes in the casing can be identified as well as casing separations.

Use of separate tubing for injection affords protection to the casing and decreases the possibility of leakage. It isolates the casing of the well from injected fluids. By preventing this contact between casing and injected fluids, the possibility of migration of contaminants through leaks in the casing is greatly diminished. For the same reason, the use of tubing and packer also lessens the chances of corrosion of casing. By monitoring the annular space between the casing and tubing, leaks in the tubing can be detected and repaired before the casing becomes faulty. Tubing and packer offers two further advantages. It isolates the annulus (between the tubing and casing) from the injection zone, facilitating detection of any leaks in the tubing. It allows for visual inspection for deterioration of the tubing during routine maintenance. Finally, wells which inject corrosive fluids should be constructed of corrosion-resistant materials. This material is intended to prolong the operating life and continued viability of the well casing.

9.2 Pathway 2: Migration of Fluids Upward Through the Annulus Between the Casing and the Well Bore

A second pathway by which contaminants can enter protected aquifers is by migrating upward through the annulus, between the drilled hole and the casing. Under usual injection conditions, injected fluids, upon leaving the well, enter a stratum in the injection zone that to some degree resists the entry of the fluids. This resistance results from friction and is inversely proportional to the size of the small openings in the stratum. Because fluids tend to take the path of least resistance, unless properly contained, they may travel upward through this annulus. If sufficient injection pressure exists, the fluids could migrate upward through the drilled hole into the overlying protected aquifer.

Leaks through holes in the well casing or upward fluid movement between the well's outer casing and the well bore are illustrated in Fig. 3.8.

Casing should be cemented to isolate the aquifers to be protected from all underlying saline aquifers and from the injection zone. Generally, two 100-ft-thick cement plugs are installed. One is located immediately below the lowermost aquifer to be protected. The other is located immediately above the injection zone. The absence of leaks and fluid movement in the well bore should be confirmed periodically using geophysical logging techniques.

9.3 Pathway 3: Migration of Fluids from an Injection Zone Through the Confining Strata

The third way by which fluids can enter a protected aquifer is through leaks in the confining strata. Upon entry into an injection zone, fluids injected under pressure will normally travel away from the well and laterally through the receiving formation. In most cases, this occurrence is expected and gives rise to no concern, but if the confining stratum that separates the injection zone from an overlying or underlying protected aquifer leaks significantly because it is either fractured or permeable, the injected fluids can migrate out of the receiving formation and into the protected aquifer.

For obvious reasons, there is no general well construction standard that can address this problem of migration of fluids through the confining strata.

Several steps should be taken to assure that fluids do not migrate through or around the confining strata (Fig. 3.9).

1. Select a deep formation as a receptor. The deeper the receptor stratum selected for injection, the greater the degree of protection.
2. Place at least 200 ft of cement about the injection zone. The thicker the cement plug placed above the injection zone, the larger the pathway of fluid movement before flow can enter the well bore above the plug.
3. Study the confining and receptor strata. Care should be taken during drilling of the test hole to determine the permeability, thickness, and other information for the confining and injection strata and the changes in fluid chemistry that can be expected as the fluids migrate through the confining bed.
4. Determine fluid movement rates at various pressures. The leakage rate versus injection pressure should be evaluated prior to operation, and the injection

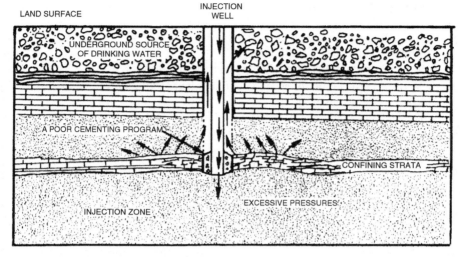

Fig. 3.9 Leakage through confining strata (Source: U.S. EPA)

pressure limited in order to avoid fluid movement through the confining strata into protected aquifers.

Frequently, when leakage out of the receptor stratum occurs, the adjacent aquifers (those leaked into) are permeable enough so that only limited vertical migration occurs. The equation for pressure buildup in the injection shell as a result of injecting into a zone with a leaky confining bed on one side is [82]:

$$P_r = P_i + P_{DL}[141.2Q\mu B]/kH \tag{3.5}$$

$$P_{DL} = \text{a function of}(t_D, \ r/B) \tag{3.6}$$

$$t_D = 6.33 \times 10^{-3}kT/\phi\mu Cr^2 \tag{3.7}$$

$$B = \sqrt{kH\text{hc}/\text{kc}} \tag{3.8}$$

The equation for determination of pressures in the injection well where both confining beds leak is the same as given above except that:

$$B = \sqrt{kHhchc'/kc'hc + kchc'} \tag{3.9}$$

where

P_r = Reservoir pressure at radius r (ft of water or psi)
P_i = Initial formation pressure (ft of water or psi)
P_{DL} = Dimensionless pressure for semiconfined reservoirs

t_D = Dimensionless time
B = Leakage factor
r = Radial distance from well bore (ft)
k = Average permeability (millidarcy)
k_c = Vertical permeability of confining layer (millidarcy)
$k_c{}'$ = Vertical permeability of second confining layer (millidarcy)
H = Reservoir thickness (ft)
h_c = Thickness of confining layer (ft)
$h_c{}'$ = Thickness of second confining layer (ft)
C = Compressibility $(psi)^{-1}$
Φ = Porosity expressed as a decimal
μ = Viscosity

For examples of solutions for Eqs. 3.5, 3.6, 3.7, 3.8, and 3.9, the reader is referred to the US EPA publication 600/2-79-170 [83].

9.4 Pathway 4: Vertical Migration of Fluids Through Improperly Abandoned or Improperly Completed Wells

Fluids from the pressurized area in the injection zone may be forced upward through nearby wells (Fig. 3.10) that penetrate the injection horizon within a zone around the injection well called the zone of endangering influence. The zone of endangering influence of a well may be defined as that radial distance from the well bore within which pressure increases because injections are sufficient to cause a potential for upward migration into freshwater zones. As shown in Fig. 3.11, the zone of endangering influence includes all that area surrounding an injection well wherein the upward pressure in the injection zone exceeds the downward pressure

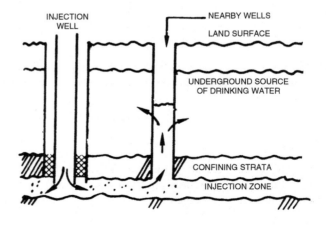

Fig. 3.10 Leakage through nearby wells (Source: U.S. EPA)

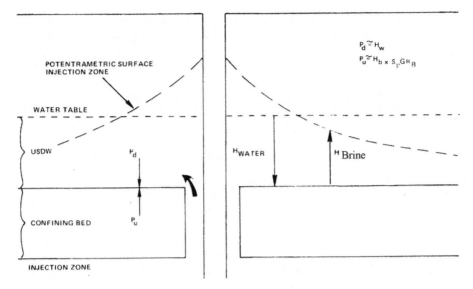

Fig. 3.11 Zone of endangering influence. Where $P_u > P_d$, then potential for endangerment exists (Source: U.S. EPA)

of freshwater when measured using the base of freshwater as the datum. In areas where before injection the upward pressures in the injection formation already exceed the downward freshwater pressure, the zone of endangering influence is infinite or very large. In such areas, an alternate fixed radius of 0.25 mile [84] was approved by the US EPA for the review of nearby wells.

Wells located within the area of pressure increase from an injection well should be examined to assure that they are properly completed and plugged. Corrective action should be taken where necessary to prevent fluids from migrating along these pathways into protected aquifers.

The key aspects influencing zone pressures during injection are:

(a)   The existing fluid pressures in the disposal aquifer
(b)   The pressure increases induced to effect waste fluid emplacement

Several methods common to the practice of engineering hydrology and reservoir analysis are applicable to the solution of the zone of endangering influence (Fig. 3.11). These methods analyze the pressure differential that exists at the base of freshwater between increased reservoir pressures because of injection and the pressure exerted downward by the freshwater column.

Consider an example. Increased formation pressures cause a column of water to rise in an open hole to a level 100 ft above the base of freshwater. Thus, at the base of freshwater, the formation fluid is exerting an upward pressure equal to 100 ft of hydrostatic head (of brine). However, if the freshwater aquifer contains, say, 200 ft of water, then a downward pressure of about 200 ft of head (of freshwater) is exerted at the base. Under these conditions if leakage were to occur, the freshwater would "leak" downward into the brine. Only when upward pressure (P_u) is greater than the downward pressure (P_d) can there be the potential for upward movement.

The zone of endangering influence around an injection well encompasses all the area within which pressure increases due to injection are sufficient to create an upward differential pressure, measured at the base of freshwater.

This type of analysis assumes "worst case" (i.e., open-hole) conditions. An analysis considering friction losses through small channels, drilling mud column displacement or seepage through beds, would yield a much smaller "zone."

The pressure change in an injection zone at distance *(r)* caused by injection volume (Qt) may be described by the equation [85] for predicting pressure increases given below:

$$s = 162.6(Qu/kH)\log(kt/70.4\phi uCr^2) \qquad (3.10)$$

where

s = Change in pressure (ft of water or psi)
Q = Flow or injection rate (barrels/d)
u = $1.87r^2S/Tt$ (centipoises = cp)
k = Average permeability (millidarcy)
H = Reservoir thickness (ft)
t = Time (d)
Φ = Porosity expressed as a decimal
C = Compressibility $(psi)^{-1}$
r = Radial distance from well bore (ft)

This pressure increase is additive to the existing formation pressure before injection began (P_2).

Step 1

Solve Eq. 3.10 for any two values of *r*, and convert s to ft of hydrostatic head by dividing (psi) by the gradient per ft of the formation fluid (Table 3.6). Add this value to P_2 (hydrostatic head of the injection zone), and plot the two values at their respective r on a semilog paper. A straight line connecting the two points establishes the pressure surface of the disposal zone as it exists in space, measured in ft of head of formation brine above the top of the injection zone.

Similarly, some pressure (P_1) exists in the basal freshwater aquifer, corresponding to the weight of a column of water in a well fully penetrating that aquifer.

Step 2

Locate the stratigraphic position of the base of the lowermost freshwater aquifer on the diagram. Convert ft of head of freshwater to ft of head as formation water measured from the base of the freshwater aquifer. Draw a horizontal line to denote the pressure surface of the freshwater as it exists in space.

Step 3

The intersection (if any) of lines P_1 and $(s + P_2)$ denotes the radius (r) of the "zone of endangering influence." That is, to the left of the intersection, the pressure in the disposal zone is sufficient to overcome the hydrostatic pressure at the base of freshwater, and an upward potential is realized.

This method of solving the "zone of endangering influence" of an injection well represents an extremely conservative viewpoint.

9.5 Pathway 5: Lateral Migration of Fluids from Within an Injection Zone into a Protected Portion of Those Strata

In most cases, the injection zone of a particular well will be physically segregated from underground sources of drinking water by impermeable material. In some instances, however, wells inject into an unprotected portion of an aquifer that is hydraulically interconnected with a protected aquifer (Fig. 3.12). In this event, there may be no impermeable layer or other barriers to prevent contact between contaminated fluids and underground drinking water.

Injection into unprotected portions of aquifers that contain drinking water in other areas must be done with great care. This type of injection can work if the predominant flow of the aquifer is such that injected fluids will tend to move away from, rather than toward, the protected part of the aquifer.

It is sometimes helpful to define the actual position of the waste front and its movement with lime.

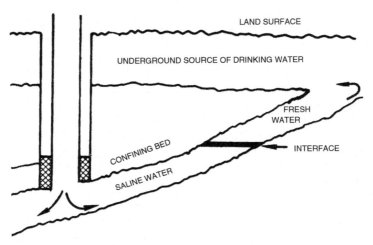

Fig. 3.12 Hydraulically interconnected aquifers (Source: U.S. EPA)

The minimum distance the waste will have traveled during injection may be described [86] by

$$r = (V/\pi H\phi)^{1/2} \tag{3.11}$$

where

r = Radial distance from well bore (ft)
V = Injected volume (ft^3)
H = Reservoir thickness (ft)
Φ = Porosity expressed as a decimal

A typical solution is given in Example 5 of Sect. 11.

In most practical situations, the minimum radial distance of travel will be exceeded because of dispersion, density segregation, and channeling through higher-permeability zones.

An estimate of the influence of dispersion can be made with the Eq. 3.12:

$$\gamma = r + 2.3\sqrt{Dr} \tag{3.12}$$

where

γ = Radial distance from well bore with dispersion (ft)
r = Radial distance from well bore (ft)
D = Dispersion coefficient

This equation is obtained by solving Eq. 10.6.5 of Bear [87] for the radial distance at which the injection front has a chemical concentration of 0.2 %. A dispersion coefficient of three represents a reasonable value for a sandstone aquifer (see Example 6 in Sect. 11).

It may be impossible to accurately predict the chemical character of the plume of waste 100 years after it has slowly moved a few hundred feet in contact with subsurface minerals. However, it is important to list some of the chemical and biological reactions that will certainly occur to degrade the waste. The rates at which these reactions will occur are only speculative, however. These reactions include [75, 76, 88–93]:

1. Dilution and dispersion
2. Biological degradation of organic compounds
3. Biochemical degradation of some species, such as nitrate, sulfate, and so on
4. Adsorption and ion-exchange reactions with clay particles
5. pH neutralization
6. Precipitation and immobilization of some constituents

It is much more important to attempt to predict the direction and ultimate location of the waste front. As was pointed out in an earlier section, the disposal reservoir is confined above and below by a number of relatively impermeable, regionally persistent clay formations. Confined by these rocks, the disposal reservoir dips and thickens toward the Gulf of Mexico, In other words, the further the waste moves coastward, the deeper and more separated from freshwater it becomes. Near the coastline, the formation exists at depths exceeding 10,000 ft. At this point, the angle of dip steepens radically, and the formation dips beneath the Gulf of Mexico to depths exceeding 20,000 ft.

Therefore, over a period of millions of years, confined above and below by clay harriers and separated from freshwater by thousands of feet of rocks, a gradually decomposed waste will move at a very slow rate and remain essentially isolated in the deep subsurface.

9.6 Pathway 6: Direct Injection of Fluids into or Above an Underground Source of Drinking Water

The last pathway of contamination of groundwater is also the most hazardous. Direct injection of contaminated fluids into or above underground sources of drinking water presents the most immediate risk to public health. Such direct injection causes an instantaneous degradation of groundwater (Fig. 3.13). The injected fluids do not benefit from natural treatment processes such as filtration and ion exchange.

Many shallow injection wells, pits, septic tanks, and other similar disposal systems are used to dispose contaminants above drinking water aquifers that need to be protected [94, 95]. The injected fluids then percolate downward into drinking water aquifers, as illustrated in Fig. 3.10. The US EPA [84] decided that wells injecting hazardous wastes directly into drinking water aquifers are to be banned. Drilling of new wells was prohibited after 1982.

Casing should be installed through all aquifers to be protected and cemented to isolate them from exposure to fluids being injected and from all underlying saline aquifers penetrated by the well bore.

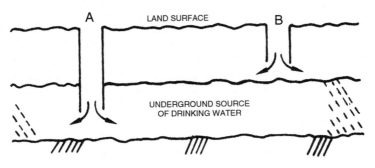

Fig. 3.13 Direct injection (Source: U.S. EPA)

10   Case Studies of Deep-Well Injection

Field studies are an important complement to geochemical modeling and to laboratory studies. Two ways to investigate interactions between injected wastes and reservoir material are (1) direct observation of the injection zone and overlying aquifers using monitoring wells and (2) backflushing the injected waste [56]. In both instances, samples of the fluids in the zone are collected at intervals to characterize the nature of geochemical reactions and to track changes over time.

1. *Monitoring wells.* Monitoring wells drilled into the injection zone at selected distances and directions from the injection well allow direct observation of formation water characteristics and the interactions that occur when the waste front reaches the monitoring well. When placed near the injection well in the aquifer above the confining layer, monitoring wells can detect the upward migration of wastes caused by casing or confining layer failure. Foster and Goolsby [96] describe detailed methods for constructing monitoring wells. Monitoring wells have several advantages: time-series sampling of the formation over extended periods is easy, and the passage of the waste front can be observed precisely. Disadvantages are the high repair cost and the potential for upward migration of wastes if monitoring well casings fail. A monitoring well at the Monsanto plant had to be plugged when unneutralized waste reached it because of fears that the casing would corrode (see Sect. 10.1.1). The two Florida case studies (Sects. 10.1 and 10.2) and the North Carolina case study (Sect. 10.3) illustrate the usefulness of monitoring wells.
2. *Backflushing of injected wastes.* Backflushing of injected wastes can also be a good way to observe waste/reservoir geochemical interactions. Injected wastes are allowed to backflow (if formation pressure is above the elevation of the wellhead) or are pumped to the surface. Backflowed wastes are sampled periodically (and reinjected when the test is completed); the last sample taken will have had the longest residence time in the injection zone. Keely [97] and Keely and Wolf [98] describe this technique for characterizing contamination of near-surface aquifers and suggest using logarithmic time intervals for chemical sampling. The two Florida studies (Sects. 10.1 and 10.2) all present results from backflushing experiments. The advantages of backflushing are reduced cost compared with that of monitoring wells and reduced sampling time (sampling takes place only during the test period). Disadvantages include less precise time- and distance-of-movement determinations and the need to interrupt injection and to have a large-enough area for backflushed fluid storage before reinjection.

10.1 Case Study 1: Pensacola, FL (Monsanto)

10.1.1 Injection Facility Overview

Monsanto operates one of the world's largest nylon plants on the Escambia River about 13 miles north of Pensacola, FL. The construction, operations, and effects of the injection well system at this site have been extensively documented by the US Geological Survey in cooperation with the Florida Bureau of Geology. Pressure and geochemical effects are reported by Goolsby [99]. Additional microbiological data are reported by Elkan and Horvath [100]. Major chemical processes observed at the site include neutralization, dissolution, biological denitrification, and methanogenesis. The geochemical fate of organic contaminants in the injected wastes, however, has not been reported.

The waste is an aqueous solution of organic monobasic and dibasic acids, nitric acid, sodium and ammonium salts, adiponitrile, hexamethylenediamine, alcohols, ketones, and esters [99]. The waste also contains cobalt, chromium, and copper, each in the range of 1–5 mg/L. Wastestreams with different characteristics, produced at various locations in the nylon plant, are collected in a large holding tank; this composite waste is acidic. The specific characteristics of the waste varied somewhat as a result of process changes. Until mid-1968, wastes were partially neutralized by pretreatment. After that, unneutralized wastes were injected. No reason was reported for suspending treatment. Goolsby [99] reports pH measurements ranging from a high of 5.6 in 1967 to a low of 2.4 in 1971 and Eh ranging from +300 mV in 1967 to + 700 mV in 1971. The chemical oxygen demand in 1971 was 20,000 mg/L.

Monsanto began injecting wastes into the lower limestone of the Floridan aquifer in 1963. In mid-1964, a second well was drilled into the formation about 1,000 ft southwest of the first. A shallow monitoring well was placed in the aquifer above the confining layer about 100 ft from the first injection well, and a deep monitoring well was placed in the injection zone about 1,300 ft south of both injection wells. The deep monitoring well (henceforth referred to as the near-deep monitoring well) was plugged with cement in 1969. In late 1969 and early 1970, two additional deep monitoring wells were placed in the injection formation, 1.5 miles south-southeast (downgradient) and 1.9 miles north-northwest (upgradient) of the site. From 1963 to 1977, about 13.3 billion gal of waste were injected. During the same period, injection pressures ranged from 125 to 235 psi.

Ten months after injection of neutralized wastes began, chemical analyses indicated that dilute wastes had migrated 1,300 ft to the nearest deep monitoring well. Injection of unneutralized wastes began in April 1968. Approximately 8 months later, unneutralized wastes reached the near-deep monitoring well, indicating that the neutralization capacity of the injection zone between the injection wells and the monitoring well had been exceeded. At this point, the monitoring well was plugged with cement from bottom to top because operators were concerned that the acidic wastes could corrode the steel casing and migrate upward [99]. The rapid

movement of the waste through the limestone indicated that most of it migrated
through a more permeable section, which was about 65 ft thick. By mid-1973,
10 years after injection began, a very dilute waste front arrived at the south
monitoring well, 1.5 miles away. As of early 1977, there was no evidence that
wastes had reached the upgradient monitoring well. The shallow monitoring well
remained unaffected during the same period.

Increases in permeability caused by limestone dissolution approximately doubled the injection index (the amount of waste that can be injected at a specified
pressure). As of 1974, the effects of the pressure created by the injection were
calculated to extend more than 40 miles radially from the injection site. An updip
movement of the freshwater/saltwater interface in the injection zone aquifer, which
lies less than 20 miles from the injection wells, was also observed.

10.1.2 Injection/Confining Zone Lithology and Chemistry

The lower limestone of the Floridan aquifer is used as the injection zone
(at 1,400–1,700 ft), and the Buckatunna clay member of the Byram Formation
(about 220 ft thick) serves as the confining layer. Figure 3.14 shows the local
stratigraphy and the monitoring well installations. The formation water in the
injection zone is a highly saline (11,900–13,700 mg/L total dissolved solids, TDS)

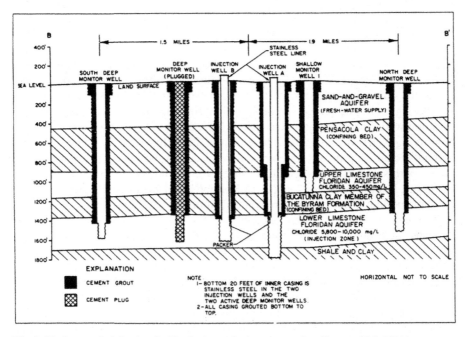

Fig. 3.14 Pensacola injection facility hydrogeologic cross section (Source: U.S. EPA)

sodium chloride solution. The Eh of samples collected from two monitoring wells located in the injection formation ranged from +23 to −32 mV, indicating reducing conditions in the injection zone that would favor anaerobic biodegradation.

The injection zone contains about 7,900 mg/L chloride, but less than 20 miles northeast of the injection site, chloride concentrations are less than 250 mg/L. Under natural conditions, water in the injection zone moves slowly south-southwestward toward the Gulf of Mexico, where it is assumed to discharge about 100 miles offshore. The preinjection hydraulic gradient was about 1.3 ft/mile.

10.1.3   Chemical Processes Observed

As a result of dissolution of the limestone by the partly neutralized acid wastes, calcium concentrations more than doubled in the near-deep monitoring well 10 months after injection started in 1963 [99]. In early 1966, however, they dropped to background levels (about 200 mg/L), possibly in response to biochemical decomposition of the waste. In September 1968, after about 300 MG of the acidic, unneutralized waste had been injected; the calcium concentration began to increase again. An abrupt increase in calcium to 2,700 mg/L, accompanied by a decrease in pH to 4.8 in January 1969, led to the decision to plug the near-deep monitoring well.

In an attempt to find out how fast the waste was reacting with limestone, a 3-h backflushing experiment, in which waste was allowed to flow back out of the injection well, yielded some unexpected results. The increase in pH of the neutralized waste could not be fully accounted for by the solution of limestone as determined from the calcium content of the backflushed liquid: the additional neutralization apparently resulted from reactions between nitric acid and alcohols and ketones in the original waste induced by increased pressure in the injection zone compared to surface conditions.

The lack of nitrates (which were present at levels of 545–1,140 mg/L in the waste) in the near-deep monitoring well, combined with the presence of nitrogen gas, indicated that degradation by denitrifying bacteria had taken place [99]. Backflushing shortly before injecting unneutralized wastes confirmed denitrification. Nitrate concentrations decreased rapidly as the backflushed waste was replaced by formation water. Similar backflushing experiments conducted after unneutralized wastes were injected, however, provided no evidence of denitrification, indicating that microbial activity was suppressed in the portion of the zone containing unneutralized wastes.

Elkan and Horvath [100] performed a microbiological analysis of samples taken from the north and south deep monitoring wells in December 1974, about 6 months after the dilute waste front had reached the south well. Both denitrifying and methanogenic bacteria were observed. The lower numbers and species diversity of organisms observed in the south monitoring well compared with those in the north well indicated suppression of microbial activity by the dilute wastes.

Chemical analyses of the north and south monitoring wells were published for the period March 1970 to March 1977. Between September 1973 and March 1977, bicarbonate concentrations increased from 282 to 636 mg/L, and dissolved organic carbon increased from 9 to 47 mg/L. These increases were accompanied by an increase in the dissolved-gas concentration and a distinctive odor like that of the injected wastes. The pH, however, remained unchanged. During the same period, dissolved methane increased from 24 to 70 mg/L, indicating increased activity by methanogenic bacteria. The observation of denitrification in the near-deep monitoring well and methanogenesis in the more distant south monitoring well fit the redox-zone biodegradation model.

Significant observations made at this site are as follows: (1) organic contaminants (as measured by dissolved organic carbon) continue to move through the aquifer even when acidity has been neutralized, and (2) even neutralized wastes can suppress microbial populations.

10.2   Case Study 2: Belle Glade, FL

10.2.1   Injection Facility Overview

The Belle Glade site, located southeast of Lake Okeechobee in south-central Florida, illustrates some of the problems that can develop with acidic-waste injection when carbonate rock is the confining layer. Contributing factors to the contamination of the aquifer above the confining zone were the dissolution of the carbonate rock and the difference in density between the injected wastes and the formation fluids. The injected waste was less dense than the groundwater because of its lower salinity and higher temperature [101].

The injected fluids include the effluent from a sugar mill and the waste from the production of furfural, an aldehyde processed from the residues of processed sugar cane. The waste is hot (about 75–93 °C) and acidic (pH 2.6–4.5) and has high concentrations of organics, nitrogen, and phosphorus. The waste is not classified as hazardous under 40 CFR 261, and the well is currently regulated by the State of Florida as a nonhazardous injection well. The organic carbon concentration exceeds 5,000 mg/L.

The well was originally cased to a depth of 1,495 ft, and the zone was left as an open hole to a depth of 1,939 ft. The depth of the zone has been increased twice. Seasonal injection (fall, winter, and spring) began in late 1966; the system was inactive during late summer. Injection rates ranged from 400 to 800 gpm, and wellhead injection pressures ranged from 30 to 60 psi. By 1973, injection had become more or less continuous. From 1966 to 1973, more than 1.1 billion gal of waste were injected [101].

At the time injection began, a shallow monitoring well was placed 75 ft south of the injection well in the upper part of the Floridan aquifer above the confining layer. A downgradient, deep monitoring well was placed in the injection zone 1,000 ft southeast of the injection well. Another shallow well, located 2 miles southeast of the injection site at the University of Florida's Everglades Experiment Station, has also been monitored for near-surface effects.

Acetate ions from the injected waste were detected in the deep monitoring well 1,000 ft. southeast of the injection well in early 1967, a matter of months after injection began. In 1971, about 27 months after injection began, evidence of waste migration was detected at a shallow monitoring well in the upper part of the Floridan aquifer. Dissolution of the carbonate confining layer by the acidic waste was the main reason for the upward migration. However, the lower density of the injected wastes compared with that of the formation waters (0.98 g/mL vs. 1,003 g/mL) served to accelerate the rate of upward migration [101]. In an attempt to prevent further upward migration, the injection well was deepened to 2,242 ft, and the inner casing was extended and cemented to 1,938 ft. When waste injection was resumed, evidence of upward migration to the shallow aquifer was observed only 15 months later. By late 1973, 7 years after injection began, the waste front was estimated to have migrated 0.6–1 mile from the injection well.

The injection well was deepened a third time, to a depth of 3,000 ft. A new, thicker confining zone of dense carbonate rock separates the current injection zone from the previous zone (see Fig. 3.15; the current injection zone is not shown). As of early 1989, the wastes were still contained in the deepest injection zone.

10.2.2 Injection/Confining Zone Lithology and Chemistry

The wastes are injected into the lower part of the carbonate Floridan aquifer, which is extremely permeable and cavernous (see Fig. 3.15). The natural direction of groundwater flow is to the southeast. The confining layer is 150 ft of dense carbonate rocks. The chloride concentration in the upper part of the injection zone is 1,650 mg/L, increasing to 15,800 mg/L near the bottom of the formation [101]. The sources used for this case study did not provide any data on the current injection zone. The native fluid was basically a sodium chloride solution but also included significant quantities of sulfate (1,500 mg/L), magnesium (625 mg/L), and calcium (477 mg/L).

10.2.3 Chemical Processes Observed

Neutralization of the injected acids by the limestone formation led to concentrations of calcium, magnesium, and silica in the waste solution that were higher than those in the unneutralized wastes. Anaerobic decomposition of the organic matter in the injected waste apparently occurred through the action of both sulfate-reducing and methanogenic bacteria. Sulfate-reducing bacteria were observed in the injected wastes that were allowed to backflow to the surface. Sulfate levels in the native groundwater declined by 45 %, and the concentration of hydrogen

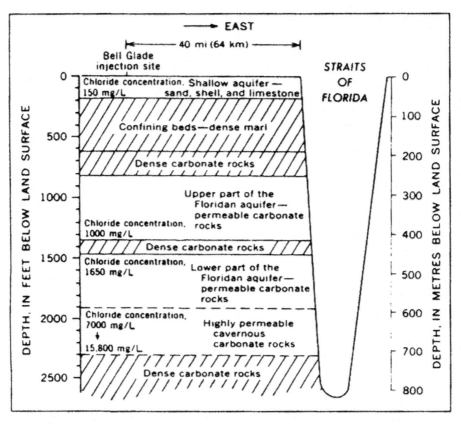

Fig. 3.15 Generalized hydrogeologic section between Belle Glade and the straits of Florida (Source: U.S. EPA)

sulfide increased by 1,600 %. Methane fermentation (reduction of CO_2 to CH_4) was also inferred from the presence of both gases in the backflow fluid, but the presence of methanogenic bacteria was not confirmed. Increased hydrogen sulfide concentrations produced by the bacteria during biodegradation and the subsequent decrease in sulfate/chloride ratio in the observation wells were taken as indicators of upward and lateral migration. Migration into the shallow monitoring well was also indicated by a decline in pH from around 7.8 to 6.5, caused by mixing with the acidic wastes.

Chemical analyses of the backflowed injected waste that had been in the aquifer for about 2.5 months (for which some dilution had occurred) indicated that chemical oxygen demand (COD) was about half that of the original waste. Samples that had been in residence for about 5 months had a COD approximately one-quarter that of the original waste (12,200 mg/L in the original waste compared with 4,166 mg/L in the samples). The percent reduction in COD resulting from bacterial action rather than dilution was not estimated.

10.3 Case Study 3: Wilmington, NC

10.3.1 Injection Facility Overview

The Hercules Chemical, Inc. (now Hercufina, Inc.) facility, 4 miles north of Wilmington, NC, attempted deep-well injection of its hazardous wastes from May 1968 to December 1972, but had to discontinue injection because of waste-reservoir incompatibility and unfavorable hydrogeologic conditions. The US Geological Survey conducted extensive geochemical studies of this site until the well was abandoned [102, 103]. Biodegradation processes were also studied [100, 104]. More geochemical fate processes affecting injected organic wastes have been documented at this site than at any other.

Hercules Chemical produced an acidic organic waste derived from the manufacture of dimethyl terephthalate, which is used in the production of synthetic fiber. The average dissolved organic carbon concentration was about 7,100 mg/L and included acetic acid, formic acid, p-toluic acid, formaldehyde, methanol, terephthalic acid, and benzoic acid. The pH ranged from 3.5 to 4.0. The waste also contained traces (less than 0.5 mg/L) of 11 other organic compounds, including dimethyl phthalate, a listed hazardous waste.

From May 1968 to December 1972, the waste was injected at a rate of about 300,000 gpd. The first injection well was completed to a depth of 850–1,025 ft (i.e., cased from the surface to 850 f. with screens placed in the most permeable sections of the injection zone to a depth of 1,025 ft). One shallow observation well was placed 50 ft east of the injection site at a depth of 690 ft. Four deep monitoring wells were also placed in the injection zone, one at 50 ft and three at 150 ft from the injection well.

The injection well became plugged after a few months of operation because of the reactive nature of the wastes and the low permeability of the injection zone. The actual plugging process was caused both by reprecipitation of the initially dissolved minerals and by plugging of pores by such gaseous products as carbon dioxide and methane. When the first well failed, a second injection well was drilled into the same injection zone about 5,000 ft north of the first, and injection began in May 1971. Nine additional monitoring wells (three shallow and six deep) were placed at distances ranging from 1,500 to 3,000 ft from the second injection well. Injection was discontinued in 1972 after the operators determined that the problems of low permeability and waste-reservoir incompatibility could not be overcome. Monitoring of the waste movement and subsurface environment continued into the mid-1970s in the three monitoring wells located 1,500-2,000 ft from the injection wells.

Within 4 months, the waste front had passed the deep observation wells located within 150 ft of the injection well. About 9 months after injection began, leakage into the aquifer above the confining layer was observed. This leakage was apparently caused by the increased pressures created by formation plugging and by the dissolution of the confining beds and the cement grout surrounding the well casing of several of the deep monitoring wells, caused by organic acids.

Eight months after injection began in the second injection well, wastes had leaked upward into the adjacent shallow monitoring well. The leak apparently was caused by the dissolution of the cement grout around the casing. In June 1972, 13 months after injection began in the second well, the waste front reached the deep monitoring well located 1,500 ft northwest of the injection well, and in August 1972, waste was detected in the well about 1,000 ft north of the injection well. Waste injection ended in December 1972. As of 1977, the wastes were treated in a surface facility [100].

10.3.2 Injection/Confining Zone Lithology and Chemistry

The injection zone consisted of multiple Upper Cretaceous strata of sand, silty sand, clay, and some thin beds of limestone (see Fig. 3.16). The clay confining layer was about 100 ft thick. The total-dissolved solid concentration in the injection zone formation water was 20,800 mg/L, with sodium chloride as the most abundant constituent.

10.3.3 Chemical Processes Observed

A number of chemical processes were observed at the site [102, 103]:

1. The waste organic acids dissolved carbonate minerals, aluminosilicate minerals, and iron/manganese oxide coatings on the primary minerals in the injection zone.
2. The waste organic acids dissolved and formed complexes with iron and manganese oxides. These dissolved complexes reprecipitated when the pH increased to 5.5 or 6.0 because of neutralization of the waste by the aquifer carbonates and oxides.
3. The aquifer mineral constituents adsorbed most waste organic compounds except formaldehyde. Adsorption of all organic acids except phthalic acid increased with a decrease in waste pH.
4. Phthalic acid was complexed with dissolved iron. The concentration of this complex decreased as the pH increased because the complex coprecipitated with the iron oxide.
5. Biochemical-waste transformation occurred at low waste concentrations, resulting in the production of methane. Additional microbial degradation of the waste resulted in the reduction of sulfates to sulfides and ferric ions to ferrous ions.

When the dilute waste front reached the monitoring well, in June 1972, microbial populations rapidly increased in this well, with methanogenesis being the major degradative process [104]. Elkan and Horvath [100] found greater numbers and species diversity of microorganisms in one of the wells, which contained dilute wastes, than in the well, which was uncontaminated. In laboratory experiments,

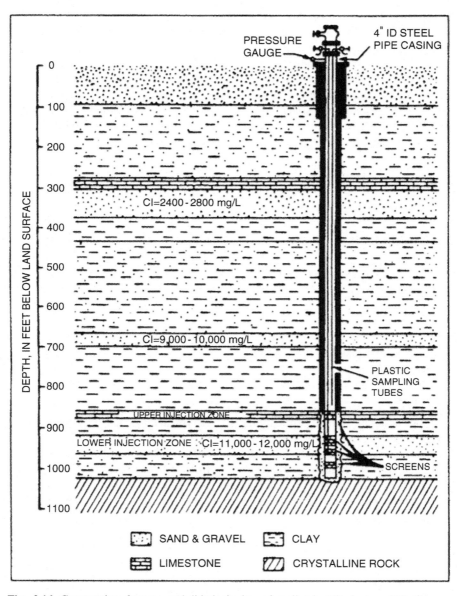

Fig. 3.16 Construction features and lithologic log of well 14, Wilmington, NC (Source: U.S. EPA)

however, DiTommaso and Elkan [104] found that bacterial growth was inhibited as the concentration of waste increased and could not decompose the waste at the rate it was being injected.

This case study illustrates the importance of dissolution/precipitation reactions in determining waste-reservoir compatibility. Adsorption was observed to immobilize most of the organic constituents in the waste except for formaldehyde. As with the Monsanto case study, biodegradation was an important process when wastes were diluted by formation waters, but the process became inhibited when undiluted waste reached a given location in the injection zone.

11 Practical Examples

11.1 Example 1

The rate of leakage through a confining bed can be determined from

$$Q = PIA \tag{3.13}$$

where

Q = the rate of leakage, ft^3/d
P = permeability, $ft^3/d/ft^2$
I = hydraulic gradient, ft/ft
A = area, ft^2

If one assumed that the permeability (P) of a typical clay confining bed averaged 0.002 gal/d/ft^2 (93×10^{-12} L/s/cm^2) or (0.00027 $ft^3/d/ft^2$), that the difference in hydraulic pressure across the confining bed is 50 ft (15.24 m), and that the confining bed is 50 ft thick (hydraulic gradient I of 1 ft/ft), then the leakage (Q) calculated through the confining bed for a circular area (A) within a 1,000-ft (305 m) radius of the injection well would be

$$P = 0.00027 \ ft^3/d/ft^2$$

$$I = 50 \ ft/50 \ ft = 1$$

$$A = \pi (1,000)^2$$

$$Q = (0.00027)(1)\left[3.142 \times (1,000)^2\right] = 848 \ ft^3/d$$

11.2 Example 2

The average velocity of fluid movement through the confining bed in Example 1 is determined from

$$v = PI/\phi \tag{3.14}$$

where

$P = $ permeability $= 0.00027$ ft^3/d/ft^2
$I = $ hydraulic gradient $= 1$ ft/ft
Assume $\phi = $ porosity $= 0.25$
Then $v = 0.00027 \times 1/0.25 = 0.001$ ft/d

At a velocity of 0.001 ft/d, the waste front would move through a 50-ft-thick confining bed in 50,000 d (137 years).

11.3 Example 3

Assume that the fluid pressure in an observation well located 500 ft (152 m) from a well injecting wastewater at a rate of 500 gpm (32 L/s) was recorded for about a week. Values of r^2/t are computed and then plotted against changes in pressure in the observation well for different values t. These are superimposed upon a second graph sheet that values of W(U) and u from Table 3.2 had previously been plotted on at the same scale. The coincident value of the match point at $u = 0.01$ and $W(u) = 1.0$ are $r^2/t = 2.8 \times 10^5$ and $s = 5.5$ ft (Fig. 3.5). Solving for T and S gives

$$T = \frac{114.60QW(u)}{s} \tag{3.15}$$

$$S = uT/\left[1.87\left(r^2/t\right)\right] \tag{3.16}$$

$$T = 114.60\,(500)\,(1.00)/(5.5) = 10,420 \text{ gal/d/ft}$$

$$S = (0.01)\,(10,420)/\left[1.87\,(2.8 \times 10^5)\right] = 0.0002$$

Having determined the T and S coefficient, long-range pressure changes with time for various distances and can be forecast by rearranging the above formulas. Thus,

$$u = 1.87\,r^2 S/(Tt) \tag{3.17}$$

$$s = 114.60\,Q\,W(u)/T \tag{3.18}$$

For example, the pressure increase at a point of 5,000 ft (1,524 m) from an injection site after injecting 800 gpm for 5 years (1,825 d) in the above-described receptor would be

u = (1 .87) (5, 000) (5, 000)(0.0002)/(10, 420 × 1, 825) = 0.0005

From Table 3.2, W(u) = 7.02

$$s = (114.60) \ (800) \ (7.02)/10,420 = 62 \ \text{ft}$$

Note: This method is not valid for u > 0.02.

11.4 Example 4

Consider the following example of how to determine pressure increases from injection which are sufficient to cause an upward gradient in the injection zone, when considered in an open hole at the base of freshwater:

1. At a distance, r = 10 ft
2. At a distance, r = 100 ft

Given that,

Q = 15,000 barrels/d
u = 1 centipoise (cp)
k = 1,127 millidarcy (md)
H = 300 ft
t = 7,300 d (20 years)
ϕ = 0.346
C = 6.5×10^{-6} psi^{-1}

The pressure change in an injection zone at distance (r) is given by s:

$$s = 162.6(Qu/kH) \log \left(kt/70.4\phi uCr^2\right) \tag{3.19}$$

The solution for pressure change at r = 10 ft and r = 100 ft in the receptor zone is given below:

$$s_{(r=10)} = 162.6 \left[(15,000) \ (1)/(1,127 \times \ 300)\right]$$

$$\times \log \left\{(1,127) \ (7,300)/\left[70.4 \ (.346) \ (1)(6.5 \times 10^{-6})(10)^2\right]\right\}$$

$$= 63 \ \text{psi}$$

$$s_{(r=100)} = 162.6\,[(15,000)\,(1)/(1,127 \times 300)]$$

$$\times \log\left\{(1,127)\,(7,300)/\left[70.4\,(.346)\,(1)(6.5 \times 10^{-6})(100)^2\right]\right\}$$

$$= 49\ \text{psi}$$

11.5 Example 5

The minimum distance (r) the waste will have traveled during injection may be described by [86]

$$r = (V/\pi H\phi)^{1/2} \tag{3.20}$$

Therefore, for

Q = 15,000 barrels/d, at 42 gal/barrel = 630,000 gpd = 84, 000 ft^3/d
H = 300 ft
ϕ = 0.346

and for t = 5 years will yield a volume V = $84,000 \times 5 \times 365 = 1.53 \times 10^8$ ft^3

$$r = \left[(1.53 \times 10^8)/(3.14 \times 300 \times 0.346)\right]^{1/2} = 685\ \text{ft}$$

and for t = 20 years will yield a volume V = $84,000 \times 20 \times 365 = 6.12 \times 10^8$ ft^3

$$r = \left[(6.12 \times 10^8)/(3.14 \times 300 \times 0.346)\right]^{1/2} = 1,370\ \text{ft}$$

11.6 Example 6

In most practical situations, the minimum radial distance of travel will be exceeded because of dispersion, density segregation, and channeling through higher-permeability zones.

An estimate of the influence of dispersion (D = 3) can be made with the equation:

$$\gamma = r + 2.3\sqrt{Dr} \tag{3.21}$$

For r = 685 ft,

$$\gamma = 685 + 2.3\sqrt{3x685} = 790.\text{ft}$$

For r = 1, 370 ft,

$$\gamma = 1,370 + 2.3\sqrt{3x1,370} = 1,517.\text{ft}$$

Appendix: US Yearly Average Cost Index for Utilities [39]

US Army Corps of Engineers civil works construction yearly average
cost index for utilities

Year	Index	Year	Index
1967	100	1991	392.35
1968	104.83	1992	399.07
1969	112.17	1993	410.63
1970	119.75	1994	424.91
1971	131.73	1995	439.72
1972	141.94	1996	445.58
1973	149.36	1997	454.99
1974	170.45	1998	459.40
1975	190.49	1999	460.16
1976	202.61	2000	468.05
1977	215.84	2001	472.18
1978	235.78	2002	486.16
1979	257.20	2003	497.40
1980	277.60	2004	563.78
1981	302.25	2005	605.47
1982	320.13	2006	645.52
1983	330.82	2007	681.88
1984	341.06	2008	741.36
1985	346.12	2009	699.70
1986	347.33	2010	720.80
1987	353.35	2011	758.79
1988	369.45	2012	769.30
1989	383.14	2013	776.44
1990	386.75	2014	790.52
		2015	803.83
		2016	819.11

References

1. U.S. EPA (2002) Technical program review: underground injection control regulations. U.S. Environmental Protection Agency, Office of Water, EPA 816-R-02-025, Dec 2002
2. U.S. EPA (2016) What is the UIC program? Summary of the history of the UIC program. U.S. Environmental Protection Agency. http://www.epa.gov/safewater/uic/whatis.html
3. U.S. EPA (2016) EPA deep wells (class I). U.S. Environmental Protection Agency. http://www.epa.gov/safewater/uic/classi.html
4. GWPC (1995) Injection well bibliography by ground water protection council, 3rd edn. Ground Water Protection Council, Oklahoma City

5. Rima DR, Chase EB, Myers BM (1971) Subsurface waste disposal by means of wells-a selected annotated bibliography. US Geological Survey, Water-supply paper 2020

6. U.S. EPA (1974) Compilation of industrial and municipal injection wells in the United States. U.S. Environmental Protection Agency, Washington, DC

7. Warner DL (1972) Survey of industrial waste injection wells vols. I, II, and III. US Department of Commerce, AD756 642, June 1972

8. Donaldson EC (1972) Injection wells and operations today. In: Underground waste management and environmental implications. The American Association of Petroleum Geologists, Tulsa

9. Warner DL, Orcutt OH (1973) Industrial wastewater-injection wells in United States-status of use and regulations. In: Underground waste management and artificial recharge, vol 2. American Association of Petroleum Geologists, Tulsa

10. Sellinger A, Aberbach SH (1973) Artificial recharge of coastal-plain aquifer in Israel. In: Underground waste management and artificial recharge, vol 2. American Association of Petroleum Geologists, Tulsa

11. Amramy A (1968) Investigation of the influence of waste disposal on groundwater qualities. Civil Eng 38(5):58

12. Donaldson EC, Bayayerd AF (1971) Reuse and subsurface injection of municipal sewage effluent-two case histories. US Bureau of Mines, Washington, Information Circular 8522

13. U.S. EPA (1973) Ocean outfalls and other methods of treated wastewater disposal in southeast Florida. U.S. Environmental Protection Agency, Final Environmental Impact Statement, 19 Mar 1973

14. The Congress of the United States of America (1972) Public law 92–500 known as the federal water pollution control act amendments of 1972, 33 USC 1151, 1972

15. Bernon RO (1970) The beneficial uses of zones of high transmissivities in the florida subsurface for water storage and waste disposal. Florida Department of Natural Resources, Tallahassee, Information Circular 70

16. U.S. EPA (2016) What is underground injection? U.S. Environmental Protection Agency. http://www.epa.gov/safewater/uic/history.html

17. Hickey JJ, Vecchioli J (1984) Subsurface injection of liquid waste with emphasis on injection practices in Florida. U.S. Geological Survey, Water supply paper 2281

18. U.S. EPA (1974) Subsurface emplacement of fluids. U.S. Environmental Protection Agency, administrators decision statement #5. Fed Reg 39(69):12922–12923, 9 Apr 1974

19. Loh HP, Loh N (2016) Hydraulic fracturing and shake gas: environmental and health impacts. In: Wang LK, Yang CT, Wang MHS (eds) Advances in water resources management. Springer, New York, pp 293–337

20. Knape B (1984) Underground injection operations in texas: a classification and assessment of underground injection activities, report 291. Texas Department of Water Resources, Austin, Dec 1984

21. U.S. EPA (2001) Class I underground injection control program: study of the risks associated with class I underground injection wells, EPA 816-R-01-007. U.S. Environmental Protection Agency, Office of Water, Mar 2001

22. CMA (1994) Class I underground injection wells: responsible management of chemical wastes. Chemical Manufacturers Association, Pamphlet, 1994

23. U.S. EPA (1989) Mid-course evaluation of the class II underground injection control program. In: Final report of the workgroup, Underground Injection Control Branch. U.S. Environmental Protection Agency, Office of Drinking Water, Aug 1989

24. Engineering Enterprises (1987) Analysis of casing and cementing compliance policy for class II wells, EPA contract No 68-03-3416. U.S. EPA, U.S. Environmental Protection Agency, Office of Drinking Water, Oct 1987

25. GAO (1989) Safeguards are not preventing contamination from injected oil and gas wastes, report. U S General Accounting Office, GAO/RCED-89-9, July 1989

26. U.S. EPA (2004) Oil and gas injection wells (class II). U.S. Environmental Protection Agency. http://www.epa.gov/safewater/uic/classii.html. Updated Oct 2004
27. U.S. EPA (2016) Mining wells (class III). U.S. Environmental Protection Agency. http://www.epa.gov/safewater/uic/classiii.html
28. U.S. EPA (2016) Shallow hazardous and radioactive injection wells (class IV). U S Environmental Protection Agency. http://www.epa.gov/safewater/uic/classiv.html. Updated Oct 2015
29. U.S. EPA (2016) EPA proposes to continue with its existing approach for managing class v injection wells. U.S. Environmental Protection Agency, EPA 816-F-01-009, Apr 2001. http://www.epa.gov/safewater/uic/classvdetermination.html
30. U.S. EPA (2016) The shallow injection wells (class V). U.S. Environmental Protection Agency. http://www.epa.gov/safewater/uic/classv.html
31. U.S. EPA (2016) State of UIC programs. U.S. Environmental Protection Agency. http://www.epa.gov/safewater/uic/primacy.html
32. Selm RP, Hulse BT (1959) Deep well disposal of industrial wastes in. Eng Bull Purdue Univ XLIV(5):566
33. Theis CV (1935) The relation between the lowering of the piezometric surface and the rate and duration of discharge of a well using ground water storage. Trans Am Geophys Union 16:519
34. U.S. EPA (1991) Underground injection control program, Annual report. U.S. Environmental Protection Agency, Office of Water, Dec 1991
35. Stewart G, Pettyjohn W (1989) Development of a methodology for regional evaluation of a confining bed integrity. EPA/600/2-89/038; U.S. EPA, U.S. Environmental Protection Agency, Office of Research and Development, July 1989
36. Davis KE, Funk RJ (1974) Experience in deep well disposal of industrial wastes. Water and Wastewater Equipment Manufacturers Association's. Industrial waste and pollution conference and exposition, Detroit, 1–4 Apr 1974
37. McKinley RM (1994) Temperature, radioactive tracer, and noise logging for injection well integrity. EPA/600/R-94/124, U.S. Environmental Protection Agency, Robert S. Kerr Research Center, July 1994
38. Temple, Barker and Sloane Inc (1979) Analysis of costs underground injection control regulations class I, class III, class IV and class V wells. EPA contract No 68-01-4778.6, May 1979
39. U.S. ACE (2016) Yearly Average Cost Index for Utilities. In: Civil works construction cost index system manual. 110-2-1304. U.S. Army Corps of Engineers, Washington, pp 44. http://www.nww.usace.army.mil/Missions/CostEngineering.aspx
40. Signor DC, Grawitz DJ, Kamn W (1970) Annotated bibliography on artificial recharge of ground water 1955–67. US Geological Survey, Water supply paper 1990
41. Winqvist G, Marelius K (1970) Thc design of artificial recharge schemes. vol. 1. England Water Research Association
42. Vocchioli J (1972) Injecting highly treated sewage into a deep sand aquifer. J New Engl Water Works Assn 136(2):87
43. Wenzel LK (1942) Methods for determining permeability of water bearing materials with special reference to discharging-well methods. US Geological Survey, Water-supply paper 887
44. Barroclouth JT (1966) Waste injection into a deep limestone in Northwestern Florida. Ground Water 3(1):13–19
45. Goolsby D (1971) Geochemical effects and movements of injected industrial waste in a limestone aquifer, AAPG-USGS. In: Symposium on underground waste management and environmental implications, Houston
46. Schicht BJ (1971) Feasibility of recharging treated sewage effluent into a deep sandstone aquifer. In: National ground water symposium, Environmental Protection Agency, 16060GR08/71

47. U.S. EPA (1972) Policy statement on water reuse. Environmental Protection Agency, 7 July 1972
48. Esmail OJ, Kinbler OK (1967) Investigation of technical feasibility of storing fresh water in saline aquifers. Water Resour Res 3(3):683
49. Sever CW (1983) A new and novel method for dewatering, compacting and disposing of sludge. U.S. Patent Office
50. Warner DL (1965) Deep-well injection in liquid waste. U.S. Department of Health, Education, and Welfare
51. U.S. EPA (1985) Report to congress on injection of hazardous waste, EPA 57019-85-003. U.S. Environmental Protection Agency, Office of Drinking Water, Mar 1985
52. GAO (1987) Hazardous waste: controls over injection well disposal operations. Report to the Chairman, Environment, Energy, and Natural Resources Subcommittee, Committee on Government Operations, House of Representatives, U.S. General Accounting Office, GAO/RCED-87-170, Aug 1987
53. Engineering Enterprises (1987) Assessment of treatment technologies available to attain acceptable levels for hazardous waste in deep injection wells. U.S. Environmental Protection Agency, Underground Injection Control Branch, UIC contract No 68-03-3416, Oct 1987
54. U.S. EPA (1991) Analysis of the effects of EPA restrictions on the deep injection of hazardous waste, EPA 570/9-91-031. U.S. Environmental Protection Agency, Office of Ground Water and Drinking Water, Oct 1991
55. Apps A, Tsang CF (1996) Deep injection disposal of hazardous and industrial waste (scientific and engineering aspects). Academic Press, San Diego
56. U.S. EPA (1990) Assessing the geochemical fate of deep-well-injected hazardous waste, a reference guide, EPA 625 6-89 025a. U.S. Environmental Protection Agency, Office of Research and Development, Washington, DC
57. Warner DL, Lehr JH (1977) An introduction to the technology of subsurface waste water injection, EPA 600/2-77-240, NTIS PB279 207
58. Langelier WF (1936) The analytical control of anti-corrosion water treatment. J Am Water Works Assoc 28:1500–1521
59. Larson TE, Buswell AM (1942) Calcium carbonation saturation index and alkalinity interpretations. J Am Water Works Assoc 34:1667–1684
60. Stiff HA, Davis LE (1952) A method for predicting the tendency of oil field waters to deposit calcium carbonate. Am Inst Mining Metall Engineers Trans Petroleum Div 195:213–216
61. Wang LK, Hung YT, Shammas NK (eds) (2005) Advanced physicochemical treatment processes. The Humana Press, Totowa
62. Wang LK, Hung YT, Shammas NK (eds) (2004) Physicochemical treatment processes. The Humana Press, Totowa
63. Ostroff AG (1965) Introduction to oil field water technology. Prentice-Hall, Englewood Cliffs
64. Watkins JW (1954) Analytical methods of testing waters to be injected into subsurface oil-productive strata. U.S. Bureau of Mines Report of Investigations 5031
65. Davis JB (1967) Petroleum microbiology. Elsevier, New York
66. Hunter JV (1971) Origin of organics from artificial contamination. In: Faust SD, Hunter JV (eds) Organic compounds in aquatic environments. Marcel Dekker, New York, pp 51–94
67. Roy WR, Mravik SC, Krapac IG, Dickerson DR, Griffin RA (1989) Geochemical interactions of hazardous wastes with geological formations in deep-well systems. Environmental Geology Notes
68. Reeder LR et al. (1977) Review and assessment of deep-well injection of hazardous wastes. EPA 600/2-77-029a-d, NTIS PB 269 001-004
69. Strycker A, Collins AG (1987) State-of-the-art report: injection of hazardous wastes into deep wells, EPA/600/8-87/01 3, NTIS PB87-1 70551
70. Warner DL, Orcutt DH (1973) Industrial wastewater-injection wells in United States-Status of use and regulation. In: Braunstein J (ed) Symposium on underground waste management and artificial recharge, Pub No 110, Int Assn of Hydrological Sciences, pp 687–697

71. Donaldson EC (1964) Subsurface disposal of industrial wastes in the United States. U.S. Bureau of Mines Information Circular 8212
72. Donaldson EC, Thomas RD, Johnston KH (1974) Subsurface waste injections in the united states: fifteen case histories. U.S. Bureau of Mines Information Circular 8636
73. U.S. EPA (1985) Report to congress on injection of hazardous wastes, EPA 570/9-85-003. U.S. Environmental Protection Agency, NTIS PB86-203056
74. U.S. EPA (1989) Injection well mechanical integrity testing, EPA 625/9-89/0 07. U.S. Environmental Protection Agency
75. Muskat M (1937) Flow of homogeneous fluids through porous media. McGraw-Hill, New York
76. Briggs LJ, McLane JW (1907) The moisture equivalent of soils. U.S. Department of Agriculture. Soils Bull 45:1–23
77. Shammas NK, Wang LK (2010) Water and wastewater engineering: water supply and wastewater removal. Wiley, Hoboken, NJ, p 824
78. U.S. EPA (1994) Underground injection wells and your drinking water. U.S. Environmental Protection Agency, Office of Water, EPA 813-F-94-001
79. U.S. EPA (1979) Guide to the underground injection control program. U.S. Environmental Protection Agency, Washington
80. Coneway PR, Coneway S (1978) The T. V. camera see what's wrong. The Johnson Drillrs J Sept–Oct 1978
81. Haire JN, Haflin JD (1977) Vertilog-A down-hole casing inspection service. 47th annual, California. regional meeting of the Society of Petroleum Engineers of AIME
82. Hantush MS, Jacob CE (1955) Solution modified for step-drawdown tests in leaky confined aquifers. Trans Am Geophys Union 36:95
83. Warner DL, Koederitz LF, Simon AD, You GM (1979) Radius of pressure influence of injection wells. U.S. Environmental Protection Agency, EPA 600/2-79-170
84. U.S. EPA (1980) Part 146 underground injection control program criteria and standards, environmental protection agency. Fed Reg 45(123):42500
85. Browning LA, Sever CW (1978) Simplified graphic analysis for calculation of the zone of endangering influence of an injection well. EPA Open Files, Dallas
86. Browning LA (1980) Hydraulic analysis of a proposed waste injection well, Adams county, Mississippi. U.S. Environmental Protection Agency open files, Atlanta, 11 Mar 1980
87. Bear J, Jacobs M (1964) The movement of injected water bodies in confined aquifers. Underground water storage study report No 13. Techpecon, Haifa
88. El-Nimir A, Salih A, Shammas NK, Khadam M, (1990) Modeling drained solid matter and salt transport through deep aquifers. In: International workshop on the application of mathematical models for assessment of changes in water quality, Tunis, pp 7–12
89. Shammas NK, Yuan P, Yang J, Hung YT (2004) Chemical oxidation. In: Wang LK, Hung YT, Shammas NK (eds) Physicochemical treatment processes. The Humana Press, Totowa, pp 229–270
90. Wang LK, Vaccari D, Li Y, Shammas NK (2004) Chemical precipitation. In: Wang LK, Hung YT, Shammas NK (eds) Physicochemical treatment processes. The Humana Press, Totowa, pp 141–198
91. Shammas NK (1986) Interactions of temperature, pH and biomass on the nitrification process. J Water Pollut Control Fed 58(1):52–59
92. Shammas NK (1982) An allosteric kinetic model for the nitrification process. In: Proceeding of the 10th annual conference of water supply improvement association, Honolulu, pp 1–30
93. Wang LK, Pereira NC, Hung YT (eds) (2005) Biological treatment processes. The Humana Press, Totowa, NJ, p 818
94. U.S. EPA (1987) Class V injection wells-current inventory, effects on ground water and technical recommendations. Report to congress, EPA 570/9-87-006, U.S. Environmental Protection Agency, Office of Water, Washington, DC

95. U.S. EPA (2016) Class V underground injection control study, office of water. EPA/816-R-99-014, 23 volumes and 5 Appendices, Sept 1999. http://www.epa.gov/safewater/uic/cl5study.html
96. Foster JB and Goolsby DA (1972) Construction of waste-injection monitoring wells near pensacola, Florida. Florida Bureau of Geology, Information Circular 74
97. Keely JF (1982) Chemical time-series sampling. Ground Water Monit Rev 2(4):29–38, Fall
98. Keely JF, Wolf F (1983) Field applications of chemical time-series sampling. Ground Water Monit Rev 3(4):26–33, Fall
99. Goolsby DA (1972) Geochemical effects and movement of injected industrial waste in a limestone aquifer. In: Cook TD (ed) Symposium on underground waste management and environmental implications. American Association of Petroleum Geology Member 18, pp 355–368
100. Elkan G and Horvath E (1977) The role of microorganisms in the decomposition of deep well injected liquid industrial waste. NSF/RA-770102, NTIS PB 268 646
101. Kaufman MI, Goolsby DA, Faulkner GL (1973) Injection of acidic industrial waste into a saline carbonate aquifer: geochemical aspects. In: Braunstein J (ed) Symposium on underground waste management and artificial recharge, Pub No 110. International Association of Hydrological Sciences, pp 526–555
102. Leenheer JA, Malcolm RL White WR (1976) Physical chemical and biological aspects of subsurface organic waste injection near wilmington, North Carolina. U.S. Geological Survey Professional Paper 987, 1976
103. Leenheer JA, Malcolm RL, White WR (1976) Investigation of the reactivity and fate of certain organic compounds of an industrial waste after deep-well injection. Environ Sci Tech 10(5):445–451
104. DiTommaso A, Elkan GH (1973) Role of bacteria in decomposition of injected liquid waste at Wilmington, North Carolina. In: Braunstein J (ed) Symposium on underground waste management and artificial recharge, Pub # 110. International Association of Hydrological Sciences, pp 585–599

Chapter 4
Treatment and Management of Industrial Dye Wastewater for Water Resources Protection

Erick Butler, Yung-Tse Hung, Mohammed Al Ahmad, and Yen-Pei Fu

Contents

1 Introduction 189
2 Description of Dyes 190
  2.1 Definition of Dye 190
  2.2 Classification of Dyes 191
  2.3 Processes Used in Dye Production 193
  2.4 Wastewater Characteristics 196
  2.5 History of Dyes 197
3 Legislation Concerning Dye Wastewater 200
  3.1 Clean Air Act (1970) 200
  3.2 Clean Water Act (1972) 200
  3.3 Resource Conservation and Recovery Act (RCRA) (1976) 201
  3.4 Emergency Planning and Community Right-to-Know Act (1986) 201
  3.5 Pollution Prevention Act (1990) 202
4 Dye Wastewater Treatment Methods 203
  4.1 Biological Treatment 203
  4.2 Advanced Oxidation Treatment 210
  4.3 Activated Carbon 213

E. Butler (✉)
School of Engineering, Computer Science and Mathematics, West Texas A&M University, Canyon, TX 79016, USA
e-mail: erick.ben.butler@gmail.com

Y.-T. Hung
Department of Civil and Environmental Engineering, Cleveland State University, Cleveland, OH 44115, USA
e-mail: yungtsehung@yahoo.com; yungtsehung@gmail.com

M.A. Ahmad
Department of Civil and Engineering, Cleveland State University, 1516 Pond Glen Way Cary, NC 27519, USA
e-mail: arabee2000@yahoo.com

Y.-P. Fu
Department of Materials Science and Engineering, National Dong Hwa University, Shou-Feng, Hualien 97401, Taiwan
e-mail: erick.ben.butler@gmail.com

© Springer International Publishing Switzerland 2016
L.K. Wang, M.-H.S. Wang, Y.-T. Hung and N.K. Shammas (eds.),
Natural Resources and Control Processes, Handbook of Environmental
Engineering, Volume 17, DOI 10.1007/978-3-319-26800-2_4

    4.4   Membrane Filtration 215
    4.5   Electrocoagulation 217
    4.6   Coagulation-Flocculation 218
    4.7   Ozone 219
  5   Conclusion 222
References 224

Abstract One can describe industries involving dye treatment as the essentials (food, beverage, and textile) simply since they involve application to entities that are used for necessity. Dye wastewater is generated from the intense application within its industry. However, it is more common to find research on the treatment and application of dyes within the textile industry, as it has been highlighted and targeted by federal regulations. Some of the more harmful components from dye wastewater include the presence of biochemical oxygen demand (BOD), chemical oxygen demand (COD), and color, to name a few. Direct discharge from the plant can cause potential harm toward a water body, impacting the lives of human health and the environment. The purpose of this chapter is to provide a survey of various treatment methods and review the literature that has been applied for the purpose of treating wastewater produced by the textile dye industry. In addition, the text will discuss some of the important preceding topics prior to treatment applications such as legislation, and definition and classification of dyes. The aim of this chapter is to be both comprehensive and concise describing some of the major components that are heavily integrated within the dyeing industry.

Keywords Textile industry • Dye wastewater treatment • Toxicity • Color removal

Nomenclature

$1/n$	Describes adsorption favorability where <1 is unfavorable, greater than 1 is favorable, and 1 is linear. (unitless)
*OH	Hydroxide radical
Am	Membrane area (ft^2)
b	Constant of net enthalpy of adsorption (L/mg)
C_e	Solute concentration within the solution (mg/L)
CO	Carbon monoxide
Fe(II) (Fe^{2+})	Ferrous (iron II) ion
Fe(III) (Fe^{3+})	Ferric (iron III) ion
Fe(III) OH_2	Ferrihydroxalate
Fe(lll)(RCO_2)^2	Ferric carboxylate complexes
H_2	Diatomic hydrogen molecule
H_2O_2	Hydrogen peroxide
h_v	Light energy
J	Flux (gfd)
K_f	K_f is a sorption capacity constant (L/g)

M	Metal
M^{n+}	Metal with charged represented by n
n^{e-}	Number of electrons (represented by n)
O	Oxygen radical
O_2	Diatomic oxygen
O_3	Ozone
OH^-	Hydroxide ion
Q	Mass of solute adsorbed per gram of activated carbon within the monolayer (mg solute/g activated carbon)
q_e	Ratio of solute that has been absorbed per gram of activated carbon (mg solute/g activated carbon)
Q_f	Feed flow (gpd)
Q_p	Filtrate flow (gpd)
R	Recovery of the membrane unit (percent)
R *	Radical

1 Introduction

One component that everyone has everyday contact with regardless of geographical and cultural differences is dye. It can be estimated that 7×10^5 million tons of dye are available worldwide [1]. Dyes are present in many essentials—for example, textile dyes are used in fabrics for clothes and bags, while coloring dyes are used for food and beverages. However, dye wastewater production from manufacturing and textile industries is problematic from an environmental aspect.

The biggest problem prevalent in dye production is high color and organic content [2]. This can be attributed to the chemical characteristics of the dyes. Many dyes consist of several aromatics such sulfo, nitro, amidocyanogen, and chloro compounds where discharge of these chemical constituents can result in major environmental problems [3]. Dye production is because dye manufacturing processes also produce various types of wastes, where the primary pollutants include volatile organic compounds, nitrogen oxides, HCl, and SO_x. Additional properties of dye wastewater from the industrial industry involve high biochemical oxygen demand (BOD), chemical oxygen demand (COD), and suspended solids (SS). For example, vast dyes contain 25 kg/kg (or 25 mg/L where 1 kg/kg $= 1 \times 10^6$ mg/L using the density of water as 1 kg/L), 85 mg/L COD, while other dyes typically have a BOD of 6 mg/L, COD of 25 mg/L, and suspended solids of 6 mg/L [4]. Without proper regulation and treatment, environmental and human health will experience various potential dangers.

One of the reasons for the potential dangers involves the interactions between discharged dyes and the surrounding world. Expounding from the previous statistic, it can be stated that out of the 7×10^5 million tons available, only 2 % of dyes are discharged as aqueous and 10 % by means of coloration. The remaining 88 % of discharged dyes retain their color due to sunlight, soils, bacteria, sweat, and degradation of microbes within the wastewater. In addition, many dyes have long half-lives. For example, Reactive Blue-19 has a half-life of 46 years. Because some dyes have very long half-lives, their impacts can further linger. Dyes can also alter photosynthesis in plants [1], specifically in plants within water bodies because of the ceasing of light penetrating within receiving waters [5]. Finally, other to the environment involve the health and vitality of stream, including undesired pollution, eutrophication, and toxicity [2].

As previously stated, it is very evident that as dye production has become very significant within human society, the characteristics of dye wastewater are the resultant of many issues for public health and the environment. Some of the issues include the toxicity, color, and biodegradability issues, halogenated organic compounds, heavy metals and surfactants, and high BOD [6].

While there is a plethora of literature that has attempted to individually describe the chemistry, classifications, legislation, and various tangible methods of dye treatment, the purpose of this text is to overlay all of these significant pieces for the purpose of truly highlighting the background for one to conclude that there is a need for proper treatment of dye wastewater. However, this text does not serve as an all-extensive guide to describe all of the contemporary methods of treating dye wastewater. Nevertheless, this text hopes to provide a starting point for those looking for background and resources that will lead them to discover more within this field of study. While there are many industries that use dyes, the majority of focus will be primarily on what is being employed within the textile dye industry.

2 Description of Dyes

2.1 Definition of Dye

A dye can be defined as "a colored substance which when applied to the fabrics imparts a permanent color and the color is not removed by washing with water, soap, or an exposure to sunlight" [7]. Dyes are within a subcategory known as colorants, entities that emit light within the visible range (400–700 nm). Most colorants can be described as being inorganic and organic chemical structures and also divided into two types—dyes or pigments. Pigments have been defined as a group of dyes that are combined with several compounds or substrates, while dyes are placed onto a compound or substrate [8].

In industry, it is important to recognize that terminology is quite important. Two terms that are necessary to define are dye baths and affinity. The places for application of dyes on a fabric are known as **dye baths**. Dye baths contribute to a significant portion of waste production. It has been estimated that dye bath solution contribution to half of the waste production from the industry. The next section discusses the processes in which dye baths are used for implementation.

Affinity can be best described as the alignment of a particular dye on a fabric. Acid dyes for example have an affinity to (applied to) nylon and wool, typically because of the formation of acid dyes and also the type of fabric wool and nylon. Acid dyes are more susceptible to be applied to wool and nylon as compared to other components. [9, 10].

2.2 Classification of Dyes

The Environmental Protection Agency Office of Toxic Substances has classified 14 different types based on chemical composition—acid, direct (substantive), azoic, disperse, sulfur, fiber reactive, basic, oxidation, mordant (chrome), developed, vat, pigment, optical fluorescent, and solvent dyes [11]. This text will describe six of the 14 major dye types—acid, direct, azo, disperse, sulfur, and fiber reactive.

Acid dyes derive from sodium salts from sulfonic or phenolic groups, where the negative ions determine the dye's color. Acid dyes are dipped into a hot constituent of the dye mixed a salt or an acid. Examples of an acid include martius yellow, orange II, and naphthol yellow [7]. Acid dyes are classified with high molecular weights resulting from the reaction between the dye and the fiber development of an insoluble color molecule on the fiber with processes at temperatures greater than 39° C. Acid dyes are further subdivided into three categories—azo, anthraquinone, and triarylmethane [11].

Direct dyes comprised of azo dyes, are a series of man made compounds adsorbed on cellulose [6]. Direct dyes are adhered to a fiber and placed in a solution that combines the fabric with the dye using an aid of ionic salts or other electrolytes. Direct dyes are known to have a high solubility in cold water [11].

Direct dyes are only applied to cellulose materials such as cotton, jute, viscose, or paper by the use of adsorption. Adsorption is accomplished by the use of electrolytes in a process known as substantivity. Substantivity happens when the application of dye molecules increase in size within open cavity spaces in the fibers too substantial to be removed by the use of water or any agents and are permanently affixed to the fabric [12].

Azo dyes consist of benzene and naphthalene groups such as chloro $[-Cl]$, methyl $[-CH_3]$, nitro $[-NO_2]$, amino $[-NH_2]$, hydroxyl $[-OH]$, and carboxyl $[-COOH]$ [13]. Typical azo dyes include cotton, rayon, silk, wool, nylon, and leather [14]. Azoic dyes form insoluble color molecules on a fiber by means of combining two soluble solutions at temperatures between 16° C and 27° C [11].

Disperse dyes are the fourth of fourteen dye classifications. These low solubility dyes affix to fabrics such as nylon and polyester in a dye bath under the process of colloidal absorption. Collodial adsorption uses high temperatures to induce a phase change from solid to gas for the purpose of being embedded into the fabric. Once this occurs, it is solidified through condensation as a colloid where it is permanently made a part of the dye [11].

Sulfur dyes are a series of dyes applied to cotton and rayon fabrics by transferring a dye from reduced water soluble form to an oxidized insoluble form, while fiber reactive dyes are a series of dyes that form covalent bonds with the fiber molecules [11]. Tables 4.1a and 4.1b have summarized the types of dyes available along with the traditional applications and fixation percentages.

The Environmental Protection Agency classifies textile dye waste in four categories —dispersive, hard to treat, high volume, and hazardous and toxic. This section will discuss disperse, hard-to-treat wastes, and high-volume. Since dyes

Table 4.1a Type of dye extraction [9, 10]

Dye class	Description	Method	Fibers type applied typically to	Typical fixation (%)	Typical pollutants associated with various dyes
Acid	Water-soluble anionic compounds	Exhaust/beck/ continuous (carpet)	Wool Nylon	80–93	Color, organics, acids, unfixed dyes
Basic	Water-soluble, applied in weakly acidic dyebaths; very bright dyes	Exhaust/beck	Acrylic Some polyesters	97–98	N/A
Direct	Water-soluble, anionic compounds; can be applied directly to cellulosics without mordants (or metals like chromium and copper)	Exhaust/beck/ continuous	Cotton Rayon Other cellulosics	70–95	Color, salts, unfixed dye; cationic fixing agents; surfactant; defoamer, leveling, and retarding agents; finish; diluents

Table 4.1b Type of dye extraction [9, 10]

Dye class	Description	Method	Fibers type applied typically to		Typical fixation (%)
Disperse	Not water soluble	High-temperature exhaust continuous	Polyester Other synthesis	Acetate	80–92
Reactive	Water-soluble anionic compounds; largest-dye class	Exhaust/beck cold pad batch/ continuous	Cotton Wool	Other cellulosics	60–90
Sulfur	Organic compounds containing sulfur or sodium sulfide	Continuous	Cotton	Other cellulosics	60–70
Vat	Oldest dyes; more chemically complex; water insoluble	Exhaust/package/continuous	Cotton Wool	Other cellulosics	80–95

are integrated into the fabric by various processes such as finishing, dye, printing, preparation, machine cleaning, pastes, and solutions collected that were not used within the processing are known as **dispersive wastes**. Dispersive wastes from a textile mill plant can be controlled by either capturing the waste streams either through various processes or the physical removal of the machines used in the plant [11]. Dispersive wastes can also be controlled by developing pollution prevention methods. Later within the chapter, pollution prevention methods will be discussed.

Hard-to-treat wastes are a series of wastes that are very difficult to remove from textile mill plant processes where at times these wastes become a hindrance toward plant productivity. These waste derive from dye and printing processes. The majority of hard-to-treat wastes are classified as being non-biodegradable or inorganic because biological processes are incapable of reducing their concentrations. When these waste streams enter through a wastewater treatment plant, they will pass through activated sludge and other wastewater treatment processing units. Some of the major wastes such as color, metals, surfactants, and toxic compounds are considered examples of hard-to-treat wastes. High-volume wastes such as wastewater, salts, and knitting oils is the category of waste that is generated in the highest quantity [11].

2.3   Processes Used in Dye Production

Regardless of industry, there are many commonalities found in the application of dye to a particular substance. For example, if one looks at the textile industry, one will find that dyeing processes can happen at various stages of textile production, including the congregation of yarn, fabric, or garments. During dye preparation, dyes are made ready to receive the application of dyes to the fabric. Fabric preparation begins with wet processing. Wet processing is when textiles receive an application of water. This strengthens the textile, allows it to last longer thereby making it more durable. At this stage in the process, fabric is usually known as being unfinished or "griege goods." After wet processes, dyeing occurs as a batch or continuous process [9, 10]. Figure 4.1 provides a process flow diagram of the textile dye industry.

In the batch dye process, 100–1000 kg of the fabric is combined with the particular dye within a machine. There are three phases of dyeing—application, fixation, and washing. **Dye fixation** is quantitative amount of dye that is applied to the fabric during the interaction between the fabric and the dye by the use of either additional chemicals or heat within the dye bath that actually is transmitted directly into the fabric in a given period of time, either in a few moments or several hours. Having had the application of dye affixed to the fabric, additional washing occurs for the purpose of removing any substances that have not fully been embedded within the fabric [9, 10].

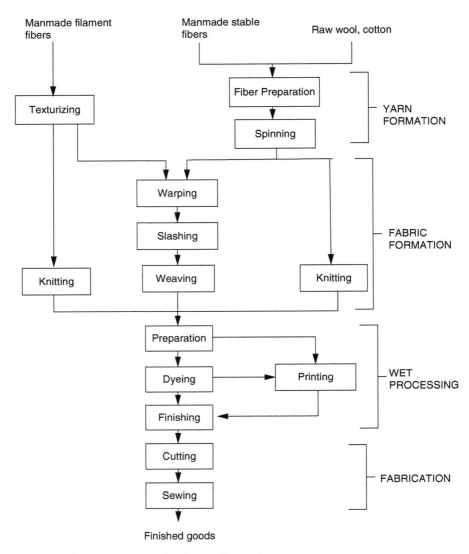

Fig. 4.1 Textile dye industry process flow diagram [9, 10]

Continuous dye application contrasts with batch dye application because of its expediency. Instead of having weighted a particular amount of fabric, the continuous process is based on length, where application is completed at rates between 50 and 250 m per min. While dyeing of fabric by the continuous dye application is similar in nature to the continuous dyeing process, 2000 m of fabric can be dyed in as little as 8 to 40 min. This method is more preferred within the textile industry

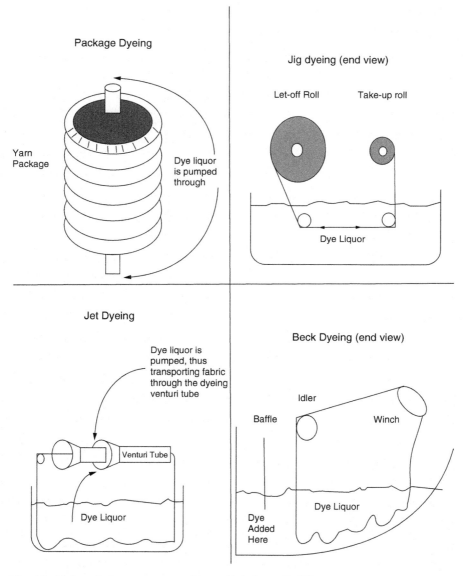

Fig. 4.2 Various dyeing processes used in textile industry [9, 10]

(continuous dyeing process occurs 60 % of the time) as compared with the batch method as it is more economical [9, 10]. Figure 4.2 provides examples of equipment that are used for textile dyeing, while Tables 4.2a and 4.2b provide the specific processes that have been employed with treatment of dyes within the textile industry.

Table 4.2a Types of coloring dye processes used in textile dyes [9, 10]

Machine	Material	Capacity		
Type	Processed	Width	Weight (kg)	Description, advantages, disadvantages
Stock	Fiber	–	500	Fiber is dyed insider perforated tubes. Same machine can be used for package and beam dyeing. Large quantities of dyed fiber can be blended for color consistency
Package	Yarn	–	550	Yarn is stationary; bath is pumped through. Same machine can be used for beam or stock dyeing
Skein	Yarn	–	100	Yarn is dyed in hanks. Used for bulky acrylic and wool yarns
Dope	Polymer melt-dyed prior to filament formation	Rope	900	Pigments are added to polymer before extrusion into fiber
Beck	Fabric	Rope	900	Very versatile—can be used almost universally. Good for repair work. Causes substantial mechanical working of goods. Can cause cracks in delicate, lightweight goods (e.g., nylon and acetone)
Jet	Fabric	Rope	500	Capable of high pressure and temperature. Fabric is handled gently. Fabric and bath are both in motion during dyeing
Beam	Fabric	Up to 5 m	1000	Fabric is handled flat, thus reducing creases and cracks in delicate goods. Optimum for lightweight, wide, and delicate goods. Fabric is stationary; bath is pumped through. Same machine can be used for stock or packaged dyeing

2.4 Wastewater Characteristics

It has been estimated that in the textile industry 90 % of wastewater or 90,000 tons goes untreated, while only 10 % is recycled. In addition, 10–15 % of a worldwide 700,000 tone of dye production is discarded as waste. The source of dyeing waste originates from dyeing processes, alkaline preparation, and also the constituents from dyeing such as salts included within some of the chemical mechanisms to develop dyes for the various processes in the industry [9].

Dyeing wastewater consists of metals, salts consisting of magnesium chloride and potassium chloride (2000 to 3000 mg/L), surfactant, biochemical oxygen demand (BOD), chemical oxygen demand (COD), total suspended solids, and toxics to name a few [10]. The highest estimated BOD concentrations are from wool scouring, complex processing, and carpeting finishing (2270, 420, and 440 mg/L, respectively). On the other hand, COD was observed as being 2 and 13 times more prevalent than the BOD concentrations—COD concentration was at

Table 4.2b Types of coloring dye processes used in textile dyes [9, 10]

Machine	Material	Capacity		
Type	Processed	Width (m)	Weight (kg)	Description, advantages, disadvantages
Jig	Fabric	2	250	Fabric is handled flat, thus reducing creases and cracks. Does not run disperse dyes very well. Too much tension for weft knits
Paddle	Fabric or product	–	100	For products such as hosiery, rugs, etc.
Garment (rotary)	Garments	–	500	Garments are dyed before cutting and sewing
Chain (tow)	Yarn		Continuous	Used to dye yarns continuously
Continuous	Fabric	Up to 3	Continuous	Best economics for long runs. Several types of fixing methods include steam, chemical reactors, Thermofix, and cold batch methods. Not effective for general repair work. The only type of dye machine that can run pigments. Too much tension for knits

7030 mg/L as compared to 2270 mg/L for wool scouring wastewater production. Research also suggests that responsible for high total suspended solid (TSS) concentration (3310 mg/L) and oil and grease (O&G) [10]. Table 4.3 provides information on the characteristics of effluent wastewater.

Wastewater toxicity varies based on constituent presence. From a test of 75 textile mills, it was observed that 38 or over 50 % of the 75 textile mills had no toxicity, while approximately 9 % had toxic components [9]. Potential sources of toxicity include salts from the dyeing status, surfactants, metals from the dyes, and organics. It was also found that 63 % of 46 tested commercial dyes have a toxicity range using the lethal concentration (LC_{50}) value, or the concentration required to kill 50 % of a given population, measured greater than 180 mg/L or having little toxicity, while only 2.2 % were considered toxic [9]. Table 4.4 summarizes the results of this test.

2.5 History of Dyes

The use of dyes has been prevalent since the beginning of time. Archeologists have found linens in Egypt that can be traced back as early as 5000 BC. Within the Bible, Moses describes to the newly freed Israelites the use of fibers and dyes for the purpose of preparing priestly garments for Aaron and the future line of priests. It also has been seen that dyes were used as extracts from various natural sources. For example, dye was created from Brazilwood in Brazil, carmine red was extracted

Table 4.3 Wastewater characteristics of effluent from textile dyeing by each category of treatment [10]

Subcategory	BOD (mg/L)	COD (mg/L)	COD/BOD	TSS (mg/L)	O&G (mg/L)	Phenol (ug/L)	Chromium (ug/L)	Sulfide (ug/L)	Color APHA Units
1. Wool scouring	2270	7030	3.1	3310	580	ID	ID	ID	ID
2. Wool finishing	170	590	3.5	60	ID	ID	ID	ID	ID
3. Low water use processing	293	692	2.4	185	ID	ID	ID	ID	ID
4. Woven fabric finishing	–	–	–	–	–	–	–	–	–
a. Simple processing	270	900	3.3	60	70	50	40	70	800
b. Complex processing	350	1060	3	110	45	55	110	100	ID
c. Complex processing plus desizing	420	1240	3	155	70	145	1100	ID	ID
5. Knit fabric finishing	–	–	–	–	–	–	–	–	–
a. Simple processing	210	870	4.1	55	85	110	80	55	400
b. Complex processing	270	790	2.9	60	50	100	80	150	750
c. Hosiery processing	320	1370	4.5	80	100	60	80	560	450
6. Carpet finishing	440	1190	2.7	65	20	130	30	180	490
7. Stock and yarn finishing	180	680	3.77	40	20	170	100	200	570
8. Nonwoven finishing	180	2360	13.1	80	ID	ID	ID	ID	ID
9. Felted fabric finishing	200	550	2.75	120	30	580	ID	ID	ID

Table 4.4 Toxicity of
46 commercial dyes [10]

Toxicity range (LC50 mg/L)	Number of dyes in range	Percent of total
>180	29	63
100–180	3	6.5
10–100	4	8.7
1–10	7	15.2
0.1–1	2	4.3
<0.1	1	2.2
Totals	46	100

from female cochineal Mexican or Peruvian cactus, and indigo for blue dyes was from Indian vegetable leaves [15]. However, author Dutfield [16] states that there are four major stages involved with the infant history of synthetic dyes. The first stage for the origin of dyes has been commonly traced to 1856 when a British student named William Perkin that accidentally discovers aniline purple, a coal tar dye having attempted to synthesize quinine. Following the discovery of aniline purple, Francoise Verguin discovers aniline red in 1859. The dye was patented in both France and Britain. Coal tar dyes were replaced by azo dyes in 1856, when two German chemists Heinrich Caro and Carl Maritus discover both Manchester yellow and brown. Three years later, the natural dye alizarin is synthesized by German scientists Carl Graebe and Carl Libermann. The fourth major stage of synthetic dye was synthetic indigo, an important stage in dyestuff, where it begins the modern development of the majority dyes present in the current dye industry [15, 16].

However, dye presence within the United States did not happen until following World War I when the German submarine *Deutchland* enters American ports trading military and money for dyes. However, the major turning point in the presence of dyes in the United States is the transferring of dye patents to the Allies via the Alien Property Custodian. Throughout the course of the twentieth century, it is documented that many companies, including DuPont, worked very meticulously in attempting to begin manufacturing in the United States. In fact, DuPont spends $43 million prior to making a profit on dye synthesis. By the 1960s, 50–60 % of all dye manufacturing production in the United States is by four major companies (Allied Chemical, American Cyanamid GAF, and Dupont), while during recent history, it has been documented that there are no current standing major dye companies present in America [17].

In summary, whenever color is present in wastewater, it can cause problems, specifically high oxygen demand and suspended solids [18]. There are many different dye wastewater types including textile wastewater because of high water presence and large quantities. One of the major components found in dye wastewater includes organic contaminants such as aromatics [19].

3 Legislation Concerning Dye Wastewater

3.1 Clean Air Act (1970)

The Environmental Protection Agency has directed several federal mandates specifically designed for this particular type of pollution. While the Clean Air Act originally passed in 1970 required limitations on exposure to ambient air by industries from primary and secondary sources, one of the core pieces of the Clean Air Act is the inclusion of the National Ambient Air Quality Standard (NAAQs) to protect human health and the environment [20].

In 1990, Title III Section 112, known as the Clean Air Act Amendments, is designated to control hazardous air pollutants. The National Emission of Hazardous Air Pollutants (NESHAPs) lists a series of amendments to include a list of 189 hazardous air pollutants (HAPs) needing to be reduced within each industry, requiring the monitoring, assessing, reporting, and potential planning with risks [21]. The Amendments play a significant role in the dye and textile industry, as many of the raw materials that are present within dyes can be found on the list of hazardous air pollutants [22]. In addition, the EPA creates legislation for the textile processing industry. Known as Maximum Achievable Control Technology Standards (MACTS), the EPA requires a control of hazardous air pollutants from textile processing plants having produced either 10 or 25 tons/year [23].

3.2 Clean Water Act (1972)

The Federal Water Pollution Control Act of 1972 (also known as the Clean Water Act) is passed to regulate the amount of pollution that was discharged into surface waters. One key component of the Clean Water Act is the administration of the National Pollution Discharge Elimination System (NPDES), which put into place for the purpose of reducing effluent that was produced from point sources by means of using available technology to remove discharge from sources other than publicly owned treatment works (POTWs) or wastewater treatment plants (WWTPs). To highlight, Section 301b designates proper control of waste production by the use of conventional treatment methods, such as secondary treatment (biological treatment) and the addition of any other advanced treatment from eight textile industries—wool scouring, finishing, process, woven and knit fabric finishing, and carpet mills, to name a few—while Sections 304b and 306b require a presentation of discharge limits and performance matrices of these discharge limits, respectively [24].

3.3 Resource Conservation and Recovery Act (RCRA) (1976)

The handling of solid and hazardous waste from the textile industry relies on the Resource Conservation and Recovery Act (RCRA) of 1976 which coins the phrase of "cradle to grave," deems responsibility over the life cycle in the production of solid and hazardous waste through its generation and production to its handling toward a final location following use. This is followed by 1984s Hazardous and Solid Waste Amendments (HSWA) which includes legislation concerning underground storage tanks. Several requirements within these two major pieces of legislation include the following: 40 CFR Part 262 puts in place requirements for the generation of hazardous waste, 40 CFR Part 261 provides the proper terminology of hazardous and solid wastes, while 40 CFR Part 280 details the design of petroleum and hazardous waste underground storage tanks [9].

RCRA and HSWA certainly are significant in the textile industry. Many of the constituents found within dye and pigments need to be treated and waste from this industry must be handled. These particular amendments were followed as a part of RCRA which highlighted wastes as being hazardous whenever generated— azo, anthraquinone, triarylmethane, dyes, pigments, and Food and Drug and Cosmetic (FD&C) colorants, sludges from triarylmethane dyes/pigments, and wastewater treatment sludge of anthraquinone dyes and pigments. These waste types have been identified as damaging to human health and the environment [25, 26].

Textile facilities should find methods of purchasing or receiving materials for waste that do not increase pollution under Section 313, Title III, of the Superfund Amendments and Reauthorization Act of 1986 (SARA), where it required, starting in 1987, a specific report of any chemical processing greater than 75,000 lbs, in 1988 and 1989 to the present, to a minimum reporting requirement of 25,000 lbs of waste each year. This is required for all facilities that produce dyes such as Disperse Yellow 3, Acid Green 3, Direct Blue 38, and dyes with copper, chromium, and cobalt-based compounds [25].

3.4 Emergency Planning and Community Right-to-Know Act (1986)

In 1986, the Environmental Protection Agency passes the Emergency Planning and Community Right-to-Know Act. This form of legislation resonates well with the textile and dye industry, specifically Section 313. This particular sanction requires manufacturing companies to publically release annual documentation on all chemical releases whether through annual activity or by accident. This is placed in conjunction with the Toxic Chemical Release Inventory rule into the *Federal*

Registrar by February 1988. This legislation requires reporting under specific sanctions of manufacturing, using, or possessing any chemicals conducting and operating with 10 or more full-time employees within a given threshold as stated in the Superfund Amendments of 1987 (SARA). The EPA provides a list of procedures for reporting and names for each chemical based on name and chemical abstract service (CAS) number [27].

3.5 Pollution Prevention Act (1990)

An alternative measure in proper removal of waste would be in pollution prevention. This is stirred by the Pollution Prevention Act of 1990 which legalizes a standard for controlling waste generation. The EPA provides a valuable series of case studies that were compiled in a publication known as "Best Management Practices for Pollution Prevention in Textile Industry." A more recent adaptation of this publication, "Profile of the Textiles," is developed in September of 1997 which includes many of the entities from this previous publication. The 1997 document lists several practices that a textile manufacturer can do to reduce pollution. These have been consolidated into 11 practices [9, 10]:

1. Textile facilities should find methods of purchasing or receiving materials for waste that do not increase pollution.
2. Prescreen materials that are purchased for various factors such as environmental impacts, handling procedures, and emergency situations.
3. Purchase materials come from reusable packages that can be resent to the merchant.
4. Waste reduction can be successful by choosing chemicals that reduce the amount of pollution and also the constituent waste.
5. Provide alternative methods outside of using chemical treatment.
6. Optimize and combine textile operation and processes.
7. Reuse dye and rinse baths.
8. Use automatic equipment that can be properly adhered to proper dye handling.
9. Use washers and ranges that consume less energy and water.
10. Have proper cleaning and housekeeping practices.
11. Provide training for workers.

Table 4.5 provides how the textile industry can improve by implementing pollution prevention methods.

Table 4.5 Example of estimated savings with textile industry [9, 10]

Description of cost/savings	Value ($)
Total costs	
Lab and support equipment	9000
Machine modification, tanks, pumps, pipes	15,000–25,000
Annual operating costs	1000–2000
Total savings (annual)	
Dye and chemicals	15,000
Water	750
Sewer	750
Energy	4500

4   Dye Wastewater Treatment Methods

4.1   Biological Treatment

4.1.1   Description of Conventional Aerobic Treatment

One of the more common methods of treatment is the use of biological treatment. The attraction to this method is the capability of microorganisms to mineralize organics naturally, i.e., providing a mutual benefit for both the treatment plants and also the organisms itself. Waste is used as a food source, while it is converted into a form suitable for discharge with little or minimum cost. In general, the major difference in treatment processes depends on whether or not molecular organic is present as in aerobic treatment or absent as in anaerobic. When deciphering between the two, one should consult the necessary treatment objectives simply because the two conditions differ based on sludge age and production, removal, and the production of additional compounds such as methane or other acids [24].

There are two major types of aerobic treatment processes—activated sludge and trickling filter. Activated sludge is the use of a suspended growth that provides intimate contact between microorganisms and organic constituents. The use of air whether through natural ventilation or artificially by a mechanical device such as a blower provides the mechanism by which intimate contact can occur between food and microorganisms. Ideally, the goal is to reduce the biochemical oxygen demand (BOD_5) developed from the production of waste from various industries. The EPA has stated that within the textile industry, activated sludge can be achieved at efficiencies as high as 95 %. For the purpose of developing nitrification or an increase in intimate contact, one can consider the use of extended aeration. Extended aeration is beyond the conventional 6–8 h sludge retention time, where it can be extended up to 3 days. Advantages of this additional time increases the metabolism of organic compounds in the reactor where more than 75 % of components can be effectively used and reduce the amount of waste generation. For effective treatment, it has been stated that treatment has a nitrogen to biochemical oxygen demand (N to BOD) ratio of 3–4 lb N/100 lb of BOD treatment and dissolved oxygen concentration is maintained around zero within aeration basins [24].

Attached growth, the antecedent to suspended growth, comes in the form of a trickling filter. Trickling filter provides media which endorses microbial growth, where examples include crushed stone, slack, or other inorganic materials. When water enters from the top of the system, it makes contact with the media, whereby initiating growth of microorganisms and removal of organic materials. Because of the structure of the trickling filter, it is necessary that a trickling filter requires the use of recirculation or the recycling of the return of treated wastewater back into the filter for the purpose of maintaining sufficient aeration and also retains moisture of the media without compromising the loss of microorganisms. One can say that removal efficiency is integrated with the BOD loading rate. Within the textile industry, 10–90 % of treatment can be achieved by a trickling filter [24].

Nevertheless, biological treatment has been criticized due to the limitation of biodegradability by microorganisms [28]. For example, Zhao and Hardin observe only 50 % color removal by microorganisms [29]. The primary reason for this is because xenobiotic components in dyes limit the effectiveness of biological treatment [28].

In addition, dyes such as azo dyes are both toxic and mutagenic and have a great difficulty using aerobic conditions because they are very hard to remove. While under anaerobic conditions, azo dyes are reduced to aromatic amines. These amines are harmful intermediates; they are toxic and create high COD [13, 30]. Unless these amines are mineralized, they can have harmful environmental effects.

However, it has been observed that the use of conventional biological methods is incapable of removing dye from wastewater due to the presence of many organic contaminants [12]. This is because organic compounds within dyes are very instable and have high resistance to organic matter decomposition [31]. The chemical composition of the dye is very difficult for the bacteria to be able to degrade down which is why it is difficult for conventional biological treatment to be used for the treatment of dye wastewater [32]. For example, Sanayei et al. through their experiments conclude that through a sequencing batch reactor for the treatment of reactive dye Cibacron Yellow FN_2R, color removal could only be achieved within the range of 31–67 % [33].

Therefore, it is best for aerobic treatment to combine alternative methods to treat dye wastewater. For example, Fernando et al. combine microbial fuel cells and an aerobic bioreactor system to decolorize Acid Orange 7 (AO7) by 99 % and 90 % COD and reduce toxicity [30]. Cui et al. apply upflow bio-electrocatalyzed electrolysis reactor (UBER) and aerobic bio-contact oxidation reactor (ABOR) to decolorize Alizarin Yellow R (AYR) by 97.5 % [13].

4.1.2 Description of Conventional Anaerobic Treatment

As stated earlier, the differences in biological treatment is contingent upon the presence or absence of molecular oxygen. With the absence of molecular oxygen, anaerobic conditions persist. Within this particular treatment method microorganisms use alternative sources of oxygen (such as sulfates, nitrates) and convert organics into organic acids and alcohols. Waste is then converted into methane and carbon dioxide. Anaerobic treatment can be considered in many instances preferable over aerobic conditions due to its ability to reduce waste and produce an entity that can be used as a valuable resource [26].

There are two types of anaerobic wastewater treatment processes—anaerobic lagoons and anaerobic digesters. An anaerobic lagoon is a depression that consists of a depth of 10–17 ft, a BOD loading rate between 15 and 20 lb BOD/1000 ft^3, and a long sludge retention time. In an anaerobic lagoon, wastewater typically flows from the bottom of the lagoon to ensure the proper entry and sustainability of food for anaerobes. Alternatively, anaerobic digestion begins within the digester at moderately high temperature (95–100° F) with BOD loadings ranging from 0.15 to 0.2 lb/ft^3 for a period of 3–12 h. This process follows entry into a gas stripper, settling within the clarifier, and then recycled back within the system. Anaerobic contact system is capable of achieving a 90–97 % removal rate of BOD and suspended solids (SS) Anaerobic digesters can be incorporated in a system known as an anaerobic contact system. This systems consists of an equalizing tank, digester with equipment for mixing, gas stripping using air or vacuum, and clarifiers [26].

Anaerobic treatment processes are capable of producing higher treatment than aerobic. Color removals have been documented to being 65 % [34, 35] to a maximum between 80 % and 90 % [36–38].

The combination of anaerobic and aerobic treatment has found more success when attempting to treat dye wastewater. One of the reasons why is because anaerobic-aerobic treatment has the potential of being able to remove pollutants while producing methane for the purpose of being used for energy. Various authors have observed more success when using this particular application than treatment using simply anaerobic or aerobic treatment by itself. For example, Karatas et al. use anaerobic-aerobic treatment of Reactive Red 24, which not only completely removes color but also produces 563 mL/day of methane [39]. Khehra et al. observe similar high color removals using an upflow anaerobic sludge blanket (UASB) to remove 98 % of color and 95 % COD from C.I. Acid Red 88 [40]. Studies completed by both Panswad and Luangdilok and An et al. determine that the success of the treatment was dependent on the type of dye [41, 42].

Overall, biological treatment processes can have higher treatment when applying aerobic and anaerobic conditions having provided opportunities to consider all bacterial types being employed in treatment. However, one of the biggest disadvantages of using biological treatment as a whole is in its long hydraulic retention times—approximately from 4 to 5.5 h [40, 43], 12 h [40], 18 h [44] to days, where

Cinar et al. complete treatment in 24 h [45], Karatas et al. record 5.76 days for their treatment [39], and Sen and Demirer observed a treatment of 128 days [38]. In addition to long hydraulic retention times, treatment is applied at lower concentrations, usually no greater than 150 mg/L [44, 46–48], but in some raw wastewater can be higher. Nevertheless, the best method of using aerobic or anaerobic treatment is to combine with nonbiological treatment methods—photocatalysis [49], membrane filtration [50–52], ozonation [53], activated carbon [54], microelectrolysis (ATCM) [55], and anaerobic upflow blanket filtration [14].

4.1.3 Alternative Biological Treatment Methods

There are many alternative biological treatment methods which have the capability to replace conventional methods. The sequencing batch reactor (SBR) is an alternative treatment method that been used to treat wastewater since the early twentieth century (1914). One can describe an SBR treatment method as a fill-and-draw technique or the process of filling the reactor and then withdrawing the following treatment. Optimum SBR treatment can occur for flows that enter in at low or intermittent values [56].

The method of sequencing batch reactor is such that all processes of aeration, sedimentation, and clarification happen within the same reactor. One cycle involve five stages—fill, react, settle, draw, and idle. Within the fill stage, the reactor begins by applying the substrate or wastewater into the reactor until approximately 25 % of the reactor volume has been achieved. Alternating presence of aeration occurs throughout this particular stage. Once the fill capacity has been achieved, the react stage, accounting for 35% of the treatment process, pertains to the point the reactor where biological reactors happen. In order to retain aerobic conditions throughout this particular point in the cycle, dissolved oxygen concentrations must remain optimum. Following the react stage, the reactor shifts to the settle period where for one to two retentions, time solids are allowed to settle inside the reactor. The final two stages inside the reactor, draw and idle, use a series of mechanical methods such as weirs to discharge clean water from the system, while idling prepares reactor transition points [57].

Some of the advantages of using sequencing batch reactors include the compartmentalization of one container which allows for cost savings. As stated previously, aeration, sedimentation, and clarification happen simultaneously [56]. In addition, this system is ideal for wastewater with high BOD such as dye wastewater, produces no return sludge, and has the capability to control undesired filamentous growth [57]. However, one must consider the potential of adding solids such as floatables to the discharged wastewater during the draw stage, an increase in maintenance demands, and meticulous design specifications may make this method disadvantageous [56].

A membrane bioreactor (MBR) is a hybrid of a biological reactor and a membrane process. In the membrane processes, wastewater passes through the membrane. This membrane is made of cellulose and has 1 µm wide diameter pores.

Fig. 4.3 Image of hollow
fiber membranes [57]

MBR treats wastewater by filtering out pollutants. This is accomplished by using a vacuum pressured system to push wastewater across the membrane surface. The system also employs self-cleaning techniques by blowing air around the membrane reversing the water back through the membrane. Any other additional components that cannot be filtered will require additional treatment.

There are two major MBR configurations—hollow fiber bundles and plate membranes. Each of the two configurations involves connectivity to a manifold—the hollow fiber configuration is connected in bundles, while the plate membrane is connected within a series of plates to provide an opportunity to add several membranes together [58]. Figure 4.3 provides detailed imagery of a hollow fiber bundle, while Fig. 4.4 shows a membrane bioreactor process in detail.

With regards to treatment performance, Yang et al. treat dye wastewater by using a microfiltration membrane that consisted of a particle size range between 1 and 50 µm, with a range in diameter between 20 and 40 µm. The authors experience removal of COD at 85 %, TOC between 85 % and 90 %, and a 70 % color removal [59]. Hai et al. treat dye wastewater using *Coriolus versicolor* using a microfiltration membrane, where it was determined that the optimum treatment parameters were a TOC concentration of 2 g/L, dye concentration of 100 mg/L, temperature of 29° C, and a pH around 4.5 and resulted in removal efficiencies of 97 % TOC and 99 % color during a 15 h hydraulic retention time (HRT) and a rate flux of 0.021 m/day [60].

A final alternative to conventional biological treatment involves the use of fungi. Fungi treatment is a viable treatment option since it can be applied on both live and dead cells [61]. However, it is more advantageous for cells to be alive because of its ability to produce enzymes such as laccases and manganese peroxidases. Fungi are capable of breaking down difficult aromatic ring structures, catalyzing the reduction of the molecular structure [62]. White-rot fungi are the most efficient fungi because they are capable of mineralizing these dye structures applying toward enzymes [61].

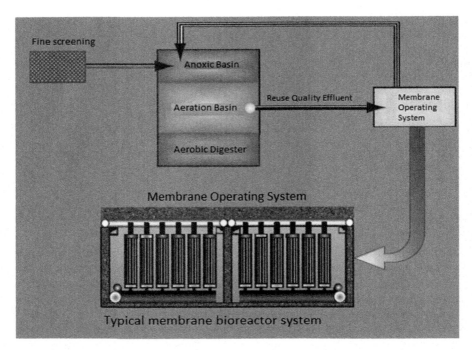

Fig. 4.4 Membrane bioreactor system [57]

On the other hand, dead cells can be used for biosorption or combined in physico-chemical methods, including adsorption, deposition, and ion exchange [61]. In order for fungi treatment to be efficient, one must consider the following growth conditions—the particular type of medium, carbon sources, nutrients (nitrogen), oxygen, pH, incubation time, and temperature [61]. One of the drawbacks of using fungal bacteria is that activity has the disadvantage of inhibiting the degradation of the molecular structure of dyes. Also, fungal treatment may not be advantageous due to the rate of reaction which may be a slower process of treatment.

There have been various adaptations throughout literature of using fungal treatment. Wesenberg et al. study the removal of dye wastewater by means of using *Clitocybila dusenii* and the laccases of manganese peroxidase (MnP) [63]. Fu and Viraraghavan remove Congo red by means of $NaHCO_3$ pretreatment along with the inclusion of *Aspergillus niger*. The authors consider the pH and kinetic and isotherm studies, where the optimum pH is 6. Following the Radke-Prausnitz model, adsorption capacities were 13.80 mg/g for granular-activated carbon and 16.81 mg/g for powder-activated carbon [61].

Cing et al. decolorize 95 % of textile dye from wastewater by the use of *Phanerochaete chrysosporium* during a treatment time of 1 day and optimum temperature of 30° C [64]. Maximo et al. [65] use *Geotrichum* sp. to treat Reactive Black 5, Reactive Red 158, and Reactive Yellow 27. Treatment time was reduced by 3/4 of the original 20 h treatment time when applying this fungal species to

200 mg/L of the dyes, specifically Reactive Black 5 dye [65]. Using the strain *Euc*-1, Dias et al. are successful in decolorizing 11 of 14 azo dyes (specifically completely decolorizing azo dye Acid Red 88) within optimum pH values of 4 and 5 [66]. Demir et al. [67] remove Remazol Red RR Gran at temperatures between 50 and 60 degrees to induce laccase activity [67]. Shin uses the white-rot fungus *Irpex lacteus* in decolorizing by shaking in 59 % and stationary at 93 % within 8 days [68]. Maximo et al. decolorize Reactive Black 5 by means of *Geotrichum* sp. CCMI 1019 within stirred tank reactors and two bubble columns, using porous plates and aeration tubes, where manganese peroxidases were found at high values within both stirred tank reactors and aeration tube bubble columns [69]. Fang et al. [70] use various fungi and bacterium and polyvinyl alcohol to decolorize Direct Fast Scarlet 4BS. The optimum conditions were 5–8 in pH, temperature range between 25 and 40° C, and a dye concentration of 100 mg/L [70]. He et al. treat Direct Fast Scarlet 4BS using white-rot fungus and *Pseudomonas* at a pH between 4 and 9, temperature range between 20 ° C and 40 °C, and a dye concentration of 1000 mg/L [71]. Amaral et al. use *Trametes versicolor* for the purpose of decolorizing textile dye by comparing looking at the application of glucose, where the decolorization was 97 % [71].

Park et al. decolorize Acid Yellow 99, Acid Blue 350, and Acid Red 114 by means of ten various fungal strains, where the optimal fungal strain was *Trametes versicolor* KCTC 16781 [73]. Yu and Wen [74] remove reactive brilliant red K-2BP using *P. rugulosa* Y-48 and *Candida krusei* G-1 at various concentrations. At 200 mg/L of dye concentration, 99 % decolorization occurred at 24 h, and at 50 mg/L dye concentration, P. rugulosa Y-48 could be removed between 22 % and 98 % and *C. krusei* G-1 between 62 % and 94 % [74].

Koseoglu and Ileri [75] use three various fungi, P. *chrysosporium*, *Trametes versicolor*, and *Pleurotus sajur-caju*. Decolorization is determined by the use of measuring at various wavelengths. *P. chrysosporium* decolorize dyes between 66 % and 77 % at a wavelength of 436 nm, 64–79 % at 525 nm, and 69–75 % at 620 nm. On the contrary, *T. versicolor* achieve 71–84 % at 436 nm, 72–85 % at 525 nm, and 70–80 % at 620 nm. while P. sajur-caju achieve 74–80 % at 436 nm, 75–81 % at 525 nm, and 72–78 % at 620 nm [75]. Payman and Mahnaz (1998) remove textile wastewater by means of using *Aspergillus niger* and a fungal species from the Georgian bay, removing 90–95 % of color from dyes [72, 76]. Other fungal species that have been used to decolorize dye wastewater include the use of white-rot fungus *Trametes versicolor* [77], *Aspergillus niger* for the treatment of basic blue 9 under various initial pH values [78], lignin peroxidase (LiP) from *Phanerochaete chrysosporium* for decolorizing about 85 % of dyes [79], and commercial laccase for the purpose of treating Remazol Brilliant Blue R (RBBR) [80]. *Aspergillus niger* is also used to for textile wastewater treatment [81].

Finally, Zhang et al. [82] combine an anaerobic reactor, a microbial electrolysis cell (MEC), and ferric (III) hydroxide ($Fe(OH)_3$) to treat brilliant red X-3B dye with sucrose. MEC combines microbial fuel cells (MFC) to produce hydrogen by degrading organic material transferring electrons outside the cell. Microbial electrolysis cells grow microorganisms on anode biofilm. Under an optimum voltage of 0.8 V increases the reduction of constituents within the wastewater [82].

4.2 Advanced Oxidation Treatment

4.2.1 Wet Air Oxidation

Wet air oxidation (WAO) is as another method for treating dye wastewater. In this advanced oxidation process, pure oxygen transforms constituents (pollutants) by oxidation, into carbon dioxide and water under high temperatures. The success of oxidation is contingent on the medium. Most common mediums include air or oxygen [83].

WAO is primarily used to treat chemical oxygen demand and total organic carbon and found to have the operative conditions of 150° C and 290° C temperature range and a partial pressure of 0.375–2.25 MPa and a temperature of 25° C temperature for oxygen [83].

The process of WAO begins by pumping wastewater at high pressures in a reactor. Air is applied using a compressor. Prior to treatment, a heating source may be necessary to increase the temperature of wastewater to meet operation conditions. After the wastewater is treated, it can be used to assist in further wastewater treatment. The liquids and gases formed in the process enter into a separator where the gas is treated using carbon adsorption. Similar to the wastewater effluent, the treated gas is capable of being reused in a turbine within the system, while the liquids are discharged or can undergo further treatment. WAO as a system is not very energy intensive and its costs are contingent on pumping power [84].

Research has been done to treat dye wastewater using WAO. Santos et al. combine the use of wet oxidation with the use of activated carbon Industrial React FE01606A for the purpose of treating orange G, methylene blue, and brilliant green [85]. Having prepared a solution of 1000 mg/L, using an oxygen flow rate of 90 mg/L, temperature of 160° C, and 16 bar, a three-phased fix-bed reactor can reduce dye between 40 % and 60 % [85]. Ma et al. use catalytic wet oxidation at 35° C, atmospheric pressure, and the $C_uOM_oO_3$-P_2O_5 catalyst for treating Methylene Blue. The authors achieve a color removal of 99.26 % within 10 min having an initial concentration and pH of 0.3 g/L and 5, respectively [86]. Liu et al. develop a catalytic wet oxidation technique using Fe_2O_3-C_eO_2-$TiIO_2$/gamma-Al_2CO_3 catalyst and capable of achieving a COD removal of 62.23 %, 50.12 % color removal, and 41.26 % TOC removal within 3 h, increasing the BOD:COD ratio from 0.19 to 0.30 [87].

4.2.2 Fenton/Photo-Fenton

The Fenton process is the chemical interaction of ferrous iron Fe (II) with hydrogen peroxide (H_2O_2) forming a reduced ferric (III) ion, hydrogen ion (OK), and a hydroxide radical (* OH). The available hydroxide radical forms an Fe(III) ion when combined with available ferrous iron (II) [88]. Equations 4.1 and 4.2 summarizes this dark reaction:

$$Fe(II) + H_2O_2 \rightarrow Fe(III) + OK + {}^*OH \qquad (4.1)$$

$${}^*OH + Fe(II) \rightarrow Fe(III) + OK \qquad (4.2)$$

The irradiation or the application of near-UV radiation and/or visible light develops the photo-Fenton reaction. The use of light emission greatly improves the removal of organics by enhancing the process of mineralization. There are three different types of reactions—the applications of photoreduction of ferrihydroxalate ($Fe(III)\ OH^{2-}$) into ferric ions and hydroxide complex. As seen in Eq. 4.2, the availability of hydroxide combined with additional ferrous ion is important. This presence is increased within photo-Fenton as compared with the dark Fenton process. The second reaction type is ferric carboxylate complexes which are irradiated to Fe(II), carbon monoxide, and radicals. Finally, photolysis of hydrogen peroxide (H_2O_2) combines light energy with hydrogen peroxide to form a hydroxide complex. Wastewater absorbance occurs at wavelength less than 300 nm [88]. Equations 4.3, 4.4, and 4.5 summarize these photo-Fenton processes [88]:

$$Fe(lll)(OH)^{+2} + hv \rightarrow Fe(ll) + {}^*OH \qquad (4.3)$$

$$Fe(lll)(RCO_2)^{+2} + hv \rightarrow Fe(ll) + CO + R^* \qquad (4.4)$$

$$H_2O_2 + \text{light energy } (hv) \rightarrow 2{}^*OH \qquad (4.5)$$

Fenton processes are advantageous because oxidation is accomplished at a lower cost and the materials needed for the process can be easily obtained [89]. The products from Fenton's reagent include low molecular weight components, along with CO_2 and H_2O [90].

There are four major conditions that must be considered to conduct a successful Fenton operation. First, high efficiency of treatment has been observed when using Fenton operated at a pH value between two and five. Evidence of this has been seen throughout literature. For example, Kuo et al.'s experiment with Fenton's reagent at a pH value is less than 3.5 [91]. Rathi et al. treated Direct Yellow 12 (DY12) using $UV/H_2O_2/Fe^{2+}$ at a pH of 4 [92]. Chen et al. treat Bromopyrogallol Red (BFR) using cobalt ions within the Fenton at a pH of 5.2 [93]. Hsieh et al. operate at an initial pH of 2 using a combined ultrasound/Fenton/nanoscale iron oxidation process [94].

Second, the Fenton reaction requires an adjustment of the molar ratio to generate the desired treatment. Kos et al. report the ideal molar ratio is 1:1 for H_2O_2 and Fe^{2+} respectively [89]. However, depending on the wastewater being treated, molar ratios may vary. Jonstrup et al. found optimum conditions when the molar ratio was 3:0.25 H_2O_2 and Fe^{2+} when using photo-Fenton for the treatment of Remazol Red RR [95]. Su et al. conclude that removal was obtained when the ratio was 0.95:3.17 for $Fe^{2+} : H_2O_2$ [96]. Koprivanac and Vujevic use a Fe^{3+}/H_2O_2 1 : 5 molar ratio when treating C.I. Direct Orange 39 [97]. Acarbabacan et al. concludes that a molar ratio of 1:5 for Fe^{2+}/H_2O_2 is optimal for treating azo dyes [98].

Third, Fenton does not always use Fenton as a metal complex. Instead of using ferrous metal, Lim et al. used the combination of H_2O_2/pyridine/Cu(II) system by means of treating Terasil Red dye [99]. Dantas et al. develop a complex combining Fe_2O_3 and carbon [100]. Torrades and Garcia-Montano apply iron-pillared Tunisian clay (Fe-PILC) to decolorize Congo red and Malachite Green at pH between 2 and 3, 8 mL of 200 mg/L H_2O_2, and 1 h UV irradiation [101].

Fourth, Fenton treatment processes have been recently combined with other treatment processes to create a high removal of wastewater components such as TOC and color. Generally, treatment is much higher when one applies a supplemental process with Fenton. Wang et al. use electro-Fenton and activated carbon fiber cathode for the purpose of the mineralization of Acid Red 14, removing 70 % total organic carbon [102]. Kusic et al. apply a series of processes involving a two-stage Fenton process treating C.I. Acid Orange 7, where 99.09 % of TOC was removed [103]. Chen et al. treat reactive brilliant orange using photo-Fenton and iron-pillared vermiculite (Fe-VT) catalyst X-GN, decolorizing at 98.7 % [104]. Fongsatitkul et al. conduct a three-stage treatment by means of biological, chemical, and a secondary biological treatment method intricately fitting Fenton within the treatment process for a textile wastewater [105]. Lucas and Peres produce high removal of Reactive Black 5 using Fenton/UV-C and ferrioxalate/ H_2O_2/solar light, where it was found that the treatment was 98.1 % and 93.2 % [106]. Gonder and Barlas combine Fenton with ion exchange [107], while Li et al. reduce Acid Brown 348 by 96 % [108].

4.2.3 Other Advanced Oxidation Methods

Another treatment method capable of being used for decolorizing dye wastewater is hydrogen peroxide/pyridine/copper (II). Hydrogen peroxide/pyridine/copper(II) [H_2O_2/pyridine/Cu(II)] is an advanced oxidation method that uses copper ions to react with a chelating agent (pyridine) for the purpose of forming anions and hydroxyl radicals. The copper ions are reduced to Cu(I) through this application.

The treatment method chemistry begins with the use of copper(II) which through the chelation of pyridine transforms and develops a pyridine copper (II) complex. Next, the reaction of hydrogen peroxide develops cuprous complex (copper I) and the emission of hydrogen. The cuprous complex is then reacted with excessive hydrogen peroxide and will break up the cuprous complex, forming copper (I), superoxide anions, and hydroxyl radicals. Through this oxidation process, dyes will be decolorized [109].

The biggest key is to transform the materials into harmless substances. H_2O_2/pyridine/copper (II) system has been used by several authors in decolorization of wastewater. Lim et al. remove COD up to 73 % when the application of H_2O_2 reached 0.059 M, pyridine at 0.087 M, and Cu(II) 0.017 [109]. Bali and Karagozoglu decolorize 90 % of the majority of dyes during the experiment, with the exception of Rifacion Yellow He-4R and Levafix Yellow Brown E-3RL, only achieving 74 % and 69 %, respectively [110]. Nerud et al. use this system and decolorize 89 % phenol red, 58 % tropaeolin 00, 95 % Evans blue, 84 % eosin yellows, and 92 % Poly B-411 within 1 h of complete treatment, maintaining the pH between 3 and 9 [111].

4.3 Activated Carbon

One of the more common methods of wastewater treatment is activated carbon. A carbon source undergoes a method such as pyrolysis and then is activated by oxidation. Pyrolysis plays an important role within the formation of activated carbon as it is completed within a high temperature range (400° C and 1000° C) without the presence of air [112].

A very important feature within activated carbon is the discussion of equilibrium isotherms. Equilibrium isotherms are used to identify the rate at which material remains within the pores of activated carbon when reaching equilibrium. There are several different isotherms. The more common are Langmuir, Brunauer-Emmett-Teller (BET), and Freundlich. The following equations summarize the three isotherm equilibrium equations.

Langmuir describes conditions where molecules from the solute can only be adsorbed on a monolayer surface [112, 113]:

$$q_e = \frac{QbC_e}{(1 + bC_e)} \tag{4.6}$$

where

q_e is the ratio of solute that has been absorbed per gram of activated carbon (mg solute/g activated carbon)

Q is the mass of solute adsorbed per gram of activated carbon within the monolayer (mg solute/g activated carbon)

C_e is the solute concentration within the solution (mg/L)

b is a constant of net enthalpy of adsorption (L/mg)

BET extends the Langmuir toward various layers where the Langmuir isotherm applies to each individual layer, while Freundlich isotherm application is applied to a heterogeneous surface [112, 113]:

$$q_e = K_f C_e^{(1/n)} \tag{4.7}$$

where

K_f is a sorption capacity constant (L/g)

$1/n$ describes adsorption favorability, where <1 is unfavorable, greater than 1 is favorable, and 1 is linear (unitless)

One of the more common adsorbents is granular-activated carbon (GAC). This adsorbent has been used to reduce heavy metal concentrations. The driving force behind GAC is adsorption or the process of retaining particulates within the pore sizes of a given material. There are various types of granular-activated carbons that are used for treating wastewater. Fresh and spent carbon sources consist of an activated carbon treatment process where wastewater is applied vertically either by the use of pressure or gravity into a fix-bed column. The column houses the carbon material and is contained at the bottom. When wastewater is applied to the reactor, organic constituents affix themselves to the walls of the pores, where the wastewater will remain within the column until the concentration reaches the water quality standards. Following such treatment, carbon is then removed and then oxidized for the purpose of removing the carbon present within the pores. An alternative to using thermal treatment is the process of applying backwash for the purpose of treating the wastewater [114].

There have been different dye wastewaters that have been treated by activated carbon. Literature within the last 3 years has suggested that either the adsorbent type has changed or the conjunction of activated carbon with another treatment method has become integral in successfully using activated carbon.

For example, Kalathil et al. [115] use a two-treatment method—granular-activated carbon microbial fuel cell (GACB-MFC) combining Nafion membrane and platinum catalyst for the purpose of treating COD, color, and toxicity. The authors achieve a decolorziation efficiency of 73 % at the anode and 77 % at the cathode and a COD removal of 71 % at the anode and 76 % at the cathode. The experiment also reduces toxicity within the first 24 h of the experiment [115]. Dong et al. reject Orange G dye wastewater using the combination of ultrafiltration (UF) and powdered activated carbon (PAC). Treatment increased by 2.29 % times higher using UF and PAC as compared with using UF [116]. Zhao et al. find success in removing Acid Orange 7 by using an activated carbon fiber-iron (ACF)/Fe anode and an activated carbon fiber-titanium ACF/Ti cathode [117]. Li et al. combine Fe-doped TiO_2 with activated carbon which has also been used for the purpose of treating dye wastewater [118].

The conventional carbon compounds have been replaced with other forms of activated carbon. Derris leaf powder was used to adsorb 93 % of the dye gray BL when the initial concentration was 25 mg/L and temperature was 300 K [18]. Others have been successful in using the combination of red mud (RM) and magnesium chloride to treat dye as well. In fact, it is considered as being a reliable system that combines powdered activated carbon and sodium hydroxide [119]. Bentonite has been proven as a successful adsorbent when attempting to treat Acid Red 18 and Acid Yellow 23 [120]. However, *P. oceanica*, a plant within the Mediterranean Sea, is capable of adsorbing CI Acid Yellow 59 within 600 min at concentrations as high as 100 mg/L, have strong retention as powdered activated carbon (PAC) [121].

Waste from other productions such as fly ash is a significant material in the treatment processes. Methylene Blue is removed using municipal solid waste incineration (MSWI) fly ash as an adsorbent to reduce color, total organic carbon, and follow Langmuir isotherm modeling techniques [122]. Biomass fly ash is used to treat Reactive Black 5, where it was found that optimum pH must range between

8.2 and 10.4. The authors conclude that the Langmuir best described the treatment process [123]. Also, Li et al. identify optimum carbonization temperature and time of 300° and 60 min and activation temperature and time of 850° C and 40 min, capable of completely removing Methylene Blue (MB) by means of sludge-based activated carbon (SAC) [124]. Wood-shaven bottom ash combined with either water (H_2O) or sulfuric acid (H_2SO_4) produces half of the adsorption capacity of activated carbon when treating synthetic Red Reactive 141 wastewater [125].

4.4 Membrane Filtration

Membrane filtration—ultrafiltration (UF), microfiltration (MF), nanofiltration (NF), and reverse osmosis (RO)—is the process of filtering desired constituents using various techniques. There are two major types of divisions within the category of membrane filtration—sieve filtration and pressure filtration. Sieve filtration is the limitation of the average pore size that is capable of removing objects that range from as small as suspended solids to as large as other higher molecular weight organic matter such as viruses and other pathogens. The key indicator within the filtration process is the diameter of the pore size for which materials can be sieved from the particular media, which in this case would be wastewater [126]. Figures 4.5 and 4.6 provide a photo and diagram of the two types of membrane filters.

Ultrafiltration is one of the three major types of membrane filtration processes. The filtration system consists of the solvent flux which is determined by using the quotient of the volume and the unit area multiplied by the unit time and solute rejection, or the difference of 1 and the ratio of the downstream and upstream concentrations (permeate and feed, respectively) of the particular entity attempting to be removed. For ultrafiltration, it can be stated that the range of pore size is

Fig. 4.5 Cross section of a hollow fiber by photomicrograph [126]

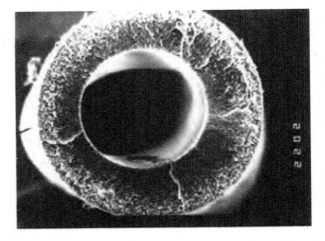

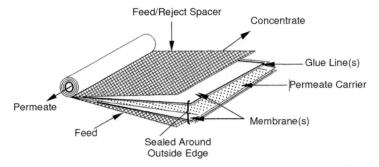

Fig. 4.6 Spiral-wound membrane module [126]

between 0.1 and 0.2 μm, while in microfiltration, the average range is 0.01 and 0.05 μm [127].

On the contrary, reverse osmosis and nanofiltration or pressure filtration are processes that heavily rely on the use of applying pressure to reverse the natural course of constituents within a semipermeable membrane. In a traditional osmotic situation, constituents gravitate toward traveling from high to low concentration. The application of a pressure (osmotic pressure) allows for the reverse effect. By doing so, the process of materials that naturally desire to gravitate to lower concentration to maintain homeostasis is reverted. For nanofiltration and reverse osmosis, the necessary pressures to achieve this state vary based on the concentration and type of constituent attempting to be removed [126].

Reverse osmosis and nanofiltration membranes are made from either cellulose acetate or polyamide material. The use of cellulose acetate limits the range (specifically the pH) at which the processes can be performed due to the biodegradability of such materials. However, the use of polyamides allow for a more suited range of pH values that do not have the limits as to what type of treatment that can be performed [126].

Some of the more considerable equations necessary include the definition of system flux and recovery which is shown below:

$$J = \frac{Q_p}{A_m} \tag{4.8}$$

where

J = flux (gfd)
Q_p = filtrate flow (gpd)
A_m = membrane area (ft^2)

$$R = \frac{Q_p}{Q_f} \qquad (4.9)$$

where

R = recovery of the membrane unit (percent)
Q_p = filtrate flow (gpd)
Q_f = feed flow (gpd)

Membrane technologies have shown some success of removing color from the wastewater. Some authors observe color removals greater than 90% [127–130]. When it came to removal of COD, it was observed that membrane technologies were capable of ranging removal between 70% and 80% [58, 128].

4.5   Electrocoagulation

Electrocoagulation-electroflotation (ECF) technology is a treatment process of applying electrical current to treat and flocculate contaminants without having to add coagulants. Shammas et al. stated that coagulation occurs with the current being applied, capable of removing small particles since direct current applied, setting them into motion. Also electrocoagulation could reduce residue for waste production [131–133]. Electrocoagulation consists of pairs of metal sheets called electrodes that are arranged in pairs—anodes and cathodes. Using the principles of electrochemistry, the cathode is oxidized (loses electrons), while the water is reduced (gains electrons), thereby making the wastewater better treated. When the cathode electrode makes contact with the wastewater, the metal is emitted into the apparatus [133].

When this happens, the particulates are neutralized by the formation of hydroxide complexes for the purpose of forming agglomerates. These agglomerates begin to form at the bottom of the tank and can be siphon out through filtration. However, when one considers an electrocoagulation-flotation apparatus, the particulates would instead float to the top of the tank by means of formed hydrogen bubbles that are created from the anode. The floated particulates can be skimmed from the top of the tank [133]. The following are chemical equations that describe this process:

Cathode:

$$M \rightarrow M^{n+} + ne^- \qquad (4.10)$$

Anode:

$$2H_2O + e^- \rightarrow 2OH^- + H_2 \qquad (4.11)$$

To consider how effective the ECF reactor can be, one must consider the following inputs or variables—wastewater type, pH, current density, type of metal electrodes (aluminum, steel, iron), number of electrodes, size of electrodes, and configuration of metals. These variables would affect the overall treatment time, kinetics, and also the removal efficiency measured. Essentially, electrocoagulation can be described as the process to produce metallic flocs, specifically the smaller particles for the purpose of treating and cleaning wastewater. One of the advantages of using electrocoagulation is that it reduces sludge production [133]. In addition, Gurses et al. indicate that electrocoagulation is capable of being a coagulant because of the sacrificial anode that is used within the treatment system [134].

Literature states that there are several optimal parameters that affect decolorization by means of electrocoagulation—current density, pH, treatment time, dye concentration, and electrolyte type. It has been an observed fact that current density is the most significant parameter affecting decolorization. The reason is such that current density determines the coagulant dosage, bubble production, size, and the floc growth [135]. The effect of pH change is significant due to hydroxide compound formation and species that are present within the water. For example, consider using aluminum electrodes. It can be stated that when the pH remains below 5, the species form $Al(OH)_3$. As the pH increases to a higher value, the aluminum electrodes form soluble Al^{+3} cations and monomeric anions [136]. When discussing species formation, another major factor that must be considered is the treatment time. This is because of the dissolution of anions resulting to coagulation species released.

As a result of the species presence from the metal ions from the electrodes and hydroxide flocs, the concentration of electrodes in the wastewater increases [130, 137]. The dye concentration determines the effectiveness of species presence within the wastewater. When hydroxides are formed within the wastewater, the decolorization occurs due to the adsorption of aluminum hydroxides. Whenever the concentration of dye is increased, the aluminum hydroxides formed within the wastewater from the electrodes have fewer sites for adsorption for the molecules from the organic dyes [137]. Finally, the electrolyte used within the treatment process is also very significant for the effectiveness of treatment. The most commonly used electrolyte, or the entity which ignites an electric current through water, is the use of table salt (NaCl). However, it has been observed that sodium sulfate (Na_2SO_4) is a more viable treatment option due to the reduction of AOX formation and also assists in the dye fixation rates [138]. Treatment efficiency using electrocoagulation has been greater than 90 % for removing color [139–142], high chemical oxygen demand greater than 70 % [143–145].

4.6   Coagulation-Flocculation

Antithetical to electrocoagulation is the use of coagulation-flocculation. This method involves the use of various coagulants, traditionally alum (aluminum sulfate), ferric chloride ($FeCl_3$), or ferrous sulfate (Fe_2SO_4). The coagulant neutralizes the charge of particulates, thereby allowing them to agglomerate and settle at the bottom of the tank [146]. This happens because coagulants form monomeric and polymeric species upon contact with water, along with metal hydroxides [147]. The chemical coagulation-flocculation parameters include the pH, mixing, and time [146]. Chemical coagulation can be very expensive depending on the volume of water treatment [146].

Recent studies, have shown that it is very difficult to find conventional use of chemical coagulation-flocculation techniques. One of the possible problems is the difficulty of being able to reduce the solubility enough for components to be able to form flocculants to be removed from the wastewater [146]. Therefore, many have resorted toward the use of combining treatment methods for better enhanced treatment [148–150].

Coagulation-flocculation combines almost every type of treatment method currently available to treat wastewater. Harrelkas et al. combine coagulation with activated carbon, specifically powdered activated carbon to drive an 80 % color removal [151]. Hassani et al. apply granular-activated carbon [152], and Yeap et al. use polyaluminum chloride-poly (3-acrylamido-isopropanol chloride) (PACl-PAMIPCl) to treat Reactive Cibacron Blue F3GA (RCB) and Disperse Terasil Yellow W-4G (DTY) [153]. Other examples of combining coagulation-flocculation include electrocoagulation [154], a natural coagulant such as chitosan that reduces color by 99 % [155], where pretreatment application of ozone reduced turbidity by 95 % [156], the fouling green algae *Enteromorpha polysaccharides* (EP) [157], and aluminum from red clay earth to treat Terasil Red R and Cibacron Red R [147].

4.7 Ozone

Ozonation is a process that dissociates one diatomic oxygen molecule to form two oxygen atoms. One oxygen molecule combines with a fully intact diatomic oxygen molecule to form ozone. While ozone occurs naturally as sunlight passes, this process can be easily replicated within a laboratory environment by either applying an electric current of low or high frequency or passing high energy radiation through either air or oxygen. One of the biggest issues with the formation of ozone is that its formation has high energy costs. It has been estimated that approximately 90 % of all ozone production is used for electrical energy, while less than 10 % that remains can be considered a consumable product [158]. The following two equations show the formation of ozone by the dissociation of a diatomic oxygen molecule:

$$O_2 \leftrightarrow 2O \qquad\qquad (4.12)$$

Table 4.6 Ozone contactors [158]

Contactor types	Advantages	Disadvantages
Spray towers	High rate of mass transfer Uniform ozone in gas phase	Requires high energy to spray liquid Solids can plus spray nozzles Short contact time
Packed columns	Wide gas/liquid operating range Small size and simplicity	Easily channeled and plugged
Plate columns	Same as packaged columns, but no channeling and broader gas/liquid operating range	Easily clogged, but easier to clean Best suited for very large installations
Spargers (bubblers)	Polymer melt-dyed prior to filament formation	Intermittent flows may cause plugging of porous media Longer contact time require larger housings Tendency to vertical channeling of gas bubbles, reducing contact efficiency
Agitators, surface aerators, injectors, turbines, static mixers	High degree of flexibility Small sizes Intimate contact and good dissolution	Require addition of energy Narrow gas/liquid operating ranges Cannot accommodate significant flow changes (injectors and static mixers), therefore require multiple contactor stages

$$2O + 2O_2 \leftrightarrow O_3 \qquad\qquad (4.13)$$

There are several important factors that must be considered when determining the amount of output ozone production for consumption desired—the electrical energy needed to run the generator, the influent gas rate, and power. The USEPA concludes that increasing the influent gas is proportionate to generation of ozone and thereby affects the use of ozone. In addition, when using ozone for treatment, one must consider the ozone contactor. There are four different types of contactors—spray towers, packed columns, plate columns, spargers, and agitators or mixers. Each of these five has their various advantages and disadvantages depending on the type of constituents treated [158]. Ozonation contactor types have been summarized in Table 4.6.

Treatment of dye wastewater by ozone membrane contactors is advantageous because of its capabilities to decrease color, COD, and increase biodegradation [159]. There have been many applications for the use of ozone, specifically in the measurement and formation of mass transfer. These applications are very significant in determining optimum treatment parameters when attempting to remove various components of wastewater. The reason for analyzing mass transfer for this particular treatment method is because one must consider that the application of the concentration of ozone dissolved within the aqueous solution differs between the application made at the equilibrium point of contact between gas and liquid. Some of the components that must be thought of including pH, type of dye, ozone dosage, dye concentration, and temperature to name a few will ultimately affect how effective the treatment is [160].

Ozone is a very popular method of treatment for dye wastewater. Consider the analysis of mass transfer has been made within the treatment of dye such as C.I. Reactive Red 120 [161], Direct Red 23, Acid Blue 113, and C.I. Reactive Red 120 [162], Acid Yellow 17 [159], and Reactive Blue 19 [163]. However, while using ozone has been used solely for the purpose of treating dye wastewater, nevertheless, the more common use of ozone is typically in conjunction with another treatment method. Treatment methods such as ultrasound [164], catalytic/wet catalytic [85, 165], UV/H_2O_2 [166], coagulation [167, 168], biological application [169, 170], and electrochemical processes [171] have been observed by means of using this treatment technology.

The treatment of dye wastewater has been very successful when using ozone. Treatment efficiencies have range depending on the optimum parameters. For example, one can treat dye wastewater at 10 min with a COD removal ranged between 56 %–59.8 % and 87.5 %–91.4 % for color [172]; complete color removal was accomplished at 90 min [173] and 25 min [174], 80 % color removal [175], and 60 % COD [176] for the purpose of using ozonation. When using combination of treatment methods, 20 % COD removal was combined with an aerated biological filter [177]; precipitation using bicarbonate, hydrogen peroxide, and powdered activated carbon achieved 99 % color [178], while color ranged between 50 % and 60 %, only 60 % COD when combining the process with coagulation [179]. Therefore, as seen in this text, literature has found ozonation to remove a higher percentage of color, but low percentages of COD, where looking at a surface level of observation, COD maxed out at around 60 %.

While ozone may remove double-bonded organic materials, it cannot mineralize refractory compound. Therefore, many applications combine ozone with other technologies including H_2O_2, UV, and TiO_2. This process is known as peroxone. While peroxone assists in mineralizing refractory compounds, there are many issues. For example, handling liquid hydrogen peroxide creates various safety hazards. Therefore, a more recent treatment method includes e-peroxone, an electrochemical process to transform the diatomic oxygen used to generate ozone into hydrogen peroxide using a carbon-polytetrafluorethylene (carbon-PTFE) cathode in acidic conditions, while forming a hydroxide radical from a hydrogen peroxide conjugate base. The equations below summarize the process of e-peroxone [180]:

$$O_2 + 2H + 2e^- \rightarrow H_2O_2 \qquad (4.14)$$

$$O_2 + H_2O + e^- \rightarrow HO_2^- + OH^- \qquad (4.15)$$

E-Peroxone reduces TOC by 95.7 % in 4 and 45 min treatment processes as compared with 55.6 and 15.3 % of TOC during ozone and electrolysis only after 90 min of treatment [180].

5   Conclusion

This chapter highlights a microcosm of treatment technologies that certainly are available and have been used for the purpose of treating dye wastewater. There are many additional treatment options that have not been discussed and if present would encompass an entire book, where one could write an entire chapter on each treatment process. However, the study and research of literature sources have made it apparent that all individual dye wastewater treatment processes have both advantages and disadvantages. With that having been said, research has now been swayed to combine treatment processes together to take advantage of the strengths for each treatment process and counterbalance the weaknesses that each possess when used alone. When considering what is available, one cannot merely suggest one treatment process over another, as it would depend on the various parameters that one desires to remove. Nevertheless, the opportunity to observe various treatment processes has led to potential success in deviating from some of the specific treatment methods available and venturing toward others. Future research and publications will be able to indicate whether dye wastewater treatment will advance on a much higher level than what is currently available at this time.

Acknowledgment With a few additions, this chapter has been taken from its entirety from Chapter 2: "Dye Wastewater Literature Review" in *Electrochemical/Electroflotation Process for Dye Wastewater Treatment* by Erick Butler [181].

Glossary of Terms Related to Dye Wastewater Treatment

Activated Carbon A treatment process where contaminants attach to the pore surfaces of carbon.

Biological Treatment A treatment method that mineralize organic material by microorganisms into a form suitable for discharge. Biological treatment can occur under aerobic (in the presence of molecular oxygen) or anaerobic (without the presence of molecular oxygen) conditions.

Chemical Coagulation-Flocculation A treatment method neutralizes the charge of particulates, allowing them to agglomerate and settle at the bottom of the tank. Traditional coagulants include alum (aluminum sulfate), ferric chloride ($FeCl_3$), or ferrous sulfate (Fe_2SO_4).

Clean Air Act The first major environmental regulation passed in 1970 by the Environmental Protection Agency (EPA) that mandates standards on ambient air quality from primary and secondary sources. This act was amended in 1990 for the purpose of controlling hazardous air pollutants.

Clean Water Act Environmental legislation passed in 1972 to regulate the amount of pollution that discharged into surface waters. The National Pollutant

Discharge Elimination System (NPDES) creates permits for effluent discharge that is produced from point sources.

Dye Any colored constituent permanently affixed to a fabric or piece of material. There are 14 major dye types—acid, direct (substantive), azoic, disperse, sulfur, fiber reactive, basic, oxidation, mordant (chrome), developed, vat, pigment, optical fluorescent, and solvent dyes.

Electrocoagulation A treatment process that applies electrical current to treat and flocculate contaminants without having to add coagulant. Contaminated particulates are neutralized by the formation of hydroxide complexes for the purpose of forming agglomerates. These agglomerates begin to form at the bottom of the tank and can be siphon out through filtration.

Emergency Planning and Community Right-to-Know Act Section 313 Legislation passed in 1983 that requires manufacturing companies to publically release annual documentation on all chemical releases whether through annual activity or by accident.

Fenton process The chemical interaction of ferrous iron Fe(II) with hydrogen peroxide (H_2O_2) forms a reduced ferric (III) ion, hydrogen ion (OK), and a hydroxide radical (*OH). The available hydroxide radical forms an additional Fe (III) ion when combined with available ferrous iron (II).

Membrane Filtration A treatment process that either limits the pore sizes for removing objects or applies pressure to reverse natural processes occurring within a membrane.

Ozone A treatment process that passes an electric current through air or oxygen. The current dissociates one diatomic oxygen molecule of oxygen to form two oxygen atoms. One of these atoms combines with a fully intact diatomic oxygen molecule.

Photo-Fenton The irradiation or application of near-UV radiation and/or visible light coupled with the Fenton process. Light emission improves organic mineralization.

Pollution Prevention Act Legislation enacted in 1990 that legalized a standard for controlling waste generation.

Resource Conservation and Recovery Act (RCRA) Legislation passed in 1976 that generates a life cycle analysis for solid and hazardous waste from its generation and production to its handling at a final location.

Toxicity A quantitative assessment that measures the impact a substance has when it comes into contact with an organism. Toxicity depends on length and dose of exposure of the organism to the substance.

Wet air oxidation An advanced oxidation process where pure oxygen transforms pollutants into carbon dioxide and water under high temperatures.

References

1. Soloman PA, Basha CA, Velan M, Ramamurthi V, Koteeswaran K, Balasubramanian N (2009) Electrochemical degradation of remazol black B dye effluent. Clean Soil Air Water 37 (11):889–900
2. Trifi B, Cavadias S, Bellakhal N (2011) Decoloration of methyl red by gliding arc discharge. Desalin Water Treat 25(1–3):65–70
3. Hu HS, Yang MD, Dang H (2005) Treatment of strong acid dye wastewater by solvent extraction. Sep Purif Technol 42(2):129–136
4. World Bank Group, United Nations Environment Programme, and United Nations Industrial Development Organization (1999) Pollution prevention and abatement handbook 1998: toward cleaner production. World Bank Group, Washington, DC
5. Yidiz YS (2008) Optimization of Bomaplex Red CR-L dye removal from aqueous solution by electrocoagulation using aluminum electrodes. J Hazard Mater 153(1–2):194–200
6. Environmental Protection Authority State Government of Victoria (1998) Environmental guidelines for the textile dyeing and finishing industry. EPA 621. Melbourne
7. Tyagi OD, Yadav M (1990) A textbook of synthetic dyes. Anmol Publishing, New Delhi
8. Zollinger H (2003) Color chemistry: syntheses, properties, and applications of organic dyes and pigments, 3rd edn. Wiley, Zurich
9. US Environmental Protection Agency (1997) Profile of the textile agency. EPA 310-R-97-009, Washington, DC, September 1997
10. US Environmental Protection Agency (1996) Best management practices for pollution prevention in the textile industry. EPA 625-R-96-004, Cincinnati, September 1996
11. US Environmental Protection Agency (1985) Project summary: textile dyes and dyeing equipment: classification, properties, and environmental aspects. EPA 600-S2-85-010. Research Triangle Park, Durham, April 1985
12. Hunger K (ed) (2003) Industrial dyes: chemistry, properties, applications. Wiley, Kelkheim
13. Cui G, Guo YQ, Lee HS, Cheng HY, Liang B, Kong FY, Wang YZ, Huang LP, Xu MY, Wang AJ (2014) Efficient azo dye removal in bioelectrochemical system and post-aerobic bioreactor: optimization and characterization. Chem Eng J 243:355–363
14. Qiu B, Dang Y, Cheng X, Sung DZ (2013) Decolorization and degradation of cationic red x-GRL by upflow blanket filter. Water Sci Technol 67(5):976–982
15. Aftalion F (2001) A history of international chemistry: from the "early days" to 2000. Chemical Heritage Foundation, Philadelphia
16. Dutfield G (2003) Intellectual property rights and the life science industries: a 20th century history (globalization and law). Ashgate Publishing Limited, Burlington
17. Kent J (ed) (2007) Kent and Riegel's handbook of industrial chemistry and biotechnology, 12th edn. Springer, New York
18. Mugugan T, Ganapathi A, Valliappan R (2010) Removal of grey BL from dye wastewater by derris (Pongamia Glabra) leaf powder by adsorption. E-J Chem 7(4):1454–1462
19. Shen ZM, Wang WH, Jia JP, Ye JC, Feng X, Peng A (2001) Degradation of dye solution by an activated carbon fiber electrode electrolysis system. J Hazard Mater 84(1):107–116
20. US Environmental Protection Agency (2012) Summary of the Clean Air Act. Available at: http://www.epa.gov/lawsregs/laws/caa.html. Last modified 27 Feb 2012
21. Bradstreet JW (1995) Hazardous air pollutants: assessment, liabilities, and regulatory compliance. Noyes Data Corporation, Park Ridge
22. Reife A, Freeman HS (1996) Environmental chemistry of dyes and pigments. Wiley, New York
23. Fung W (2002) Coated and laminated textiles. CRC Press, Cambridge
24. US Environmental Protection Agency (1974) Treatment processes use: development document for effluent limitations guidelines and new source performance standards for the textile mills point source category. EPA 440-174-022A, June 1974

25. Hazardous waste management system; identification and listing of hazardous waste; dye and pigment industries; land disposal restrictions for newly identified wastes; CERCLA hazardous substance designation and reportable quantities; proposed rule. 40 CFR Parts 148, 261 et al. (23 July 1999). pp. 40191–40230
26. US Environmental Protection Agency (2011) Dyes and pigments: uly 23, 1999. Available at: http://www.epa.gov/osw/hazard/tsd/ldr/dyes.htm. Last accessed 27 Feb 2012
27. US Environmental Protection Agency (1988) Title III Section 313 Release Reporting Guidance: estimating chemical releases from textile dyeing. EPA 560/4-88-004h. Washington, DC, February 1988
28. Lu X, Ma L, Wang Z, Huang M (2010) Application of polymerase chain reaction-denaturing gradient gel electrophoresis to resolve taxonomic diversity in white rot fungus reactors. Environ Eng Sci 27(6):493–503
29. Zhao X, Hardin IR (2007) HPLC and spectrophotometric analysis of biodegradation of azo dyes by Pleurotusostreatus. Dyes Pigm 73(3):322–325
30. Fernando E, Keshavarz T, Kyazze G (2014) Complete degradation of the azo dye acid orange-7 and bioelectricity generation in an integrated microbial fuel cell, aerobic two-stage bioreactor system in continuous flow mode at ambient temperature. Bioresour Technol 156:155–162
31. Hu TL, Lin CF, Chiang KY (2004) Simulations of organic carbon transformation in dye wastewater and treatment suggestions. Adv Environ Res 8(3–4):493–500
32. Zuriaga-Agusti E, Iborra-Clar MI, Mendoza-Roca JA, Tancredi M, Alcaina-Miranda MI, Iborra-Clar A (2010) Sequencing batch reactor technology coupled with nanofiltration for textile wastewater reclamation. Chem Eng J 161(1–2):122–128
33. Sanayei Y, Ismail N, Teng TT, Morad N (2009) Biological treatment of reactive dye (cibacron yellow FN_2R) by sequencing batch reactor performance. Aust J Basic Appl Sci 3(4):4071–4077
34. Firmino PIM, da Silva MER, Mota FSB, dos Santos AB (2011) Applicability of anthraquinone-2,6-disulfonate (AQDS) to enhance colour removal in mesophilic UASB reactors treating textile wastewater Rid E-7728-2011. Brazil J Chem Eng 28(4):617–623
35. Haroun M, Dris A (2009) Treatment of textile wastewater with an anaerobic fluidized bed reactor. Desalination 237(1–3):357–366
36. Wijetunga S, Li X, Jian C (2010) Effect of organic load on decolourization of textile wastewater containing acid dyes in upflow anaerobic sludge blanket reactor. J Hazard Mater 177(1–3):792–798
37. Somasiri W, Li X, Ruan W, Jian C (2008) Evaluation of the efficacy of upflow anaerobic sludge blanket reactor in removal of colour and reduction of COD in real textile wastewater. Bioresour Technol 99(9):3692–3699
38. Sen S, Demirer G (2003) Anaerobic treatment of synthetic textile wastewater containing a reactive azo dye. J Environ Eng-ASCE 129(7):595–601
39. Karatas M, Dursun S, Argun ME (2011) Methane production from anaerobic-aerobic sequential system treatment of azo dye reactive red 24. Environ Prog Sustainable Energy 30(1):50–58
40. Khehra MS, Saini HS, Sharma DK, Chadha BS, Chimni SS (2006) Biodegradation of azo dye CI Acid Red 88 by an anoxic-aerobic sequential bioreactor. Dyes Pigm 70(1):1–7
41. Panswad T, Luangdilok W (2000) Decolorization of reactive dyes with different molecular structures under different environmental conditions. Water Res 34(17):4177–4184
42. An H, Qian Y, Gu X, Tang W (1996) Biological treatment of dye wastewaters using an anaerobic-oxic system. Chemosphere 33(12):2533–2542
43. Li Y, Xi D (2004) Decolorization and biodegradation of dye wastewaters by a facultative-aerobic process. Environ Sci Pollut Res 11(6):372–377
44. Kapdan I, Tekol M, Sengul F (2003) Decolorization of simulated textile wastewater in an anaerobic-aerobic sequential treatment system. Process Biochem 38(7):1031–1037

45. Cinar O, Yasar S, Kertmen M, Demiroz K, Yigit NO, Kitis M (2008) Effect of cycle time on biodegradation of azo dye in sequencing batch reactor. Process Saf Environ Prot 86 (B6):455–460

46. Jin Y, Wu M, Zhao G, Li M (2011) Photocatalysis-enhanced electrosorption process for degradation of high-concentration dye wastewater on TiO$_2$/carbon aerogel. Chem Eng J 168 (3):1248–1255

47. Ibrahim Z, Amin MFM, Yahya A, Aris A, Muda K (2010) Characteristics of developed granules containing selected decolourising bacteria for the degradation of textile wastewater RID A-6870-2011. Water Sci Technol 61(5):1279–1288

48. Isik M, Sponza DT (2008) Anaerobic/aerobic treatment of a simulated textile wastewater. Sep Purif Technol 60(1):64–72

49. Harrelkas F, Paulo A, Alves MM, El Khadir L, Zahraa O, Pons MN, van der Zee FP (2008) Photocatalytic and combined anaerobic-photocatalytic treatment of textile dyes. Chemosphere 72(11):1816–1822

50. Hai FI, Yamamoto K, Nakajima F, Fukushi K (2011) Bioaugmented membrane bioreactor (MBR) with a GAC-packed zone for high rate textile wastewater treatment. Water Res 45 (6):2199–2206

51. Debik E, Kaykioglu G, Coban A, Koyuncu I (2010) Reuse of anaerobically and aerobically pre-treated textile wastewater by UF and NF membranes. Desalination 256(1–3):174–180

52. Zylla R, Sojka-Ledakowicz J, Stelmach E, Ledakowicz S (2006) Coupling of membrane filtration with biological methods for textile wastewater treatment. Desalination 198 (1–3):316–325

53. Ledakowicz S, Solecka M (2001) Influence of ozone and advanced oxidation processes on biological treatment of textile wastewater. Ozone-Sci Eng 23(4):327–332

54. Kuai L, De Vreese I, Vandevivere P, Verstraete W (1998) GAC-amended UASB reactor for the stable treatment of toxic textile wastewater. Environ Technol 19(11):1111–1117

55. Huang LH, Sun GP, Yang T, Zhang B, He Y, Wang XH (2013) A preliminary study of anaerobic treatment coupled with micro-electrolysis for anthraquinone dye wastewater. Desalination 309:91–96

56. Environmental Protection Agency (1999) Wastewater technology fact sheet: sequencing batch reactor. EPA 832-F-99-073. Washington, DC, September 1999

57. Environmental Protection Agency (1986) Sequencing batch reactor. EPA 625-8-86-011, Cincinnati, October 1986

58. Environmental Protection Agency (2007) Membrane management fact sheet: membrane bioreactors. Available at http://water.epa.gov/scitech/wastetech/upload/2008_01_23_mtb_etfs_membrane-bioreactors.pdf. Last accessed 28 Feb 2012

59. Yang Q, Shang H, Wang J (2009) Dye wastewater treatment by using ceramic membrane bioreactor. Int J Environ Pollut 38(3):267–279

60. Hai F, Yamamoto K, Fukushi K (2006) Development of a submerged membrane fungi reactor for textile wastewater treatment. Desalination 192(1–3):315–322

61. Fu Y, Viraraghavan T (2002) Removal of congo red from an aqueous solution by fungus Aspergillus niger. Adv Environ Res 7(1):239–247

62. Hardin I, Cao H, Wilson S, Akin D (2000) Decolorization of textile wastewater by selective fungi. Text Chem Color Am Dyest Rep 32(11):38–42

63. Wesenberg D, Buchon F, Agathos S (2002) Degradation of dye-containing textile effluent by the agaric white-rot fungus Clitocybula dusenii. Biotechnol Lett 24(12):989–993

64. Cing S, Asma D, Apohan E, Yesilada O (2003) Decolorization of textile dyeing wastewater by Phanerochaete chrysosporium. Folia Microbiol (Praha) 48(5):639–642

65. Maximo C, Amorim M, Costa-Ferreira M (2003) Biotransformation of industrial reactive azo dyes by Geotrichum sp CCMI 1019. Enzyme Microb Technol 32(1):145–151

66. Dias A, Bezerra R, Lemos P, Pereira A (2003) In vivo and laccase-catalysed decolourization of xenobiotic azo dyes by a basidiomycetous fungus: characterization of its ligninolytic system. World J Microbiol Biotechnol 19(9):969–975

67. Demir G, Borat M, Bayat C (2004) Decolorization of Remazol Red RR Gran by the white rot fungus Phanerochaete chrysosporium. Fresenius Environ Bull 13(10):979–984
68. Shin K (2004) The role of enzymes produced by white-rot fungus Irpex lacteus in the decolorization of the textile industry effluent. J Microbiol 42(1):37–41
69. Maximo C, Lageiro M, Duarte A, Reis A, Costa-Ferreira M (2004) Different bioreactor configurations for the decolourisation of the azo dye reactive black 5 by Geotrichum sp. CCMI 1019. Biocatal Biotransfor 22(5–6):307–313
70. Fang H, Hu W, Li Y (2004) Investigation of isolation and immobilization of a microbial consortium for decoloring of azo dye 4BS. Water Res 38(16):3596–3604
71. He F, Hu W, Li Y (2004) Biodegradation mechanisms and kinetics of azo dye 4BS by a microbial consortium. Chemosphere 57(4):293–301
72. Amaral P, Fernandes D, Tavares A, Xavier A, Cammarota M, Coutinho J, Coelho M (2004) Decolorization of dyes from textile wastewater by Trametes versicolor. Environ Technol 25 (11):1313–1320
73. Park C, Lee Y, Kim T, Lev B, Lee J, Kim S (2004) Decolorization of three acid dyes by enzymes from fungal strains. J Microbiol Biotechnol 14(6):1190–1195
74. Yu Z, Wen X (2005) Screening and identification of yeasts for decolorizing synthetic dyes in industrial wastewater. Int Biodeterior Biodegrad 56(2):109–114
75. Koseoglu G, Ileri R (2006) Decolorization of dying textile wastewater by advanced activated sludge in a sequencing batch reactor. Fresenius Environ Bull 15(9B):1111–1114
76. Payman M, Mahnaz M (1998) Decolourization of textile effluent by Aspergillus niger (marine and terrestrial). Fresenius Environ Bull 7(1–2):1–7
77. Srinivasan S, Murthy D (2000) Fungal treatment for wastewater containing recalcitrant compounds. Battle Press, Columbus
78. Fu Y, Viraraghavan T (2000) Removal of a dye from an aqueous solution by the fungus Aspergillus niger. Water Qual Res J Can 35(1):95–111
79. Ferreira V, Magalhaes D, Kling S, da Silva J, Bon E (2000) N-demethylation of methylene blue by lignin peroxidase from Phanerochaete chrysosporium—stoichiometric relation for H_2O_2 consumption. Appl Biochem Biotechnol 84(6):255–265
80. Soares G, de Amorim M, Costa-Ferreira M (2001) Use of laccase together with redox mediators to decolourize remazol brilliant blue R. J Biotechnol 89(2–3):123–129
81. Assadi M, Jahangiri M (2001) Textile wastewater treatment by Aspergillus niger. Desalination 141(1):1–6
82. Zhang JX, Zhang YB, Quan X, Chen S, Afzal S (2013) Enhanced anaerobic digestion of organic contaminants containing diverse microbial population by combined microbial electrolysis cell (MEC) and anaerobic reactor under Fe(III) reducing conditions. Bioresour Technol 136:273–280
83. Lei L, Hu X, Chen G, Porter J, Yue P (2000) Wet air oxidation of desizing wastewater from the textile industry. Ind Eng Chem Res 39(8):2896–2901
84. Fu J, Kyzas GS (2014) Wet air oxidation for the decolorization of dye wastewater: an overview of the last two decades. Chin J Catal 35(1):1–7
85. Santos A, Yustos P, Rodriguez S, Garcia-Ochoa F, de Gracia M (2007) Decolorization of textile dyes by wet oxidation using activated carbon as catalyst. Ind Eng Chem Res 46 (8):2423–2427
86. Ma H, Zhuo Q, Wang B (2007) Characteristics of CuO-MoO_3-P_2O_5 catalyst and its catalytic wet oxidation (CWO) of dye wastewater under extremely mild conditions. Environ Sci Technol 41(21):7491–7496
87. Liu Yan, Sun De-zhi, Cheng Lin, Li Yan-ping (2006) Preparation and characterization of Fe2O3-CeO2-TiO2/gamma-Al2O3 catalyst for degradation dye wastewater. J Environ Sci 18 (6):1189–1192
88. Environmental Protection Agency (1998) Advanced photochemical oxidation processes. EPA 625-R-98-004, Washington, DC, December 1998

89. Kos L, Michalska K, Perkowski J (2010) Textile wastewater treatment by the Fenton method. Fibres Text East Eur 18(4):105–109

90. Zaharia C, Suteu D, Muresan A, Muresan R, Popescu A (2009) Textile wastewater treatment by homogeneous oxidation with hydrogen peroxide. Environ Eng Manage J 8(6):1359–1369

91. Kuo W (1992) Decolorizing dye wastewater with Fenton reagent. Water Res 26(7):881–886

92. Rathi A, Rajor H, Sharma R (2003) Photodegradation of direct yellow-12 using UV/H_2O_2/Fe^{2+}. J Hazard Mater 102(2–3):231–241

93. Chen J, Liu M, Zhang J, Xian Y, Jin L (2003) Electrochemical degradation of bromopyrogallol red in presence of cobalt ions. Chemosphere 53(9):1131–1136

94. Hsieh L, Kang H, Shyu H, Chang C (2009) Optimal degradation of dye wastewater by ultrasound/Fenton method in the presence of nanoscale iron. Water Sci Technol 60 (5):1295–1301

95. Jonstrup M, Punzi M, Mattiasson B (2011) Comparison of anaerobic pre-treatment and aerobic post-treatment coupled to photo-Fenton oxidation for degradation of azo dyes. J Photochem Photobiol A-Chem 224(1):55–61

96. Su C, Pukdee-Asa M, Ratanatamskul C, Lu M (2011) Effect of operating parameters on the decolorization and oxidation of textile wastewater by the fluidized-bed Fenton process. Sep Purif Technol 83:100–105

97. Koprivanac N, Vujevic D (2007) Degradation of an azo dye by Fenton type processes assisted with UV irradiation. Int J Chem React Eng 5:A56

98. Acarbabacan S, Vergili I, Kaya Y, Demir G, Barlas H (2002) Removal of color from textile wastewater containing azo dyes by Fenton's reagent. Fresenius Environ Bull 11 (10B):840–843

99. Lim CL, Morad N, Teng TT, Ismail N (2009) Treatment of terasil red R dye wastewater using H_2O_2/pyridine/Cu(II) System RID G-1915-2010 RID F-6428-2010. J Hazard Mater 168(1):383–389

100. Dantas T, Mendonca V, Jose H, Rodrigues A, Moreira R (2006) Treatment of textile wastewater by heterogeneous Fenton process using a new composite Fe_2O_3/carbon RID F-7465-2010. Chem Eng J 118(1–2):77–82

101. Torrades F, Garcia-Montano J (2014) Using central composite experimental design to optimize the degradation of real dye wastewater by Fenton and photo-Fenton reactions. Dyes Pigm 100:184–189

102. Wang A, Qu J, Ru J, Liu H, Ge J (2005) Mineralization of an azo dye acid red 14 by electro-Fenton's reagent using an activated carbon fiber cathode. Dyes Pigm 65(3):227–233

103. Kusic H, Koprivanac N, Srsan L (2006) Azo dye degradation using Fenton type processes assisted by UV irradiation: a kinetic study. J Photochem Photobiol A-Chem 181 (2–3):195–202

104. Chen Q, Wu P, Dang Z, Zhu N, Li P, Wu J, Wang X (2010) Iron pillared vermiculite as a heterogeneous photo-Fenton catalyst for photocatalytic degradation of azo dye reactive brilliant orange X-GN. Sep Purif Technol 71(3):315–323

105. Fongsatitkul P, Elefsiniotis P, Yamasmit A, Yamasmit N (2004) Use of sequencing batch reactors and Fenton's reagent to treat a wastewater from a textile industry. Biochem Eng J 21 (3):213–220

106. Lucas MS, Peres JA (2007) Degradation of reactive black 5 by Fenton/UV-C and ferrioxalate/H_2O_2/solar light processes. Dyes Pigm 74(3):622–629

107. Gonder B, Barlas H (2005) Treatment of coloured wastewater with the combination of Fenton process and ion exchange. Fresenius Environ Bull 14(5):393–399

108. Li JT, Lan RJ, Bai B, Song YL (2013) Ultrasound-promoted degradation of acid brown 348 by Fenton-Like processes. Asian J Chem 25(4):2246–2250

109. Lim CI, Morad N, Teng TT, Norli I (2011) Chemical oxygen demand (COD) reduction of a reactive dye wastewater using H_2O_2/pyridine/Cu (II) system. Desalination 278:26–30

110. Bali U, Karagozoglu B (2007) Decolorization of remazol-turquoise blue G-133 and other dyes by Cu(II)/pyridine/H_2O_s system. Dyes Pigm 73(2):133–140

111. Nerud F, Baldrian P, Gabriel J, Ogbeifun D (2001) Decolorization of synthetic dyes by the Fenton reagent and the Cu/pyridine/H_2O_2 system RID A-9170-2009. Chemosphere 44 (5):957–961
112. US Environmental Protection Agency (1980) Preparation and evaluation of powdered activated carbon from lignocellulosic material. EPA 600-2-80-123. Cincinnati, August 1980
113. US Environmental Protection Agency (2014) Drinking water treatability database: GAC isotherm. Available at: http://iaspub.epa.gov/tdb/pages/treatment/treatmentOverview.do? treatmentProcessId = -979193564. Accessed 12 July 2014
114. US Environmental Protection Agency (2000) Wastewater technology fact sheet: granulated activated carbon and regeneration. EPA 832-F-00-017, Washington, DC, September 2000
115. Kalathil S, Lee J, Cho MH (2011) Granular activated carbon based microbial fuel cell for simultaneous decolorization of real dye wastewater and electricity generation. New Biotechnol 29(1):32–37
116. Dong Yanan, Su Yanlei, Chen Wenjuan, Peng Jinming, Zhang Yan Z, Jiang Zhongyi (2011) Ultrafiltration enhanced with activated carbon adsorption for efficient dye removal from aqueous solution. Chin J Chem Eng 19(5):863–869
117. Zhao H, Sun Y, Xu L, Ni J (2010) Removal of Acid Orange 7 in simulated wastewater using a three-dimensional electrode reactor: removal mechanisms and dye degradation pathway RID E-8694-2010. Chemosphere 78(1):46–51
118. Li Y, Chen J, Liu J, Ma M, Chen W, Li L (2010) Activated carbon supported TiO(2)-photocatalysis doped with Fe ions for continuous treatment of dye wastewater in a dynamic reactor. J Environ Sci (China) 22(8):1290–1296
119. Wang Q, Luan Z, Wei N, Li J, Liu C (2009) The color removal of dye wastewater by magnesium chloride/red mud (MRM) from aqueous solution. J Hazard Mater 170 (2–3):690–698
120. Qiao S, Hu Q, Haghseresht F, Hu X, Lua GQ (2009) An investigation on the adsorption of acid dyes on bentonite based composite adsorbent RID A-6057-2010. Sep Purif Technol 67 (2):218–225
121. Guezguez I, Dridi-Dhaouadi S, Mhenni E (2009) Sorption of Yellow 59 on Posidonia oceanica, a non-conventional biosorbent: comparison with activated carbons. Ind Crops Prod 29(1):197–204
122. Liu Q, Zhou Y, Zou L, Deng T, Zhang J, Sun Y, Ruan X, Zhu P, Qian G (2011) Simultaneous wastewater decoloration and fly ash dechlorination during the dye wastewater treatment by municipal solid waste incineration fly ash. Desalin Water Treat 32(1–3):179–186
123. Pengthamkeerati P, Satapanajaru T, Chatsatapattayakul N, Chairattanamanokorn P, Sananwai N (2010) Alkaline treatment of biomass fly ash for reactive dye removal from aqueous solution. Desalination 261(1–2):34–40
124. Li W, Yue Q, Gao B, Wang X, Qi Y, Zhao Y, Li Y (2011) Preparation of sludge-based activated carbon made from paper mill sewage sludge by steam activation for dye wastewater treatment. Desalination 278(1–3):179–185
125. Leechart P, Nakbanpote W, Thiravetyan P (2009) Application of 'waste' wood-shaving bottom ash for adsorption of azo reactive dye. J Environ Manage 90(2):912–920
126. US Environmental Protection Agency (2005) Membrane filtration guidance manual. EPA 815-R-06-009, November 2005
127. Al-Bastaki N, Banat F (2004) Combining ultrafiltration and adsorption on bentonite in a one-step process for the treatment of colored waters. Resour Conserv Recycl 41(2):103–113
128. Gholami M, Nasseri S, Alizadehfard M, Mesdaghinia A (2003) Textile dye removal by membrane technology and biological oxidation. Water Qual Res J Can 38(2):379–391
129. Han R, Zhang S, Xing D, Jian X (2010) Desalination of dye utilizing copoly(phthalazinone biphenyl ether sulfone) ultrafiltration membrane with low molecular weight cut-off. J Membr Sci 358(1–2):1–6

130. Nora'aini A, Norhidayah A, Jusoh A (2009) The role of reaction time in organic phase on the preparation of thin-film composite nanofiltration (TFC-NF) membrane for dye removal. Desalin Water Treat 10(1–3):181–191
131. Fersi C, Gzara L, Dhahbi M (2009) Flux decline study for textile wastewater treatment by membrane processes. Desalination 244(1–3):321–332
132. Shammas NK, Pouet M, Grasmick A (2010) Wastewater treatment by electrocoagulation–flotation. In: Wang L (ed) Flotation technology. Springer, New York, pp 99–124
133. Butler E, Hung Y-T, Yeh R-L, Suleiman Al Ahmad M (2011) Electrocoagulation in wastewater treatment. Water 3(2):495–525
134. Gurses A, Yalcin M, Dogar C (2002) Electrocoagulation of some reactive dyes: a statistical investigation of some electrochemical variables. Waste Manage 22(5):491–499
135. Akbal F, Kuleyin A (2011) Decolorization of levafix brilliant blue E-B by Electrocoagulation Method. Environ Prog Sustainable Energy 30(1):29–36
136. Hmida ESBH, Mansour D, Bellakhal N (2010) Treatment of lixiviate from Jebel Chakir-Tunis by electrocoagulation. Desalin Water Treat 24(1–3):266–272
137. Aoudj S, Khelifa A, Drouiche N, Hecini M, Hamitouche H (2010) Electrocoagulation process applied to wastewater containing dyes from textile industry. Chem Eng Process 49 (11):1176–1182
138. Arslan-Alaton I, Kabdasli I, Vardar B, Tuenay O (2009) Electrocoagulation of simulated reactive dyebath effluent with aluminum and stainless steel electrodes. J Hazard Mater 164 (2–3):1586–1594
139. Chafi M, Gourich B, Essadki AH, Vial C, Fabregat A (2011) Comparison of electrocoagulation using iron and aluminium electrodes with chemical coagulation for the removal of a highly soluble acid dye. Desalination 281:285–292
140. Dalvand A, Gholami M, Joneidi A, Mahmoodi NM (2011) Dye removal, energy consumption and operating cost of electrocoagulation of textile wastewater as a clean process. Clean-Soil Air Water 39(7):665–672
141. Phalakornkule C, Polgumhang S, Tongdaung W, Karakat B, Nuyut T (2010) Electrocoagulation of blue reactive, red disperse and mixed dyes, and application in treating textile effluent. J Environ Manage 91(4):918–926
142. Cerqueira A, Russo C, Marques MRC (2009) Electroflocculation for textile wastewater treatment. Braz J Chem Eng 26(4):659–668
143. Balla W, Essadki AH, Gourich B, Dassaa A, Chenik H, Azzi M (2010) Electrocoagulation/electroflotation of reactive, disperse and mixture dyes in an external-loop airlift reactor. J Hazard Mater 184(1–3):710–716
144. Daneshvar N, Ashassi-Sorkhabi H, Tizpar A (2003) Decolorization of orange II by electrocoagulation method. Sep Purif Technol 31(2):153–162
145. Merzouk B, Gourich B, Sekki A, Madani K, Vial C, Barkaoui M (2009) Studies on the decolorization of textile dye wastewater by continuous electrocoagulation process. Chem Eng J 149(1–3):207–214
146. Verma AK, Dash RR, Bhunia P (2012) A review on chemical coagulation/flocculation technologies for removal of colour from textile wastewaters. J Environ Manage 93 (1):154–168
147. Alkarkhi AFM, Lim HK, Yusup Y, Teng TT, Abu Bakar MA, Cheah KS (2013) Treatment of terasil red R and cibacron red R wastewater using extracted aluminum from red earth: factorial design. J Environ Manage 122:121–129
148. Nabi Bidhendi GR, Torabian A, Ehsani H, Razmkhah N, Abbasi M (2007) Evaluation of industrial dyeing wastewater treatment with coagulants. Int J Environ Res 1(3):242–247
149. Wong PW, Teng TT, Norulaini NARN (2007) Efficiency of the coagulation-flocculation method for the treatment of dye mixtures containing disperse and reactive dye RID F-6428-2010. Water Qual Res J Can 42(1):54–62
150. Georgiou D, Aivazidis A, Hatiras J, Gimouhopoulos K (2003) Treatment of cotton textile wastewater using lime and ferrous sulfate. Water Res 37(9):2248–2250

151. Harrelkas F, Azizi A, Yaacoubi A, Benhammou A, Pons MN (2009) Treatment of textile dye effluents using coagulation-flocculation coupled with membrane processes or adsorption on powdered activated carbon. Desalination 235(1–3):330–339
152. Hassani AH, Seif S, Javid AH, Borghei M (2008) Comparison of adsorption process by GAC with novel formulation of coagulation—flocculation for color removal of textile wastewater. Int J Environ Res 2(3):239–248
153. Yeap KL, Teng TT, Poh BT, Morad N, Lee KE (2014) Preparation and characterization of coagulation/flocculation behavior of a novel inorganic–organic hybrid polymer for reactive and disperse dyes removal. Chem Eng J 243:305–314
154. Merzouk B, Gourich B, Madani K, Vial C, Sekki A (2011) Removal of a disperse red dye from synthetic wastewater by chemical coagulation and continuous electrocoagulation. A comparative study. Desalination 272(1–3):246–253
155. Szygula A, Guibal E, Arino Palacin M, RuizA M, Maria Sastre A (2009) Removal of an anionic dye (acid blue 92) by coagulation-flocculation using chitosan RID B-1045-2008. J Environ Manage 90(10):2979–2986
156. Barredo-Damas S, Iborra-Clar M, Bes-Pia A, Alcaina-Miranda M, Mendoza-Roca J, Iborra-Clar A (2005) Study of preozonation influence on the physical-chemical treatment of textile wastewater. Desalination 182(1–3):267–274
157. Zhao S, Gao B, Yue Q, Wang Y, Li Q, Dong H, Yan H (2014) Study of Enteromorpha polysaccharides as a new-style coagulant aid in dye wastewater treatment. Carbohydr Polym 15(103):179–186
158. US Environmental Protection Agency (1980) Ozone for industrial water and wastewater treatment: a literature survey. EPA-600-2-80-060. Ada, April 1980
159. Lackey LW, Mines RO Jr, McCreanor PT (2006) Ozonation of acid yellow 17 dye in a semi-batch bubble column. J Hazard Mater 138(2):357–362
160. Wu JN, Wang TW (2001) Effects of some water-quality and operating parameters on the decolorization of reactive dye solutions by ozone. J Environ Sci Health A 36(7):1335–1347
161. Atchariyawut S, Phattaranawik J, Leiknes T, Jiraratananon R (2009) Application of ozonation membrane contacting system for dye wastewater treatment. Sep Purif Technol 66 (1):153–158
162. Bamperng S, Suwannachart T, Atchariyawut S, Jiraratananon R (2010) Ozonation of dye wastewater by membrane contactor using PVDF and PTFE membranes. Sep Purif Technol 72 (2):186–193
163. Tehrani-Bagha AR, Mahmoodi NM, Menger FM (2010) Degradation of a persistent organic dye from colored textile wastewater by ozonation. Desalination 260(1–3):34–38
164. Zhou X, Guo W, Yang S, Ren N (2012) A rapid and low energy consumption method to decolorize the high concentration triphenylmethane dye wastewater: operational parameters optimization for the ultrasonic-assisted ozone oxidation process. Bioresour Technol 105:40–47
165. Dong Y, He K, Zhao B, Yin Y, Yin L, Zhang A (2007) Catalytic ozonation of azo dye active brilliant red X-3B in water with natural mineral brucite. Catal Commun 8(11):1599–1603
166. Shu H, Chang M (2005) Pre-ozonation coupled with UV/H_2O_2 process for the decolorization and mineralization of cotton dyeing effluent and synthesized CI Direct Black 22 wastewater. J Hazard Mater 121(1–3):127–133
167. Arslan-Alaton W (2003) Novel catalytic photochemical and hydrothermal treatment processes for acid dye wastewater. J Environ Sci Health A 38(8):1615–1627
168. Hsu Y, Yen C, Huang H (1998) Multistage treatment of high strength dye wastewater by coagulation and ozonation. J Chem Technol Biotechnol 71(1):71–76
169. Turan-Ertas T (2001) Biological and physical-chemical treatment of textile dyeing wastewater for color and COD removal. Ozone-Sci Eng 23(3):199–206
170. Qi L, Wang X, Xu Q (2011) Coupling of biological methods with membrane filtration using ozone as pre-treatment for water reuse. Desalination 270(1–3):264–268

171. Leshem E, Pines D, Ergas S, Reckhow D (2006) Electrochemical oxidation and ozonation for textile wastewater reuse. J Environ Eng-ASCE 132(3):324–330

172. Avsar Y, Batibay A (2010) Ozone application as an alternative method to the chemical treatment technique for textile wastewater. Fresenius Environ Bull 19(12):2788–2794

173. Turhan K, Turgut Z (2009) Decolorization of direct dye in textile wastewater by ozonization in a semi-batch bubble column reactor. Desalination 242(1–3):256–263

174. Turhan K, Turgut Z (2007) Reducing chemical oxygen demand and decolorization of direct dye from synthetic-textile wastewater by ozonization in a batch bubble column reactor. Fresenius Environ Bull 16(7):821–825

175. Chu L, Xing X, Yu A, Sun X, Jurcik B (2008) Enhanced treatment of practical textile wastewater by microbubble ozonation. Process Saf Environ Prot 86(B5):389–393

176. Gharbani P, Tabatabaii SM, Mehrizad A (2008) Removal of congo red from textile wastewater by ozonation. Int J Environ Sci Technol 5(4):495–500

177. Fu Z, Zhang Y, Wang X (2011) Textiles wastewater treatment using anoxic filter bed and biological wriggle bed-ozone biological aerated filter. Bioresour Technol 102(4):3748–3753

178. Oguz E, Keskinler B (2008) Removal of colour and COD from synthetic textile wastewaters using O_3, PAC, H_2O_2 and HCO_3^-. J Hazard Mater 151(2–3):753–760

179. Meric S, Selcuk H, Belgiorno V (2005) Acute toxicity removal in textile finishing wastewater by Fenton's oxidation, ozone and coagulation-flocculation processes. Water Res 39 (6):1147–1153

180. Bakheet B, Yuan S, Li ZX, Wang HJ, Zuo JN, Komarneni S, Wang YJ (2013) Electro-peroxone treatment of orange II dye wastewater. Water Res 47(16):6234–6243

181. Butler E (2013) Electrochemical/electroflotation process for dye wastewater treatment. Doctoral dissertation, Cleveland State University

Chapter 5
Health Effects and Control of Toxic Lead in the Environment

Nancy Loh, Hsue-Peng Loh, Lawrence K. Wang, and Mu-Hao Sung Wang

Contents

1   Introduction 236
2   Lead Exposure and Health Effects 239
    2.1   Occupational Lead Exposure Criteria 240
    2.2   General Health Effects of Lead Exposure on People of All Ages 241
    2.3   Health Effects of Lead Exposure on Children 243
    2.4   Health Effects of Lead Exposure on Women and Infants 245
    2.5   Health Effects of Lead Exposure on Workers 247
3   Lead Hazard Abatement Methods 248
    3.1   Abatement by Replacement 250
    3.2   Abatement Using Enclosure 250
    3.3   Abatement by Encapsulation 251
    3.4   Abatement by Lead Paint Removal 252
4   Lead Hazard Abatement Operational Procedures 257
    4.1   General Lead Hazard Abatement Process Procedures 258
    4.2   Building Component Replacement Procedures 259
    4.3   Enclosure Installation Procedures 260
    4.4   Lead Paint Removal Procedures 260
    4.5   Encapsulation Procedures 262
5   Lead Hazard Abatement Practices 263
6   Risk Management 265
    6.1   Risk Assessment 265
    6.2   Risk Evaluation Options 266
    6.3   Lead Hazard Screen 266

N. Loh (✉)
Wenko Systems Analysis, 2141 Eye Street, NW, Suite 801, Washington, DC 20037, USA
e-mail: nancyloh1@gmail.com

H.-P. Loh
Wenko Systems Analysis, 230 Beverly Road, Pittsburgh, PA 15216, USA
e-mail: hp.loh2@gmail.com

L.K. Wang • M.-H.S. Wang
Lenox Institute of Water Technology, Newtonville, NY 12128-0405, USA

Rutgers University, New Brunswick, NJ, USA
e-mail: lawrencekwang@gmail.com; lenox.institute@gmail.com

© Springer International Publishing Switzerland 2016
L.K. Wang, M.-H.S. Wang, Y.-T. Hung and N.K. Shammas (eds.),
Natural Resources and Control Processes, Handbook of Environmental
Engineering, Volume 17, DOI 10.1007/978-3-319-26800-2_5

6.4   Lead-Based Paint Inspection 267
7   Lead Hazard Control Planning 269
7.1   Long-Term or Short-Term Response 269
7.2   Lead Hazard Control Planning Steps 270
8   Rules and Regulations for Lead Poisoning Prevention and Environmental Control 270
8.1   Worldwide Awareness and Regulations 270
8.2   OSHA's Lead Regulations 271
8.3   Summary of All US Federal Lead Standards 273
9   New York State Lead Poisoning Prevention and Control Programs 274
9.1   New York State (NYS) Lead Poisoning Prevention Program 274
9.2   NYS Advisory Council on Lead Poisoning
       Prevention Program 276
9.3   Lead Screening of Children and Pregnant Women
       by NYS Health-Care Providers 277
9.4   Lead Screening of Child Care or Preschool Enrollees in NYS 278
9.5   Reporting Lead Exposure Levels in NYS 278
9.6   Manufacture and Sale of Lead-Painted Toys and Furniture in NYS 279
9.7   Use of Leaded Paint in NYS 279
9.8   Abatement of Lead Poisoning Conditions in NYS 279
9.9   Enforcement Agencies in NYS 280
9.10   Sale of Consumer Products Containing Lead or Cadmium in NYS 281
References 281

Abstract Lead is a heavy metal that occurs naturally in the Earth's crust. Lead does not break down in the environment; if left undisturbed, lead is virtually immobile. However, once mined and transformed into man-made products dispersed throughout the environment, lead becomes highly toxic to humans. Lead is ubiquitous in older American homes in house paint and is common in certain industrial workplaces, due to its widespread use over the course of the past century.

The health effects of lead poisoning can be quite serious, and there are numerous regulations regarding its use. The Consumer Product Safety Commission bans the use of lead-based paint in residences. The Occupational Safety and Health Administration sets limits on permissible exposure to lead in the workplace. However, because lead-based paint inhibits the rusting and corrosion of iron and steel, lead continues to be used on bridges, railways, ships, lighthouses, and other steel structures. This is the case even though substitute lead-free coatings are available.

Lead is most commonly absorbed into the body by inhalation, in the form of lead-contaminated dust or mist, often generated from old lead-based paint that has begun to chip. A significant portion of inhaled or ingested lead is absorbed into the bloodstream. Once in the bloodstream, lead circulates through the body and is stored in various organs and body tissues. As exposure continues, the amount stored will increase and accumulate in the body, where it can slowly cause irreversible damage, first to individual cells, then to organs and whole body systems.

Lead is toxic to both male and female reproductive systems. Lead can lead to miscarriage and stillbirth in women exposed to lead. Children born to parents who were exposed to excess lead levels are more likely to have birth defects, mental retardation, or behavioral disorders and are more likely to die during the first year of childhood. Lead is much more harmful to children than adults because it can affect children's brains and nervous systems, which are still developing. Children are more vulnerable to permanent damage; for example, learning disabilities, behavioral abnormalities, attention deficit problems, and insomnia.

Children are also at higher risk because they are more likely to unknowingly inhale or ingest lead. Exposure to lead in house dust tends to be highest for young children, due to their frequent and extensive contact with floors, carpets, window areas, and other surfaces where dust gathers, as well as their frequent hand-to-mouth activity. It is common for young children to put everything, including hands, pacifiers, toys, and other small objects, into their mouths.

Workers involved in iron work, demolition work, painting, lead-based paint abatement, plumbing, heating and air conditioning, and carpentry are also potentially at risk for high lead exposure.

There have been significant efforts over the past few decades to reduce lead exposure in the USA, but lead poisoning is still an important public health issue. To address potential lead poisoning, a risk evaluation of a building or home will determine the risk and extent of the lead hazard present. Once a lead risk is confirmed, the building should undergo abatement, the process of eliminating or mitigating the lead hazard. There are several approaches to abatement that can be taken.

This chapter will discuss the health effects of lead on children, women, workers, and their protection. It will then detail the types of abatement and how they can be implemented. The risk evaluation and the US federal and New York State rules and regulations on lead will also be discussed.

Keywords Heavy metal • Lead • Lead poisoning • Lead-based paint • Lead exposure • Lead hazard • Lead-contaminated dust • Occupational lead exposure criteria • Blood lead level (BLL) • Permissible exposure limit (PEL) • Health effect • Worker • Children • Infant • Women • Male infertility • Paint replacement • Enclosure installation • Encapsulation • Paint removal • Abatement procedures • Risk management • Hazard screen • Hazard control • Rules • Regulations • US EPA • New York State

Acronym

ABLES	New Jersey Adult Blood Lead Epidemiology and Surveillance System
ACCLPP	Advisory Committee on Childhood Lead Poisoning Prevention
ACGIH	American Conference of Government Industrial Hygienists
ASTM	American Society for Testing and Materials
ATSDR	Agency for Toxic Substances and Disease Registry
BLL	Blood lead level
CDCP	Centers for Disease Control and Prevention
CPSC	Consumer Product Safety Commission
FDA	Food and Drug Administration
HEPA	High-efficiency particulate air

HUD   US Department of Housing and Urban Development
LBP   Lead-based paint
MCLG   Maximum contaminant level goal
MSDS   Material safety data sheet
NAAQS   National Ambient Air Quality Standard
NHANES   National Health and Nutrition Examination Survey
NIOSH   National Institute for Occupational Safety and Health
NLLAP   National Lead Laboratory Accreditation Program
NTP   National Toxicology Program
NYCRR   New York Codes, Rules, and Regulations
NYS   New York State
NYSDOH   New York State Department of Health
OSHA   Occupational Safety and Health Administration
PEL   Permissible exposure limit
PPE   Personal protection equipment
PPM   Parts per million
REL   Recommended exposure limit
TLV   Threshold limit value
TSCA   Toxic Substances Control Act
TWA   Time-weighted average
US EPA   US Environmental Protection Agency
µg/dL   Microgram per deciliter
µg/g   Microgram per gram
µg/L   Microgram per liter
µg/m^3 Microgram per cubic meter

1   Introduction

Lead is a malleable, blue-gray, heavy metal that occurs naturally in the Earth's crust, with trace amounts found in soil and plants. Since lead is a natural element, it does not break down in the environment. If left undisturbed, lead is virtually immobile. However, once mined and transformed into man-made products that are dispersed throughout the environment, lead becomes highly toxic.

Lead is used to produce a variety of industrial and consumer products and is frequently used in construction and electrical conduits. In plumbing, soft solder, which is used chiefly for soldering, is an alloy of lead and tin. Lead is ubiquitous in older American homes painted with lead-based paint. Lead exposures in industrial workplaces or construction sites are common because of the widespread use, during the past century, of lead compounds in paints, gasoline, and various industrial materials.

Soft solder has been banned for many uses in the USA. In addition, the Consumer Product Safety Commission bans the use of lead-based paint in residences. Because lead-based paint inhibits the rusting and corrosion of iron and steel, however, lead continues to be used on bridges, railways, ships, lighthouses, and other steel structures, although substitute coatings are available. Today lead is still found in:

1. House paint made before 1978
2. Toys and furniture made before 1976
3. Painted toys and decorations
4. Lead bullets, fishing sinkers, and curtain weights
5. Plumbing, pipes, and faucets
6. Soil contaminated by decades of car exhaust and lead-based gasoline spills
7. Dust from years of house paint scraping
8. Hobbies involving soldering, stained glass
9. Children's paint sets and art supplies
10. Pewter pitchers and dinnerware
11. Storage batteries

For centuries, exposure to high concentrations of lead has been known to pose health hazards. Lead's toxicity was recognized as early as 2000 BC [1].

Lead is most commonly absorbed into the body by inhalation, in the form of lead-contaminated dust or mist. Once the dust or mist is inhaled, the lungs and upper respiratory tract absorb it into the body. Lead can also be absorbed through the digestive system if it enters the mouth and is ingested.

A significant portion of inhaled or ingested lead gets absorbed into the bloodstream. Once in the bloodstream, lead circulates through the body and is stored in various organs and body tissues. As exposure continues, and if the body absorbs more lead than it excretes, the amount stored will increase and accumulate in the body. The lead stored in the tissue can slowly cause irreversible damage, first to individual cells, then to organs and whole body systems.

There are many possible symptoms of lead poisoning. Some of the common symptoms include:

1. Loss of appetite
2. Constipation
3. Nausea
4. Excessive tiredness
5. Headaches
6. Fine tremors
7. Severe abdominal pain
8. Metallic taste in the mouth
9. Weakness
10. Nervous irritability
11. Hyperactivity
12. Muscle and joint pain or soreness
13. Anxiety
14. Pallor
15. Insomnia
16. Numbness
17. Dizziness

Lead is toxic to both male and female reproductive systems. Lead can alter the structure of sperm cells; and there is evidence of increased miscarriage and stillbirth in women exposed to lead or whose partners have been exposed to lead. Children born to parents who were exposed to excess lead levels are more likely to have birth defects, mental retardation, or behavioral disorders and are more likely to die during the first year of childhood.

Because of lead's ubiquitous presence in industrial societies, there are many sources and pathways of lead exposure in children. These include lead-based paint in industrial, commercial, and residential buildings and equipment, lead-based paint in art, dust and soil contaminated by lead, presence of lead in drinking water distribution systems, and current use of lead in the manufacture of some products like toys.

It is common for young children to put everything, including hands, pacifiers, toys, and other small objects, into their mouths. Anything which contains lead, from small dust particles to large paint chips, can cause harm if swallowed. Exposure to lead in house dust tends to be highest for young children, due to their frequent and extensive contact with floors, carpets, window areas, and other surfaces where dust gathers, as well as their frequent hand-to-mouth activity.

Lead poisoning commonly contributes to problems which may become permanent in young children. Learning disabilities, behavior abnormalities, attention deficit problems, and insomnia are common symptoms. Lead is much more harmful to children than adults because it can affect children's brains and nervous systems, which are still developing.

Children less than 6 years of age are of special concern because their developing brains and bodies can easily be damaged by lead and because they are more likely to put the lead-contaminated dust, loose paints, etc., into their mouths.

Unborn children are the most vulnerable. Possible complications in behavior or attention problems include:

1. Failure at school
2. Hearing problems
3. Kidney damage
4. Reduced IQ
5. Slowed body growth

Workers potentially at risk for lead exposure include those involved in iron work, demolition work, painting, lead-based paint abatement, plumbing, heating and air conditioning, and carpentry, among others.

Lead-based paint is paint containing lead pigment. Chrome yellow is a yellow pigment containing lead chromate $PbCrO_4$. Lead white is a white pigment containing lead carbonate $PbCO_3$. Lead pigment is added to paint liquid for speeding up the drying process, increasing paint's durability, resisting corrosion, and maintaining an attractive appearance. Lead-based paint is one of the main environmental and health hazards.

Once discovered, lead hazards must be eliminated and/or reduced, a process called abatement. There are four general approaches to abatement depending on the existence, nature, severity, and location of lead-based hazards: replacement, enclosure, encapsulation, and paint removal.

1. Replacement is removal of lead-based painted components and replacing them with new "lead-free" components.
2. Encapsulation is the coating of a lead-based painted surface with rigid materials that rely on adhesion to that surface instead of being mechanically fastened.
3. Enclosure involves covering the lead-based paint with a solid, dust-tight barrier so that it is completely enclosed.
4. Paint removal is removal of lead-based paint from building components, leaving the subsurface intact.

Once abatement is completed, the property must be cleared—that is, it must be inspected to ensure that it clears acceptable standards for lead hazards.

There are numerous regulations and laws relevant to lead poisoning and abatement. Some of the federal agencies that oversee or regulate lead in the USA include the US Environmental Protection Agency (US EPA), Centers for Disease Control and Prevention (CDCP), and the National Institute for Occupational Safety and Health (NIOSH), which is part of the CDC. The Occupational Safety and Health Administration (OSHA) sets guidance on lead exposure in workplace settings; and the Department of Housing and Urban Development (HUD) does so with regard to housing. There are other rules and regulations from state and local authorities.

A risk evaluation is usually performed to identify the lead risk and to determine the best approach for abatement. There are four ways to perform risk evaluation: risk assessment, lead hazard screen, lead-based paint inspection, and combination inspection/risk assessment. Risk assessment is on-site investigations to determine the existence, nature, severity, and location of lead-based paint hazards; lead hazard screen identifies lead-based paint hazard and other potential lead hazards. Lead-based paint inspection involves conducting a visual assessment, analyzing dust and soil samples, and interviewing property owners or residents.

The issues to be addressed in the following sections include:

1. Health effects on children, women, and lead-based paint workers
2. Abatement approaches
3. Risk evaluation
4. Federal rules and lead standards

2 Lead Exposure and Health Effects

The toxic health effects of lead in both children and adults are well documented. Lead exposure in adults can damage the central nervous system, cardiovascular system, reproductive system, hematological system, and the kidneys. Lead exposure in adults can also harm the development of their children. Lead has been shown to be an animal carcinogen as well, and toxic lead compounds are listed under various categories by the International Agency for Research on Cancer (IARC) (www.cancer.org).

The most vulnerable groups are children, nursing and pregnant women, and workers who work with lead-based paint or other lead-based materials. Lead is much more harmful to children than adults because it can affect children's brains and nervous systems, which are not yet fully developed. The younger the child, the more harmful lead can be.

The well-known danger of lead poisoning prompted government action in 1978, when the Occupational Safety and Health Administration (OSHA) promulgated a lead standard to protect workers in general industry.

Because of these national efforts to reduce environmental lead exposures, general population lead exposures in the USA have dropped significantly in the past two decades. Lead exposures in the workplace, however, continue to be a significant public health problem. One complicating factor is the nature of the symptoms of lead poisoning; lead poisoning often goes undetected since many of the symptoms, such as stomach pain, headaches, anxiety, irritability, and poor appetite, are nonspecific and may not be immediately recognized as symptoms of lead poisoning.

Research studies on lead toxicity in humans indicate that current OSHA standards should prevent the most severe symptoms of lead poisoning, but these standards do not fully protect workers and their developing children from all of the adverse effects of lead.

2.1 Occupational Lead Exposure Criteria

Human lead exposure occurs when dust and fumes are inhaled or when lead is ingested via lead-contaminated hands, food, water, cigarettes, and clothing. Once exposure occurs, lead entering the respiratory and digestive systems is released into the bloodstream and distributed throughout the body. More than 90 % of the total body burden of lead is accumulated in the bones, where it can remain for decades. In addition, secondary exposure may occur; lead that has accumulated in bones may be subsequently released into the blood and reexpose organ systems long after the original environmental exposure. In pregnant women, this process can also expose the fetus to lead.

There are several biological indices of lead exposure. Lead concentrations in the blood, urine, teeth, and hair can be used as indicators to measure the level of lead exposure. At present, the best available method for monitoring biological exposure to lead is measurement of the blood lead level (BLL), usually measured in micrograms/deciliter (μg/dL) of blood.

Under the OSHA general industry lead standard (29 CFR 1910.1025), the permissible exposure limit (PEL) for personal exposure to airborne inorganic lead was set in 1978, at a maximum BLL of 50 μg/m^3 (micrograms per cubic meter of air). Specifically, the OSHA set a PEL of 50 μg/m^3 averaged over an 8-h period (as an 8-h time-weighted average or TWA), when respirators are used, and a PEL of 30 μg/m^3 when respirators are not used. The 1970s saw the elimination or restriction of the use of leaded gasoline and lead-based paint in the USA. Maintaining the concentration of airborne particles of lead in the work environment below the PEL is a preventive measure intended to protect workers from excessive exposure. Much progress has been made in reducing general lead exposures; more than 90 % of adults now have a BLL < 10 μg/dL, and more than 98 % have a BLL < 15 μg/dL. However, exposures in the workplace continue to be a significant public health problem. Even with the federal regulations, thousands of adults with BLLs of at least 25 μg/dL are reported each year to NIOSH by states participating in a NIOSH surveillance program.

In the 1978 general industry standard, OSHA advised that men or women planning to have children should limit their exposure to maintain a BLL less than 30 μg/dL.

Research studies on lead toxicity in humans indicate that while current OSHA standards should prevent the most severe symptoms of lead poisoning, they do not protect workers and their children from all of the adverse effects of lead. Even lower BLLs, within the permissible limit, can cause damage. In recognition of this problem, standards and public health goals, set up by various agencies, have established lower exposure limits for workers exposed to lead, to offer increased protection for those workers and their children.

2.2 General Health Effects of Lead Exposure on People of All Ages

The health effects of lead have been previously extensively documented by the federal public health agencies: Agency for Toxic Substances and Disease Registry (ATSDR), Centers for Disease Control and Prevention (CDCP), and National Institute for Occupational Safety and Health (NIOSH) [2–4].

Excessive exposure to lead can lead to several types of health effects, such as neurotoxic effects, hematological and renal effects, cardiovascular effects, and reproductive problems. These are described in greater detail below.

2.2.1 Neurotoxic Effects

One of the major targets of lead toxicity is the nervous system, including the central and peripheral nervous systems. Lead damages the blood-brain barrier and, subsequently, contaminated blood can harm brain tissues. Severe exposures resulting in BLLs > 80 μg/dL may cause coma, encephalopathy, or even death.

Because of the improved control of occupational lead exposures in recent decades, such occurrences of extreme lead toxicity are rare today in the USA.

Occupational lead exposures allowable under the current OSHA lead standards will not produce these obvious neurologic clinical symptoms. However, lead exposure levels permissible under the OSHA standards may still be harmful to the central nervous system. Workers with BLLs of 40–50 µg/dL may experience fatigue, irritability, insomnia, headaches, and subtle evidence of mental and intellectual decline [5, 6]. BLLs as low as 30–40 µg/dL decrease motor nerve conduction velocity in workers, although these lead exposure levels are not associated with clinical symptoms [7]. These subclinical symptoms represent early stages of neurologic damage to the central and peripheral nervous system.

2.2.2 Hematologic and Renal Effects

Anemia, a blood iron deficiency, is one of the most characteristic symptoms of high and prolonged human exposures to lead, associated with BLLs > 80 µg/dL. For children, a significant association was found for mild and severe anemia, even at the BLL range of 10–20 µg/dL. The anemia results from the damaging effects of lead on the formation and functioning of red blood cells.

Chronic high exposure to lead, above the OSHA permissible exposure limits (PEL) of 50 µg/m³, may cause chronic nephropathy and, in extreme cases, kidney failure. There is substantially less evidence of kidney disease at lower exposures to lead [8].

2.2.3 Reproductive and Developmental Effects

Historical studies indicate that high exposures to lead can produce stillbirths and miscarriages [9]. Several studies conducted in the USA and abroad have indicated that exposures to lower concentrations of lead, with BLLs at or below 15 µg/dL, may result in adverse pregnancy outcomes, such as shortened time of gestation and decreased fetal mental development and growth [10, 11].

The developing nervous system of the fetus is particularly vulnerable to lead toxicity. Neurological toxicity is observed in children of exposed female workers, a result of the ability of lead to cross the placental barrier and to cause neurological impairment in the fetus [12].

BLLs of 60 µg/dL or higher may be associated with male infertility [13]. Studies in male workers also indicate that exposures to lead resulting in BLLs as low as 40 µg/dL may cause decreased sperm count and abnormal sperm morphology [14, 15].

Several reports indicate that decreased sperm quality and hormonal changes can occur among male workers exposed to lead with BLLs of 30–40 µg/dL [16, 17].

2.2.4   Cardiovascular Effects

Chronic high exposures to lead that occurred earlier in the last century were associated with an increased incidence of hypertension and cardiovascular disease [18]. Today these severe effects of lead exposure are rarely observed in the USA [19]. Studies conducted in the general population, where lead exposures are much lower, have also indicated that increased BLLs are associated with small increases in blood pressure. This relationship appears to extend to BLLs below 10 µg/dL [20–23].

2.2.5   Carcinogenic Effects

Results from two studies indicate that lead may increase the risk of cancer among workers exposed to high levels of lead [24, 25].

2.3   Health Effects of Lead Exposure on Children

While people of all ages can suffer from excessive lead exposures, the groups most at risk are fetuses, infants, and children under 6 years of age, because they are more vulnerable to the effects of lead poisoning and because they are more likely to unknowingly expose themselves to lead.

Because children's brains, nervous systems, and other body systems are still developing, they sustain the greatest impact of lead exposure. The first 3 years of life are characterized by major growth and developmental events in the nervous system, so young children have increased susceptibility to the neurodevelopmental effects of lead [26]. Once absorbed, some lead goes into the blood and can be eliminated more quickly, but most of the lead is stored in bones, where it can stay many years.

Lead poisoning commonly contributes to a variety of problems that may ultimately become permanent in young children. Lead can have neurotoxic effects even at low levels, causing reductions in attention span, reading and learning disabilities, hyperactivity, insomnia, and behavioral problems and abnormalities [27]. The National Toxicology Program (NTP) has concluded that childhood lead exposure is associated with reduced cognitive function [28]. Children with higher blood lead levels generally have lower scores on IQ tests [29–35] and reduced academic achievement [28]. In addition to the effects on IQ and school performance, research on the effects of lead has increasingly been addressing the effects of lead on behavior.

Lead poisoning affects children across all socioeconomic strata and in all regions of the country. However, because lead-based paint hazards are most severe in older, dilapidated housing, the poor in inner cities are disproportionately affected.

Despite steady and impressive progress since the 1970s in reducing blood lead levels (BLL) among the US population, childhood lead poisoning remains a major, but preventable, environmental health problem in the USA. Children continue to be exposed to lead due to the widespread distribution of lead in the environment.

2.3.1 Lead in Surface Dust

In the USA, the major current source of early childhood lead exposure is lead-contaminated house dust ingested by normal hand-to-mouth and toy-to-mouth activity. The major contributor to lead in house dust is deteriorated or disrupted lead-based paint [36–40] most typically found in older houses. Leaded dust is generated as lead-based paint deteriorates over time, damaged by moisture, abraded on friction and impact surfaces, or disturbed in the course of renovation, repair, or abatement projects. Deteriorating paint can also chip and fall to floors or window sills, creating lead-based paint chips that children may ingest. The likelihood, extent, and concentration of lead-based paint increase with the age of the building. Because the greatest risk of paint deterioration is in dwellings built before 1950, older housing generally commands a higher priority for lead hazard controls [41].

2.3.2 Lead in Soil

Children can be exposed to lead from direct contact with lead-contaminated soil or soil tracked in from outside the home. The high levels of lead in soil typically come from deteriorating exterior lead-based paint around the foundation of a house. Other known sources of lead in soil include historical airborne emissions of leaded gasoline and emissions from industrial sources such as smelters [26, 42–45].

2.3.3 Other Causes of Lead Poisoning

Children can also be exposed to lead in drinking water or by inhaling lead in ambient air. Contamination of drinking water can occur from corrosion of lead pipes and other elements of water distribution systems. Exposure via drinking water may be particularly high among very young children who consume baby formula prepared with water that is contaminated by leaching lead pipes [40, 46, 47].

Other sources and pathways of lead poisoning in children can include point sources, ceramics, toys, children's jewelry, lead brought home from a parent's workplace, imported candy and its candy wrappers, home and folk remedies, cosmetics, and hobby supplies [48, 49]. These sources may account for a small amount of children's exposure; for most children, paint, dust, and soil are the primary sources of lead poisoning.

Mothers who are exposed to lead can also transfer lead to the fetus during pregnancy and to the child while breast feeding [48, 49].

While infants and very young children are at greatest risk, lead is toxic to children older than 5 years as well; they are also susceptible to the neurodevelopmental effects of lead.

In October 1991, CDCP formally revised its statement on Preventing Lead Poisoning in Young Children [2] reducing its "level of concern" for childhood lead poisoning from the previous threshold of 25 μg/dL to 10 μg/dL.

Until recently, CDCP had defined this blood lead level of 10 micrograms per deciliter (μg/dL) as "elevated"; this definition was used to identify children for blood lead case management [50, 51]. CDCP now specifically notes that "no level of lead in a child's blood can be specified as safe," and the NTP has concluded that there is sufficient evidence of adverse health effects in children at blood lead levels less than 5 μg/dL [28, 36].

CDCP recommends that sources of lead in children's environments be controlled or eliminated before children come into contact with them and are poisoned, i.e., "primary prevention" [52, 53].

2.4 Health Effects of Lead Exposure on Women and Infants

Lead exposure remains a concern for pregnant and lactating women. There is increasing awareness that unintended exposures to environmental contaminants such as lead adversely affect maternal and infant health, including the ability to become pregnant, maintain a healthy pregnancy, and have a healthy baby.

However, guidance for clinicians and prenatal health-care providers regarding screening and managing pregnant and lactating women exposed to lead has not kept pace with the scientific evidence. There are currently no national recommendations by any medical, obstetric, family practice, or pediatric nursing professional association that covers lead risk assessment and management during pregnancy and lactation.

The CDCP has not identified an allowable exposure level, level of concern, or any other bright line intended to connote a safe or unsafe level of lead exposure for either mother or fetus. In other words, there is no apparent threshold below which adverse effects of lead do not occur. Instead, CDCP is applying public health principles of prevention, i.e., recommending follow-up blood lead testing and interventions when prudent. These guidelines recommend follow-up activities and interventions beginning at BLLs of ≥5 μg/dL in pregnant women.

There is evidence that a significant number of pregnant women and presumably their infants are being exposed to lead in the USA today. Lead exposure remains a public health problem for subpopulations of women of childbearing age and for the developing fetus and nursing infant as prenatal lead exposure has known influences on both maternal health and infant birth and neurodevelopmental outcomes [54].

Since bone lead stores persist for decades, women and their infants may be at risk for continued exposure long after initial exposure to external environmental sources has been terminated.

High levels of lead exposure can result in delirium, seizures, stupor, coma, or even death. Other overt signs and symptoms may include hypertension, peripheral neuropathy, ataxia, tremor, headache, loss of appetite, weight loss, fatigue, muscle and joint aches, changes in behavior and concentration, gout, nephropathy, lead colic, and anemia. In general, symptoms tend to increase with increasing blood lead levels.

2.4.1 Impact on Sexual Maturation and Fertility

Although studies are limited, there is some suggestion that blood lead at relatively low levels may lead to alterations in the onset of sexual maturation and reduced fertility. These findings underscore the importance of considering sensitive markers of human fecundity in relation to lead exposure and should be confirmed in studies that can address the methodological limitations of previous research [54].

2.4.2 Impact on Maternal/Gestational Hypertension

Lead is an established risk factor for hypertension in adults [22, 55]. Hypertension is also one of the most common complications of pregnancy.

Gestational hypertension has been associated with adverse maternal and perinatal outcomes. Lead exposure has been associated with increased risk for gestational hypertension, but the magnitude of the effect, the exposure level at which risk begins to increase, and whether risk is most associated with acute or cumulative exposure remain uncertain.

2.4.3 Impact on Pregnancy Outcomes

Overall, increased risk for spontaneous abortion (miscarriage) appears to be associated with blood lead levels ≥ 30 µg/dL. Limited evidence suggests that maternal blood lead levels less than 30 µg/dL could also increase the risk for spontaneous abortion, although these findings remain to be confirmed in further research. Maternal lead exposure may increase the risk for preterm delivery and low birth weight, although data are limited and a blood lead level at which these risks begin to increase has not been determined. The available data are inadequate to establish the presence or absence of an association between maternal lead exposure and major congenital anomalies in the fetus [54].

2.4.4 Impact on Infant Growth and Neurodevelopment

Numerous studies on the association between prenatal lead exposure and infant growth have been conducted, but data is limited and thus it is difficult to make sweeping conclusions.

Two studies suggest an association between maternal lead exposure and decreased growth. In one study, maternal bone lead levels were negatively associated with infant weight at 1 month of age and with postnatal weight gain between birth and 1 month [56]. In another study, the postnatal linear growth rate was negatively related to prenatal blood lead level, although only when infants' postnatal lead exposure was also elevated [57].

The findings of recent cohort studies suggest that prenatal lead exposure at maternal blood lead levels below 10 μg/dL is inversely related to neurobehavioral development independent from the effects of postnatal exposure.

As previously noted, CDCP has not identified any threshold below which adverse effects of lead do not occur for either mother or fetus, instead, applying public health principles of prevention to intervene when prudent. Specific recommendations are presented throughout the rest of these guidelines.

2.5 Health Effects of Lead Exposure on Workers

Workers who work with any lead-based materials—possible examples are construction or demolition work—are at higher risk for lead exposure. The potential for worker exposure to lead (as well as to other hazardous substances, safety hazards, and physical agents) also exists during all lead hazard control projects, efforts meant to remove or minimize lead hazards. Due to the recognized adverse health effects of lead, employers should minimize worker lead exposures as much as possible. Employers should refer directly to the OSHA construction lead standard (Sect. 5) for complete requirements.

Families of construction workers, including those involved in lead-based paint (LBP) activities, may also be exposed to lead brought home from the workplace. Studies suggest that construction workers' occupational lead exposures combined with ineffective hygiene practices can result in lead contamination of their cars and homes [58]. NIOSH and the New Jersey Department of Health conducted a surveillance study in 1993 and 1994 involving the voluntary participation of 46 construction workers' families. The workers, who had reported BLLs of 25 μg/dL, were identified from the 510 construction workers in the New Jersey ABLES registry [59]. BLL testing of young children indicated that the workers' children, particularly those under age six, were at greater risk of having elevated BLLs (10 μg/dL) than children in the general population. Higher percentages of workers' children in age categories 1–2 and 3–5 years had elevated BLLs than national averages for these ages.

There is also potential for take-home lead exposures among families of renovation and remodeling workers. Exposure to lead in construction activities can result in workers' vehicles being contaminated and a significant amount of lead being transported into the home. A NIOSH study of lead-exposed residential renovation and repair workers found higher surface lead levels in 20 full-time workers' vehicles (arithmetic mean: 3300 $\mu g/m^3$) than in those of 11 part-time volunteers (1500 $\mu g/m^3$), although the difference did not reach statistical significance [60].

3   Lead Hazard Abatement Methods

Once a lead hazard is discovered or determined, it should be removed or mitigated. The process for doing so is called abatement. According to the Lead-Based Paint Hazard Reduction Act of 1992, "abatement" refers to the methods used to permanently eliminate lead-based paint hazards or to make lead-based paint unavailable. HUD has defined "permanent" as lasting at least 20 years. US EPA defines a "paint-lead hazard" as:

1. Lead-based paint (LBP) on any friction surface that rubs against another surface and creates a dust-lead hazard
2. Lead-based paint that is damaged or deteriorated on any impact surface
3. Any chewable lead-painted surface on which there is evidence of teeth marks
4. Any other deteriorated lead-based paint on the inside or outside of any residential building or child-occupied facility

The term "abatement" includes a number of other ancillary activities that are not directly related to the removal of lead itself, but that must be included in the overall effort for the abatement to be successful. These activities include lead hazard evaluation, planning, cleaning, clearance, and waste disposal. When abatement is performed inadequately, or without sufficient protection, it can be even more harmful, causing lead exposures to increase.

Because lead abatement work is dangerous, all lead-based paint abatement contractors and firms must be certified to perform this type of work, and all abatement workers and supervisors must be trained and certified. Certification of abatement contractors and completion of clearance examinations by independent, certified risk assessors, lead-based paint inspectors, or sampling technicians ensures that abatement work is conducted properly and safely. The US EPA's regulations are generally implemented through state, tribal, or territorial programs.

There are four basic methods of lead abatement (Table 5.1):

1. Replacement—removing the building part coated with lead-based paint and replacing it with a new one
2. Enclosure—covering the lead-based paint with a solid barrier
3. Encapsulation—coating the lead-based painted surface so that it is not accessible
4. Paint removal

Table 5.1 Comparison of four abatement methods

Approach	Pros	Cons
Replacement		
Removing lead-based painted components and replacing them with new "lead-free" components	Quick way to remove lead-based paint (LBP) Permanent solution Can improve building through upgrades New component can lower heating bills and maintenance costs	May involve demolition work Can create a lot of dust Personal protection equipment (PPE) may be necessary
Enclosure		
A rigid, mechanically affixed barrier Surface preparation is necessary "Source" problems must be fixed LBP surfaces must be labeled PPE may be needed	Uses locally available materials Durable and long lasting Low generation of waste and dust	LBP is still present LBP may be disturbed during routine work on enclosure Enclosed surfaces must be monitored for damage
Encapsulation		
Uses a liquid, paint-like material Surface preparation is critical Must be strong but flexible Must provide complete coverage over old paint PPE may be needed	Little dust is generated Lower cost than other abatement options Many choices are available	Not appropriate for use on friction surfaces Durability depends on condition of previous paint layers Periodic monitoring and maintenance is required Susceptible to water damage Some systems may contain toxic ingredients
Paint removal		
Takes off lead-based paint Dust generation must be controlled Wet scraping Wet planning Electric heat guns Local exhaust hand tools Chemical stripping	LBP is gone Useful for historic preservation projects or detailed components Many options are available	Tedious and time consuming Dust is generated Strippers create hazardous waste Surface must be properly prepared for new surface

Source: US EPA model worker course. March 2004 [61]

Enclosure and encapsulation do not remove the lead-based paint; they abate or mitigate the lead-based paint hazard. The four methods are described in detail below.

3.1 Abatement by Replacement

Replacement entails removing the lead-painted building component (such as a window) and replacing it with a new one that is not painted with lead-based paint. This method is mostly recommended for windows, doors, and other woodwork that has been coated with lead-based paint.

Replacement is the easiest and quickest way to get rid of lead-based paint. Replacement is a permanent solution and removes lead-based paint forever. When combined with overall modernization, replacement can upgrade the building itself.

Replacement is expensive and can be a lot of work. Skilled carpenters are often needed to put in the new parts—especially windows and doors. Surfaces next to the part being removed may also get damaged in the process. Replacement can involve manual demolition work which can create a lot of dust.

3.2 Abatement Using Enclosure

Enclosure involves covering the lead-based paint with a solid, dust-tight barrier so that it is completely enclosed. An enclosure keeps the lead-based paint away from the rest of the building and its building occupants.

The materials used to enclose the lead-painted surface must be durable. Common materials used to build enclosures include aluminum, fiber board, vinyl, plywood, drywall, tiles, and acrylic sheets.

Wallpaper and contact paper are not suitable enclosure materials because they are not dust-tight.

Enclosures are made using locally available construction materials and are usually affixed to the substrate using screws or nails or some other mechanical fastening system. Enclosures are durable and, if done right, don't create much waste or dust. They should typically last for at least 20 years under normal conditions.

Enclosure does not permanently remove the lead-based paint as the lead source still remains underneath the covering. Any subsequent renovation or repair work to enclosed surfaces will likely disturb the lead-based paint and release lead dust that has collected behind the enclosure barrier. Therefore, enclosed surfaces and joints of the enclosures must be periodically monitored for damage and deterioration.

3.3   *Abatement by Encapsulation*

Encapsulants are coatings of rigid materials that rely on adhesion to a lead-based painted surface instead of being mechanically fastened to the substrate. Encapsulation should not be confused with enclosure (described above), which involves mechanical fastening of rigid materials as the primary method of attachment.

Encapsulation is a process that makes lead-based paint inaccessible by providing a barrier between the lead-based paint and the environment. This barrier is formed using a liquid-applied coating or an adhesively bonded covering material. While encapsulant systems may also be attached to a surface by using supplemental mechanical fasteners, the primary means of attachment for an encapsulant is bonding of the coating or covering to the painted surface (either by itself or through the use of an adhesive).

Effective encapsulation depends upon a strong bond between the surface of the existing paint film and the encapsulant. However, this condition alone is not sufficient for encapsulation system success. All layers of the existing paint film must adhere well to each other, as well as to the base substrate. If not, the encapsulation system may fail. Thus, proper assessment of the suitability of the surface and substrate for encapsulation is essential prior to the application and installation of the product.

If the assessment shows significant surface deterioration, then encapsulation cannot be done, or the surface must undergo preparation for encapsulation; patch testing of the product should be done in the field before application. In addition to proper completion of preparation and application procedures, the encapsulated surface should be monitored by the owner and residents on an ongoing basis, with periodic reevaluation by a certified risk assessor.

Encapsulation technologies can offer safe and effective control of lead-based paint hazards and may be one of the only alternatives that can be used in certain situations because it is often less expensive.

Encapsulants may also be used in combination with other methods. Unless there is significant surface deterioration, encapsulants typically generate low amounts of leaded dust. However, if the encapsulation system fails, repairing the damage, as well as covering the exposed lead-based paint surfaces, may result in high maintenance costs.

Encapsulation's durability depends on the condition of previous paint layers. Field compatibility testing of the encapsulant with the particular lead-based painted surface is essential. Encapsulant system's success depends on proper surface preparation. Application may be weather and temperature dependent and may require several coats. Encapsulation can also be susceptible to water damage.

In recent years, encapsulation has been used less often than other abatement methods. The disadvantages appear to have outweighed the advantages in many cases.

3.4   Abatement by Lead Paint Removal

Paint removal is the removal of lead paint from building components, leaving the substrate intact. More than any other abatement method, on-site paint removal involves the greatest degree of disturbance and dust generation. Therefore, on-site removal of lead-based paint from a substrate should be carried out only if abatement is required and if no other abatement method is feasible. Paint removal may increase the level of lead in household dust and make effective cleaning more difficult. Even if dust clearance standards are met, any increase in leaded dust levels, no matter how small, over baseline levels means some increase in exposure. Furthermore, all paint removal methods leave behind residues embedded in the substrate, which could continue to pose a hazard if the surface from which the paint is removed is later disturbed. Therefore, paint removal is the most invasive of abatement methods and should be avoided if possible.

In some cases, off-site paint removal can be an option, i.e., where the painted element is removed from the site and treated in a commercial or more controlled setting. This is a safer option for potential residents or neighbors of the building.

There are three methods of paint removal: mechanical, heat, and chemical.

3.4.1   Mechanical Methods

Mechanical methods (scraping, sanding, and blasting) are abrasive methods for removing paint from lead-based surfaces. They have two major drawbacks. They produce high levels of dangerous residue, and they have the potential to damage the underlying substrate.

3.4.1.1   Scraping

Traditional scraping involves using hand tools to scrape the paint off the surface/ substrate. It is a slow process that can remove lead coatings only with great effort. Aggressive scraping can gouge wooden surfaces and molding profiles. In general, scraping is best suited for use on completely flat surfaces.

Two kinds of scraping are available: wet and dry. Wet scraping entails misting loose paint with water before scraping it and continuing to mist with water while scraping. It is usually preferable to use wet scraping, as dry scraping creates a huge amount of dust. Wet scraping tends to create less dust than many other on-site paint removal methods; paint chips removed by scraping can be controlled much more easily than the more minute residue generated by other methods.

3.4.1.2 Sanding

Sanding can be done manually with sandpaper or using power sanding tools (Fig. 5.1); both hand sanding and power sanding generate extremely high levels of paint dust. Rotary, disk, and belt sanding procedures, in addition to the problem of lead dust generation, can destroy carved work and molding profiles altogether. For these reasons, power sanding procedures are usually not appropriate for on-site surface preparation, especially in occupied buildings. The dust generated from sanding can permeate the building and endanger residents.

If the elements to be treated can be safely removed and taken to a controlled site for treatment, some sanding methods may be appropriate. Belt sanders can be used on flat surfaces, and orbital sanders, if used with great care, can be used with damage kept to a minimum. If off-site treatment is considered, proper care must be taken to remove features without damaging them. High-efficiency particulate air (HEPA)-equipped sanders, often used by professionals, help control the amount of lead dust put into the air. If power sanding is to be considered, HEPA equipment is strongly recommended. In addition, workers should also wear the HEPA mask shown in Fig. 5.2 for protection.

Fig. 5.1 A typical power sanding tool

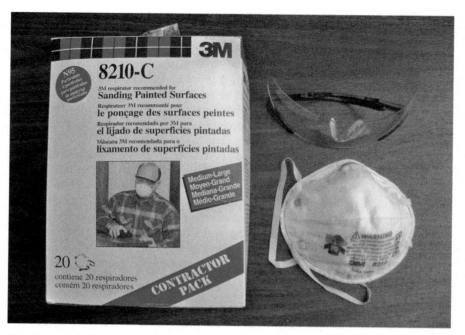

Fig. 5.2 A typical HEPA filtration mask

3.4.1.3 Blasting

All methods of abrasive blasting involve semi-controlled pulverization of the surface being treated. Under most conditions, the damage to the substrate will be irreparable [62]. It does not matter whether the blasting is dry or wet, fine grit or coarse grit, low pressure or high pressure, or whether the aggregate is sand, crushed walnut shells, glass beads, water, or air; abrasive blasting is destructive. Abrasive blasting will remove lead-based paint, but it will also create tremendous amounts of lead dust. The dust will permeate interior spaces, and in exterior applications it can be carried over a wide area, contaminating neighboring properties. Because of these potent hazards, abrasive blasting should never be used to remove lead paint.

3.4.2 Heat Methods

Lead-based paint can also be removed using a variety of heat (thermal) methods. Heat is applied to the paint, which softens it; the loosened paint can then be scraped off with hand tools. Certain tools, like blowtorches and other open-flame techniques, are extremely dangerous and should never be used [63]. In addition to the very real danger of burning down the building and the threat to human lives,

Fig. 5.3 A typical heat gun

open-flame techniques produce very high levels of toxic lead dust and fumes that can be easily inhaled.

Heat guns (Fig. 5.3) and heat plates generate lower levels of heat than do torches. Electric heat guns may be used to force warmed air onto a painted surface. However, even at these lower levels of heat, there is still a danger of generating lead fumes and of igniting sawdust, construction debris, or other materials. Heat guns and heat plates should only be used under carefully controlled circumstances. Heat guns that generate heat of 1,100 °F or more are prohibited.

3.4.3 Chemical Methods

Standard paint-stripping chemicals include solvents (methylene chloride, Fig. 5.4) and caustics (sodium hydroxide or potassium hydroxide, Fig. 5.5). Methylene chloride is toxic and must be handled with appropriate precautions.

Methylene chloride is the basis for most solvent-based paint strippers. Methylene chloride is particularly effective on wooden surfaces because it causes little damage to wooden surfaces; it is often used for older buildings.

Caustic paint removers tend to be less toxic, but they have a greater tendency to pit wooden surfaces and raise wood grain.

Fig. 5.4 A typical solvent-based paint stripper containing methylene chloride

Fig. 5.5 A typical caustic paint stripper containing sodium hydroxide

Chemical methods generally generate much less paint dust than abrasive or thermal methods. Chemical stripping can be done on-site or off-site. Each procedure has its own problems and recommended precautions.

3.4.3.1   On-Site Chemical Stripping

On-site chemical stripping must be done with great care, since residents or neighbors may be exposed to toxic chemicals. Good ventilation, protective clothing, and respirators are essential when using methylene chloride, since it can cause severe burns, liver and heart damage, and possibly cancer. There is also a danger of inhaling chemical fumes. Strict control over the worksite is essential when using this toxic chemical stripper.

The runoff from chemical stripping procedures is often hazardous as it can include lead paint residue, so even greater care must be taken to treat it accordingly and to dispose of it properly.

3.4.3.2   Off-Site Chemical Stripping

In many cases, painted elements can be removed from buildings and treated at commercial stripping joints. Elements are immersed in tanks of chemical remover, either solvent based or caustic, scrubbed down, and then returned to the building. Because the work is done completely off-site, this method will not create the dust or fumes that are the chief hazard of on-site treatment, and the toxic chemicals are restricted to controlled areas. In fact, with respect to the health of workers and residents, any off-site method will always be safer than on-site removal. However, there are challenges associated with off-site treatment of affected architectural elements that are not an issue with on-site treatment. These include the need to limit the damage to the elements as they are removed, the proper reinstallation of the treated elements in their correct locations and configurations, and the limitations to the size of elements that can be removed for treatment.

4   Lead Hazard Abatement Operational Procedures

HUD published a report titled "Guidelines for the Evaluation and Control of Lead-Based Paint Hazards in Housing" (HUD 2012) that includes how-to-do-it recommendations for the abatement process, replacement of building components, enclosure, paint removal, and encapsulation [64]. These following checklists include each step that should be considered when undertaking any of these projects.

4.1   General Lead Hazard Abatement Process Procedures

1. **Arrange for risk assessment or paint inspection.** Have a lead hazard risk
   assessment or lead-based paint inspection performed by a certified risk assessor
   or a certified inspector who is independent of the abatement contractor.
2. **Develop a hazard control plan.** Develop a site-specific lead hazard control
   plan based on the hazards (risk assessment) or lead-based paint (inspection)
   identified and financing available. Avoid high-dust jobs and procedures.
3. **Obtain waste permits.** Have the contractor obtain any necessary building or
   waste permits; notify local authorities if the local jurisdiction requires it.
4. **Select needed materials.** Together with the contractor (or designer or risk
   assessor), select specific building component replacement items, enclosure
   materials, paint removal equipment and/or chemicals, tools, and cleaning
   supplies. Consider waste management and historic preservation implications
   of the selected treatment.
5. **Develop specifications** (usually for large projects only).
6. **Schedule other construction work.** Schedule other needed construction work
   first so that leaded surfaces are not inadvertently disturbed and unprotected
   workers are not placed at risk. Include time for clearance examinations and
   laboratory dust sample analysis in the scheduling process.
7. **Select a contractor.** Select a certified abatement contractor using the lowest
   qualified bidder.
8. **Conduct preconstruction conference.** Conduct a preconstruction conference to
   ensure the contractor fully understands the work involved (for large projects only).
9. **Notify residents.** Notify residents of the dwelling and adjacent dwellings of the
   work and the date when it will begin. Implement relocation (if appropriate).
10. **Correct housing conditions that might impede work.** Correct any existing
    conditions that could impede the abatement work like trash removal or struc-
    tural deficiencies.
11. **Post warning signs.** Post warning signs and restrict entry to authorized per-
    sonnel only. Implement the worksite preparation procedures.
12. **Consider a pilot project.** For large projects only, consider conducting a pilot
    project to determine if the selected abatement method will actually work.
13. **Consider collecting soil samples for quality control.** As an optional quality
    control procedure, consider collecting pre- and post-abatement soil samples.
    Once collected, post-abatement soil samples should be analyzed and compared
    to clearance standards. The pre-abatement samples need not be analyzed if
    post-abatement soil levels are below the applicable limit. Soil sampling is not
    required by US EPA regulations as part of clearance. This is an optional
    activity.
14. **Execute construction work.** Execute abatement work. See step-by-step sum-
    maries for building component replacement, enclosure, paint removal, and
    encapsulation.

15. **Store waste.** Following completion of abatement work, be sure to store all waste in a secure area.
16. **Cleanup.** Conduct cleanup of the work site at the end of each day and final cleanup. Execute waste disposal.
17. **Arrange for clearance.** Have an independent certified inspector technician or risk assessor conduct a clearance examination. The clearance examination should be done after waiting at least 1 h after cleanup has been completed to allow dust to settle.
18. **Repeat cleaning if clearance fails.** If clearance is not achieved, repeat cleaning and/or complete abatement work before repeating the clearance examination. Once clearance is achieved, obtain any required formal release, if required by HUD or local authorities.
19. **Notify residents.** Notify residents of affected dwellings of the nature and results of the abatement work.
20. **Pay contractors.** Pay contractor and clearance examiner.
21. **Conduct periodic monitoring.** Following successful abatement, conduct periodic monitoring and reevaluation of enclosure or encapsulation systems (if applicable) or lead-based paint that was not abated. Maintain records of all abatement, monitoring, reevaluation, and maintenance activities, and turn them over to any new owner upon sale of the property as part of lead disclosure. Provide proper disclosure and notification to tenants.

4.2 Building Component Replacement Procedures

1. Prepare work area and plan how the new component installation will be installed. Whenever possible, use new, energy-efficient window, door, and insulating systems.
2. Prepare lead-painted building component for removal. Turn off and disconnect any electrical circuits inside or near the building component to be removed.
3. Lightly mist the component to be removed.
4. Score all painted seams with a sharp knife.
5. Remove or bend back any screws, nails, or fasteners.
6. Dry component. Use a flat pry instrument (crowbar) and hammer to pry the component from the substrate.
7. Wrap and seal bulk components in plastic and take them to a covered truck for disposal or to secured waste storage area.
8. Vacuum any dust or paint chips in the area where the component was located.
9. Replace component (optional).
10. Conduct cleanup of the work site.
11. Arrange for clearance and reclean if necessary.

4.3 Enclosure Installation Procedures

1. Post warnings on affected components. Stamp, label, or stencil all lead-based painted surfaces that will be enclosed with a warning approximately every 2 ft (0.61 m) both horizontally and vertically on all components. The warning should read: "Danger: Lead-Based Paint." Deteriorated paint should not be removed from the surface to be enclosed.
2. Identify enclosure. Attach a durable drawing to the utility room or closet showing where lead-based paint has been enclosed in the dwelling.
3. Plan for annual monitoring of the enclosure by the owner.
4. Repair substrates. Repair any unsound substrates and structural elements that will support the enclosure, if necessary.
5. Select enclosure material. Select appropriate enclosure material. Acceptable choices include drywall or fiberboard, wood paneling, laminated products, rigid tile and brick veneers, vinyl, aluminum, or plywood.
6. Prepare electrical fittings. Install extension rings for all electrical switches and outlets that will penetrate the enclosure.
7. Clean floors. If enclosing floors, remove all dirt with a vacuum cleaner to avoid small lumps in the new flooring.
8. Seal seams. Seal and back-caulk all seams and joints. Back-caulking is the application of caulk to the underside of the enclosure.
9. Anchor enclosures. When installing enclosures directly to a painted surface, use adhesive and then anchor with mechanical fasteners (nails or screws).
10. Conduct cleanup.
11. Arrange for clearance. Have a certified risk assessor or inspector technician conduct clearance testing and provide documentation.

4.4 Lead Paint Removal Procedures

1. Decide on paint removal method—use only approved removal methods. Avoid the following prohibited methods: (a) open-flame burning or torching, (b) heat guns operating above 1100 °F, (c) machine sanding or grinding without a HEPA vacuum exhaust tool, (d) abrasive blasting or sandblasting without a HEPA vacuum exhaust tool, (e) paint stripping in a poorly ventilated space using volatile stripper, and (f) dry scraping (except for limited areas).
2. Ensure safe use of heat guns (Fig. 5.3)—for heat gun work, provide fire extinguishers in the work area and ensure that adequate electrical power is available. Use for limited areas only. Train workers to avoid gouging or abrading the substrate.
3. When using mechanical tools, use only HEPA-equipped tools—vacuum blasting and needle guns should not be used on wood, plaster, drywall, or other soft substrates. Observe the manufacturer's directions for the amount of vacuum airflow required.

4. Use a spray bottle or wet sponge for wet scraping to keep the surface wet while scraping—apply enough water to moisten the surface completely, but not so much that large amounts of water run onto the floor or ground. Do not moisten areas near electrical circuits.

5. Use off-site chemical stripping facilities, if feasible—for chemical paint removers, determine if the building component can be removed and stripped off-site. Off-site stripping is generally preferred to on-site paint removal. Observe all manufacturers' directions for use of paint removers.

6. Remove components carefully—score the component edges with a knife or razor blade to minimize damage to adjacent surfaces. Inform the off-site paint remover that lead-based paint is present before shipping. Wrap the component in plastic and send to the off-site stripping location.

7. Test effectiveness of on-site stripper, if used—for on-site paint removal, first test the product on a small area to determine its effectiveness. Chemical paint removers may not be effective or desirable on exterior, deteriorated wood surfaces, aluminum, and glass.

8. Stripping/removal—provide neoprene, nitrile, rubber, or polyvinyl chloride (PVC) gloves (or other types of glove recommended by the manufacturer), face shields, respirators with combination filter cartridges (Fig. 5.6) for leaded dust and organic vapors (if appropriate), and chemical-resistant clothing. Be sure to select the right type of organic vapor filter cartridge, gloves, or clothing for the specific chemical being used. Portable eyewash stations capable of providing a 15-min flow must be on-site. Apply the chemicals and wait the

Fig. 5.6 A typical protection gear including face shield, respirator, and filter cartridges

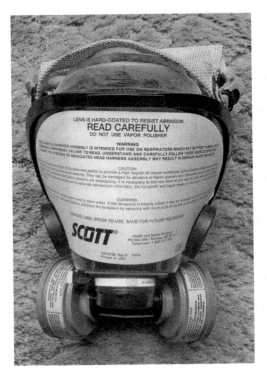

required period of time. Maintain security overnight to prevent passersby from coming into contact with the chemicals. For caustic chemical paint removers (Fig. 5.5), neutralize the surface before repainting using glacial acetic acid (not vinegar).

9. Repaint with lead-free paint.
10. Dispose of waste properly.
11. Conduct cleanup.
12. Arrange for clearance—have a certified risk assessor or lead-based paint inspector conduct a clearance examination and provide documentation.

4.5   Encapsulation Procedures

1. Determine if encapsulants can be used. Do not encapsulate the following surfaces: (a) friction surfaces, such as window jambs and doorjambs, (b) surfaces that fail patch tests, (c) surfaces with substrates or existing coatings that have a high level of deterioration, (d) surfaces in which there is a known incompatibility between two existing paint layers, (e) surfaces that cannot support the additional weight stress of encapsulation due to existing paint thickness, and (f) metal surfaces that are prone to rust or corrosion.
2. Conduct field tests of surfaces to be encapsulated for paint film integrity.
3. Consider special use and environmental requirements (e.g., abrasion resistance and ability to span base substrate cracks).
4. Examine encapsulant performance test data supplied by the manufacturer.
5. Conduct at least one on-site test patch on each type of building component where encapsulant will be used.
6. Prepare the surface selected for the complete job. For both non-reinforced and reinforced coatings, use a 6- by 6-in. (15.24×15.24 cm) test patch area. Prepared surfaces for patch testing should be at least 2 in. larger in each direction than the patch area.
7. Use a 3- by 3-in. (7.62×7.62 cm) patch for fiber-reinforced wall coverings. For rigid coatings that cannot be cut with a knife, conduct a soundness test.
8. Allow coating to cure and then assess results of the patch test. For liquid coating encapsulant, visually examine it for wrinkling, blistering, cracking, bubbling, or other chemical reaction with the underlying paint. Carry out the appropriate adhesion tests.
9. Record the results of all patch tests and decide which one to use.
10. Develop job specifications.
11. Implement a proper worksite preparation level.
12. Repair all building components and substrates as needed, e.g., caulk cracks, and repair sources of water leaks.
13. Prepare surfaces. Remove all dirt, grease, chalking paint, mildew and other surface contaminants, remnants of cleaning solutions, and loose paint. All surfaces should be deglossed as needed.

14. Ventilate the containment area whenever volatile solvents or chemicals are used.
15. Monitor temperature and humidity during encapsulant application or installation. For liquid coatings, monitor coating thickness to ensure that the encapsulant manufacturer's specifications are met.
16. Conduct cleanup and clearance.
17. Have the owner monitor the condition of the encapsulant after the first 6 months and at least annually thereafter to ensure it remains intact. Repairs should be made as necessary.
18. Provide information to owners and/or tenants on how to care for the encapsulation system properly and complete repairs safely and quickly.
19. Maintain accurate records. Make sure the records include: exact detailed locations of encapsulant applications, concentration of lead in the paint underneath the encapsulant, patch test specifications and results, reevaluations, product name, contractor, and date of application or installation, along with a copy of the product label and a material safety data sheet (MSDS) for the product. Record failures and corrective measures and signs of wear and tear.

5 Lead Hazard Abatement Practices

Below are the practices recommended by US Department of Housing and Urban Development with regard to lead abatement [64].

1. Develop a written compliance plan and designate a competent person to oversee worker protection efforts (usually an industrial hygienist or a certified lead abatement supervisor)—ensure that worker exposure to airborne lead during residential lead-related work does not exceed the permissible exposure limit (PEL) set by OSHA (50 $\mu g/m^3$ averaged over an 8-h period).
2. Conduct an exposure assessment for each job classification in each work area. Monitoring current work is the best means of conducting exposure assessments—perform air sampling of work that is representative of the exposure for each employee in the workplace who is exposed to lead. Alternatively, if working conditions are similar to previous jobs by the same employer within the past 12 months, previously collected exposure data can be used to estimate worker exposures. Finally, objective data (as defined by OSHA) may be used to determine worker lead exposures in some cases. Exposures to airborne leaded dust greater than 30 $\mu g/m^3$ (8-h, time-weighted average) trigger protective requirements.
3. Use specific worker protection measures—if lead hazard control will include manual demolition, manual scraping, manual sanding, heat gun use, or use of power tools such as needle guns that generate greater dust exposure, then specific worker protection measures are required until an initial exposure assessment is completed. If the initial exposure assessment indicates exposures

are less than 30 $\mu g/m^3$, the protection measures are not legally required, although exposure to lead should be kept as low as possible at all times.

4. Implement engineering, work practice, and administrative controls to bring worker exposure levels below the PEL—examples of such controls include the use of wet abatement methods, ventilation, and the selection of other work methods that generate less dust.

5. Supplement the use of engineering and work practice controls with appropriate respirators and implement a respiratory protection program where needed—provide a respirator to any employee who requests one, regardless of the degree of exposure, to prevent inhalation of lead-contaminated dust.

6. Conduct medical exams and fit testing for all workers who will be required to wear respirators—before work begins, medical exams should be arranged for each worker who will be required to wear a respirator. The exam entails fit testing, which will indicate whether the worker is physically capable of wearing a respirator safely.

7. Provide protective clothing and arrange for proper disposal or laundering of work clothing and proper labeling of containers of contaminated clothing and equipment.

8. Provide handwashing facilities with showers if exposures are over the PEL.

9. Implement a medical surveillance program that includes blood lead monitoring under the supervision of a qualified physician pursuant to OSHA regulations—initial blood testing for lead exposure is required by OSHA for workers performing certain tasks, such as manual scraping, whenever an exposure determination has not been completed, and for any worker who may be exposed to greater than 30 $\mu g/m^3$ of lead on any day.

10. Ensure that workers are properly trained in the hazards of lead exposure, the location of lead-containing materials, the use of job-specific exposure control methods (such as respirators), the use of hygiene facilities, and the signs and symptoms of lead poisoning—OSHA requires all lead hazard control workers to be trained and to be given (communicated) specific information on lead hazards for the specific job they are doing. Employers are responsible for training their employees to comply with all of OSHA's construction standards, not just the lead standard; this training needs to be worksite specific.

11. Post lead hazard warning signs around work areas. Also, post an emergency telephone number in case an on-the-job injury occurs.

12. Conduct work as specified.

13. Conduct worker decontamination before all breaks, before lunch, and at the end of each shift—decontamination of workers performing abatement usually consists of:

   (a) Cleaning all tools in the work area or a specially designated area in the restricted work area (end of the shift only);
   (b) HEPA vacuuming all protective clothing if visibly contaminated with paint chips or dust before entering the decontamination area;
   (c) Entering the decontamination area (dirty side);

(d) Removing protective clothing by rolling outward (without removing the respirator), removing work shoes, and placing them in a plastic bag;

(e) Entering shower or washing facility;

(f) Washing hands and then removing respirator;

(g) Taking a shower, if available, using plenty of soap and water, and washing hair, hands, fingernails, and face thoroughly (before lunch and at the end of the shift only); and

(h) Entering the clean area and putting on street clothing and shoes

14. Review and maintain exposure assessment and medical surveillance records continuously for 30 years—notify workers of air sampling and blood lead level results within 5 working days after receiving the results. Provide each worker with a copy of the written medical opinion from their examining physician. Employers must maintain all records of exposure monitoring for 30 years and all medical records for the duration of each worker's employment plus 30 years.

6 Risk Management

6.1 Risk Assessment

Risk assessment is a procedure for determining the existence, nature, severity, and location of lead-based paint hazards in or on a residential property; it also includes reporting the findings of the assessment and the options for controlling or abating the hazards that are found.

Legally, risk assessments are on-site investigations by licensed professionals to determine the existence, nature, severity, and location of lead-based paint hazards accompanied by a report explaining the results and options for reducing lead-based paint hazards (40 CFR 745.227(d)(11)). A lead-based paint hazard is any condition that causes exposure to lead from dust-lead hazards, soil-lead hazards, or lead-based paint that is deteriorated or present in chewable surfaces, friction surfaces, or impact surfaces that would result in adverse human health effects. A risk assessment may be conducted in any residential property, regardless of occupancy. However, in the case of an environmental investigation of the home of a child with an elevated blood lead level (EBL), the standard risk assessment should be supplemented with additional questions and activities.

Activities that are required by US EPA or HUD regulations are identified in this section as being "required" or as actions that "must" be done. Activities that are not required by US EPA or HUD regulations but are recommended are identified as being "recommended" or as actions that "should" be done. Note that there may be state, tribal, or local laws and regulations that must also be followed, especially if they are more stringent or protective than the federal requirements. Activities that may be done at the discretion of the owner or manager are identified as "optional." Section 6.4 discusses risk assessment further.

6.2   Risk Evaluation Options

This section offers owners, planners, and risk assessors guidance on choosing the most appropriate evaluation method for specific housing situations. There are other types of evaluation aside from a risk assessment. Except where regulations specifically require a risk assessment or a lead-based paint inspection, there are no simple rules for choosing an evaluation method.

A property owner has a choice of the following evaluation options, except where regulations limit or determine the choice:

1. A risk assessment, which identifies lead-based paint hazards, as defined by US EPA regulations
2. A lead hazard screen (for properties in good physical condition)
3. A lead-based paint inspection, which identifies all lead-based paint, whether hazardous or not
4. A combination risk assessment/paint inspection, which provides complete information on lead-based paint and lead-based paint hazards

Tables 5.2 and 5.3 provide an overview comparing two options: (a) risk assessment and lead hazard screen and (b) lead-based paint inspection and combination inspection/risk assessment.

6.3   Lead Hazard Screen

A second type of lead-based paint evaluation is the lead hazard screen. This evaluation method identifies lead-based paint hazards like the risk assessment does, but it also identifies other potential lead hazards. It is an abbreviated form of evaluation and generally is available at a lower cost than a full risk assessment. However, this method should be used only in dwellings in good condition where the probability of finding lead-based paint hazards is low. A screen employs limited sampling (soil sampling is usually not conducted) and, as a trade-off, more sensitive hazard identification criteria. The protocol for a lead hazard screen is described later in this chapter. If a screen indicates that lead hazards may be present, the owner should have a full risk assessment performed. All lead hazard screens must be performed by risk assessors certified or licensed by US EPA or an US EPA-authorized state, tribe, or territory.

Table 5.2 Comparison of risk assessment and lead hazard screen

Analysis, content, or use	Risk assessment	Lead hazard screen
Paint	Deteriorated paint and intact paint on friction and impact surfaces only	Deteriorated paint only
Dust	Yes	Yes
Soil	Yes	No
Water	Optional	No
Air	No	No
Maintenance status	Optional	No
Management plan	Optional	No
Status of any current child lead poisoning cases	If information is available	If information is available
Review of previous paint testing	Yes	Yes
Typical applications	1. Interim controls 2. Building nearing the end of expected life 3. Sale of property or turnover 4. Insurance (documentation of lead-safe status) 5. Remodeling and repainting 6. Lead Safe Housing Rule compliance	Post-1960 housing in good condition for which a risk assessment is required
Final report	Location of lead-based paint hazards and options for acceptable hazard control methods or certification that no lead-based paint hazards were found. Also includes interim controls and abatement measures if hazards are found	Probable existence of lead-based paint hazards (based on more stringent standards used for screen) or the absence of lead-based paint hazards

Source: Guidelines for the evaluation and control of lead-based paint hazards in housing [64]

6.4   Lead-Based Paint Inspection

The third type of evaluation is a paint inspection. It evaluates all painted surfaces to detect any presence of lead-based paint and must be done by a certified paint inspector.

HUD [64] recommends the following step-by-step process for a comprehensive risk assessment:

1. Determine scope. Determine if the client is requesting a risk assessment, a lead-based paint inspection, lead hazard screen, or a combination of the two. Reach an agreement on costs and scope of effort. If the dwelling is in good condition, a lead hazard screen may be conducted to determine if a full risk assessment is needed.

Table 5.3 Comparison of lead-based paint inspection and combination inspection/risk assessment

Analysis, content, or use	Lead-based paint inspection	Combination inspection/risk assessment
Paint	Surface-by-surface	Surface-by-surface
Dust	Yes	Yes
Soil	No	Yes
Water	No	Optional
Air	No	No
Maintenance status	No	Optional
Management plan	No	Optional
Status of any current child lead poisoning cases	No	If information is available
Review of previous paint testing	Yes	Yes
Typical applications	1. Abatement 2. Renovation work 3. Weatherization 4. Sale of property or turnover	Renovation
Final report	Lead concentration for each painted building component	Combination of risk assessment and inspection report content

Source: Guidelines for the evaluation and control of lead-based paint hazards in housing [64]

2. Interview residents and/or owners. For individual residences, interview residents about family use patterns, especially of young children (if any).
3. Survey building condition. Perform a brief building condition survey to identify any major deficiencies that may affect the success of lead hazard controls.
4. Determine whether units will be sampled and, if so, select units. Visual assessments and environmental sampling should be conducted in each dwelling if assessing individual dwelling units, fewer than five rental units, or multiple rental units where the units are not similar.
5. Conduct visual assessment. Perform a visual assessment of the building and paint condition, using the forms and protocols in this chapter, and select dust sampling and paint testing locations based on use patterns and visual observations.
6. Conduct dust sampling: (a) in individual dwelling units, dust samples are typically collected in the entryway and at least four living areas where children under age 6 are most likely to come into contact with dust (such as the kitchen, the children's principal playroom, and children's bedrooms); (b) in multifamily properties, dust samples are also collected from the common areas, including main entryway, stairways, and hallways, and other common areas frequented by a young child; (c) submit dust samples to a laboratory recognized for the analysis of lead in dust by US EPA through the National Lead Laboratory Accreditation Program (NLLAP).
7. Conduct soil sampling. Collect a composite soil sample from bare soil in each of the three following area types: (a) each play area with bare soil, (b) nonplay

areas in the foundation area, and (c) nonplay areas in the rest of the yard (including gardens).

8. Conduct paint testing as needed. Conduct testing of deteriorated paint and intact paint on friction surfaces. Lead in deteriorated paint can be measured with a portable x-ray fluorescence (XRF) analyzer if there is a large enough flat surface with all layers present.

9. Sample tap water (optional). At the client's request, collect optional water samples to evaluate lead exposures that can be corrected by the owner (leaded service lines, fixtures). Water sampling is not recommended for routine risk assessments of lead-based paint hazards, since drinking water hazards are outside the scope of lead-based paint hazards, and US EPA has another program that regulates drinking water. US EPA has a protocol, including specific sample collection procedures and when to collect the samples, which should be followed.

10. Interpret the laboratory results. Interpret the results of the environmental testing in accordance with applicable regulations.

11. Analyze data and discuss with client. Integrate the laboratory results with the visual assessment results, any XRF measurements, and other maintenance and management data to determine the presence or absence of lead-based paint hazards, as defined under applicable statutes or regulations.

12. Prepare a report listing any hazards identified and acceptable control measures, including interim control and abatement options.

13. Discuss all of the safe and effective lead hazard control options, and provide recommendations, for specific lead hazards with the owner.

7 Lead Hazard Control Planning

7.1 Long-Term or Short-Term Response

Owners have a wide range of options for lead hazard evaluation and control that include both long- to short-term solutions.

Complete and permanent elimination of all known or presumed lead-based paint through abatement is a long-term approach. It can be effective and safe provided that:

1. All types of lead hazards are addressed, including lead-contaminated dust and soil.
2. Workers and residents are not adversely affected during the work.
3. The process is properly controlled so that new lead hazards are not created.
4. Cleanup is adequate as determined by clearance testing.

Risk assessment followed by abatement of specific lead-based paint hazards is a more focused long-term approach. It focuses treatment resources on specific hazards. If encapsulation or enclosure is performed, the condition of these treatments should be periodically monitored through a lead-safe maintenance program. Short-term solutions are only appropriate as an interim measure.

7.2   Lead Hazard Control Planning Steps

1. Review of existing conditions and preliminary determination of lead hazard control strategy, including historic preservation considerations
2. Evaluation of lead-based paint and/or lead-based paint hazards
3. Preparation of notice of evaluation for the presence of lead to residents, if required
4. Selection of specific lead hazard control methods
5. Selection of level of resident protection and worksite preparation level
6. Initiation of pilot project (not necessarily required in single-family dwellings)
7. Scheduling of other related construction work
8. Selection of lead hazard control contractors. Notifications to state/local jurisdictions, if required
9. Lead-safe correction of preexisting conditions that could impede lead hazard control work
10. Ongoing monitoring of the work and cleanup process
11. Clearance (and certification if required by the local jurisdiction)
12. Preparation of format for notice of lead hazard control activities to residents, if required
13. Arrangement of ongoing monitoring and reevaluation after completion of lead control project

8   Rules and Regulations for Lead Poisoning Prevention and Environmental Control

8.1   Worldwide Awareness and Regulations

Lead-laden paints are still widely sold around the world due to their durability, corrosion resistance capability, and low cost. However, the European Union (EU) has passed a directive controlling the use of lead-based paint. The Canadian government has also established rules and regulations on surface-coating materials, which came into force in 2005, limiting lead content to its background level for both interior and exterior paints sold to consumers [49].

In the USA, the US EPA, HUD, and OSHA all have rules and regulations for lead in their respective areas (i.e., environmental concerns, housing, and the workplace). As previously mentioned, states may also have their own individual regulations.

The US Congress passed the Residential Lead-Based Paint Hazard Reduction Act of 1992, also known as Title X, to protect families from exposure to lead from paint, dust, and soil. Section 1018 of this law directed HUD and US EPA to require the disclosure of known information on lead-based paint and lead-based paint hazards before the sale or lease of most housing built before 1978. Before ratification of a contract for housing sale or lease, sellers and landlords must:

1. Supply a US EPA-approved information pamphlet on identifying and controlling lead-based paint hazards.
2. The seller or landlord must disclose any known information concerning lead-based paint or lead-based paint hazards, such as the location of the lead-based paint and/or lead-based paint hazards, and the condition of the painted surfaces.
3. Provide any records and reports on lead-based paint and/or lead-based paint hazards that are available to the seller or landlord (For multi-unit buildings, this requirement includes records and reports concerning common areas and other units obtained as a result of a building-wide evaluation).
4. Include an attachment to the contract or lease (or language inserted in the lease itself) that includes a Lead Warning Statement and confirms that the seller or landlord has complied with all notification requirements. This attachment is to be provided in the same language used in the rest of the contract. Sellers or landlords, and agents, as well as homebuyers or tenants, must sign and date the attachment.
5. Sellers must provide homebuyers a 10-day period to conduct a paint inspection or risk assessment for lead-based paint or lead-based paint hazards. Parties may mutually agree, in writing, to lengthen or shorten the time period for inspection. Homebuyers may waive this inspection opportunity.

The Lead Disclosure Rule (the identical 24 CFR 35, subpart A, and 40 CFR 745, subpart F) was jointly issued by HUD and US EPA in 1996 (61 FR 9063-9088, March 6, 1996) as part of Title X. As of 2011, HUD and US EPA had issued three interpretive guidance documents about the Lead Disclosure Rule; these are available from both agencies' websites on the rule.

8.2   OSHA's Lead Regulations

OSHA's lead regulations are described at OSHA's main lead regulation web page at: http://www.osha.gov/SLTC/lead/. In addition, as of 2014, 25 states, Puerto Rico, and the Virgin Islands had OSHA-approved State Plans and had adopted their own standards and enforcement policies. For the most part, these states adopted standards similar to OSHA's. However, some states have adopted different standards or have different enforcement policies.

OSHA has two lead standards, one specifically for construction and one for general industry. The two standards complement each other. The first covers construction work (construction, alteration, repair, painting, and/or decorating (29 CFR 1926.10, (a))), while the second covers work that is not related to construction work (such as maintenance work). Employers are responsible for determining which standard applies to their workers on a particular project.

8.2.1 OSHA's Lead in Construction Standard

OSHA's Lead in Construction Standard (29 CFR 1926.62) applies to all construction work where an employee may be occupationally exposed to lead. OSHA has published a 332-page booklet on this regulation (OSHA 3142-09R 2003), posted at http://www.osha.gov/Publications/osha3142.pdf. OSHA has also posted an online interactive expert system (compliance advisor) on the Lead in Construction Standard at http://www.dol.gov/elaws/oshalead.htm.

The Lead in Construction Standard applies to any source or concentration of lead to which workers may be exposed as a result of construction work. OSHA standards are not limited to lead-based paint as defined by HUD or US EPA or lead-containing paint as defined by the Consumer Product Safety Commission (CPSC). All work related to construction, alteration, or repair, including painting and decorating, is included. Under this standard, construction includes but is not limited to:

1. Demolition or salvage of structures where lead or materials containing lead are present
2. Removal or encapsulation of materials containing lead
3. New construction, alteration, repair, or renovation of structures, substrates, or portions or materials containing lead
4. Installation of products containing lead
5. Lead contamination from emergency cleanup
6. Transportation, disposal, storage, or containment of lead or materials containing lead where construction activities are performed
7. Maintenance operations associated with these construction activities

8.2.2 OSHA's Lead in General Industry Standard

The OSHA's Lead in General Industry Standard (29 CFR 1910.1025) covers the use of lead in general industry. This industry includes nonconstruction-related maintenance work, as well as lead smelting, manufacturing, and the use of lead-based pigments contained in inks, paints, and other solvents in addition to the manufacturing and recycling of lead batteries.

Maintenance work associated with construction, alteration, or repair activities is covered separately by the Construction Standard (29 CFR 1926.62, subsection (a), as discussed below). Nonconstruction-related maintenance work (or if lead is a component of any product that workers make or use) is covered by the General Industry Standard (29 CFR 1910.1025(e)(3)(ii)(A)). Construction activities do not include routine cleaning and repainting where there is insignificant damage, wear, or corrosion of existing lead-containing paint and coating or substrates.

8.3 Summary of All US Federal Lead Standards

8.3.1 Lead Content Standards

OSHA, US EPA, CDC, NIOSH, FDA, CPSC, and ACGIH have issued various lead standards for blood (OSHA, CDC, ACGIH), air (OSHA, CDC/NIOSH, ACGIH, US EPA), water (US EPA), soil (US EPA), paint (CPSC), and food (FDA). Table 5.4 is a summary of the standards for different media, agencies, allowable levels, and comments for applicability. For example, Table 5.4 shows that (a) the CDC advisory's lowest blood level is 10 µg/dL for individual management, (b) the US EPA regulated NAAQS for the lowest ambient air lead level is 0.15 µg/m^3 (3-month average), (c) the US EPA's residential play area's soil lead level is limited at 400 ppm, and (d) the action level for public water supplies is set at 15 µg/L by the US EPA.

In the USA, Consumer Product Safety Commission (CPSC) formally banned lead-based paint (LBP) in residential properties and public buildings along with toys and furniture containing lead paint in 1977, because children may ingest lead paint chips or peelings and be poisoned [66]. The CPSC also instituted the

Table 5.4 Standard and regulation for lead content

Media	Agency	Level	Comments
Blood	OSHA	40 µg/dL 60 µg/dL	Regulation Regulation, cause for medical removal from exposure
Blood	CDC	10 µg/dL	Advisory, level for individual management
Blood	ACGIH	30 µg/dL	Advisory, indicates exposure at the TLV
Air (workplace)	CDC/ NIOSH	100 µg/m^3	REL (non-enforceable)
Air (workplace)	OSHA	50 µg/m^3 30 µg/m^3	Regulation, PEL (8-h average) (general industry) Action level
Air (ambient)	US EPA	0.15 µg/m^3	Regulation, NAAQS, 3-month average
Air (workplace)	ACGIH	150 µg/m^3 50 µg/m^3	TLV/TWA guideline for lead arsenate TLV/TWA guideline for other forms of lead
Water (drinking)	US EPA	15 µg/L 0 µg/L	Action level for public supplies Non-enforceable goal, MCLG
Soil (residential)	US EPA	400 ppm (play areas) 1200 ppm (nonplay areas)	Soil screening guidance level, requirement for federally funded projects only
Paint	CPSC	600 ppm (0.06 %)	Regulation, by dry weight. There is a new standard for lead in children's jewelry
Food	FDA	Various	Action levels for various foods, for example, lead-soldered food cans now banned

Source: Case Studies in Environmental Medicine (CSEM) lead toxicity (2012) [65]
http://www.atsdr.cdc.gov/csem/lead/docs/lead.pdf

Consumer Product Safety Improvement Act (CPSIA) of 2008 for manufacturers. The CPSIA changed the cap on lead content in paint from 0.06 % to 0.009 % starting August 14, 2009.

8.3.2 US EPA Lead Renovation, Repair, and Painting Rule

US EPA's 2008 Lead-Based Paint Renovation, Repair and Painting (RRP) Rule (as amended in 2010 and 2011), aims to protect the public from lead-based paint hazards associated with renovation, repair, and painting activities. These activities can create hazardous lead dust when surfaces with lead paint, even from many decades ago, are disturbed. The rule requires workers to be certified and trained in the use of lead-safe work practices and requires renovation, repair, and painting firms to be US EPA certified. These requirements became fully effective on April 22, 2010. Specifically the RRP Rule requires that (a) all renovators working in homes built before 1978 and disturbing more than 6 ft^2 (0.557 m^2) of lead paint inside the home or 20 ft^2 (1.858 m^2) outside the home be certified; (b) firms performing renovation, repair, and painting projects that disturb lead-based paint (LBP) in homes and any child-occupied facility (child care facilities, preschools, etc.) built before 1978 be certified by US EPA; (c) the certified renovators be trained by the US EPA-approved training providers; and (d) only the certified renovators be used for lead remediation projects [67].

9 New York State Lead Poisoning Prevention and Control Programs

Almost all states in the USA have established their own lead poisoning prevention and control programs [68–70] with collaboration of the US Federal government [71–72]. As a model example, this section introduces the New York State's lead programs that are part of the New York Codes, Rules, and Regulations (NYCRR). The information contained in this section is not the official, final version of the NYCRR's compilation. No representation is made as to its accuracy. Furthermore, the NYCRR is only valid in the State of New York and is subject to revision periodically. To ensure accuracy and for evidentiary purposes, readers should obtain the most current and site-specific version from the government which has direct the jurisdiction.

9.1 New York State (NYS) Lead Poisoning Prevention Program

1. The New York State Department of Health (NYSDOH) has established a lead poisoning prevention program, which is responsible for establishing and coordinating activities to prevent lead poisoning and to minimize risk of exposure to

lead. NYSDOH exercises any and all authority which may be deemed necessary and appropriate to effectuate the provisions of this title.

2. NYSDOH:

   (a) Promulgates and enforces regulations for screening children and pregnant women, including requirements for blood lead testing, for lead poisoning, and for follow-up of children and pregnant women who have elevated blood lead levels;

   (b) Enters into interagency agreements to coordinate lead poisoning prevention, exposure reduction, identification and treatment activities, and lead reduction activities with other federal, state, and local agencies and programs;

   (c) Establishes a statewide registry of lead levels of children provided such information is maintained as confidential except for (c-i) disclosure for medical treatment purposes; (c-ii) disclosure of non-identifying epidemiological data; and (c-iii) disclosure of information from such registry to the statewide immunization information system; and

   (d) Develops and implements public education and community outreach programs on lead exposure, detection, and risk reduction

3. NYSDOH identifies and designates areas in the state with significant concentrations of children identified with elevated blood lead levels as communities of concern for purposes of implementing a childhood lead poisoning primary prevention program and may, within amounts appropriated, provide grants to implement approved programs. The commissioner of health of a county or part-county health district, a county health director or a public health director, and, in the City of New York, the commissioner of the New York City Department of Health and Mental Hygiene shall develop and implement a childhood lead poisoning primary prevention program to prevent exposure to lead-based paint hazards for the communities of concern in their jurisdiction. NYSDOH provides funding to the New York City Department of Health and Mental Hygiene or County Health Departments to implement the approved work plan for a childhood lead poisoning primary prevention program. The work plan and budget, which shall be subject to the approval of the department, shall include but not be limited to:

   (a) Identification and designation of an area or areas of high risk within communities of concern;

   (b) A housing inspection program that includes prioritization and inspection of areas of high risk for lead hazards, correction of identified lead hazards using effective lead-safe work practices, and appropriate oversight of remediation work;

   (c) Partnerships with other counties or municipal agencies or community-based organizations to build community awareness of the childhood lead poisoning primary prevention program and activities, coordinate referrals for services, and support remediation of housing that contains lead hazards;

(d) A mechanism to provide education and referral for lead testing for children and pregnant women to families who are encountered in the course of conducting primary prevention inspections and other outreach activities; and

(e) A mechanism and outreach efforts to provide housing inspections for lead hazards upon request

9.2 NYS Advisory Council on Lead Poisoning Prevention Program

9.2.1 Designees of NYS Advisory Council

The New York State (NYS) advisory council on lead poisoning prevention is established in the NYSDOH, to consist of the following or their designees: the commissioner, the commissioner of labor, the commissioner of environmental conservation, the commissioner of housing and community renewal, the commissioner of children and family services, the commissioner of temporary and disability assistance, the secretary of state, the superintendent of insurance, and 15 public members appointed by the governor. The public members shall have a demonstrated expertise or interest in lead poisoning prevention, and at least one public member shall be representative of each of the following: local government, community groups, labor unions, real estate, industry, parents, educators, local housing authorities, child health advocates, environmental groups, and professional medical organizations and hospitals. The public members of the council shall have fixed terms of 3 years, except that five of the initial appointments shall be for 2 years and five shall be for 1 year. The council shall be chaired by the commissioner or his or her designee. Members of the advisory council shall serve without compensation for their services, except that each of them may be allowed necessary and actual expenses which he or she shall incur in the performance of his or her duties under this article.

9.2.2 Powers and Duties of NYS Advisory Council

The council shall meet as often as may be deemed necessary to fulfill its responsibilities. The council shall have the following powers and duties:

1. To develop a comprehensive statewide plan to prevent lead poisoning and to minimize the risk of human exposure to lead.
2. To coordinate the activities of its member agencies with respect to environmental lead policy and the statewide plan.
3. To recommend the adoption of policies with regard to the detection and elimination of lead hazards in the environment.
4. To recommend the adoption of policies with regard to the identification and management of children with elevated lead levels.

5. To recommend the adoption of policies with regard to education and outreach strategies related to lead exposure, detection, and risk reduction.
6. To comment on regulations of the department under this title when the council deems appropriate.
7. To make recommendations to ensure the qualifications of persons performing inspection and abatement of lead through a system of licensure and certification or otherwise.
8. To recommend strategies for funding the lead poisoning prevention program, including but not limited to ways to enhance the funding of screening through insurance coverage and other means, and ways to financially assist property owners in abating environmental lead, such as tax credits, loan funds, and other approaches.
9. To report on or before the first of December of each year to the governor and the legislature concerning the previous year's development and implementation of the statewide plan and operation of the program, together with recommendations it deems necessary and the most currently available lead surveillance measures, including the actual number and estimated percentage of children tested for lead in accordance with New York state regulations, including age-specific testing requirements, and the actual number and estimated percentage of children identified with elevated blood lead levels. Such report shall be made available on the department's website.

9.3   Lead Screening of Children and Pregnant Women by NYS Health-Care Providers

1. NYSDOH is authorized to promulgate regulations establishing the means by which and the intervals at which children and pregnant women shall be screened for elevated lead levels. The department is also authorized to require screening for lead poisoning in other high-risk groups.
2. Every physician or other authorized practitioner who provides medical care to children or pregnant women shall screen children or refer them for screening for elevated lead levels at the intervals and using the methods specified in such regulations. Every licensed, registered, or approved health-care facility serving children including but not limited to hospitals, clinics, and health maintenance organizations shall ensure, by providing screenings or by referring for screenings, that their patients receive screening for lead at the intervals and using the methods specified in such regulations.
3. The health practitioner who screens any child for lead shall give a certificate of screening to the parent or guardian of the child.
4. The department shall establish a separate level of payment, subject to the approval of the director of the budget, for payments made by governmental agencies for screenings performed pursuant to this section by hospitals.

9.4   Lead Screening of Child Care or Preschool Enrollees in NYS

1. Except as provided pursuant to regulations of the NYSDOH, each child care provider, public and private nursery school, and preschool licensed, certified, or approved by any state or local agency shall, prior to or within 3 months after initial enrollment of a child under 6 years of age, obtain from a parent or guardian of the child evidence that said child has been screened for lead.
2. Whenever there exists no evidence of lead screening as provided for in subdivision one of this section or other acceptable evidences of the child's screening for lead, the child care provider, principal, teacher, owner, or person in charge of the nursery school or preschool shall provide the parent or guardian of the child with information on lead poisoning in children and lead poisoning prevention and refer the parent or guardian to a primary care provider or the local health authority.
3. (a) If any parent or guardian to such child is unable to obtain lead testing, such person may present such child to the health officer of the county in which the child resides, who shall then perform or arrange for the required screening. (b) The local public health district shall develop and implement a fee schedule for households with incomes in excess of 200 % of the federal poverty level for lead screening pursuant to section six hundred six of this chapter, which shall vary depending on patient household income.

9.5   Reporting Lead Exposure Levels in NYS

1. Every physician or authorized practitioner shall give notice of elevated lead levels as specified by the commissioner pursuant to regulation, to the health officer of the health district wherein the patient resides, except as otherwise provided.
2. The commissioner may, by regulation, provide that cases of elevated lead levels which occur (a) in health districts of less than 50,000 population not having a full-time health officer or, (b) in state institutions, shall be reported directly to the department or its district health officer.
3. Whenever an analysis of a clinical specimen for lead is performed by a laboratory or a physician or authorized practitioner, the director of such laboratory or such physician or authorized practitioner shall, within such period specified by the commissioner, report the results and any related information in connection therewith to the local and state health officer to whom a physician or authorized practitioner is required to report such cases, pursuant to this section.
4. The person in charge of every hospital, clinic, or other similar public or private institution shall give notice of every child with an elevated blood lead level coming under the care of the institution to the local or state health officer to

whom a physician or authorized practitioner is required to report such cases, pursuant to this section.

5. The notices required by this section shall be in a form and filed in such time period as shall be prescribed by the commissioner.

9.6   Manufacture and Sale of Lead-Painted Toys and Furniture in NYS

1. No person shall manufacture, sell, or hold for sale a children's toy or children's furniture having paint or other similar surface-coating materials thereon containing more than 0.06 of one per centum of metallic lead based on the total weight of the contained solids or dried paint film.

2. The commissioner of health may waive the provisions of this section in whole or in part upon a finding by the commissioner in a particular instance that there is no significant threat to public health; with respect to miniatures the commissioner shall do so, on terms and conditions he or she shall establish, upon a final judicial or administrative finding that there is no immediate public health threat in that instance.

9.7   Use of Leaded Paint in NYS

No person shall apply paint or other similar surface-coating materials containing more than 0.06 of one per centum of metallic lead based on the total weight of the contained solids or dried paint film to any interior surface, window sill, window frame, or porch of a dwelling.

9.8   Abatement of Lead Poisoning Conditions in NYS

1. Whenever the commissioner or his representative shall designate an area of high risk, he may give written notice and demand, served as provided herein, for the discontinuance of a paint condition conducive to lead poisoning in any designated dwelling in such area within a specified period of time.

2. Such notice and demand shall prescribe the method of discontinuance of a condition conducive to lead poisoning which may include the removal of paint containing more than one-half of one per centum of metallic lead based on the total weight of the contained solids or dried film of the paint or other similar surface-coating materials from surfaces specified by the commissioner or his/her representative under such safety conditions as may be indicated, and the refinishing of such surfaces with a suitable finish which is not in violation of

section one thousand three hundred seventy-two of this title, or the covering of such surfaces with such material or the removal of lead-contaminated soils or lead pipes supplying drinking water, as may be deemed necessary to protect the life and health of occupants of the dwelling.

3. In the event of failure to comply with a notice and demand, the commissioner or his/her representative may conduct a formal hearing upon due notice in accordance with the provisions of section twelve-a of this chapter and, on proof of violation of such notice and demand, may order abatement of a paint condition conducive to lead poisoning upon such terms as may be appropriate and may assess a penalty not to exceed 2,500 US dollars for such violation.

4. A notice required by this section may be served upon an owner or occupant of the dwelling or agent of the owner in the same manner as a summons in a civil action or by registered or certified mail to his last known address or place of residence.

5. The removal of a tenant from or the surrender by the tenant of a dwelling with respect to which the commissioner or his representative, pursuant to subdivision one of this section, has given written notice and demand for the discontinuance of a paint condition conducive to lead poisoning shall not absolve, relieve, or discharge any persons chargeable therewith from the obligation and responsibility to discontinue such paint condition conducive to lead poisoning in accordance with the method of discontinuance prescribed therefor in such notice and demand.

9.9 Enforcement Agencies in NYS

1. The commissioner's designee having jurisdiction, county and city commissioners of health and local housing code enforcement agencies designated by the commissioner's designee having jurisdiction, or county or city commissioner of health shall have the same authority, powers, and duties within their respective jurisdictions as has the commissioner under the provisions of this title.

2. The commissioner or his/her representative and an official or agency specified in subdivision one of this section may request and shall receive from all public officers, departments, and agencies of the state and its political subdivisions such cooperation and assistance as may be necessary or proper in the enforcement of the provisions of this title.

3. Nothing contained in this title shall be construed to alter or abridge any duties and powers now or hereafter existing in the commissioner, county boards of health, city and county commissioners of health, the New York City department of housing preservation and development, and the department of health, local boards of health or other public agencies or public officials, or any private party.

9.10   Sale of Consumer Products Containing Lead or Cadmium in NYS

1. In the absence of a federal standard for a specific type of product, the commissioner shall establish the maximum quantity of lead or cadmium (and the manner of testing therefor) which may be released from glazed ceramic tableware, crystal, china, and other consumer products. Such maximum quantity shall be based on the best available scientific data and shall insure the safety of the public by reducing its exposure to lead and cadmium to the lowest practicable level. The commissioner may amend such maximum quantity (and the manner of testing therefor) where necessary or appropriate for the safety of the public. Until such maximum quantity of lead or cadmium established by the commissioner is effective, no glazed ceramic tableware shall be offered for sale which releases lead in excess of 7 parts per million or cadmium in excess of 0.5 parts per million.
2. The commissioner is empowered to order the recall of or confiscation of glazed ceramic tableware, crystal, china, or other consumer products offered for sale which do not meet the standards set forth in or pursuant to this section.
3. The commissioner of health may waive the provisions of this section in whole or in part upon a finding by the commissioner in a particular instance that there is no significant threat to the public health; with respect to miniatures the commissioner shall do so, on terms and conditions he or she shall establish, upon a final judicial or administrative finding that there is no immediate public health threat in that instance.

References

1. Needleman H (2014) History of lead poisoning in the World. http://www.org.au/history_of_lead_poisoning_in_the_world
2. ATSDR (1990) Toxicological profile for lead. U.S. Department of Health and Human Services, ATSDR Publication No. TP–88/17
3. CDCP (1991) Preventing lead poisoning in young children, US Department of Health and Human Services, Centers for Disease Control and Prevention, Atlanta
4. NIOSH (1978) Criteria for a recommended standard: occupational exposure to inorganic lead, Revised Criteria—1978. U.S. Department of Health, Education, and Welfare, Public Health Service, Center for Disease Control, National Institute for Occupational Safety and Health, DHEW (NIOSH) Publication No. 78–158
5. Mantere P, Hanninen H, Hernberg S, Luukkonen R (1984) A prospective follow-up study on psychological effects in workers exposed to low levels of lead. Scand J Work Environ Health 10:43–50
6. Hogstedt C, Hane M, Agrell A, Bodin L (1983) Neuropsychological test results and symptoms among workers with well-defined long-term exposure to lead. Br J Ind Med 40:99–105
7. Seppäläinen AM, Hernberg S, Vesanto R, Kock B (1983) Early neurotoxic effects of occupational lead exposure: a prospective study. Neurotoxicology 4(2):181–192
8. Goyer RA (1989) Mechanisms of lead and cadmium nephrotoxicity. Toxicol Lett 46:153–162

9. Rom W (1976) Effects of lead on the female and reproduction: a review. Mt Sinai J Med 43:542–552
10. Andrews KW, Savitz DA, Hertz-Picciotto I (1994) Prenatal lead exposure in relation to gestational age and birth weight: a review of epidemiologic studies. Am J Ind Med 26:13–32
11. Schwartz J (1994) Low-level lead exposure and children's IQ: a meta analysis and search for a threshold. Environ Res 65:42–55
12. Zi-quiang C, Qi-ing C, Chin-chin P, Jia-yian Q (1985) Peripheral nerve conduction velocity in workers occupationally exposed to lead. Scand J Work Environ Health 11(4):26–28
13. Fisher-Fischbein J, Fischbein A, Melnick HD, Bardin W (1987) Correlation between biochemical indicators of lead exposure and semen quality in a lead-poisoned firearms instructor. JAMA 257(6):803–805
14. Lancranjan I, Popescu HI, Gavanescu O, Klepsch I, Serbanescu M (1975) Reproductive ability of workmen occupationally exposed to lead. Arch Environ Health 30:396–401
15. Alexander BH, Checkoway H, van Netten C, Muller CH, Ewers TG, Kaufman JD, Mueller BA, Vaughan TL, Faustman EM (1996) Semen quality of men employed at a lead smelter. Occup Environ Med 53:411–416
16. Braunstein GD, Dahlgren J, Loriaux DL (1978) Hypogonadism in chronically lead-poisoned men [Abstract]. Infertility 1(1):33–51
17. Ng TP, Goh HH, Ng YL, Ong HY, Ong CN, Chia KS, Chia SE, Jeyaratnam J (1991) Male endocrine functions in workers with moderate exposure to lead. Br J Ind Med 48:485–491
18. Dingwall–Fordyce I, Lane RE (1963) A follow-up study of lead workers. Br J Ind Med 20:313–315
19. Schwartz J (1991) Lead, blood pressure, and cardiovascular disease in men and women. Environ Health Perspect 91:71–75
20. Pocock SJ, Shaper AG, Ashby D, Delves HT, Clayton BE (1988) The relationship between blood lead, blood pressure, stroke, and heart attacks in middle-aged British men. Environ Health Perspect 78:23–30
21. Pirkle JL, Schwartz J, Landis JR, Harlan WR (1985) The relationship between blood lead levels and blood pressure and its cardiovascular risk implications. Am J Epidemiol 121 (2):246–258
22. Hertz-Picciotto I, Croft J (1993) Review of the relation between blood lead and blood pressure. Epidemiol Rev 15(2):352
23. Schwartz J (1995) Lead, blood pressure and cardiovascular disease in men. Environ Health 50:51
24. Anttila A, Heikkilä P, Nykyri E, Kauppinen T, Hernberg S, Hemminki K (1995) Excess lung cancer among workers exposed to lead. Scand J Work Environ Health 21:460–469
25. Steenland K, Sevelan S, Landrigan P (1992) Mortality of lead smelter workers: an update. Am J Public Health 82(12):1641–1644
26. US EPA (2006) Air quality criteria for lead. Volume I of II. US Environmental Protection Agency. EPA/600/R-5/144aF
27. Davis JM, Elias RW, Grant LD (1993) Current issues in human lead exposure and regulation of lead. Neurotoxicology 14(2–3):1528
28. National Toxicology Program (2012) NTP monograph on health effects of low-level lead. National Institute of Environmental Health Sciences, National Toxicology Program, Research Triangle Park, Durham. http://ntp.niehs.nih.gov/go/36443
29. Bellinger D, Sloman J, Leviton A, Rabinowitz M, Needleman HL, Waternaux C (1991) Low-level lead exposure and children's cognitive function in the preschool years. Pediatrics 87(2):219–227
30. Canfield RL, Henderson CR Jr, Cory-Slechta DA, Cox C, Jusko TA, Lanphear BP (2003) Intellectual impairment in children with blood lead concentrations below 10 microg per deciliter. N Engl J Med 348(16):1517–1526
31. Jusko TA, Henderson CR, Lanphear BP, Cory-Slechta DA, Parsons PJ, Canfield RL (2008) Blood lead concentrations < 10 microg/dL and child intelligence at 6 years of age. Environ Health Perspect 116(2):243–248

32. Lanphear BP, Dietrich K, Auinger P, Cox C (2000) Cognitive deficits associated with blood lead concentrations <10 microg/dL in US children and adolescents. Public Health Rep 115 (6):521–529
33. Lanphear BP, Hornung R, Khoury J, Yolton K, Baghurst P, Bellinger DC, Canfield RL, Dietrich KN, Bornschein R, Greene T et al (2005) Low-level environmental lead exposure and children's intellectual function: an international pooled analysis. Environ Health Perspect 113(7):894–899
34. Schnaas L, Rothenberg SJ, Flores MF, Martinez S, Hernandez C, Osorio E, Velasco SR, Perroni E (2006) Reduced intellectual development in children with prenatal lead exposure. Environ Health Perspect 114(5):791–797
35. Surkan PJ, Zhang A, Trachtenberg F, Daniel DB, McKinlay S, Bellinger DC (2007) Neuropsychological function in children with bloodLead levels <10 microg/dL. Neurotoxicology 28 (6):1170–1177
36. CDCP (2005) Preventing lead poisoning in young children. U.S. Department of Health and Human Services, Centers for Disease Control and Prevention, Atlanta
37. Lanphear BP, Hornung R, Ho M, Howard CR, Eberly S, Knauf K (2002) Environmental lead exposure during early childhood. J Pediatr 140(1):40–47
38. Lanphear BP, Roghmann KJ (1997) Pathways of lead exposure in urban children. Environ Res 74(1):67–73
39. Rabinowitz M, Leviton A, Needleman H, Bellinger D, Waternaux C (1985) Environmental correlates of infant blood lead levels in Boston. Environ Res 38(1):96–107
40. Levin R, Brown MJ, Kashtock ME, Jacobs DE, Whelan EA, Rodman J, Schock MR, Padilla A, Sinks T (2008) Lead exposures in U.S. Children, 2008: implications for prevention. Environ Health Perspect 116(10):1285–1293
41. McElvaine MD, DeUngria EG, Matte TD, Copley CG, Binder S (1992) Prevalence of radiographic evidence of paint chip ingestion among children with moderate to severe lead poisoning, St Louis, Missouri, 1989 through 1990. Pediatrics 89(4 Pt 2):740–742
42. Lanphear BP, Matte TD, Rogers J, Clickner RP, Dietz B, Bornschein RL, Succop P, Mahaffey KR, Dixon S, Galke W et al (1998) The contribution of lead-contaminated house dust and residential soil to children's blood lead levels. Environ Res 79(1):51–68
43. Ter Har G, Aronow R (1974) New information on lead in dirt and dust as related to the childhood lead problem. Environ Health Perspect 7:83–89. http://www.ncbi.nlm.nih.gov/pubmed/4831152
44. Linton RW, Natush DFS, Solomon RL, Evans CA (1980) Physicochemical characterization of lead in urban dusts: a microanalytical technique to lead tracing. Environ Sci Technol 14:159–164
45. Mielke HW, Reagan PL (1998) Soil is an important pathway of human lead exposure. Environ Health Perspect 106(Suppl 1):217–229
46. Edwards M, Triantafyllidou S, Best D (2009) Elevated blood lead in young children due to lead-contaminated drinking water: Washington, DC, 2001–2004. Environ Sci Tech 43 (5):1618–1623
47. Miranda ML, Kim D, Hull AP, Paul CJ, Galeano MA (2007) Changes in blood lead levels associated with use of chloramines in water treatment systems. Environ Health Perspect 115 (2):221–225
48. Ettinger AS, Tellez-Rojo MM, Amarasiriwardena C, Gonzalez-Cossio T, Peterson KE, Aro A, Hu H, Hernandez-Avila M (2004) Levels of lead in breast milk and their relation to maternal blood and bone lead levels at one month postpartum. Environ Health Perspect 112(8):926–931
49. Health Canada (2013) Final human health state of the science report on lead. Health Canada, Canada, 102 pages
50. Centers for Disease Control and Prevention (2002) Managing elevated blood lead levels among young children: recommendations from the advisory committee on childhood lead poisoning prevention. CDC, Atlanta

51. Centers for Disease Control and Prevention (1997) Screening young children for lead poisoning: guidance for state and local public health officials. Centers for Disease Control and Prevention, National Center for Environmental Health, U.S. Department of Health and Human Services, Public Health Service, Atlanta

52. CDCP, Atlanta (2007) Interpreting and managing blood lead levels < 10 µg/dL in children and reducing childhood exposures to lead: recommendations of the centers for disease control and prevention advisory committee on childhood lead poisoning prevention. Recomm Rep 56:1–14

53. CDCP, Atlanta (2012) Low level lead exposure harms children: a renewed call of primary prevention. www.cdc.gov/

54. Bellinger DC (2005) Teratogen update: lead and pregnancy. Birth Defects Res A Clin Mol Teratol 73(6):409–420

55. Kosnett MJ, Wedeen RP, Rothenberg SJ, Hipkins KL, Materna BL, Schwartz BS et al (2007) Recommendations for medical management of adult lead exposure. Environ Health Perspect 115(3):463–471

56. Sanin LH, Gonzalez-Cossio T, Romieu I, Peterson KE, Ruiz S, Palazuelos E et al (2001) Effect of maternal lead burden on infant weight and weight gain at one month of age among breastfed infants. Pediatrics 107(5):1016–1023

57. Shukla R, Bornschein RL, Dietrich KN, Buncher CR, Berger OG, Hammond PB et al (1989) Fetal and infant lead exposure: effects on growth in stature. Pediatrics 84(4):604–612

58. Piactelli GM, Whelan EA, Sieber WK, Gerwel M (1997) Elevated lead contamination in homes of construction workers. Am Ind Hyg Assoc J 58:447–454

59. CDCP (2016) Adult Blood Lead Epidemiology and Surveillance (ABLES). Centers for Disease Control and Prevention, Atlanta. Jan. 26, 2016

60. NIOSH (1997) Hazards evaluation and technical assistance report: people working cooperatively. NIOSH report no. HETA-0818-2649

61. US EPA Model Worker Course (2004) US Environmental Protection Agency, Washington, DC

62. Grimmer AE (1979) Preservation Brief no. 6, "Dangers of Abrasive Cleaning to Historic Buildings." National Park Service. Washington, DC. June 1979

63. Burning the Paint Off: The Dangers Associated with Torches, Heat Guns, and other Thermal Devices for Paint Removal (FYI 10). Updated July 2000. Ttp//www.dhr.virginia.gov/pdf/burningpaint1.pdf

64. HUD (2012) Guidelines for the evaluation and control of lead-based paint hazards in housing, 2nd edn. Office of Healthy Homes and Lead Hazards Control, Washington, DC

65. Case Studies in Environmental Medicine (CSEM) Lead Toxicity (2012) http://www.atsdr.cdc.gov/csem/lead/docs/lead.pdf

66. CPSC (2011) Ban of lead-containing paint and certain consumer products bearing lead-containing paint (16 CFR 1303). US Consumer Product Safety Commission, Washington, DC

67. US EPA (2008 amended 2011) Lead renovation, repair and painting program rules. US Environmental Protection Agency, Washington, DC

68. NYSDOH (2014) NYS regulations for lead poisoning prevention and control. NY State Department of Health, Albany

69. VDOH (2014) Childhood lead poisoning prevention. Virginia Department of Health, Richmond, www.virginia.gov

70. CDPH (2014) Childhood lead poisoning prevention branch. California Department of Public Health, Sacramento. www.cdph.ca.gov

71. Chuang HY, Schwartz J, Gonzales-Cossio T, Lugo MC, Palazuelos E, Aro A, Hu H, Hernandez-Avila M (2001) Interrelations of lead levels in bone, venous blood, and umbilical cord blood with exogenous lead exposure through maternal plasma lead in peripartum women. Environ Health Perspect 109(5):527–532

72. US EPA (2015) Evaluating and eliminating lead-based paint hazards. US Environmental Protection Agency, Washington, DC. Www2.epa.gov/lead

Chapter 6
Municipal and Industrial Wastewater Treatment Using Plastic Trickling Filters for BOD and Nutrient Removal

Jia Zhu, Frank M. Kulick III, Larry Li, and Lawrence K. Wang

Contents

1   Trickling Filter Introduction 288
  1.1   Process Description 288
  1.2   Evolution of Trickling Filters 289
  1.3   Key Components of Modern Trickling Filters 291
2   Trickling Filter Applications 295
  2.1   Trickling Filter Loading Rates 295
  2.2   Trickling Filter Classification 296
3   Design of Plastic Trickling Filters 297
  3.1   Process Considerations 297
  3.2   Plastic Media Selection 304
  3.3   Single-Stage and Multistage Trickling Filters 306
  3.4   Design Models for BOD Removal Filters 307
  3.5   Design Procedures for Combined BOD Removal and Nitrification Filters 309
  3.6   Design Procedure for Nitrification Trickling Filters (NTFs) 310
4   Combined Trickling Filter/Activated Sludge Processes 314
  4.1   Trickling Filter/Solids Contact (TF/SC) 314
  4.2   Roughing Filter/Activated Sludge (RF/AS) 316
5   Trickling Filter and Nutrient Removal 317
  5.1   Process Modifications to Achieve Total Nitrogen Removal 317
  5.2   Coarse Media Denitrification Filter 319
  5.3   Phosphorus Removal in Trickling Filter Plants 320

J. Zhu (✉) • L. Li
Water Group, Brentwood, 500 Spring Ridge Drive, Reading, PA 19610, USA
e-mail: Jia.Zhu@brentwoodindustries.com; Larry.Li@brentwoodindustries.com

F.M. Kulick III
Research and Development Department, Brentwood, 500 Spring Ridge Drive,
Reading, PA 19610, USA
e-mail: Frank.Kulick@brentwoodindustries.com

L.K. Wang
Lenox Institute of Water Technology, Newtonville, NY 12128-0405, USA

Rutgers University, New Brunswick, NJ, USA
e-mail: lawrencekwang@gmail.com

© Springer International Publishing Switzerland 2016
L.K. Wang, M.-H.S. Wang, Y.-T. Hung and N.K. Shammas (eds.),
Natural Resources and Control Processes, Handbook of Environmental
Engineering, Volume 17, DOI 10.1007/978-3-319-26800-2_6

285

6   Trickling Filter Plant Upgrades 320
7   Trickling Filters for Industrial Applications 322
8   Construction Consideration of Trickling Filters 326
     8.1   Media Integrity 326
     8.2   Miscellaneous Considerations 328
9   Operation and Maintenance Considerations of Plastic Trickling Filters 329
     9.1   Uniform Distribution Across Media Surface 330
     9.2   Recirculation and Hydraulic Wetting 330
     9.3   Distribution Rate and Flushing Operation 330
     9.4   Ventilation 330
     9.5   Operation of Associated Processes 331
     9.6   Macrofauna Control 331
     9.7   Cold Weather Operation 331
     9.8   Odor Control 332
     9.9   Plastic Media Replacement 332
10   Design Examples 333
     10.1 Design Example 1: Design a Plastic Trickling Filter System
           for Domestic Wastewater Carbonaceous BOD Removal
           Using Germain Formula 333
     10.2 Design Example 2: Design a Plastic Trickling Filter for Combined
           BOD Removal and Nitrification 336
     10.3 Design Example 3: Design a Plastic Trickling Filter for Tertiary
           Nitrification Using Albertson and Okey's Procedure 339
References 345

Abstract A trickling filter (TF) is one of the oldest biological wastewater technologies that has been used for over 100 years. With years of evolution, modern trickling filters can be built up to 12 m (40 ft) tall with the use of high-performance plastic media. Modern flow distribution and ventilation systems can also be incorporated to improve TF performance. The treatment capacity of modern filters has been significantly increased in comparison with that of old rock filters. The application of trickling filter technologies has also been greatly expanded from early BOD roughing to incorporate complete carbon oxidizing and nitrification. In many cases, trickling filters are designed in "stages" to achieve different treatment levels.

This chapter presents trickling filter technology fundamentals including history, general system configurations, design methodologies, and considerations. In addition to trickling filter design for BOD removal and nitrification, necessary modifications to the process configuration to achieve total nitrogen removal are also discussed. General guidelines for industrial trickling filter design are addressed as well.

Keywords Trickling filters • Biological treatment • Plastic media • Wastewater • Carbon oxidizing • Nitrification • Denitrification • Nutrient removal • Biofilm • Practical design examples • Municipal wastewater treatment • Industrial wastewater treatment

Nomenclature

A	Trickling filter media plan view area, m^2 or ft^2
A_S	Trickling filter media-specific surface area, m^2/m^3 or ft^2/ft^3
AR_{20}	Airflow rate at 20 °C and 1.0 atm, m^3/min or scfm
AR_T	Airflow rate corrected for temperature T, m^3/min or scfm
$AR_{T>20}$	Airflow rate at temperature of greater than 20 °C, m^3/min or scfm
a	Number of distributor arms
D	Trickling filter media depth, m or ft
D_{Air}	Natural draft driving pressure, mm of water
DR	Dosing rate, mm/pass or inch/pass
g	Acceleration of gravity, 9.8 m/s^2
K_V	First-order rate constant (Velz), 1/day
$k_{C,\ 20}$	Treatability coefficient at 20 °C, $(L/s)^{0.5}/m^2$ or $gpm^{0.5}/ft^2$
k_G	Wastewater treatability and packing coefficient (Germain), $(L/s)^{0.5}/m^2$ or $gpm^{0.5}/ft^2$
$k_{n,T}$	Zero-order nitrification rate corrected for temperature, g $N/m^2 \bullet day$
k_n'	First-order nitrification rate, g $N/m^2 \bullet day$
k_T	Kinetic rate constant at temperature T
K_{20}	Kinetic rate constant at 20 °C
$k_{20,\ MV}$	Overall treatability coefficient at 20 °C (Modified Velz), $(L/s)^{0.5}/m^2$ or $gpm^{0.5}/ft^2$
L	Liquid loading rate, kg/h
n	Exponent characteristic of plastic media dimensionless
N	Tower resistance head loss in the tower
N_e	Trickling filter effluent NH_4-N concentration, mg/L
N_P	Media packing head loss in terms of the number of velocity heads
N_T	Transition ammonia-nitrogen concentration, mg/L
NLR	Nitrogen loading rate, g ammonium nitrogen/m^2 media surface$\bullet$day or lb/1000 $ft^2 \bullet day$
NO_x	Oxidized nitrogen concentration, mg/L
OFR	Clarifier overflow rate, m/h or gpd/ft^2
OLR	Trickling filter organic loading rate, kg BOD/m^3 media volume$\bullet$day or lb/1000 $ft^3 \bullet day$
ΔP	Total head loss, kPa
PF	Peaking factor for flow rate
Q_{in}	Influent flow to trickling filter, m^3/day or million gallons per day (mgd)
Q_R	Trickling filter recirculation flow rate, m^3/day or mgd
R	Recirculation ratio (recycled flow/influent flow)
R_o	Oxygen supply, kg O_2/kg BOD_5 applied
S_D	Removable BOD concentration at depth D, mg/L
S_e	Settled effluent BOD_5 concentration, mg/L
$S_{e,s}$	Effluent soluble BOD_5 concentration, mg/L
S_o	Influent BOD_5 concentration, g/m^3 or mg/L

$S_{N, in}$ Influent ammonium nitrogen concentration, g/m^3 or mg/L

$S_{o,s}$ Influent soluble BOD_5 concentration mg/L, excluding recirculation

S_{TKN} Influent total Kjeldahl nitrogen (TKN) concentration, mg/L

SWD   Side water depth, m or ft

T   Wastewater temperature, °C

T_C Cold air temperature, K

THL   Trickling filter hydraulic loading rate including recirculation, m^3/m^2•day or gpm/ft^2

TKN_{OX} Nitrification rate, g N/m^2•day

T_H Hot air temperature, K

T_1 Higher air temperature inside a filter, K

T_2 Lower air temperature inside a filter, K

T_m Log mean of air temperature inside TF, K

T_0 Ambient air temperature, °C

θ Temperature correction constant

ω Distributor arm rotational speed, rev/min

V_M Trickling filter media volume, m^3 or ft^3

v   Superficial air velocity, m/s

WHL   Trickling filter hydraulic loading rate excluding recirculation, m^3/m^2 •day or gpm/ft^2

1   Trickling Filter Introduction

1.1   *Process Description*

A trickling filter is an aerobic process that is used to remove organic matter and ammonium nitrogen from wastewater. A trickling filter consists of fixed bed media through which wastewater trickles for treatment. The term "filter" can be misleading as no physical filtration actually occurs in a trickling filter system. Rather the filter serves as a host for microorganisms that grow on the media to form biofilms. Trickling filters are essentially a solid-liquid-gas system in which the wastewater (liquid) flows over the biofilm (solids) and contacts with air (gas), and at the same time, organic matter and nitrogen pollutants are absorbed and subsequently degraded by the microorganisms in the biofilm as shown in Fig. 6.1.

During the operation of a trickling filter, the biofilm tends to grow in thickness as a result of microbial growth. Portions of the biofilm will detach when they lose the ability to remain attached to the media, which is called "sloughing." In modern trickling filters with rotary distributor arms, biofilm sloughing can be effectively controlled through a flushing operation, during which the shear force of the wastewater flow refreshes the biofilm. The sloughed biofilm together with the treated wastewater will drain to the filter bottom and then be conveyed to the downstream solid separation units such as clarifiers for settling. Advanced filtration such as sand filters may be used following secondary clarifiers or nitrifying trickling filters where higher effluent quality is desired.

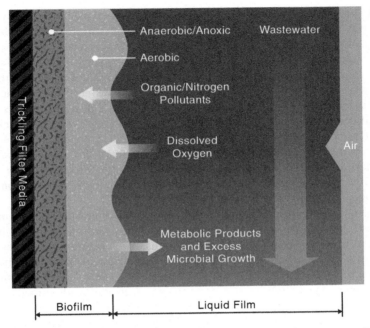

Fig. 6.1 Cross section schematic of the wastewater, biofilm, and media in a trickling filter

Trickling filters are an attached growth processes, where the biomass is attached to the fixed media. Unlike the activated sludge processes, recycling of settled biomass is typically not required. The other major difference between the two types of processes is the aeration method. With trickling filters, air is provided passively either through natural air draft or by forced ventilation systems with low-capacity blowers, while an activated sludge process requires active diffusion aeration with high-capacity blowers or aerators. Therefore, it is generally accepted that trickling filters are more energy efficient than activated sludge processes for similar levels of treatment.

Another known advantage of the trickling filter process over activated sludge processes is the ability to handle shock loads without seriously upsetting the performance. Trickling filters are also considered to be less affected by toxic compounds than other biological systems [1].

1.2   Evolution of Trickling Filters

Trickling filters are one of the oldest wastewater treatment processes used in the USA and around the world. Trickling filters were first initiated in the late 1800s in the USA and UK. The first experimental trickling filter plant in the USA was built in 1901 in Madison, Wisconsin, and the first major trickling filter installation was constructed in 1907 in Columbus, Ohio. By the 1960s, more than 3,500 trickling filter plants were in operation in the USA [2, 3].

Early trickling filters used beds of rocks or other natural media. Plastic synthetic media was developed in the 1950s to provide more surface area for biomass attachment. This reduced clogging and odor problems and has been the dominant media choice for constructing new filters or upgrading rock filters. The specific surface area of the plastic media is two to four times greater than that of the rock media, while the weight is only 2–3 %. The void to volume ratio of the plastic media is also much higher which promotes ventilation. The lightweight of the plastic media also allows for ease of installation and construction of deep bed filters.

Fixed nozzle distribution was employed in most of the early filters. In the 1940s, rotary flow distributors were introduced to provide even distribution of wastewater to the top of the filters, which improved treatment performance. Motor driven distributors were introduced in the 1980s which allow more flexible control of wastewater dosing and enhance biofilm control.

Modern trickling filters can be built up to 12 m (40 ft) tall with high-performance structured sheet plastic media, rotary distributor arms, and forced ventilation systems. The treatment capacity of these filters has been significantly increased in comparison with the earlier filters with rock or other natural media. The applications of the trickling filter technology have also been greatly expanded beyond biochemical oxygen demand (BOD) roughing to include secondary carbon oxidation and tertiary nitrification. Figure 6.2 shows a modern trickling filter in operation.

In recent decades, processes combing trickling filters and activated sludge systems have been developed to take advantage of the benefits of each system and achieve

Fig. 6.2 A modern tertiary nitrifying trickling filter in operation, constructed with high-performance structured sheet PVC Media and a rotary distributor arm system (Courtesy of Brentwood Industries, Inc.)

higher level of overall treatment. For example, the trickling filter/solids contact (TF/SC) process consists of a trickling filter and a very small aeration tank downstream to improve BOD removal and sludge settling ability. The TF/SC process has been widely used in both new facilities and retrofitting existing trickling filter facilities since the first successful full-scale demonstration in the late 1970s. Other combined processes include roughing filter/activated sludge (RF/AS) and activated biofilter (ABF). RF/AS consists of a heavily loaded trickling filter and an aeration tank. The ABF process involves recirculating mixed liquor to the trickling filter.

The most recent research on trickling filters involves using trickling filters to remove toxic and volatile organics and metals such as manganese. Some findings appear to be promising [4].

1.3 Key Components of Modern Trickling Filters

A typical trickling filter consists of a flow distribution system, a filter media, an under drain system, a ventilation system, a containment structure, and, in some cases, a dome. Figure 6.3 displays the components of a modern trickling filter.

Fig. 6.3 Key components of a modern trickling filter (Courtesy of Brentwood Industries, Inc.)

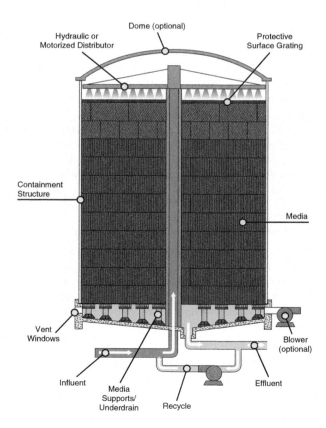

1.3.1 Flow Distribution System

The system provides distribution of wastewater over the media. Fixed nozzle distribution was used in early trickling filters. Modern trickling filters commonly use rotary distributor arms, which are basically two to four horizontal pipes suspended approximately 0.15 m (0.5 ft) or more above the filter media. Wastewater is distributed through the orifices along the horizontal pipes (arms) as the arms rotate. Figure 6.2 shows a modern rotary distributor.

The rotary distributor arms can be hydraulically driven utilizing the hydrodynamic force of the wastewater flowing out of the orifices. Back spray or reversing orifices are used in modern hydraulic rotary distributors to reduce the rotational speed and maintain a desired dosing rate. Rotary distributors can also be motor driven, with which the rotary speed can be controlled independent of flow and therefore allows enhanced wastewater dosing control. Reducing rotational speed to allow for high instantaneous dosing in the distributor mechanism is particularly valuable in systems that have high organic loads. In nitrifying filters, reduced rotational speed is used to flush predators such as snails from the filter.

1.3.2 Filter Media

Rock, wood, and plastic media have been used as the biomass carrier in trickling filters. Comparisons of characteristics of most commonly used media are presented in Table 6.1. In the USA, modular media blocks made from thermoformed plastic sheets are used almost exclusively for new trickling filters due to superior compression strength and ability to be stacked as high as 12 m (40 ft). Random dump plastic media, on the other hand, requires lateral support from the containment structure and typically limits filter media depth to 3 m (10 ft) generally due to compressibility of the media itself. Media consisting of hanging plastic strips are sometimes used for high organic load roughing applications but are not typical. Figure 6.4 shows the most commonly used plastic modular media of cross flow and vertical flow type.

Table 6.1 Physical properties of trickling filter media

Media type	Nominal size, cm[a]	Unit weight, kg/m^3 [b]	Specific surface area, m^2/m^3 [c]	Void space, %
Rock (slag)	5–10	1620	46	60
Rock (river)	2.5–7.5	1442	62	50
Plastic modular – cross flow	$61 \times 61 \times 122$	24–54	102–223	95
Plastic modular – vertical flow	$61 \times 61 \times 236$	24–54	98–131	95
Plastic random	Varies	27.2	98	95

[a]cm/2.54 = in.
[b]kg/m^3/16.02 = lb/ft^3
[c]m^2/m^3/3.281 = ft^2/ft^3

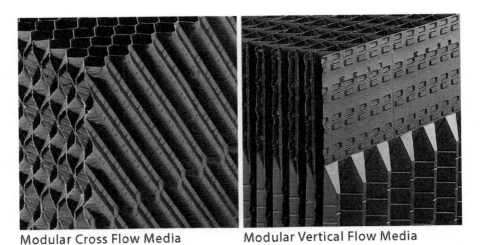

Modular Cross Flow Media   Modular Vertical Flow Media

Fig. 6.4 Commonly used plastic modular media (Courtesy of Brentwood Industries, Inc.)

1.3.3   Containment Structure

Trickling filter containment structures vary in construction depending on the media used. Rock media and random dump plastic media are not self-supporting and therefore require lateral structural support to contain the media within the filter, while modular plastic media is self-supporting and does not require lateral structural support from the containment. Containment structures are typically precast or formed concrete tanks. When self-supporting media is used, materials such as wood, fiberglass, and welded and bolted steel can also be used as containment structures. Some concrete containment structures are specially designed to provide flooding ability allowing for macrofauna and predator control.

1.3.4   Dome

Trickling filters may be covered by a dome for a few reasons. In cold climates, the dome serves to reduce temperature loss in winter and shield the system from strong winds that could interfere with ventilation. Trickling filters may also be covered if odor is a concern. The dome is used to create an enclosed environment where trickling filter-spent air can be collected and channeled to odor scrubbers for treatment.

1.3.5   Underdrain System

The trickling filter underdrain system serves three purposes: (1) collect treated wastewater for conveyance, (2) provide support to the media, and (3) create a plenum that allows air circulation through the media bed. For trickling filters with

Fig. 6.5 Underdrain system, consisting of plastic stanchions and FRP gratings (Courtesy of Brentwood Industries, Inc.)

rock media, clay or concrete underdrain blocks are typically used. A variety of concrete support systems including concrete beams, concrete column with beams, and concrete column with reinforced fiberglass grating have been used to support wood or plastic media. In recent decades, an underdrain support system consisting of field-adjustable plastic stanchions and fiberglass reinforced gratings, as shown in Fig. 6.5, has often been used as an alternative to conventional underdrains. This media support system typically offers improved drainage and ventilation and in many cases is more economical than a conventional concrete support system.

1.3.6   Ventilation

BOD removal and nitrification in a trickling filter are aerobic processes that rely on airflow, which provides sufficient dissolved oxygen concentration through contact of the air with the wastewater being treated. In trickling filters with natural draft ventilation, air density gradients created by temperature and humidity difference between inside and outside of the filter are utilized to drive circulation. Forced ventilation with a low-pressure blower system is sometimes used to provide more consistent and controlled airflow, especially for nitrifying trickling filters and covered trickling filters.

2 Trickling Filter Applications

2.1 Trickling Filter Loading Rates

Trickling filters are used to remove organic matter (BOD) and ammonium nitrogen from wastewater.

Organic loading is typically expressed as $kg/m^3 \cdot day$ (or $lb/1000 \ ft^3 \cdot day$) of filter media as BOD_5. The organic loading rate (OLR) may be calculated using Eq. 6.1:

$$OLR = \left(\frac{BOD_5 \ Applied}{Media \ Volume} \right) = \frac{Q_{in} \cdot S_o}{V_M} \frac{1kg}{1000g} \tag{6.1}$$

where

V_M = trickling filter media volume, m^3
Q_{in} = influent flow to trickling filter, m^3/day
S_o = influent BOD_5 concentration, g/m^3

Nitrifying trickling filter loading rate is normally expressed as grams of ammonium nitrogen per square meter of media surface area, e.g., $g/m^2 \cdot day$ (or $lb/1000 \ ft^2 \cdot day$). The ammonium nitrogen loading rate (NLR) is calculated using Eq. 6.2:

$$NLR = \frac{Ammonium \ Nitrogen \ Applied}{Media \ Surface} = \frac{Q_{in} \cdot S_{N,in}}{V_M \cdot A_s} \tag{6.2}$$

where

V_M = trickling filter media volume, m^3
A_s = trickling filter media-specific surface area, m^2/m^3
Q_{in} = influent to trickling filter, m^3/day
$S_{N,in}$ = influent ammonium nitrogen concentration, g/m^3, determined by influent TKN − dissolved organic nitrogen in the effluent. Dissolved organic nitrogen concentration in the effluent is reported to be in the range of $0.5–2.5g/m^3$ [5].

Hydraulic loading rate or wetting rate to a trickling filter is a critical design and operational parameter. The hydraulic loading rate can significantly impact retention time of wastewater, wetting efficiency of media, and biofilm thickness. Hydraulic loading rate excluding recirculation (WHL) can be calculated using Eq. 6.3. In many cases, trickling filter effluent is recycled to combine with the influent forward flow to increase the wetting rate of the filter, returns dissolved oxygen (DO) to top of filter and also helps control the biofilm thickness. The recirculation ratio is calculated based on recycle flow to forward flow and typically ranges from 0.5 to 4.0. Total hydraulic loading rate to a trickling filter includes the forward flow and the recycled flow and can be calculated using Eq. 6.4:

$$\text{WHL} = \frac{Q_{in}}{A} \tag{6.3}$$

where

WHL = hydraulic loading rate excluding recirculation, $m^3/day \bullet m^2$
Q_{in} = trickling filter influent, m^3/day
A = trickling filter media plan view area, m^2

$$\text{THL} = \frac{Q_{in} + Q_R}{A} \tag{6.4}$$

where

THL = hydraulic loading rate including recirculation, $m^3/d \bullet m^2$
Q_{in} = trickling filter influent, m^3/day
Q_R = trickling filter recirculation flow, m^3/day
A = trickling filter media plan view area, m^2

2.2   Trickling Filter Classification

Modern trickling filters are classified as carbonaceous BOD (cBOD) roughing, carbon oxidizing (cBOD), combined cBOD oxidizing and nitrification, and tertiary nitrification. Table 6.2 summarizes the characteristics of each type of trickling filter [6].

Roughing Filter Roughing filters receive organic loadings of more than 1.6 kg/m^3 •day (100 lb/1000 ft^3•day) and hydraulic loadings up to 180 m^3/day•m^2 (3.0 gal/min•ft^2). Plastic vertical flow modular media is typically used in roughing filters to facilitate flushing under high growth rate conditions. Roughing filters remove 50–80 % of soluble BOD$_5$ and 40–70 % of total BOD$_5$ with proper clarification. Roughing filters are widely used in industrial and municipal wastewater treatment. Treated wastewater may discharge to secondary treatment or to public sewer systems. Energy consumption of a roughing filter includes pumping the influent wastewater and recirculation flows and, in some cases, providing forced ventilation using low-pressure blowers, which is significantly less than that required for aeration and pumping in activated sludge processes. The energy requirement for a roughing filter was estimated to be 2–4 kg BOD applied/kWh, while that for activated sludge, it was estimated to be 1.2–2.4 kg BOD$_5$/kWh [7].

Carbon-Oxidizing Filter Carbon-oxidizing filters receive organic loadings from 0.32 to 0.96 kg/m^3•day (20–60 lb/1,000 ft^3•day). With proper clarification, treated effluent concentrations of 15–30 mg/L of BOD$_5$ and TSS can be achieved.

Combined Carbon-Oxidizing and Nitrification Filter For a combined carbon-oxidizing and nitrification filter, the nitrification efficiency is dependent on the volumetric organic loading [7]. At low organic loadings from 0.08 to 0.24 kg/m^3

Table 6.2 Trickling filter classification [6]

Design parameter	Roughing	Carbon oxidizing (cBOD)	cBOD and nitrification	Nitrifying
Media[a]	VF	RA, RO, XF, or VF	RA, RO, or XF	RA or XF
Wastewater source	Primary effluent	Primary effluent	Primary effluent	Secondary effluent
Hydraulic loading m^3/day•m^2 (gpm/ft^2)	52.8–178.2 (0.9–2.9)	13.7–88.0 (0.25[b]–1.5)	13.7–88.0 (0.25[b]–1.5)	35.2–88.0 (0.6–1.5)
Organic loading, kgBOD$_5$/m^3•day (lb BOD$_5$/day/1000 ft^3)	1.6–3.52 (100–220)	0.32–0.96 (20–60)	0.08–0.24 (5–15)	N/A
Nitrogen loading, kg NH$_3$-N/m^2•day (lb NH$_4$-N/day/1000 ft^2)	N/A	N/A	0.2–1.0 (0.0–0.2)	0.5–2.4 (0.1–0.5)
Effluent quality	40–70 % BOD$_5$ conversion	15–30 mg/L BOD$_5$ and TSS	<10 mg/L BOD$_5$ <3 mg/L NH$_4$-N	0.5–3 NH$_4$-N
Predation	No appreciable growth	Beneficial	Detrimental (nitrifying biofilm)	Detrimental
Filter flies	No appreciable growth	No appreciable growth	No appreciable growth	No appreciable growth
Depth, m (ft)	0.91–6.10 (3–20)	≤12.2 (40)	≤12.2 (40)	≤12.2 (40)

[a]Media used: *VF* vertical flow media, *XF* cross flow media, *RA* random media, *RO* rock media
[b]Applicable to shallow filters

•day (5–15 lb/1000 ft^3•day), significant nitrification occurs in the trickling filter, and treated effluent concentrations of less than 3 mg/L of NH$_4$ − N and 10 mg/L of BOD$_5$ can be achieved.

Tertiary Nitrifying Filter For tertiary nitrification applications, very low BOD loading is applied, and a thin biofilm consisting of predominantly nitrifying bacteria develops on the media surface. With proper design, tertiary nitrifying filters can achieve effluent NH$_4$ − N concentration of 0.5–3 mg/L.

3 Design of Plastic Trickling Filters

3.1 Process Considerations

The design of a trickling filter for municipal wastewater treatment typically relies on empirical models. For industrial applications, pilot studies are recommended to optimize trickling filter design.

Typical factors to be considered for a trickling filter design include wastewater characteristics and treatability, temperature and pH, pretreatment, trickling filter media type, media depth, recirculation, hydraulic and organic loadings, and ventilation. In addition, a solid separation system following a trickling filter needs to be designed properly to achieve expected levels of total BOD removal efficiency.

Wastewater Characteristics and Treatability. Wastewater characteristics to be considered include concentrations of 5-day biochemical oxygen demand (BOD_5), soluble BOD_5, chemical oxygen demand (COD), soluble COD (sCOD), total suspended solids (TSS), total Kjeldahl nitrogen (TKN), ammonium nitrogen ($NH_4 - N$), etc. The data generated from these tests is used in conjunction with the design flow rate to calculate organic and nitrogen load to a trickling filter.

Organic pollutants in a wastewater are in the forms of soluble and suspended/colloidal substances. Treatability of wastewater partly depends on the suspended/colloidal to soluble organic concentration ratio. In a trickling filter, soluble organics of small molecular size are easily removed through biochemical oxidation. On the other hand, suspended/colloidal organics are primarily removed by combined processes of biological flocculation and adsorption. Therefore, the trickling filter is a very efficient treatment means for industrial wastewater containing high percentage of soluble organic matter such as food-processing wastewaters [1].

Temperature and pH. The temperature effect on filter performance for BOD removal has been expressed by the following relationship shown in Eq. 6.5 [8]. This equation is applicable to the multiple empirical models discussed later in this chapter:

$$k_T = k_{20}\theta^{T-20} \qquad (6.5)$$

where

T = wastewater temperature, °C
θ = *temperature correction constant*, commonly accepted as 1.035; values ranging from 1.00 to 1.049 were reported [1].
k_T = kinetic rate constant at temperature T
k_{20} = kinetic rate constant at 20 °C

The temperature effect on nitrifying filters has been inconclusive. Studies [9, 10] suggested an equivalent θ value of 1.018 for nitrification rate correction at temperature range of 10–20 °C, while other studies [11, 12] found little correlation between filter performance and temperature.

As with other biochemical oxidizing processes, wastewater pH should be in the range of 6.5–8.5 in order to achieve optimum performance in a trickling filter. For nitrifying trickling filters, it was reported that higher pH of 8.4 in the feed stream favored nitrification, but this conclusion has not been well accepted by the industry [1].

Pretreatment. Trickling filter performance can be affected by the degree of pretreatment. For domestic wastewater, primary clarification is typically considered to provide adequate pretreatment for a trickling filter. In some cases, fine screens are used as an alternative to primary clarifiers for plastic trickling filters. However, decreased capacity and efficiency were reported for a trickling filter receiving screened wastewater compared to clarified wastewater [13]. It is recommended that fine screening should not be used for rock media, random media, and cross flow media with specific surface area of greater than $102 \ m^2/m^3$ ($31 \ ft^2/ft^3$) due to the increased potential of media fouling [1]. A well-designed primary clarifier should be able to remove at least 55 % of the TSS and 25–30 % of BOD_5. Chemical precipitation to remove phosphorus in the primary clarifier is commonly used in the trickling filter plants with a phosphorus discharge limit. In such cases, TSS and BOD_5 removal increase as a result of chemical addition, and trickling filter performance is improved.

Recirculation Recirculation is a process in which the filter effluent is recycled and brought into contact with the biofilm more than once. Recirculation can be beneficial in several ways. Firstly, it improves flushing and keeps biofilm refreshed therefore reduces odor and ponding; secondly, treated effluent dilutes the influent wastewater pollutant concentrations and helps distribute the load evenly through the depth of the filter; thirdly, recirculation brings back dissolved oxygen (DO) from the filter effluent and helps with treatment performance. Dissolved oxygen is consumed at a higher rate near the top of the TF, while the filter influent is typically unaerated. Recirculation brings back DO-rich effluent to raise the DO concentration near the top of the tower.

In many cases, forward flow alone cannot provide adequate wetting efficiency; therefore, recirculation is designed to achieve the proper total hydraulic loading rate (or wetting rate) to the filter. The minimum hydraulic loading rate recommended by Dow Chemical Company is $0.5 \ L/m^2 \cdot s$ ($0.75 \ gpm/ft^2$) to achieve a maximum BOD_5 removal efficiency [1]. Hydraulic loading rate to a trickling filter can be in a range of $0.17–2 \ L/m^2 \cdot s$ ($0.25–3 \ gpm/ft^2$) for various applications and media configurations [6].

A typical recirculation rate ranges from 0.5 to 4 times of the average influent flow, but for high-strength industrial wastewater, a recirculation ratio of ten or more can be used to achieve adequate wetting and flushing of the media.

Dosing Rate Dosing rate (DR) is the depth of water applied on top of the filter media during each pass of the distributor arm. Dosing rate is also referred to as the SK rate or *Spulkraft*, a term used in Germany to define dosing intensity. Dosing rate is expressed in Eq. 6.6:

$$DR = \frac{(\text{THL})\left(1,000\,\frac{mm}{m}\right)}{(a)(\omega)\left(1,440\,\frac{min}{day}\right)} \tag{6.6}$$

where

DR = dosing rate, mm/pass
THL = total hydraulic loading rate (including forward flow and recycle flow), $m^3/m^2 \cdot day$
a = number of distributor arms
ω = distributor rotational speed, rev/min

At a fixed hydraulic loading rate, slowing down the rotary distributor speed can increase the DR. Studies indicate that slowing down the rotation speed can provide greater wetting efficiency, improve biofilm control, reduce odor, and help control predators [14, 15]. It is also suggested that a higher dosing rate may also compensate for low recirculation rate to a certain level and save energy [1].

In North America, rotary distributors are typically operated in the range of 2–10 mm/pass and are much lower than the dosing rate of 20–750 mm/pass recommended by some researchers. It has been reported that plants adopting recommended dosing rate control achieved better or more stable performances [1]. Higher dosing rates are recommended for filters with higher loads because excessive biofilm growth occurs in these filters and high dosing help with reducing biofilm thickness therefore preventing filters from ponding and other operational issues. When possible, it is recommended to conduct a periodic flushing operation with an extra high dosing rate (i.e., low rotation speed). The flushing operation can be conducted once a day for an hour. Table 6.3 summarizes the recommended operating and flushing dosing rates for filters with different loads [6, 16]. Data presented in the table provides a general guideline for dosing rates; however, optimal dosing rates are best determined from field operation. For this reason, distributor design should include the flexibility of providing a range of dosing rates for filter performance optimization.

Ventilation The oxygen required to maintain aerobic conditions in a trickling filter is obtained from the atmosphere, and therefore, an adequate flow of air is critical to the successful operation of a trickling filter.

Table 6.3 Recommendation of trickling filter operating and flushing dosing rates [6, 16]

Organic loading, kg/m³·day (lb BOD₅/day/1,000 ft³)	Operating dosing rate, mm/pass (in./pass)	Flushing dosing rate, mm/pass (in./pass)
<0.4(<25)	25–75 (6–3)	100(4)
0.8 (50)	50–150 (2–6)	150(6)
1.2 (75)	75–225 (3–9)	225(9)
1.7(100)	100–300 (4–12)	300(12)
2.4(150)	150–450 (6–18)	450 (18)
3.2(200)	200–600(8–24)	600(24)

Natural draft has been the primary means for providing airflow through a trickling filter, which is driven by the temperature and humidity difference between air in the trickling filter media and ambient air. If the wastewater is colder than the ambient air, the direction of airflow will be downward and vice versa. A downward airflow in natural draft is desirable for trickling filters as the top portion of a trickling filter receives the heaviest loading and the dissolved oxygen is consumed in a higher rate than that in the lower portion of the trickling filter. Natural draft-driving pressure as a result of temperature differences may be determined by Eq. 6.7 developed by Schroeder and Tchobanoglous [17]. The inside air temperature is conservatively determined by the log mean temperature between higher temperature and lower temperature as Eq. 6.8. Effects of humidity difference on natural draft are considered less significant [18]:

$$D_{\text{Air}} = 353 \left(\frac{1}{T_c - \frac{1}{T_h}} \right) D \qquad (6.7)$$

$$T_m = \frac{T_1 - T_2}{\ln(T_1/T_2)} \qquad (6.8)$$

where

D_{Air} = natural draft driving pressure, mm of water
T_c = cold air temperature, K
T_h = hot air temperature, K
T_m = log mean of air temperature inside a filter, K, substituted for T_c or T_H in 6.7
D = media depth, m
T_1 = higher air temperature inside a filter, K
T_2 = lower temperature inside a filter, K

Natural draft is considered to be subject to environmental changes which can negatively impact trickling filter performances from time to time. In many areas, during spring and fall, there could be periods where little to no airflow would occur through the filter because of insufficient temperature differences between wastewater and air inside a filter and ambient air, which is called "stagnation." Less air supply during the stagnation can reduce treatment efficiency or capacity.

Forced ventilation is a low-cost measure to provide trickling filters with reliable and controlled airflows [18]. The power consumption for a forced ventilation system to treat 3,800 m³/day (1 MGD) wastewater flow was estimated to be 0.15 Kw (0.2 hp) [7]. Forced ventilation is also recommended for heavily loaded trickling filters and nitrifying filters, where treatment can be limited by oxygen availability from natural draft. For filters suffering from natural draft stagnation during certain periods of the year, forced ventilation equipment, if available, can be operated to compensate for the required airflow.

Forced ventilation is required for domed trickling filters, where natural draft is significantly limited. During winter, forced ventilation airflow through the trickling filter can be controlled to minimize or eliminate freezing in the filter.

Axial-flow, radial-flow, and centrifugal, low-head fans can be used. Air distribution header is usually needed to control air distribution from forced ventilation.

The amount of oxygen required for optimizing trickling filter performance has been less studied. Studies by the Dow Chemical Company suggested that at least a supply of 20 kg oxygen/kg of oxygen demand, or 5 % transfer efficiency, should be used to determine the minimum airflow to the filter. Accordingly, oxygen requirements for BOD removal and combined BOD removal and nitrification trickling filters can be estimated based on Eqs. 6.9 and 6.10, respectively.

BOD removal only:

$$R_o = \left(\frac{20\text{kg}}{\text{kg}}\right)[0.8e^{-90LR} + 1.2e^{-0.17OLR}]PF \tag{6.9}$$

Combined BOD removal and nitrification:

$$R_o = \left(\frac{40\text{kg}}{\text{kg}}\right)\left[0.8e^{-90LR} + 1.2e^{-0.17OLR} + 4.6\frac{N_{OX}}{BOD_5}\right]PF \tag{6.10}$$

where

R_o = oxygen supply, kg O_2/kg BOD_5 applied
OLR = BOD loading to trickling filter, kg/m^3•day
N_{OX}/BOD_5 = ratio of oxidized nitrogen to influent BOD_5
N_{OX} = oxidized nitrogen, calculated by: Influent TKN – Effluent TKN – Net Yield Organic Nitrogen
Net yield organic nitrogen = 0.07 (VSS yield)
BOD_5 = influent BOD_5, mg/L
PF = peaking factor, typically 1.3 for large plants and 1.75 for small plants [1]

The ventilation air rate is based on oxygen requirement and can be determined by Eq. 6.11. Actual ventilation rate should be corrected based on temperature as expressed in Eq. 6.12. Where the wastewater temperature is greater than 20 °C, a further correction based on Eq. 6.13 is recommended in order to compensate for the low-oxygen saturation concentration in water at higher temperatures:

$$AR_{20} = \frac{(R_o)(Q_{in})(S_o)(3.5N\ \text{m}^3/\text{kg oxygen})}{\left(\frac{1,000g}{\text{kg}}\right)\left(1,440\frac{\text{min}}{\text{day}}\right)} \tag{6.11}$$

$$AR_T, = AR_{20}\left(\frac{273 + T_0}{273}\right) \tag{6.12}$$

$$AR_{T>20} = AR_T + AR_T(1\%)(T - 20) \tag{6.13}$$

where

AR_{20} = airflow rate at 20 °C and 1.0 atm, m^3/min
Q_{in} = wastewater flow, m^3/day

S_o = TF influent BOD_5 concentration, g/m^3
AR_T = actual airflow corrected for temperature, m^3/min
T_0 = ambient air temperature, °C
$AR_{T>20}$ = actual airflow at temperature of greater than 20 °C

Pressure drop through plastic media can be calculated based on Eq. 6.14:

$$\Delta P = N \left(\frac{v^2}{2g}\right) \tag{6.14}$$

where

ΔP = total head loss, kPa
v = superficial air velocity, m/s
g = acceleration of gravity, 9.8 m/s^2
N = tower resistance, head loss in the tower

The tower resistance N represents the sum of all the individual head losses from airflow. Dow Chemical Company developed a formula (Eq. 6.15) for vertical flow media with specific surface area of 89 m^2/m^3 to determine the packing loss:

$$N_p = 10.33D\,e^{(1.36\times10^{-5})(L/A)} \tag{6.15}$$

where

N_p = media packing head loss in terms of velocity heads
D = media depth, m
A = tower top surface area, m^2
L = liquid loading rate, kg/h

As there are no further data on other types of media, it is suggested to use the correction factors in Table 6.4 to estimate Np based on calculated Np per Eq. 6.15 [1]

Clarification Properly designed clarifiers are critical in achieving desired levels of TSS and BOD_5 removal in the trickling filter effluent. Historically, shallow clarifiers with a side water depth of 2.1 m (7 ft) and relatively high-average overflow rates based on recommendations from the Recommended Standards for Sewage Works, also known as the Ten States Standards [19], were designed and resulted in relatively poor effluent quality. The updated Ten States Standards [20, 21] increased the minimum side water depth to 3.0 m (10 ft), which may still not be adequate for

Table 6.4 Correction factors for head loss in trickling filters with non-vertical flow media [1]

Media	Specific surface area, m^2/m^3 (ft^2/ft^3)	Correction factor
Plastic cross flow	100 (30)	1.3
Plastic cross flow	138 (42)	1.6
Plastic random	100 (30)	1.6
Rock	39–49 (12–15)	2.0

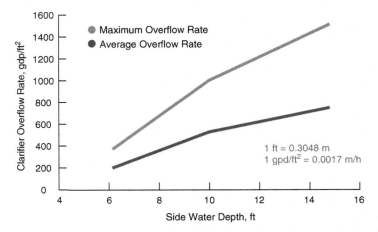

Fig. 6.6 Recommended trickling filter clarifier overflow rate as a function of clarifier side water depth

higher overflow rates. WEF [1] recommends that overflow rates of clarifiers be a function of the side water depth for clarifiers with a floor slope greater than 1:12 as shown in Fig. 6.6, which was developed based on Eqs. 6.16, 6.17, 6.18, and 6.19.

For TF clarifiers with floor slope $\geq$1:12, overflow rates (OFRs) can be determined by:

$$\text{Maximum OFR (m/h)} \leq 0.182\,\text{SWD}^2 \; (\text{SWD} = 1.83 \text{ to } 0.35 \text{ m}) \qquad (6.16)$$

$$\text{Average OFR (m/h)} \leq 0.092\,\text{SWD}^2 (\text{SWD} = 1.83 \text{ to } 3.05\text{m}) \qquad (6.17)$$

$$\text{Maximum OFR (m/h)} \leq 0.556\,\text{SWD} \; (\text{SWD} = 3.05 \text{ to } 4.57 \text{ m}) \qquad (6.18)$$

$$\text{Average OFR (m/h)} \leq 0.278\,\text{SWD} (\text{SWD} = 0.305 \text{ to } 4.57\text{m}) \qquad (6.19)$$

3.2 Plastic Media Selection

The most commonly used modular media types are cross flow media (XF), which is made of sheets formed with alternating corrugations at 60° to horizontal and vertical flow media (VF), which has vertical channels throughout the module. CH2M Hill [22] conducted a comprehensive study comparing different types of trickling filter media, and some of the results are shown in Table 6.5. Sixty degree (60°) cross flow media were concluded to be the most efficient media for secondary treatment and nitrification applications, and the vertical flow media is more suitable for high-loading applications.

Table 6.5 Relative comparison of commonly used trickling filter media [22]

			60° Cross flow	Vertical flow	Random	Rock
Effluent quality	Low load	Soluble BOD	Best	Good	Good	Average
		TSS	Best	Average	Average	Average
		Total BOD	Best	Average	Average	Average
		Nitrification	Best	Good	Average	Worst
	High load	Soluble BOD	Good	Best	Average	Worst
		TSS	Best	Average	Poor	Average
		Total BOD				
Operating characteristics		Hydraulic retention	Average	Worst	Good	Best
		Oxygen transfer	Average	Average	Average	Worst
		Wetting	Average	Average	Worst	Good
		Filter flies	Good	Best	Average	Worst
		Biogrowth	Average	Best	Poor	Average
		Plugging	Average	Best	Average	Worst
Qualitative factors		Ease of installation	Good	Good	Average	Worst
		Air ventilation	Good	Poor	Average	Worst
		Media strength	Good	Average	Poor	Best
		Tower construction	Average	Average	Poor	Worst
		Media height	Good	Good	Average	Worst
		Media surface area	Good	Good	Good	Worst
		Media void space	Good	Good	Good	Worst
		Weight of media	Good	Good	Good	Worst

Among the various plastic media, modular cross flow media is most widely used in new trickling filter installations and rock filter retrofits. Low-density cross flow media with specific surface area of 101 m^2/m^3 (31 ft^2/ft^3) is typically used in carbonaceous BOD removal filters; medium-density media has also been used in shallow rock filter retrofits, and good performance was reported [23, 24]. For combined carbon-oxidizing and nitrification filters, both low-density cross flow and medium-density cross flow media have been used in full-scale applications. For tertiary nitrification applications, medium-density cross flow media was recommended based on several studies [25–27]. High-density cross flow media with specific surface area of 223 m^2/m^3 (68 ft^2/ft^3) has also been used in full-scale tertiary nitrification filters since the 1980s. Li et al. [28] reported several full-scale nitrifying trickling filters with rotary distributors, and high-density cross

Table 6.6 TF application and plastic modular media selection

Application	Media type	Specific surface area, m^2/m^3 (ft^2/ft^3)
Roughing filter	Vertical flow	89–98 (27–30)
	Combined cross flow and vertical flow	101 (30.8)[a]
Carbon oxidizing trickling filter (secondary treatment filter)	Cross flow (low density)	102 (31)
	Cross flow (medium density)[b]	138–157 (42–48)
Combined carbon oxidizing and nitrification trickling filter	Cross flow (low or medium density)	102 (31)
		138–157 (42–48)
Tertiary nitrification	Cross flow (medium or high density)	138–157 (42–48)
		223 (68)

[a]Actual surface area varies with media depth
[b]For shallow filter of less than 1.8–2.4 m (6–8 ft)

flow media achieved excellent nitrification performance, and the surface nitrification rates are comparable to those of medium-density cross flow media reported from full-scale applications. When a fixed nozzle distribution system was used, the high-density nitrifying filter performance was poorer as a result of media plugging and insufficient wetting. Vertical flow media with a specific surface area of 98 m^2/m^3 (30 ft^2/ft^3) is often used for high-strength wastewater roughing. In the past two decades, a mixed media combination of cross flow media at the top of the filter and vertical flow media at the bottom has also been used for roughing applications. General guidelines for plastic modular media selection based on their application are summarized in Table 6.6. Figure 6.7 displays a modular mixed media using cross flow in top layers and vertical flow media in bottom layers.

3.3 Single-Stage and Multistage Trickling Filters

Trickling filters can be operated in single or multistages as shown in Fig. 6.8. In many cases, the second- or third-stage trickling filters are built as a plant upgrade to expand, meet more stringent BOD limits, and achieve nitrification.

Benefits of intermediate clarification vary based on the application. If the wastewater contains a high portion of soluble BOD and produces a large quantity of biomass in the first-stage filter, intermediate clarification may avoid the adverse impact of high TSS to the second-stage filter performance; intermediate clarification can also improve nitrification performance in the second-stage filters. For other two-stage domestic trickling filter plants, negligible benefits of intermediate clarifier have been reported [29].

Fig. 6.7 Modular mixed media includes cross flow media in the top two layers for enhanced distribution and *vertical* flow media in lower layers to resist plugging (Courtesy of Brentwood Industries, Inc.)

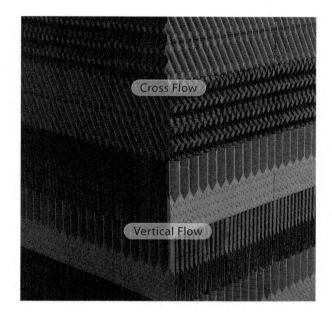

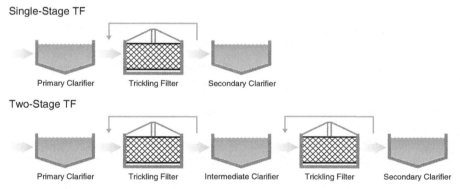

Fig. 6.8 Common flow diagrams of single- and two-stage trickling filter processes

3.4 Design Models for BOD Removal Filters

3.4.1 Velz Formula

The first process design approach to use fundamental principles was published by Velz in 1948. His equation expressed BOD removal (Eq. 6.20) as a first-order function of filter depth:

$$\frac{S_D}{S_o} = 10^{-K_V D} \qquad (6.20)$$

where

S_o = influent BOD$_5$, mg/L
S_D = *removable* BOD at depth D, mg/L
D = media depth, m
K_V = Velz first-order constant, d^{-1}

3.4.2 Germain Formula

Schultz modified the Velz equation to account for hydraulic loading rate, and later Germain applied Schultz's formula to plastic trickling filter media as Eq. 6.21 [1]:

$$\frac{S_e}{S_o} = e^{-k_G D/\text{WHL}^n} \tag{6.21}$$

where

S_o = influent BOD$_5$, excluding recirculation, mg/L
S_e = settled effluent BOD$_5$, mg/L
D = media depth, m
WHL = hydraulic loading rate, excluding recirculation, L/m^2•s
n = exponent characteristic of plastic media, typically assumed to be 0.5
k_G = wastewater treatability and packing coefficient, (L/s)$^{0.5}$/m^2

The value of k is related to wastewater characteristics, media depth, media surface area, and media configuration. Germain (1966) reported that the value of k for a plastic media filter 6.6 m (21.5 ft) deep treating domestic wastewater was 0.24 (L/s)^n/m^2 (0.088 gpm^n/ft^2). The value of n is normally assumed to be 0.50 for different types of plastic media [29]. Media used in Germain's study was vertical flow media with a specific surface area of 89 m^2/m^3 (27 ft^2/ft^3). The k values were evaluated for different types of wastewater based on more than 140 pilot studies by the Dow Chemical Company and showed wide variations. This suggests that for nonmunicipal wastewater treatment a pilot study is the only reliable means of determining the *k* value and consequently sizing the trickling filter properly [29].

The Germain equation does not include the effect of recirculation based on the pilot studies conducted in which no statistically significant differences were found. However, it is noted that the relatively high media depth of 6.6 m (21.5 ft) was used in the study resulting in relatively high-application loading which ensures adequate wetting of the media. In shallow filters with lower THL, recirculation should benefit the system performance as a result of increase media wetting.

3.4.3   Modified Velz Formula

Eckenfelder and Barnhart [30, 31] proposed design formulas that account for the effect of media surface area. Later, an equation termed modified Velz formula [29] was developed for soluble BOD removal that accounts recirculation as follows (Eq. 6.22):

$$S_{e,s} = \frac{S_{o,s}}{(R+1)\exp\left\{\frac{k_{20,MV}A_s D\theta^{T-20}}{\text{WHL}[(R+1)^n]}\right\} - R} \tag{6.22}$$

where

$S_{o,s}$ = influent soluble BOD_5, mg/L, excluding recirculation
$S_{e,s}$ = effluent soluble BOD_5, mg/L
$k_{20,MV}$ = overall treatability coefficient at 20 °C, $(L/s)^{0.5}/m^2$
A_s = clean media surface area, m^2/m^3
D = depth of media, m
θ = temperature correction coefficient, 1.035
WHL = hydraulic loading rate excluding recirculation, L/s•m^2
R = recirculation ratio (recycled flow/influent flow)
n = exponent characteristic of plastic media, typically assumed to be 0.5

3.4.4   Normalization of Treatment Coefficient

Published k values are normally developed based on data from different tower configurations and wastewater strengths and therefore are not directly comparable. In order to make them comparable, the k values can be generalized or normalized to a specified depth of 6.1 m (20 ft) and influent BOD concentration of 150 mg/L using Eq. 6.23 [29]:

$$k_{c20} = k_{20}\left(\frac{D}{6.1}\right)^{0.5}\left(\frac{S_o}{150}\right)^{0.5} \tag{6.23}$$

For shallow filters with cross flow media of less than 4 m (13 ft) high, it is suggested that a coefficient exponent of 0.3 instead of 0.5 should be used to correct the k value in terms of depth $D/6.1^{0.3}$ [1].

3.5   Design Procedures for Combined BOD Removal and Nitrification Filters

Combined BOD removal and nitrification filters are also referred as "dual-purpose trickling filters." In a dual-purpose trickling filter, BOD removal primarily takes place in the upper portion of the tower and nitrification in the lower portion. This is because nitrifier growth can only become dominant when most soluble organic substrate has been removed by the upstream fast-growing heterotrophs and therefore less competition for available oxygen and media surface. Recirculation may increase nitrification capacity due to reduced BOD concentration in the filter.

The kinetics of dual-purpose trickling filters are complex due to the complexity of biofilm and limited technical understanding of the influencing factors. Therefore, design methods of dual-purpose trickling filters have been empirical [1].

Several researchers have reported that soluble BOD concentration of less than 20 mg/L are required to initiate nitrification [32, 33]. High levels of nitrification were reported in multiple full-scale and pilot installations at organic loadings of less than 0.2 kg/m^3-day (12.6 lb/1,000 ft^3-day) [34, 35]. It was also reported that cross flow media outperformed vertical flow media when used in dual-purpose trickling filters [36]. Design and performance data of three full-scale dual-purpose trickling filters are summarized in Table 6.7. All three trickling filters were followed with a solids contact basin, and the effluent data shown in Table 6.7 represent solids contact basin effluent characteristics. The trickling filter/solids contact process will be discussed later in this chapter.

WEF [1] described a design procedure based on the study by Okey and Albertson [11]. Extensive data from four plants was compared based on the influent BOD-to-TKN ratio. A relationship between the TKN removal rate and influent BOD-to-TKN ratio was derived as Eq. 6.24. Higher nitrification rates were observed at temperatures of 9–20 °C:

$$TKN_{OX} = 1.086 \left(\frac{S_o}{S_{TKN}} \right)^{-0.44}$$

(6.24)

$$\text{median } \bar{y} = 0.460 \pm 0.175; \ \bar{x} = 11.081$$

where

TKN_{OX} = nitrification rate, g/m^2•day at approximately 15 °C
S_o = influent BOD$_5$, mg/L
S_{TKN} = influent total Kjeldahl nitrogen (TKN) concentration, mg/L

3.6 Design Procedure for Nitrification Trickling Filters (NTFs)

Trickling filters have been used in tertiary nitrification applications since the 1980s and have proven to be a reliable and cost-effective technology to control ammonium nitrogen. A properly designed and operated NTF can achieve effluent NH$_4$-N concentration of less than 1 mg/L during warm weather and 1–4 mg/L [7] or 2–3 mg/L during cold weather [6]. Studies focusing on the kinetics of tertiary nitrification filters revealed that the performance of the filters is dependent on many factors including influent cBOD$_5$ and TSS levels, oxygen availability, ammonium nitrogen concentration, temperature, pH, media type, and configuration and hydraulics.

Table 6.7 Design and performance of full-scale dual-purpose trickling filters [1]

	Wauconda, Illinois	Buckeye Lake, Ohio	Chemung County, New York
Design basis			
Number of units	2	2	2
Diameter, m (ft)	15 (50)	14 (45)	40 (130)
Depth, m (ft)	8.5 (28)	13 (42)	4 (13)
Volume, m^3 (ft^3)	3110 (109,900)	3780 (133,528)	9770 (344,930)
Media	Cross flow	Cross flow	Cross flow
Specific surface area, m^2 (ft^2)	98 (30)	98 (30)	98 (30)
Flow, L/s (mgd)	61 (1.4)	48 (1.1)	526 (12)
Primary effluent BOD$_5$, mg/L	117	156	73
Primary effluent soluble BOD$_5$, mg/L	80	–	–
Primary effluent TSS, mg/L	68	90	40
Primary effluent TKN, mg/L	30	36	18
Primary effluent NH$_4$-N, mg/L	20	26	–
Temperature, °C	7–20	10–25	10–20
Permit criteria			
BOD$_5$, mg/L	12	20	15
TSS, mg/L	10	30	–
NH$_4$-N, mg/L	1.5/4.0	3.0	5.0
Design rates			
BOD$_5$, kg/m^3•day (lb/1000 ft^3•day)	0.20 (12.5)	0.17 (10.6)	0.34 (21.2)
NH$_4$-N, g/m^2•day(lb/1000 ft^2•day)	0.37 (0.075)	0.32 (0.0655)	0.86 (0.176)
TF influent BOD$_5$/TKN	3.9	4.3	4.1
TF/SC effluent quality			
cBOD$_5$, mg/L	6	2	7.8
TSS, mg/L	10	5	11.7
NH$_4$-N, mg/L	<1.0	0.3	1.1

NTF Influent cBOD$_5$ and TSS Increased carbonaceous BOD$_5$ (cBOD$_5$) concentration in the NTF influent can negatively impact NTF performance due to the bacterial competition for oxygen and media surface. NTF influent BOD$_5$: TKN ratio typically ranges from 0.3 to1.0. Negative impacts on nitrification capacity from high influent TSS concentrations have also been reported [27]. Influent TSS concentrations of less than 10 mg/L were suggested for maximizing NTF performance [1]. An increase in organic load may result in reduced nitrification performance due to heterotrophic competition for available surface with slower growing nitrifiers.

Oxygen Availability It is generally accepted that in the upper portion of an NTF, the nitrification rate is limited by the oxygen availability and oxygen diffusion rate to the biofilm [7]. Therefore, forced ventilation is typically employed in NTFs to improve the oxygen availability.

Table 6.8 Reported nitrification rates

Plant	Full scale or pilot	Media type	Media surface area, m^2/m^3 (ft^2/ft^3)	Nitrification rates, g/m^2·day	Reference
ANNP, AZ	Pilot	Vertical flow	89 (27)	3.2	[11]
Zurich, Switzerland	Pilot	Vertical flow	92 (28)	1.6	[1]
Zurich, Switzerland	Pilot	Cross flow	223 (68)	1.1–1.2	[1]
Central Valley, UT	Pilot	Cross flow	138 (42)	2.1–3.2	[1]
Lima, OH	Full scale	Vertical flow	89 (27)	1.2–1.8	[1]

NH$_4$-N Concentration At higher NH$_4$-N concentrations, nitrification is considered to be limited by oxygen availability and becomes zero order with respect to nitrogen. When the NH$_4$-N concentration reaches a lower level, NH$_4$-N concentration then becomes the limiting factor, and nitrification becomes a first-order reaction. Okey and Albertson [11] proposed an approach to determine the boundary NH$_4$-N concentration, or the so-called transitional NH$_4$-N concentration based on temperature and oxygen availability as shown in Table 6.9. A wide variation of nitrification rates have been reported and summarized in Table 6.8.

Temperature Several studies [9, 10] showed an approximate 20 % increase in nitrification rates at a temperature increase of 10–20 °C. Other studies [11, 12] reported little correlation between temperature and nitrification rate, and rather the impact of oxygen availability was found to be significant. A recent study [28] reported that NTF performance was relatively independent of temperature, for temperatures between 11 °C and 28.6 °C as long as rapid temperature changes did not occur. A rapid temperature decrease of more than 4–5 °C within 1–2 days period appeared to have negative impact on nitrification performance.

Hydraulics Hydraulic loading rates of NTFs range from 0.4 to 1.0 L/s·m^2 (0.6–1.5 gpm/ft^2). Higher application rates above 1.0 L/s·m^2 (1.5 gpm/ft^2) provide better NTF performances but are not energy efficient [1]. Filter depths of 6.1 m (20 ft) or greater are typically designed to achieve high hydraulic rates and maximize the zero-order kinetic rates.

Media Selection Medium-density cross flow media with specific surface areas of 138–157 m^2/m^3 (42–48 ft^2/ft^3) and high-density cross flow media with specific surface areas of 200–223 m^2/m^3 (62–68 ft^2/ft^3) have been used in full-scale NTFs. High-density media was not recommended based on findings from pilot studies [25, 26]. It was claimed that high-density media had poorer wetting efficiencies and higher clogging potential than medium-density cross flow media which could result in reduced performance. However, a recent study [28] summarized the performance of three NTF plants with high-density cross flow media and showed that NTFs with

Table 6.9 Zero-order to first-order nitrification transitional NH$_4$-N concentration as a function of temperature [11, 37]

Oxygen saturation (%)	Transitional NH$_4$-N concentration, mg/L		
	10 °C	20 °C	30 °C
25	1.0	0.9	0.8
50	2.2	1.8	1.5
75	3.3	2.6	2.1
100	4.3	3.5	2.9

high-density media and modern design features such as rotary distributor arms with speed control, forced air ventilation, and macrofauna control systems can achieve comparable nitrification rate performed to a sustained loading up to 1.14 g NH$_4$-N/ m^2/day as NTFs with medium-density cross flow media. The nitrification rates of these NTFs with high-density cross flow media are consistent with nitrification rates reported in earlier pilot studies [1].

WEF [6] described an empirical design procedure for the NTF using medium-density cross flow media based on Albertson and Okey's work [38] which includes four steps:

1. Determine the transition NH$_4$-N concentration (N_T) from Table 6.9 for the boundary between zero-order and first-order nitrification reaction.
2. Determine media volume required for zero-order region using the maximum rate of 1.2 g N/m^2•day over a temperature range of 10–30 °C. Below 10 °C, the rate ($k_{n,T}$) is adjusted based on Eq. 6.25:

$$k_{n,T} = 1.2 \frac{\text{g}}{\text{m}^2} \cdot \text{day} \times 1.045^{T-10} \qquad (6.25)$$

3. Determine media volume required for first-order performance using the rate which is calculated based on Eq. 6.26. No temperature correction is required between 7 °C and 30 °C:

$$k_n' = 1.2 \left(\frac{N_e}{N_T} \right)^{0.75} \qquad (6.26)$$

where

k_n' = first-order nitrification rate, g N/m^2•day
N_e = trickling filter effluent NH$_4$-N concentration
N_T = transitional NH$_4$-N concentration, mg/L, determined from Table 6.9

4. Sum up the media volume for zero-order and first-order performance for total media volume requirement

Albertson and Okey's design procedure can be used for designing NTFs when the following conditions are met [6]:

- Influent $BOD_5/TKN \leq 1.0$
- Influent soluble $BOD_5 < 12$ mg/L
- Overall hydraulic loading rate (THL) ≥ 0.54 L/m^2·s $\left(0.8 \text{ gpm/ft}^2\right)$
- Influent $cBOD_5$ and TSS ≤ 30 mg/L for medium-density media (138 m^2/m^3 or 42 ft^2/ft^3)
- Forced ventilation
- Distributor control to provide an operating SK number of 25–75 mm/pass and a flushing SK number of 300 mm/pass

4   Combined Trickling Filter/Activated Sludge Processes

Combined trickling filter/activated sludge processes are also referred as dual biological processes. In these processes, trickling filters and activated sludge reactors are operated in series to take advantage of the benefits of both. Trickling filters have higher resistance to shock loads and lower energy consumption, while activated sludge processes typically produce higher effluent quality with lower effluent BOD and TSS concentrations as a result of bioflocculation of the suspended biomass.

The most commonly used combined processes are the trickling filter/solids contact (TF/SC) process and the roughing filter/activated sludge (RF/AS) process. In both processes, effluent from trickling filters is discharged to downstream activated sludge basins without clarification. Returned activated sludge (RAS) from the secondary clarifiers is recirculated back to the activated sludge aeration basins and is included to maintain a proper MLSS concentration. Figure 6.9 shows typical flow schematics of the TF/SC and RF/AS processes. Combined trickling filter and activated sludge processes have been widely used in the USA and provide flexibility in trickling filter plant upgrades. Table 6.10 shows several representative combined trickling filter/activated sludge plants.

4.1   Trickling Filter/Solids Contact (TF/SC)

The TF/SC process includes a trickling filter followed by a small solids contact aeration basin that has only 5–20 % of the volume that would be otherwise required by a stand-alone activated sludge process for similar treatment. The trickling filter is designed to remove a majority of the soluble BOD in the wastewater, and the solids contact basin is designed to bioflocculate the non-readily settleable, colloidal, and particulate organic matters and to polish the residual soluble BOD. Many TF/SC plants also include a sludge reaeration tank which helps improve sludge quality before recirculation and slightly increases system sludge retention time

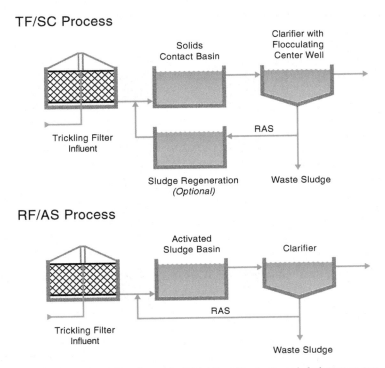

Fig. 6.9 Typical flow schematics of combined trickling filter/activated sludge processes

Table 6.10 Example plants using combined trickling filter/activated sludge processes

Plant	Process	Plant flow, L/s (MGD)	Secondary effluent BOD/TSS/NH$_4$-N, mg/L	Source
Honouliuli, HI	TF/SC	568 (13)	10/6/7	[39]
Duck Creek, TX	TF/SC	1311 (30)	6–12/10–14/0.2–2.4	[40]
Rowlett, TX	TF/AS	700 (16)	7–10/7–10/1.4–5.0	[40]
Corvallis, OR	TF/SC	424 (9.7)	4/5/5	[41]
Richmond, CA	RF/AS	263 (6.0)	30/–/–	[1]
Medford, OR	RF/AS	144.6 (3.3)	5/–/–	[1]
Muscatine, IA	RF/AS	267.4 (6.1)	34/–/–	[1]

(SRT). Flocculating center wells are commonly employed in the clarifier to further promote bioflocculation and improve effluent quality. TSS and BOD concentrations of a TF/SC process are typically less than 15 mg/L and can be as low as 10 mg/L [42, 43]. Typical design criteria for TF/SC processes are summarized in Table 6.11.

It is generally accepted that the organic loading to the trickling filter in a TF/SC process should be limited to 1.2 kg/m^3/day (75 lb/1000 ft^3/day). However, TF/SC processes with higher organic loadings have been reported. Kelly et al. [45]

Table 6.11 Typical design criteria for TF/SC processes [29, 44]

Parameters (based on modular plastic media)	Design criteria	
	Range	Commonly used
Solids production (mg VSS/mg BOD₅ removed)	0.7–0.9	0.7
Trickling filter hydraulic load, m/h (gpm/ft^2)	0.24–4.8 (0.1–2.0)	2.4 (1.0)
Trickling filter organic load, kg/m^3/day (lb/1,000 ft^3/day)	0.32–1.20 (20–75)	0.80 (50)
Solids contact basin HRT at average flow, min	45–120	60
Solids contact basin HRT at peak flow, min	15–30	30
Solids contact basin SRT (d)	1.0–2.0	1.0
Solids contact basin MLSS concentration, mg/L	1,500–3,000	2,000
Clarifier basin overflow rate at average flow m/h (gpd/ft^2)	0.85–1.69 (500–1,000)	1.35 (800)
Clarifier underflow concentration, % total solids	0.6–1.2	0.8

reported that the world's largest TF/SC plant in Vancouver, BC, Canada, with a capacity of 772 MLD (207 MGD) was operated at a loading of approximately 1.6 kg/m^3/day (100 lb/1000 ft^3/day), achieved in excess of 70 % of soluble BOD removal through the trickling filter and maintained consistently low SVI through the solids contact basin.

It has been reported that nitrification can also occur in TF/SC processes with lower organic loads [39, 40]. Daigger et al. [40] reported that a lightly loaded TF/SC system with a TF total BOD₅ loading of 23.2–34.2 lb/kcf•day and soluble BOD₅ loading of 10.6–12.5 lb/kcf•day consistently achieved low effluent NH₄-N concentration of 0.2–2.4 mg/L. A seeding effect was also observed where nitrifiers sloughed from the trickling filter and seeded the solids contact basin that enabled the latter to achieve nitrification at a lower than theoretically required SRT.

4.2 Roughing Filter/Activated Sludge (RF/AS)

In an RF/AS process, the roughing filter is designed to remove 40–70 % of the BOD. The residual BOD will be removed in the aeration basin following the roughing filter. In some cases, the AS basins are also designed to achieve nitrification. The major difference between the TF/SC and RF/AS processes is the organic load to the trickling filter. In the RF/AS process, high organic loads are designed so that a large portion of the carbon oxidation takes place in the activated sludge basin. Another difference is that in contrast to the TF/SC process, regular secondary clarifiers are typically sufficient for the RF/AS process instead of flocculating-type clarifiers due to the nature of the process and solids characteristics. Typical design criteria for RF/AS (or so-called TF/AS) processes are summarized in Table 6.12.

Table 6.12 Typical design criteria for RF/AS processes [29, 44]

Parameters (based on modular plastic media)	Design criteria	
	Range	Common
Solids production (mg VSS/mg BOD₅ removed)	0.8-1.2	1.0
Trickling filter hydraulic load, m/h (gpm/ft²)	2.0–12 (0.8–5.0)	2.4 (1.0)
Trickling filter organic load, kg/m³/day (lb/1,000 ft³/day)	1.20–4.8 (75–300)	2.4 (150)
Activated sludge basin HRT at average flow, min	120–480	240
Activated sludge basin HRT at peak flow, min	40–120	90
Activated sludge basin SRT (d)	1.0–12.0	8.0
Activated sludge basin MLSS concentration, mg/L	1,500–6,000	3,000
Clarifier basin overflow rate at average flow m/h (gpd/ft²)	0.85–1.69 (500–1,000)	1.35 (800)
Clarifier underflow concentration, % total solids	0.6–1.2	0.8

5   Trickling Filter and Nutrient Removal

Trickling filters have been widely used in nitrification applications since the 1980s. In recent decades, many states are requiring total nitrogen and phosphorus removal for effluent being discharged to certain sensitive receiving waters. Therefore, trickling filter plants are facing upgrade challenges to meet the nutrient removal requirements.

5.1   Process Modifications to Achieve Total Nitrogen Removal

Achieving denitrification in trickling filters has been of interest to many researchers as it can be an economical means of upgrading an existing trickling filter process for achieving total nitrogen removal. Parker and Richards [36] observed significant denitrification within a lightly loaded (less than 20 lb/1,000 ft³/day) pilot trickling filter at the city of Garland, TX, when internal recycling was applied. The denitrification was believed to occur when nitrified effluent was brought back in contact with the high-soluble BOD laden influent in the upper portion of the tower. Pearce [46] reported that total nitrogen reduction of 0 to over 50 % can be achieved across the trickling filter processes, and 26–63 % of total nitrogen removal can be obtained if primary treatment is included. Dai et al. [47] reported an over 60 % total nitrogen removal across a lightly loaded rock filter process when internal recycle to the primary clarifier was applied and further enhanced denitrification was achieved by increasing the recycle flow and maintaining adequate sludge retention time in the primary clarifier. Biesterfeld et al. [48] conducted a study in quantifying denitrification potential in carbonaceous trickling filters. During the study, biofilm samples

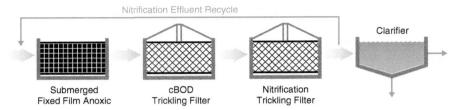

Fig. 6.10 Conceptual process schematic of integrated submerged fixed film trickling filter process

collected from the carbonaceous trickling filters at Littleton-Englewood Wastewater treatment plant in Colorado were placed in a bench-scale reactor and spiked with nitrate. Denitrification rates of 3.09–5.55 g NO_3-N/m^2-day were observed. Based on the finding of the study, internal recycle of tertiary nitrification trickling filter effluent to the primary clarifier has been employed at the Littleton-Englewood Wastewater Treatment Plant (WWTP), and 50 % total inorganic nitrogen removal was obtained [49].

All of these studies indicate that partial total nitrogen removal can be achieved by recycling nitrified effluent, and this approach may be used to meet less stringent total nitrogen discharge limits. This may also be combined with postdenitrification processes to reduce the postdenitrification system size and external carbon consumption. Pearce [46] suggested that achievable reliable process performance through this approach is limited to 30–50 % total nitrogen removal, which is consistent with the findings of the above studies. Another limitation of this approach is that denitrification performance is limited by hydraulic constraints at a given facility.

Submerged denitrification filters have been used in several trickling filter plants to achieve postdenitrification where external carbon sources are added. At Littleton-Englewood WWTP, both pre-denitrification and postdenitrification processes are involved in meeting the total nitrogen removal requirements. Treatment processes at this plant will be discussed further later in this chapter.

Coupling activated sludge denitrification and trickling filter processes has been studied by several researchers, and significant nitrogen removal was observed [47]. However, this process would involve intermediate clarifiers between the activated sludge tanks and the trickling filters, which may increase the system capital cost and operational complexity.

In recent years, submerged fixed film processes such as moving bed biofilm reactor (MBBR) and fixed bed biofilm reactor (FBBR) have become popular in wastewater treatment. As such systems utilize submerged fixed film for treatment and do not require sludge recirculation and backwash, integrating the systems with trickling filters to achieve total nitrogen removal appears to be a promising solution for upgrading trickling filter plants to achieve high levels of total nitrogen removal and is under investigation. Figure 6.10 shows a conceptual process schematic.

5.2   Coarse Media Denitrification Filter

Submerged denitrification filters have been commonly used in trickling filter plants to achieve total nitrogen removal. In a denitrification filter, nitrate and nitrite are reduced to nitrogen gas through the action of facultative heterotrophic bacteria [50–55]. Coarse media denitrification filters are attached growth biological processes in which nitrified wastewater is passed through submerged beds containing synthetic plastic media or equivalent. The systems may be pressurized or driven by gravity. Minimum filter media particle diameter or media flute size is about 15 mm. Anoxic conditions are maintained in the submerged bed, and since the nitrified wastewater is usually deficient in carbonaceous materials and contains a high residual DO concentration, a supplemental carbon source (usually methanol) may be required to maintain the attached denitrifying biofilm [50–52].

Because of the high-percent void volume and low specific surface area characteristic of low-density coarse denitrification trickling filters, biofilm or attached slime continuously sloughs off the media. As a result, the denitrification filter effluent is usually moderately high in suspended solids (20–40 mg/L), requiring a final polishing step, such as sand filtration or membrane filtration.

A wide variety of media types may be used as long as high void volume and low specific volume are maintained. Both random dump plastic media and corrugated plastic sheet media may be used.

Backwashing is infrequent and is usually done to control effluent suspended solids rather than pressure drop. Alternate carbon sources such as sugars, volatile acids, ethanol, or other organic compounds, as well as nitrogen deficient materials such as brewery wastes, may be used. In some installations where the filters are used as pre-anoxic reactor, the BOD or chemical oxygen demand (COD) concentration in the denitrification filter influent may serve as the carbon source.

The denitrification filter process is capable of converting nearly all nitrate in a nitrified secondary effluent to gaseous nitrogen with an overall nitrogen removal of 70–90 %. Although the process is well developed at full scale, it is not in widespread use. Currently, the process is used almost exclusively to denitrify domestic sewage that has undergone carbon oxidation and nitrification. It may also be used to reduce nitrate in industrial effluents. Under controlled pH, temperature, nitrogen load, and chemical feed, the process may achieve high levels of reliability. The denitrification filter can be installed either above ground [50, 51] or underground [55].

The required amount of methanol, the most common energy source, may be estimated using the following values per mg/L of the nitrogen species and dissolved oxygen in the inlet to the denitrification process:

(a) $2.47 \, mg/L \, CH_3OH$ per $1 \, mg/L$ nitrate $- N$
(b) $1.53 \, mg/L \, CH_3OH$ per $1 \, mg/L$ nitrite $- N$
(c) $0.87 \, mg/L \, CH_3OH$ per $1 \, mg/L$ dissolved oxygen

Table 6.13 Recommended design criteria and operational conditions for coarse media denitrification filter

Design/operation parameters	Value
pH	6.5–7.5
Void ratio	70–96 %
Specific surface area	65–274 ft^2/ft^3 (213–899 m^2/m^3)
Nitrogen loading rate at 5 °C	0.5×10^{-4} lb NO$_3$-N/ft^2 filter media surface area/day (2.44×10^{-4} kg/m^2/day)
at 15 °C	Up to 0.8×10^{-4} lb NO$_3$-N /ft^2 filter media surface area/day (3.91×10^{-4}kg/m^2/day)
at 25 °C	Up to 1.3×10^{-4} lb NO$_3$-N/ft^2 filter media surface area/day (6.35×10^{-4} kg/m^2/day)
Hydraulic surface loading rate for plant of 0.3 MGD (1.14 MLD)	2.5 gal/ft^2/day (101.85 L/m^2/day)
For plant of 0.5 MGD (1.89 MLD)	4.1 gal/ft^2/day (167.04 L/m^2/day)

The expected sludge production rate is 0.6–0.8 lb/lb (or 0.8 kg/kg) NO$_3$-N reduced.

The recommended design criteria and operational conditions are summarized in Table 6.13.

5.3 Phosphorus Removal in Trickling Filter Plants

Phosphorus removal in a trickling filter plant is typically achieved through chemical precipitation in the primary or secondary clarifiers. Chemical precipitation to remove phosphorus in the primary clarifier is commonly used in the trickling filter plants with a phosphorus discharge limit. The biological P removal concept in a trickling filter-related process has not been well studied.

6 Trickling Filter Plant Upgrades

Trickling filters have been the workhorse of the wastewater treatment industry for over 100 years, and today many trickling filter plants are facing the challenges of meeting stricter limits and achieving higher levels of treatment goals. In recent decades, various synthetic media have been developed for different treatment goals. In addition, process modifications around trickling filters have been well studied and introduced into practice which allow for cost-effective retrofits of trickling filter plants to meet the ever tighter limits. For example, replacing rock media with high-efficiency structured sheet media has proven to increase the volumetric BOD removal capacity than the original rock filters [23, 56]. A well-designed TF/SC can

Table 6.14 Typical design concepts for upgrading trickling filter to meet various goals

Existing	Purposes of upgrade	Possible upgrades
Rock or plastic media	Capacity increase	Replace rock with high-efficiency plastic media Increase plastic TF height or construct more TFs Incorporate motorized distribution and forced ventilation
	Lower effluent BOD/TSS	TF/SC process Chemical enhanced secondary clarifier Modified secondary clarifier
	Nitrification	TF/AS process Construct second/third TF Increase TF height for a dual-purposes TF
	P removal	Chemical addition to primary clarifier or secondary clarifier
	TN removal	Post denitrification filter following final clarifier Recycle nitrified effluent to primary clarifiers[a, b]

[a]Only partial TN removal can be achieved
[b]Evaluation on wastewater characteristics, existing process configuration, and primary clarifier capacity shall be conducted

achieve high-quality effluent with BOD and TSS concentrations of less than 10 mg/L without tertiary filtration [43, 57]. Combined carbon-oxidizing filters, nitrification filters, and tertiary nitrification filters have also been widely used for ammonium nitrogen removal as discussed earlier in this chapter. Post denitrification filters following nitrifying filters have been employed by several trickling filter facilities to achieve total nitrogen removal. In addition, it has been reported that partial nitrogen removal can be achieved through pre-anoxic processes in which nitrified effluent is recycled back to the primary clarifiers. Denitrification occurs in either the primary clarifiers or the upper portion of the cBOD-oxidizing trickling filters [47, 49]. However, to implement such process modification, careful evaluation of plant wastewater characteristics, existing process configuration, and individual unit capacity should be conducted.

Typical design concepts for trickling filter upgrade are summarized in Table 6.14.

The Littleton-Englewood Wastewater treatment plant in Colorado is an example of a contemporary trickling filter plant built to today's configuration through several expansions and upgrades. Figure 6.11 shows the liquid process scheme of the plant. This 50-MGD wastewater treatment facility consists of headworks, primary clarifiers, trickling filters, solids contact basins, secondary clarifiers, tertiary nitrification filters, and denitrification sand filters. Treatment has proven to be stable and reliable: the annual average secondary effluent BOD and TSS concentrations are 5 mg/L and 6 mg/L, respectively, and the secondary sludge volume index (SVI) is less than 100 ml/g. Annual average NH_4-N concentration is well below the discharge limits. Total nitrogen removal is achieved through both pre-denitrification and post denitrification processes. Nitrified effluent is recycled back to the headworks and encourages pre-denitrification in the upper portion of the

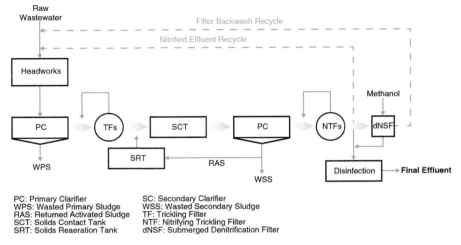

Fig. 6.11 Liquid flow scheme of Littleton-Englewood Wastewater treatment plant

carbonaceous trickling filters; postdenitrification sand filters are designed to polish the effluent to meet TN limits. The plant reported an approximate 50 % of total inorganic nitrogen removal through the pre-denitrification process. In addition, the recycled nitrified effluent eliminates hydrogen sulfide (H_2S) odor in the primary clarifier [49].

7   Trickling Filters for Industrial Applications

Trickling filters are used in industrial wastewater pretreatment or integrated with activated sludge for higher levels of treatment. There have been no established design models for industrial wastewater treatment as the treatability varies significantly. In addition, trickling filter tower configuration and operating parameters such as ventilation, organic load, and hydraulic load can also impact the trickling filter performance. Dow Chemical Company conducted extensive pilot studies on different types of industrial wastewater treatment using vertical flow media with a specific surface area of 89 m^2/m^3 (27 ft^2/ft^3). The treatability coefficient k_{20} for the Germain model was evaluated based on the performance data and operating parameters. Some data is presented in Table 6.15. As shown, the k_{20} values varied significantly among different facilities. Even for wastewaters from the same industry, treatability varied significantly. For example, pulp and paper wastewater showed a wide range of treatability as the wastewater came from different processes. Therefore, pilot studies are highly recommended for industrial trickling filter applications.

Table 6.15 Industrial wastewater Germain k values [1, 58]

Type of waste and location	Flow – gpm/ft² [a]		Depth, ft [b]	BOD load, Lb/1,000 ft³/day [c]	BOD conc., mg/L	BOD removal, %	Average temperature °C	Germain k_{20}, gpm$^{0.5}$/ft² [d]	Normalized Germain k_{20}, gpm$^{0.5}$/ft² [e, f]
	Feed	Recycle							
Relatively readily biodegradable wastewater									
Cereal (dry mill), IL	0.29	0.71	21.5	155	956	75.2	32	0.023	0.060
	0.43	0.71	21.5	230	956	69.9	32	0.024	0.063
Meat packing, NC	2.00	2.00	21.5	2089	1870	63	38	0.035	0.129
	6.43	0.00	21.5	5291	1473	42	38	0.035	0.112
Wet corn milling, IN	1.00	0.57	21.5	232	415	40	20	0.024	0.041
Synthetic dairy, MI	0.15	1.37	21.5	16	191	65	20	0.054	0.063
Synthetic dairy, CA	0.11	0.22	7.2	39	215	72	20	0.059	0.042
Meat packing, IA	0.46	0.74	31.5	289	1645	69	32	0.017	0.069
	1.00	1.50	31.5	448	1175	67	32	0.023	0.082
	1.31	1.00	31.5	1144	2290	49	32	0.016	0.079
Frozen foods, VA	0.60	0.00	21.5	488	1456	52	20	0.026	0.085
	0.65	0.00	21.5	452	1248	50	20	0.026	0.078
Fruit canning, CA	1.00	1.50	21.5	398	381	49	20	0.031	0.071
	2.00	1.00	21.5	1103	712	30	20	0.023	0.062
Sugar processing, CA	1.00	0.00	21.5	274	491	50	20	0.032	0.060
	2.50	1.00	21.5	666	477	36	20	0.033	0.061
Potato processing, ID	0.72	2.00	21.5	861	2140	59	20	0.035	0.138
Relatively slowly biodegradable wastewater									
Kraft mill, AL	0.57	0.71	21.5	51	160	76.2	37	0.028	0.030
	1.50	0.71	21.5	134	160	72.4	38	0.039	0.042
	2.00	2.00	21.5	179	160	64.6	38	0.037	0.039
	3.70	1.29	21.5	331	160	59.5	41	0.039	0.042

(continued)

Table 6.15 (continued)

Type of waste and location	Flow – gpm/ft² [a] Feed	Recycle	Depth, ft [b]	BOD load, Lb/1,000 ft³/day [c]	BOD conc., mg/L	BOD removal, %	Average temperature °C	Germain k₂₀, gpm^0.5/ft² [d]	Normalized Germain k₂₀, gpm^0.5/ft² [e,f]
Meat packing (ana. pond eff.), NC	0.43	2.00	21.5	27	114	61	30	0.020	0.018
	3.00	2.00	21.5	204	122	41	30	0.030	0.028
Refinery, CA	2.00	0.00	21.5	80	72	60	40	0.030	0.022
	1.50	0.50	21.5	78	93	54	36	0.026	0.021
Textile mill, VA	0.86	0.47	21.5	94	196	78	42	0.031	0.036
	1.50	0.71	21.5	195	233	76	42	0.038	0.049
Paper co. CAN	0.43	1.00	21.5	67	279	58.6	27	0.021	0.030
	0.70	1.00	21.5	96	244	73.7	36	0.030	0.040
	0.72	1.00	21.5	55	136	73.5	32	0.035	0.034
	0.86	1.00	21.5	62	129	67.8	38	0.026	0.025
	1.21	1.57	21.5	88	129	34.8	35	0.013	0.013
	1.50	1.00	21.5	152	181	40.2	35	0.017	0.019
Paper co. NY	1.35	1.35	21.5	88	116	64.7	–	0.056	0.051
	1.35	2.00	21.5	100	133	60.9	–	0.051	0.050
	2.50	0.86	21.5	222	159	45.3	–	0.044	0.047
Hardboard mill, OR	0.43	1.0	21.5	184	765	49.5	11	0.028	0.065
Tannery (pigskin), MI	0.29	1.00	21.5	115	727	84	33	0.029	0.063
Kraft mill, New England	0.33	0.67	40.0	50	251	81.3	–	0.024	0.044
	0.65	0.70	40.0	89	228	85.1	–	0.038	0.067
	1.00	0.50	40.0	178	296	77.7	–	0.038	0.075
	1.02	0.68	40.0	167	274	70.1	–	0.031	0.058
	1.34	0.66	40.0	207	258	57.8	–	0.025	0.054
	1.66	0.84	40.0	246	247	62.4	–	0.032	0.057
	2.00	1.00	40.0	384	320	65.6	–	0.038	0.078
	2.34	1.16	40.0	577	411	55.7	–	0.031	0.073

Textile mill, VA	0.33	0.67	21.6	28	154	81.2	–	0.044	0.047
	1.00	1.00	21.6	68	124	72.6	–	0.060	0.057
	1.56	1.56	21.6	118	137	60.6	–	0.054	0.053
Creosote mill, MS	0.035	1.00	21.5	76	3914	75.2	–	0.012	0.064
	0.07	1.00	21.5	66	1698	91.9	–	0.031	0.108
	0.14	1.00	21.5	85	1089	80.7	–	0.029	0.080

[a] gpm/ft^2 × 0.6791 = L/m^2•s

[b] ft × 0.3048 = m

[c] lb/day/1,000 ft^3 × 0.01602 = kg/m^3/day

[d] Corrected for temperature when available by $k_{G,20} = \dfrac{\ln\left(\frac{S_e}{S_o}\right)(WHL)^{0.5}}{D\left(1.035^{T-20}\right)}$

[e] gpm$^{0.5}$/ft^2 × 2.704 = (L/s)$^{0.5}$/m^2

[f] Normalized to D = 20 ft and S_o = 150 mg/L by $k_{G,20} = \dfrac{\ln\left(\frac{S_e}{S_o}\right)(WHL)^{0.5}}{D\left(1.035^{T-20}\right)}\left(\frac{D}{20}\right)^{0.5}\left(\frac{S_o}{150}\right)^{0.5}$

The data also indicates that trickling filter process is more efficient per unit volume in treating industrial wastes containing soluble and readily biodegradable matters such as food-processing wastes. Different from municipal wastewaters where nutrients required for biological synthesis are present, most industrial wastewaters are nutrient deficient and require nutrient addition. pH adjustment is also required for certain industrial wastewater sources. Proper design of pretreatment prior to the trickling filters and clarification after the trickling filter is also an important factor to be considered.

Trickling filters have been widely used in industrial roughing applications, particularly in the food industries such as dairy, soft drink, fruit canning, and potato chips. However, general design guidelines for industrial roughing filters have not been well established. Design engineers should base their designs on pilot studies or refer to successful and similar full-scale installations to determine design parameters. Published information regarding industrial roughing filters is very limited due to either inadequate sampling from the industrial plant owners or lack of interest in having their plant performance data published. Soluble BOD reduction of 50–80 % from full-scale installations is generally expected for these applications depending on loading and treatability of the wastewaters.

8 Construction Consideration of Trickling Filters

8.1 Media Integrity

Structural integrity of plastic media is an important aspect to consider when designing trickling filters to prevent structural failure. Parker [59] reported 12 trickling filter catastrophic failures between year 1990 and 1998 and seven of which were attributed to weak media strength or poor installation. Polyvinyl chloride (PVC) is the most common plastic used for trickling filters. The properties of the PVC material actually used for a filter are determined using ASTM procedures. Specific tests and recommended values for PVC are summarized in Table 6.16.

Table 6.16 Recommended tests and properties of standard PVC

Property	ASTM	Units	Value
Specific gravity	D792	g/cm^3	1.4–1.6
Tensile strength	D638/D882	psi	6,000 min
Flexural modulus	D790	psi	425,000 min
Flexural strength	D790	psi	11,000 min
Elastic modulus	D638/D882	psi	360,000 min
Impact resistance	D5420	in. lbs/mil	0.8 min
Heat deflection	D648	Temperature °F under load of 264 psi	158 min
Flammability	D635		<5 sec

Fig. 6.12 Typical modular media deflection curve from short-term test

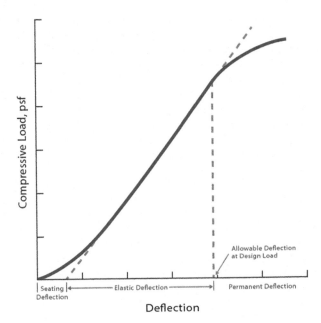

Structural testing of media modules is critical to ensure the media will satisfy structural load requirements during the lifespan of the filter media. Plastic media structural failure is a result of a continuous load which causes long term media deformation, or creep. Creep can result in loss of structure moment of inertia and consequently loss of product design strength. Creep is accelerated at higher temperatures and is generally related to the modulus of material. It is therefore important that the plastic material is properly designed to resist creep uner the load applied which includes the weight of media, water and biomass.

Each trickling filter tower is designed layer by layer for required strength based on the depth of the tower. Deeper trickling filters must have higher compressive strength media at the bottom to withstand the cumulative load from above. Layers higher up in the filter have less weight to support and hence have lower strength requirements.

A short-term test procedure is used in the industry to evaluate media load capacity. During the test, loads are gradually applied to the media packs, and corresponding deflections (i.e., media height change) are measured. Figure 6.12 represents a typical media performance curve obtained from a short-term structural test. The capacity of the media should be defined at the upper point of departure of deflection from the straight line or elastic region as shown. Media failure usually occurs between media deflections of 0.75–1.5 % as reported [1].

The short-term compressive strength of PVC media decreases at higher temperatures; therefore, media testing should be performed at the maximum sustained operating wastewater temperature in the trickling filter to ensure that the media used will satisfy the load requirement during operation. For high-temperature applications, heat-elevated PVC formulations exhibiting higher heat deflection properties are required to achieve media long-term structural performance.

Typical service life of structured sheet PVC media is expected to be 20 years or more. As PVC media will lose strength over time, media conditions should be evaluated and replacement planned before a plastic trickling filter reaches its service life. Random dump polypropylene (PP) media requires lateral structural support and the media height is typically less than 3.05 m (10 ft). Media strength of random dump PP media is considered to be lower than that of PVC structured sheet media [22].

8.2 Miscellaneous Considerations

8.2.1 Containment Structure

Containment structures for rock or random dump plastic media are typically precast or panel-type concrete tanks. For self-supporting plastic modular media, fiberglass and coated steel tanks may also be used. If a trickling filter is designed to allow flooding operation to control macrofauna, watertight walls must be used. For filters that are covered, the domes are often constructed from aluminum or fiber glass. Forced ventilation requires a sealed containment construction; otherwise wastewater leaks from the structural seams as it is pushed out by the air pressure.

8.2.2 Subfloor and Drainage

The floor of trickling filters should provide sufficient strength to support the media, attached biomass, underdrain system, and wastewater load. The trickling filter floor is typically constructed with steel-mesh-reinforced concrete. The floor typically slopes to the drainage channels at a 1–5 % grade.

The drainage channel should be adequately designed to convey treated water and sloughed biomass and macrofauna to the downstream clarifier and should be sized to produce a minimum velocity of 0.6 m/s (2 ft/s) [4].

8.2.3 Ventilation

Plastic trickling filters typically have multiple vents placed in the periphery wall to provide ventilation. For municipal wastewater, plastic media manufacturers have recommended 0.1 m^2 (1 ft^2) of ventilating area for each 3–4.6 m (10–15 ft) of

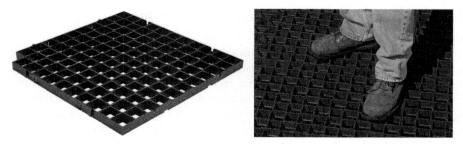

Fig. 6.13 Media-protective grating panel and installed grating as walking surface (Courtesy of Brentwood Industries, Inc.)

trickling filter tower periphery and 1–2 m^2 of ventilation area in the underdrain area per 1000 m^3 of media [29].

8.2.4 Media Protection Grating

For uncovered trickling filter, the top layer of media can suffer from UV degradation and hydraulic impact, foot traffic, and trash accumulation. In recent decades, plastic gratings have been used to protect the media. The plastic gratings are typically made from polypropylene (PP) or polyethylene (PE) and are specially designed for wastewater application. These gratings are placed on the top of the media to protect the media from UV (uncovered trickling filters) and reduce the hydraulic impact on the media. The gratings also provide a nonskid walking surface for the operators for trash cleaning and distributor arm servicing to distribute the point source footprint across a larger plan view surface area of the media. The PP or PE gratings are considered economical solutions for media protection. Fiber glass-reinforced plastic (FRP) gratings have been occasionally used for similar purposes but have been gradually replaced by the PP and PE gratings due to the relatively high cost of FRP grating. Figure 6.13 shows a typical protective grating panel and installed grating as walking surface.

9 Operation and Maintenance Considerations of Plastic Trickling Filters

There are certain practices in the operation of trickling filters that are of significant importance in maintaining optimum treatment capacity and quality of effluent, which are discussed as follows.

9.1   Uniform Distribution Across Media Surface

The uniformity of wastewater distribution is essential to a successful trickling filter operation. It is desirable to have a consistent dosing rate across the entire surface of the media tower. Fixed nozzles are not recommended for new construction filter installations. Rotary distributor arms should be used to provide adequate flushing and control of dosing and hydraulic loading rates for optimum performance based on specific conditions encountered at each facility. The uniformity of distribution can be evaluated by a pan test in which wastewater is collected during one rotation in a series of pans at various radial positions on the top of the filter.

9.2   Recirculation and Hydraulic Wetting

As discussed earlier, recirculation offers benefits of increasing dissolved oxygen and improving wetting efficiency of the media. Recirculation is often used to maintain minimum hydraulic wetting rates for adequate distribution and wetting of internal media surfaces. Typical wetting rates for different trickling filter applications are summarized in Table 6.2.

9.3   Distribution Rate and Flushing Operation

In North America, most trickling filters employ rotary distributor arms that are either hydraulically driven or equipped with mechanical or other speed controls. Speed-controlled distributors typically allow rotary speeds of 4–10 min per revolution under normal operation and allow for higher dosing rates during flushing operation.

Biofilm thickness control is an important aspect of trickling filter operation, especially for trickling filters with high organic loads. WEF [16] recommends a once per week flushing operation for highly loaded or roughing filters. Recommendations on trickling filter dosing rates under normal and flushing operations are summarized in Table 6.3. However, the optimum dosing rates and schedule may vary among filters and should be established based on field performance.

9.4   Ventilation

Trickling filters are primarily aerobic processes. Aerobic conditions should be maintained at all times if possible. During the seasons that stagnation may occur, forced ventilation should be used if available to compensate for the lack of oxygen supply. This is especially important for highly loaded roughing filters or nitrification filters that are more likely to suffer from oxygen limitation.

9.5   Operation of Associated Processes

Proper control and maintenance of other treatment processes associated with the trickling filter system are of importance for a successful overall system performance, which include:

- Proper maintenance and operation of primary-treatment facilities to ensure minimum solids and BOD load to downstream processes. This includes proper removal of accumulating solids, maintenance of the mechanical equipment, and maintenance of any chemical feed systems required to maintain optimum performance.
- Proper and effective operation of any primary biological treatment systems preceding the trickling filters, such as extended aeration systems, lagoons, roughing trickling filters, or suspended solids treatment units
- Proper maintenance and operation of secondary clarifiers
- Maintenance of sludge disposal operations, including controlled return of sidestreams
- Maintenance of information records pertaining to influent characteristics (wastewater and atmospheric temperature, pH values, influent characteristics, effluent characteristics).

9.6   Macrofauna Control

Macrofauna, such as snails, occur in trickling filters and may cause nuisance problems most of which can be controlled by flushing. A proven method for snail control is raising the wastewater pH to approximately 10 which will result in free ammonia that is toxic to snails. Snail traps have also been used in some facilities to remove accumulated snails. Snail traps should be prior to recirculation as recirculation of snails exacerbates the issue.

9.7   Cold Weather Operation

Icing on a trickling filter can occur under cold climates and can cause several problems including damage to the media and possibly to containment structure, poor distribution, and reduce the effective media volume (e.g., bypass freezing media). General practices that may help solve the problems include reducing recirculation, closing ventilation ports, constructing wind breakers or filter covers, using high pressure water stream to remove ice from distributor orifice, and, if possible, flooding the trickling filter [1, 60].

9.8   Odor Control

Odor generated from a trickling filter process is mainly a result of anaerobic condition in the trickling filter. Primary causes of the anaerobic condition include excessive organic loads, media plugging from macrofauna, or biofilm overgrowth and ventilation system failure such as ventilation port/underdrain clogging. Other causes include improper recirculation (e.g., recirculation from bottom of downstream clarifiers) or poor primary clarification.

If caused by excessive load, reducing load by increasing BOD removal in the upstream treatment units can be a remedy; if caused by media plugging or ventilation system failure, checking the air ports and/or ventilation system, increasing hydraulic loading rate, flushing the filter at high dosing rate, and cleaning the effluent channels would improve the oxygen supply and reduce odor.

9.9   Plastic Media Replacement

The service life of plastic media ranges from 20 to 30 years. As a trickling filter approaches the end of its expected service life, it is prudent to plan for media replacement in order to avoid poor treatment or catastrophic filter structural failure [61]. Factors to be considered include:

- Filter performance: Extended periods of decreased performance, especially after ruling out most possible causes such as temperature changes, loading changes, etc., and is likely an indicator of plugged media, which results in less surface area for biofilm, decreased ventilation, and decreased treatment.
- Filter appearance: Ponding on the top or uneven water drainage in the underdrain area may indicate plugging of media or underdrains.
- Wastewater type: Nitrifying filters typically last longer than industrial roughing filters, which have high organic loadings and heavy biomass growth that may accelerate filter aging.

10 Design Examples

10.1 Design Example 1: Design a Plastic Trickling Filter System for Domestic Wastewater Carbonaceous BOD Removal Using Germain Formula

Given	
Average flow, m^3/day (MGD)	18,927 (5)
Flow rate peaking factor	1.8
Wastewater temperature, °C	12 °C
Raw wastewater BOD_5, g/m^3	200
Raw wastewater TSS, g/m^3	300
Primary clarifier efficiency, TSS, %	55 %
Primary clarifier efficiency, BOD_5, %	30 %
Clarified effluent BOD_5, g/m^3	25
Exponent characteristic, n	0.5
Media type	Cross flow PVC modular
Media-specific surface area, m^2/m^3 (ft^2/ft^3)	102 (31)
Plastic media k_{20}, $(L/s)^{0.5}/m^2$ $[(gpm)^{0.5}/ft^2]$	0.21 [0.078]
Minimum THL, $L/m^2 \cdot s$ (gpm/ft^2)	0.5 (0.75)
Number of tricking filter in parallel	3
Number of clarifier in parallel	3
Number of distributor arms	2
Trickling filter media depth, m (ft)	4.88 (16)
Secondary clarifier side water depth, m (ft)	3.66 (12)

Determine
- Trickling filter diameter
- Secondary clarifier diameter
- Recirculation rate
- Dosing rates in normal and flushing operation
- Distributor rotational speed

Solution:

1. Determine primary effluent (TF influent) BOD_5 and TSS concentrations:

$$\text{Pri. Eff. } BOD_5 = \text{Raw } BOD_5 \times (1 - PC_{BOD}\%) = 200 \times (1 - 30\%)$$
$$= 140 \text{ g/m}^3$$

$$\text{Pri. Eff. TSS} = \text{Raw TSS} \times (1 - PC_{TSS}\%) = 300 \times (1 - 55\%)$$
$$= 135 \text{ g/m}^3$$

2. Determine k_{20} for the design conditions using Eq. 6.23:

   a. Solve for k_{c20}

$$k_{c20} = k_{20}\left(\frac{D}{6.1}\right)^{0.4}\left(\frac{s}{150}\right)^{0.5} = 0.21\left(\frac{6.1}{4.88}\right)^{0.5}\left(\frac{150}{140}\right)^{0.5} = 0.243(\text{L/s})^{0.5}/\text{m}^2$$

b. Correct k_{c20} for temperature effect using Eq. 6.5:

$$k_{c12} = k_{c20}(1.035)^{T-20} = 0.243(1.035)^{(12-20)} = 0.185(\text{L/s})^{0.5}/\text{m}^2$$

c. Determine hydraulic loading rate WHL by rearranging Eq. 6.21:

$$\frac{S_e}{S_o} = e^{-kD/\text{WHL}^2}$$

$$\text{WHL} = \left(\frac{kD}{\ln\left(\frac{S_o}{S_e}\right)}\right)^{1/n} = \left(\frac{0.185 \times 4.88}{\ln\left(\frac{140}{25}\right)}\right)^{\frac{1}{0.5}} = 0.275\text{L/s}\bullet\text{m}^2$$

d. Determine the filter area A.
   Flow to each filter $Q_{TF} = 18,927/3 = 6,309\text{m}^3/\text{day}$

$$A = \frac{Q_{TF}}{\text{WHL}} = \frac{6,309(\text{m}^3/\text{day})\left(\frac{1,000\,L}{\text{m}^3}\right)\left(\frac{1\,\text{day}}{1,440\,\text{min}}\right)\left(\frac{1\,\text{min}}{60s}\right)}{0.275\text{L/s} \cdot m^2} = 265.5\text{m}^2$$

$$\text{Filter Dia.} = \left(\frac{265.5}{\pi}\right)^{0.5} \times 2 = 18.4 \text{ m}$$

e. Determine the filter volume:

$$V = AD = 265.5 \text{ m}^2 \times 4.88\text{ft} = 1,296 \text{ m}^3$$

$$\text{Total Media Volume} = 1296 \times 3 = 3,888 \text{ m}^3$$

f. Determine organic loading rate using Eq. 6.1:

$$\text{OLR} = \frac{QS_o}{V} = \frac{18,922(\text{m}^3/\text{day}) \times 140\text{g/m}^3 \times \frac{1\text{kg}}{1,000\text{g}}}{3,888\text{m}^3} = 0.681 \text{ kg/m}^3 \cdot \text{day}$$

3. Determine recirculation rate.
   The minimum hydraulic loading rate is 0.5 L/s•m^2:

$$\text{THL} = \text{WHL} + q_r = 0.5\text{L/s} \cdot \text{m}^2$$

$$q_r = 0.5 - 0.275 = 0.225 \text{ L/s} \cdot \text{m}^2$$

$$R = \frac{qr}{\text{WHL}} = \frac{0.225}{0.275} = 0.818$$

Recirculation rate $= 0.818 \times 18{,}927$ m^3/day $= 15{,}485$ m^3/day $= 645$ m^3/h

4. Determine the dosing rates at normal and flushing operation.

From Table 6.3, select dosing rates at design organic loading.

Normal operation DR $= 75$ mm/pass

Flushing operation DR $= 150$ mm/pass.

5. Determine distributor rotational speeds during normal and flushing operation using Eq. 6.6:

$$\text{THL} = 0.5\frac{L}{s} \cdot \text{m}^2 = 1.8 \, \text{m}^3/\text{m}^2 \cdot \text{h}$$

$$\omega_{\text{normal}} = \frac{(\text{THL})\left(1{,}000 \, \frac{\text{mm}}{\text{m}}\right)}{(a)(\text{DR}_{\text{normal}})\left(60 \, \frac{\text{min}}{\text{h}}\right)} = \frac{1.8\left(\frac{\text{m}^3}{\text{m}^2} \cdot \text{hr}\right)\left(1{,}000 \, \frac{\text{mm}}{\text{m}}\right)}{2\left(\frac{75 \, \text{mm}}{\text{pass}}\right)\left(60 \, \frac{\text{min}}{\text{h}}\right)}$$

$$= 0.2 \, \text{rev/min} \cdot (5 \cdot \text{min/rev})$$

$$\omega_{\text{flushing}} = \frac{(\text{THL})\left(1{,}000 \, \frac{\text{mm}}{\text{m}}\right)}{(a)(\text{DR}_{\text{flushing}})\left(60 \, \frac{\text{min}}{\text{h}}\right)} = \frac{1.8\left(\frac{\text{m}^3}{\text{m}^2} \cdot \text{hr}\right)\left(1{,}000 \, \frac{\text{mm}}{\text{m}}\right)}{2\left(\frac{150 \, \text{mm}}{\text{pass}}\right)\left(60 \, \frac{\text{min}}{\text{h}}\right)}$$

$$= 0.1 \, \text{rev/min} \cdot (10 \cdot \text{min/rev})$$

6. Determine clarifier diameter

From Fig. 6.6 and Eqs. 6.18 and 6.19, determine overflow rates at average and maximum flow.

Average OFR $= 0.278 \times 3.66 = 1.017$ m/h

Maximum OFR $= 0.556 \times 3.66 = 2.035$ m/h

Because the peaking factor is 1.8, average OFR controls the clarifier design.

Clarifier surface area $= (6309$ m^3/day$)(1$ day/24 h$)/1.017$ m/hr $= 258$ m^2

Clarifier diameter $= (258$ m$^2/\pi)^{0.5} \times 2 = 18.1$ m

10.1.1 Design Summary

Design parameter, unit	Value
Number of TF in parallel	3
TF dimension, dia. m × media depth m (ft × ft)	18.4 × 4.88 (60.5 × 16)
Media volume/TF, m^3 (1000 ft^3)	1,296 (45.75)
Total media volume, m^3 (1000 ft^3)	3,888 (137.25)
Organic loading rate, kg/m^3•day (lb/1000 ft^3•day)	0.681 (42.5)
Total hydraulic loading rate, L/s•m^2 (gpm/ft^2)	0.5 (0.75)
Recirculation rate, m^3/h (MGD)	645 (4.09)

(continued)

Design parameter, unit	Value
Number of distributor arms	2
Rotational speed during normal operation, min/rev.	5
Rotational speed during flushing operation, min/rev.	10
No. of secondary clarifier	3
Secondary clarifier side water depth, m (ft)	3.66 (12)
Secondary clarifier diameter, m	18.1 (59.4)

10.2 Design Example 2: Design a Plastic Trickling Filter for Combined BOD Removal and Nitrification

Given	
Average flow, m³/day (MGD)	3785 (1)
Wastewater temperature, °C	15 °C
Primary effluent BOD$_5$, g/m³	120
Primary effluent TSS, g/m³	70
Primary effluent TKN, g/m³	30
Primary effluent NH$_4$-N, g/m³	22
Clarified effluent cBOD$_5$, g/m³	10
Clarified effluent TKN, g/m³	5
Clarified effluent NH$_4$-N, g/m³	2
Clarified effluent TSS, g/m³	15
Number of distributor arms	4
Media type	Cross flow PVC modular
Media specific surface area, m²/m³	102 (31)
No. of tricking filter in parallel	1
Trickling filter media depth, m (ft)	6.7 (22)
Minimum THL, L/m²•s (gpm/ft²)	0.5 (0.75)

Determine
- Trickling filter diameter
- Recirculation Rate
- Forced ventilation airflow at high temperature of 25 °C
- Dosing rates in normal and flushing operation
- Distributor rotational speed

Solution:

1. Determine TKN removal rate using Eq. 6.24:

$$TKN_{OX} = 1.086 \left(\frac{s_o}{S_{TKN}} \right)^{-0.44} = 1.086 \left(\frac{120}{30} \right)^{-0.44} = 0.59 \text{ g/m}^2 \cdot \text{day}$$

Chose standard deviation of −1 for high probability of success; therefore

$$\text{Design TKN}_{OX} = (0.59 - 0.175)\frac{g}{m^2}.\text{day} = 0.415 \text{ g/m}^2 \cdot \text{day}$$

2. Determine TKN removal requirement:

$$\text{TKN}_R = (\text{TKN}_{inf} - \text{TKN}_{eff})Q = \left(\frac{30g}{m^3} - \frac{5g}{m^3}\right)\left(3,785\frac{m^3}{day}\right) = 94,625 \text{ g/day}$$

3. Determine surface area requirement:

$$A_S = \frac{\text{TKN}_R}{\text{TKN}_{OX}} = \frac{94,625 \text{ g/day}}{0.415 \text{ g/m}^2 \cdot \text{day}} = 228,012 \text{ m}^2$$

4. Determine media volume requirement:

$$V = \frac{228,102 \text{ m}^2}{101 \text{ m}^2/\text{m}^3} = 2,258 \text{ m}^3$$

5. Determine filter surface area:

$$A = \frac{2258 \text{ m}^3}{6.7 \text{ m}} = 337 \text{ m}^2$$

6. Determine filter diameter:

$$\text{Dia.} = \left(\frac{337 \text{ m}^2}{\pi}\right)^{0.5} \times 2 = 20.7\text{m}$$

7. Determine filter organic loading rate (OLR) using Eq. 6.1:

$$\text{OLR} = \frac{\left(3,785\frac{m^3}{day}\right)\left(120\frac{g}{m^3}\right)\left(\frac{kg}{1,000g}\right)}{2,258\text{m}^3} = 0.2\text{kg/m}^3 \cdot \text{day}$$

8. Determine filter WHL Eq. 6.3:

$$\text{WHL} = \frac{Q_{in}}{A} = \frac{\left(3,785\frac{m^3}{day}\right)\left(\frac{1,000L}{m^3}\right)\left(\frac{1day}{1,440 \text{ min}}\right)\left(\frac{1min}{60s}\right)}{337 \text{ m}^2} = 0.13 \text{ L/m}^2 \cdot \text{s}$$

9. Determine recirculation rate by rearranging Eq. 6.4:

$$Q_R = \left(\frac{\text{THL}}{\text{WHL}} - 1\right)Q_{in}$$

$$= \left(\frac{0.5}{0.13} - 1\right)\left(3,785\frac{m^3}{day}\right) = 10,772 \text{ m}^3/\text{day} = 449 \text{ m}^3/\text{h}$$

10. Determine oxygen supply requirement using Eqs. 6.10, 6.11, 6.12, and 6.13:

$$R_o = \left(\frac{40\text{kg}}{\text{kg}}\right)\left[0.8e^{-90LR} + 1.2e^{-0.170LR} + 4.6\frac{N_{ox}}{BOD_5}\right]PF$$

N_{OX} = influent TKN − net yield organic nitrogen effluent TKN
Assume sludge yield = 0.5 g VSS/g BOD (typical for TF process)
VSS Yield = 0.5 g VSS/g BOD × (120 − 10) mg/L = 55 mg/L
Net yield organic nitrogen = 0.07 × VSS Yield = 3.85 mg/L:

$$N_{OX} = 30 - 3.85 - 5 = 21.15 \text{ mg/L}$$

Assume PF = 1.5:

$$R_o = \left(\frac{40\text{kg}}{\text{kg}}\right)\left[08e^{-9\times0.2} + 1.2e^{-0.17\times0.2} + 4.6\frac{21.15}{120}\right] \times 1.5$$

$$= 126.2\text{kgO}_2/\text{kg BOD}_5$$

$$R_o = \left(126.2\frac{\text{kg O}_2}{\text{kg BOD}_5}\right)\left(3,788\frac{\text{m}^3}{\text{day}}\right)\left(120\frac{\text{g}}{\text{m}^3}\right)\left(\frac{1\text{kg}}{1,000\text{g}}\right) = 57,365 \text{ kg O}_2/\text{day}$$

$$AR_{20} = \frac{\left(\text{Oxygen Supply, }\frac{\text{kg}}{\text{d}}\right)\left(3.5N\frac{\text{m}^3}{\text{kg}}\text{ oxygen}\right)}{1,440 \text{ mi n/day}}$$

$$= \frac{(57,365\text{kg } O_2/\text{day})\left(3.5 \, N\frac{\text{m}^3}{\text{kg}}O_2\right)}{1,440 \text{ min/day}} = 139.4\text{m}^3/\text{min}$$

Actual airflow rate using Eq. 3.7

$$AR_T = N\left(\frac{273 + T_0}{273}\right) = \left(139.4\frac{\text{m}^3}{\text{min}}\right)\left(\frac{273 + 25}{273}\right) = 152.2 \text{ m}^3/\text{min}$$

Future correct for high temperature

$$AR_{25} = 152.2 + 152.2(1\%)(25 - 20) = 159.6 \text{ m}^3/\text{min}$$

11. Determine the dosing rates at normal and flushing operation.
    From Table 6.3, select dosing rates at design organic loading.
    Normal operation DR = 25 mm/pass
    Flushing operation DR = 100 mm/pass
12. Determine distributor rotational speeds during normal and flushing operation by rearranging Eq. 6.6:

$$THL = 0.5\frac{L}{s} \cdot m^2 = 1.8 m^3/m^2 \cdot h$$

$$\omega_{flushing} = \frac{THL\left(1,000\,\frac{mm}{m}\right)}{(a)(DR_{flushing})\left(60\,\frac{min}{h}\right)} = \frac{1.8\left(\frac{m^3}{m^2} \cdot hr\right)\left(1,000\,\frac{mm}{m}\right)}{4\left(\frac{1,000mm}{pass}\right)\left(60\,\frac{min}{h}\right)}$$

$$= 0.075 \text{ rev/min} \cdot (13.4 \text{min/rev})$$

10.2.1 Summary of Design

Design parameter, unit	Value
Number of TF in parallel	1
TF dimension, dia. m × media depth m (ft × ft)	20.7 × 6.7 (68 × 22)
Media volume/TF, m^3 (1,000 ft^3)	2258 (79.7)
Organic loading rate, kg/m^3•day (lb/kcf•day)	0.2 (12.5)
THL, L/s•m^2 (gpm/ft^2)	0.5 (0.75)
Recirculation flow rate, m^3/h (MGD)	449 (2.84)
Maximum forced ventilation rate, m^3/hr (cfm)	159.6 (5634)
Number of distributor arms	4
Rotational speed during normal operation, min/rev.	3.3
Rotational speed during flushing operation, min/rev.	13.4

10.3 Design Example 3: Design a Plastic Trickling Filter for Tertiary Nitrification Using Albertson and Okey's Procedure

Given	
Average flow, m^3/day (MGD)	3,785 (1)
Wastewater temperature, °C	10 °C
Secondary effluent BOD_5, g/m^3	20
Secondary effluent soluble BOD_5, g/m^3	10
Secondary effluent TSS, g/m^3	20
Secondary effluent TKN, g/m^3	25
Secondary effluent NH_4-N, g/m^3	22
Tertiary effluent $cBOD_5$, g/m^3	10
Tertiary effluent TKN, g/m^3	4
Tertiary effluent NH_4-N, g/m^3	1

(continued)

Tertiary effluent TSS, g/m^3	15
Media type	Cross flow PVC modular
Media-specific surface area, m^2/m^3 (ft^2/ft^3)	138 (42)
Number of tricking filters	1
Trickling filter media depth, m (ft)	6.1 (20)
Minimum THL, L/m^2•s (gpm/ft^2)	0.6 (0.89)

Determine
- Trickling filter diameter and media volume requirement
- Recirculation rate, if any

Solution:

1. Determine transitional NH$_4$-N concentration using Table 6.9.

   Assume that dissolved oxygen saturation 75 %, transitional NH$_4$ − N is approximately 3.3 mg/L (3.3 g/m^3) at temperature of 10 °C.

2. Determine NH$_4$ − N removal through zero-order nitrification:

$$N_{OX} = (Q_{in})(NH_4-N_{inf}- NH_4-N_T)$$

$$= \left(3,785\,\frac{m^3}{day}\right)(22 - 3.3)\left(\frac{g}{m^3}\right) = 70,780 \text{ g/day}$$

3. Determine media surface area required for zero-order nitrification. $k_n = 1.2$ g/m^2 •day as per Albertson and Okey [38]:

$$A_s = \frac{NH_4 - N \text{ removed}}{k_n} - \frac{70,780\frac{g}{day}}{1.2 \text{ g/m}^2 \cdot \text{day}} = 58,983\,\text{m}^2$$

4. Determine NH$_4$ − N removal through first-order nitrification:

$$N_{OX1st} = (Q_{in})(NH_4 - N_{inf}- NH_4 - N_T) = \left(3,785\,\frac{m^3}{day}\right)(3.3 - 1)\left(\frac{g}{m^3}\right)$$

$$= 8,706 \text{ g/day}$$

5. Determine first-order nitrification rate using Eq. 6.26:

$$k_n' = 1.2\left(\frac{1}{3.3}\right)^{0.75} = 0.49 \text{g/m}^2 \cdot \text{day}$$

6. Determine media surface area required for first-order nitrification $\left(k_n' = 0.49 \text{ g/m}^2\text{•day}\right)$:

$$A_s = \frac{NH_4 - N \text{ removed}}{k_n'} = \frac{8,706\frac{g}{day}}{0.49\text{g/m}^2 \cdot \text{day}} = 17,762 \text{ m}^2$$

7. Determine overall nitrification rate:

$$\overline{k}_n = \frac{\text{Total NH}_4 - \text{N removed}}{\text{Total Surface Area}} = \frac{(70,780 + 8,706)\text{g/day}}{(58,983 + 17,762)\text{m}^2} = 1.04\text{g/m}^2 \cdot \text{day}$$

8. Determine total media volume required:

$$V = \frac{58,983\,\text{m}^2 + 17,762\,\text{m}^2}{138\ \text{m}^2/\text{m}^3} = 556\ \text{m}^3$$

9. Determine filter top surface area and diameter:

$$A = \frac{556\text{m}^3}{6.1\text{m}} = 91.2\text{m}^2$$

$$\text{Dia.} = \left(\frac{91.2\text{m}^2}{\pi}\right)^{0.5} \times 2 = 10.77\text{m}$$

10. Determine filter WHL using Eq. 6.3:

$$\text{WHL} = \frac{Q_{in}}{A} = \frac{\left(3,785\,\frac{\text{m}^3}{\text{day}}\right)\left(\frac{1,000L}{1\text{m}^3}\right)\left(\frac{1\text{day}}{1,440\ \text{min}}\right)\left(\frac{1\text{min}}{60\text{s}}\right)}{91.2\text{m}^2} = 0.48L/\text{s} \cdot \text{m}^2$$

11. Determine recirculation rate by rearranging Eq. 6.4:

$$R = \frac{\text{THL}}{\text{WHL}} - 1 = \frac{0.6}{0.48} - 1 = 0.25$$

$$Q_R = Q_{in}R = \left(3,785\,\frac{\text{m}^3}{\text{day}}\right) \times 0.25 = 946\,\frac{\text{m}^3}{\text{day}} = 39.4\text{m}^3/\text{h}$$

10.3.1 Summary of Design

Design parameter, unit	Value
Number of TF in parallel	1
TF dimension, dia. m × media depth m (ft × ft)	10.77 × 6.1 (35.5 × 20)
Media volume/TF, m³ (1,000 ft³)	556 (19.6)
Recirculation flow rate, m³/h	39.4

Practical Design Notes The design methodology was developed for medium-density media with specific surface area of 138 m^2/m^3 based on pilot studies. For full-scale design, a safety factor of 1.2–1.5 on overall nitrification rates may be considered, especially when medium-density media with higher surface area (157 m^2/m^3) or high-medium density media (223 m^2/m^3) are used.

Glossary of Trickling Filter Processes Terms

Activated Sludge It is a biological treatment process using a mass of suspended microorganisms and air to remove pollutant from wastewater. Activated sludge processes are typically used for oxidizing carbonaceous matter and ammonium nitrogen. A further solids separation (sedimentation, floatation, filtration) stage is required to separate the sludge.

Aerobic It is a condition that free oxygen (O_2) is available for oxidizing organic matters and ammonium nitrogen.

Anaerobic It is a condition that is lack of free oxygen (O_2) or bound oxygen (e.g., NO_3^- or NO_2^-).

Anoxic It is a condition that is lack of free oxygen (O_2), but bound oxygen (e.g., NO_3^- or NO_2^-) is available.

Attached Growth Attached growth processes use a medium to retain and grow microorganisms that are used to remove pollutants from wastewater. Trickling filter (TF) and rotating biological contactors (RBCs) are two common types of attached growth systems.

Biochemical Oxygen Demand (BOD) It is a measure of the quantity of oxygen consumed by microorganisms during the decomposition of organic matter sometimes including oxidizing ammonium nitrogen (typically within 5 days of period). Therefore, BOD is a sum of carbonaceous biochemical oxygen demand (cBOD) and nitrogenous biochemical oxygen demand (NBOD).

Biofilm It is a group of microorganisms in which cells stick to each other on a surface. In trickling filter processes, the biofilm attached to the media serves to remove pollutants from wastewater that is trickled through the biofilm on the surface of the media.

Carbon-Oxidizing Filter It is a trickling filter that removes organic pollutants and achieves effluent BOD and TSS of less than 30 mg/L after clarification. Organic loading to a carbon-oxidizing filter typically ranges from 20 to 60 lb/1,000 ft^3/day.

Clarification It refers to a process following trickling filters to separate sludge and treated wastewater.

Combined cBOD and Nitrification Filter It is a trickling filter that oxidizes carbonaceous matter and ammonium nitrogen and achieves an effluent BOD concentration of less than 10 mg/L and an effluent ammonium nitrogen concentration of less than 3 mg/L after clarification.

Denitrification It is a biological process in which nitrate (NO_3^-) is converted to nitrogen gas (N_2).

Distributor It provides even distribution of wastewater over the trickling filter media. Modern trickling filters commonly use rotary distributor arms, which are basically two to six horizontal pipes suspended approximately 6 in. or more above the filter media. The distributor arms can be hydraulically or electrically driven.

Dosing Rate on a Trickling Filter It is the depth of liquid discharged on top of the media for each pass of the distributor. Dosing rate depends on hydraulic loading to the filter and distributor rotation speed.

Dual Biological Processes The processes typically consist of an attached growth reactor and a suspended growth reactor in series. Commonly used dual biological processes include trickling filter/solids contact (TF/SC) and roughing filter/activated sludge (RF/AS).

Fine Screening It is a unit operation that is used to remove solids from wastewater in preliminary or primary treatment. Openings of fine screens typically vary from 0.2 to 6 mm. Fine screens sometimes are used as a substitute for primary clarifier upstream of a trickling filter.

Forced Ventilation It provides a reliable supply of oxygen to trickling filter using low-pressure fan to blow air into trickling filters. Forced ventilation is a low-cost means to improve trickling filter performance and is typically required in highly loaded filters, nitrification filters, and cold climates.

Flushing It is the operation in which an intermittent high dose rate is applied to trickling filter to control the biofilm thickness and solids inventory.

Hydraulic Loading It is a measure of wastewater flow applied to the top of a trickling filter. If recirculation is excluded, the hydraulic loading is referred as "WHL" (wastewater hydraulic loading), and if recirculation is included, the hydraulic loading is referred as "THL" (total hydraulic loading), in this book.

Natural Draft It provides ventilation to a tickling filter without the aid of mechanical means. Natural draft occurs by the air temperature differences inside and outside of the filter.

Nitrification It is a biological process in which ammonium nitrogen (NH_4-N) is oxidized to nitrate (NO_3^-).

Nitrifying Filter It is a trickling filter that oxidizes ammonium nitrogen (NH_4-N) to nitrate (NO_3^-). The organic loading rate to a nitrifying filter is less than 5–10 lb/1,000 ft^3/day.

Organic Loading It is measure of organic pollutant (BOD) mass loading to a unit volume of trickling filter media.

Plastic Media It is used in modern trickling filters to support biofilm growth. Commonly used plastic media include structured sheet cross flow and vertical flow media and random dump media.

Primary Sedimentation It is used to remove readily settleable solids and floating material. Primary clarification is typically used upstream of a trickling filter to reduce TSS and BOD in the influent wastewater. A well-designed and operated primary sedimentation should remove 50–70 % of TSS and 25–40 % of BOD.

Recirculation (Recycle) It is a process in which the filter effluent is recycled and brought into contact with the biofilm more than once.

Roughing Filter/Activated Sludge (RF/AS) It is a dual biological process that consists of a high organically loaded trickling filter (roughing filter) and an activated sludge basin in series. The roughing filter is designed to remove 40–70 % of the BOD load by mass and where the residual particulate and soluble BOD will be removed in the subsequent aeration basin through oxidation, bioflocculation, and adsorption.

Secondary Treatment It refers to biological wastewater treatment that achieves effluent BOD_5 and TSS of less than 30 mg/L and $cBOD_5$ of less than 25 mg/L after clarification (as per minimum US National Standard).

Sloughing During the operation of a trickling filter, the biofilm tends to grow in thickness as a result of steady-state microbial growth. Portions of the biofilm will fall off when they lose the ability to stay attached to the media, which is called "sloughing."

Specific Surface Area of Trickling Filter Media It is the total surface area of the media per unit of bulk volume, typically in units of m^2/m^3 or ft^2/ft^3.

Stagnation It refers to the phenomenon that little airflow would occur through the filter because of insufficient temperature differences between air inside and outside of a trickling filter. Stagnation can negatively impact the trickling filter performance due to a period of low oxygen transfer rate.

Submerged Fixed Film System It is a type of attached growth system utilizing submerged media to support biofilm growth and achieve biological treatment. Commonly used systems include moving bed biofilm reactor (MBBR), fixed bed biofilm reactor (FBBR), and biological aerated filter (BAF).

Suspended Growth Processes The processes utilize microorganisms that are suspended in the wastewater to remove pollutants. Typical suspended growth processes include activated sludge and aerated lagoons.

Trickling Filter It is an attached growth process utilizing media to support biofilm growth. Trickling filters are essentially a three-phase system in which the wastewater flows over the biofilm and contacts with air, and at the same time organic matter and ammonium are absorbed and subsequently degraded by the microorganisms in the biofilm.

Trickling Filter/Solids Contact (TF/SC) It is a dual biological process consists of a trickling filter and a small aeration basin in series. The trickling filter is designed to remove a majority of the soluble BOD in the wastewater, and the solids contact basin is designed to bioflocculate the non-readily settleable, colloidal, and particulate organic matters and to polish the residual soluble BOD.

Underdrain System in a Trickling Filter It collects wastewater for conveyance, provides support to the media, and creates a plenum that allows air circulation through the media bed. The trickling filter underdrain system can be constructed with clay or concrete blocks, concrete beams, concrete posts with beams, concrete piers with reinforced fiberglass gratings, or PVC stanchions with fiberglass gratings.

Wetting Rate to a Trickling Filter It is the same as total hydraulic loading (THL).

References

1. WEF (2000) Aerobic fixed-growth reactor, *Special Publication*. Water Environment Federation, Alexandria
2. Albertson OE (1992) Milestones in trickling filter development
3. Shriver LE, Bowers DM (1975) Operational practice to upgrade trickling filter plant performance. J Water Pollut Control Fed 47(11):2640–2651
4. Wang LK, Wu Z, Shammas NK (2009) Trickling filters. In: Wang LK, Pereira NC, Hung YT (eds) Biological treatment processes. Human Press, New York, pp 371–434
5. US EPA (2009) Nutrient control design manual, EPA/600/R-09/012. US Environmental Protection Agency, Washington, DC
6. WEF (2009) Design of municipal wastewater treatment plants, manual of practice 8, 5th edn. Water Environment Federation, Alexandria
7. Metcalf and Eddy (2002) Wastewater engineering: treatment and reuse, 4th edn. McGraw-Hill, New York
8. Onda K, Takeuchi H, Okumoto Y (1968) Mass transfer coefficients between gas and liquid phases in packed columns. J Chem Eng Jpn 1:56–62
9. Parker DS, Lutz MP, Pratt AM (1990) New trickling filter applications in the U.S.A. Water Sci Technol 22(1–2):215–226
10. Paulson C (1989) Nitrification for the 90's. Water Eng Manag 136(9):57–59
11. Okey RW, Albertson OE (1989) Evidence for oxygen-limiting conditions during tertiary fixed-film nitrification. J Water Pollut Control Fed 61:510
12. Okey RW, Albertson OE (1989) Diffusion's role in regulating rate and masking temperature effects in fixed film nitrification. J Water Pollut Control Fed 61:510
13. Sarner E (1984) Effect on filter medium, substrate concentration and substrate and hydraulic loading on trickling filter performance. Paper presented at 2nd international conference on fixed film biological processes, Arlington
14. Albertson OE, Davies G (1984) Analysis of process factors controlling performance of plastic bio-media. Paper presented at 57th Annual Conference of Water Pollution Control Federation, New Orleans
15. Albertson OE (1995) Excessive biofilm control by distributor-speed modulation. J Environ Eng NY 121(4):330–336
16. WEF (2007) Operation of municipal wastewater treatment plants, manual of practice 11, 5th edn. Water Environment Federation, Alexandria
17. Schroeder ED, Tchobanoglous G (1976) Mass transfer limitations on trickling filter design. J Water Pollut Control Fed 48(4):771–775
18. Albertson OE, Okey RW (1988) Trickling filter need to breath too. Paper presented at Iowa Water Pollution Control Federation meeting, Des Moines
19. Great Lakes – Upper Mississippi River Board of State Sanitary Engineering (1971) Recommended standards for Sewage works. Health Education Service, Albany
20. Great Lakes – Upper Mississippi River Board of State Sanitary Engineering (1990) Recommended standards for Sewage works. Health Education Service, Albany
21. Great Lakes – Upper Mississippi River Board of State Sanitary Engineering (2004) Recommended standards for Sewage works. Health Education Service, Albany
22. CH2MHILL (1984) A comparison of trickling filter media. CH2MHILL, Denver
23. Wood D, Drury D, Middleton B (1989) Evaluation shows plastic media to be more effective than rock trickling filters. John Carollo Engineers reports, Phoenix
24. Drury DD, Carmona J III, Delgadillo A (1986) Evaluation of high density cross flow media for rehabilitating an existing trickling filter. J Water Pollut Control Fed 58(5):364–367
25. Gujer W, Boller M (1986) Operating experience with plastic media tertiary trickling filters for nitrification. In: Von der Ende W, Tench HB (eds) Design and operation of large wastewater treatment plants. Pergamon Press, Oxford, pp 201–213

26. Parker DS, Lutz M, Dahl R, Bernkopf S (1989) Enhancing reaction rates in nitrifying trickling filters through biofilm control. J Water Pollut Control Fed 61:618

27. Parker DS, Lutz M, Andersson B, Aspegren H (1995) Effect of operating variables on nitrification rates in trickling filters. Water Environ Res 67(7):1111–1118

28. Li H, Daigger GT, Boltz JP, Kulick FM, Zhu J (2013) Practical experience with high-density cross-flow media in nitrifying trickling filters: a basis for the reform of generally accepted design criteria. Paper presented at 2013 IWA Biofilm Conference, Paris

29. WEF (1998) Design of municipal wastewater treatment plants, manual of practice 8, 4th edn. Water Environment Federation, Alexandria

30. Enkenfelder WW (1963) Trickling filter design and performance. Trans Am Soc Civ Eng 128 (Part III):371–384

31. Enkenfelder WW, Barnhart W (1963) Performance of a high-rate trickling filter using selected media. J Water Pollut Control Fed 35:1535

32. Harremoes P (1982) Criteria for nitrification in fixed film reactors. Water Sci Technol 14:167

33. Wanner J, Gujer W (1984) Competition in biofilms. Paper presented at 12th Annual Conference of International Association on Water Pollution Research and Control, Amsterdam

34. US EPA (1975) Process design manual for nitrogen control. U.S. EPA Techno. Transfer, EPA-625/1-77-007. US Environmental Protection Agency, Washington, DC

35. Lin CS, Heck G (1987) Design and performance of the trickling filter/solids contact processes for nitrification in a cold climate. Paper presented at 60th Annual Conference of Water and Pollutant Control Federation, 27, p 805

36. Parker DS, Richards T (1986) Nitrification in trickling filters. J Water Pollut Control Fed 58 (9):896–902

37. Chen WF, Richard Liew JY (2002) The civil engineering handbook, 2nd edn. CRC Press, Boca Raton, Florida

38. Albertson OE, Okey RW (1989) Design procedure for tertiary nitrification. Prepared for American Surfpac Inc., West Chester

39. Harrison JR (2012) Honouliuli WWTP – TFSC & water reclamation., Fixed Film Forum, www.fixedfilmforum.com

40. Daigger GT, Norton LE, Watson RS, Crawford D, Sieger RB (1993) Process and kinetic analysis of nitrification in coupled trickling filter/activated sludge processes. Water Environ Res 65(6):750–758

41. Harrison JR (2012) Corvallis wastewater treatment plant-TFSC Fixed Film Forum. www.fixedfilmforum.com

42. Harrison JR, Daigger GT, Filtert W (1984) A survey of combined trickling filter and activated sludge processes. J Water Pollut Control Fed 56:1073–1079

43. Norris DP, Parker DS, Daniels ML, Owens EL (1982) High quality trickling filter effluent without tertiary treatment. J Water Pollut Control Fed 54(7):1087–1098

44. Daigger GT, Boltz JP (2011) Trickling filter and trickling filter-suspended growth process design and operation: a state-of-the-art review. Water Environ Res 83(5):388–404

45. Kelly R, Melcer H, Krugel S, Hystad B (2010) Ten years gone: an operational update of Annacis Island, the world's largest trickling filter solids contact treatment facility. Paper presented at Biofilm Technology conference, Portland, 14–18 Aug 2010

46. Pearce P (2004) Trickling filters for upgrading low technology wastewater plants for nitrogen removal. Water Sci Technol 49(11–12):47–52

47. Dai Y, Constantinou A, Griffiths P (2013) Enhanced nitrogen removal in trickling filter plants. Water Sci Technol 67(10):2273–2280

48. Biesterfeld S, Farmer G, Figueroa L, Parker D, Russell P (2003) Quantification of denitrification potential in carbonaceous trickling filters. Water Res 37:4011–4017

49. Farmer G (2013) Littleton/Englewood WWTP. A case study presented at Trickling Filter Media Webinar, March, 2013, Honolulu, Hawaii

50. Shammas NK, Wang LK (2009) Emerging attached-growth biological processes. In: Wang LK, Shammas NK, Hung YT (eds) Advanced biological treatment processes. Humana Press, Totowa, pp 649–682
51. Wang LK (1984) Investigation and design of a denitrification filter. Civ Eng Pract Des Eng 3:347–362
52. US EPA (2007) Wastewater management fact sheet denitrifying filters. EPA 832-F-07-014. US Environmental Protection Agency, Washington, DC
53. US EPA (2015) Consumer fact sheet nitrates/nitrites. US Environmental Protection Agency, Washington, DC
54. Zhu J (2014) Everything you need to know about trickling filters. Clear Waters 44(2):16–19
55. Hirst J, Anderson D (2015) Backyard BNR – a passive nitrogen reduction system. Water Environ Technol 27(3):40–43
56. Gorder PJ, Steele JD, Tobel T (1990) Major trickling filter facility retrofit with synthetic media saves costs and increases plant capacity. Paper presented at 63rd Water Pollution Control Federation Annual Conference & Exposition, Washington, DC
57. USACE (1988) The trickling filter/solids contact process: application to army wastewater plants, USA-CERL technical report N-88/14. US Army Corp of Engineers, Washington, DC
58. Albertson OE (1990) Evaluation of the wastewater treatability in trickling filters, www.orrisealbertson.com
59. Parker DS (1998) Trickling filter mythology.In: ASCE 1998 National Conference on Environment Engineering, Chicago
60. US EPA (2000) Wastewater technology fact sheet trickling filters EPA 832-F-00-014. US Environmental Protection Agency, Washington, DC
61. Harrison JR (2013) TFSC update or abandon practice, Fixed Film Forum Webinar, March, 2013, Honolulu, Hawaii

Chapter 7
Chlorides Removal for Recycling Fly Ash from Municipal Solid Waste Incinerator

Fenfen Zhu, Masaki Takaoka, Chein-Chi Chang, and Lawrence K. Wang

Contents

1 Overview of Existing Intermediate Treatment and Recycling for Fly Ash from Municipal Solid Waste Incinerator 350
  1.1 Melting 351
  1.2 Chemical Stabilization 351
  1.3 Chemical Extraction 353
  1.4 Cement Solidification 353
  1.5 Sintering or Calcining 354
  1.6 Recycling 355
2 Chlorides in Fly Ash 357
  2.1 The Importance of Studying Chlorides in Fly Ash 357
  2.2 Chloride Reduction Characteristics by Washing 358
  2.3 Chloride Speciation in Fly Ash 359
  2.4 Chloride Behavior in Washing Experiments 362
3 Conclusion, Further Investigations, and Recent Advances 362
References 364

F. Zhu (✉)
Department of Environmental Science and Engineering, Renmin University of China,
Beijing, China
e-mail: zhufenfen@ruc.edu.cn; zhufenfen78@hotmail.com

M. Takaoka
Environmental Engineering Department, Graduate School of Engineering,
Kyoto University, Kyoto, Japan
e-mail: takaoka.masaki.4w@kyoto-u.ac.jp

C.-C. Chang
DC Water and Sewer Authority, 5000 Overlook Avenue, SW, Washington, DC 20032, USA

University of Maryland, Baltimore, MD, USA
e-mail: cchang@dcwater.com; chang87@gmail.com

L.K. Wang
Lenox Institute of Water Technology, Newtonville, NY 12128-0405, USA

Rutgers University, New Brunswick, NJ, USA
e-mail: lawrencekwang@gmail.com

© Springer International Publishing Switzerland 2016
L.K. Wang, M.-H.S. Wang, Y.-T. Hung and N.K. Shammas (eds.),
Natural Resources and Control Processes, Handbook of Environmental
Engineering, Volume 17, DOI 10.1007/978-3-319-26800-2_7

Abstract Both the incineration residues of the municipal solid waste (MSW) and the residues of air pollution control facilities are frequently classified as hazardous wastes if they do not pass the toxicity characteristic leaching procedure (TCLP) testing. As the hazardous wastes, they may be either properly disposed of with a cost or recycled for reuse with a value.

If the above-described residues or fly ashes are mainly composed of calcium and silicon compounds, they have the potential for recycling. To utilize the waste residues and reduce their negative environmental impact, three types of pretreatment prior to recycling may be needed: (a) separation processes, (b) solidification or stabilization processes, and (c) thermal processes. The major problem preventing fly ashes from being recycled for reuse is their high content of soluble salts, such as chloride. Water extraction process is a feasible pretreatment of fly ashes for extracting many problematic soluble salts from the ash matrix, which opens the windows for possible resources recovery. This chapter provides an overview of existing intermediate treatment and recycling for fly ash from MSW incinerator. These pretreatment and recycling technologies include melting, chemical stabilization, chemical extraction, cement solidification, sintering or calcining, and recycling. Environmental significance of chlorides in fly ash is emphasized. Special topics covered in this chapter are (a) chloride reduction characteristics by washing, (b) chloride speciation in fly ash, and (c) chloride behavior in washing experiments. The authors also introduce further investigations and recent advances by other researchers.

Keywords Municipal solid waste • Incinerator • Fly ash • Recycling • Melting • Chemical stabilization • Chemical extraction • Cement solidification • Sintering • Calcining • Chloride reduction • Washing • Speciation • Chloride behavior • Recent advances

1 Overview of Existing Intermediate Treatment and Recycling for Fly Ash from Municipal Solid Waste Incinerator

Incineration or combustion is one of the main methods to treat municipal solid waste besides landfill and compost. It has many advantages. First, it can effectively reduce the volume of the waste and the volume reduction rate can reach even more than 90 %. Second, it occupies much less area than landfill and compost, which is very significant to the narrow place. Third, it can recover heat produced during incineration to generate electricity or to get hot water. Those are the reasons why the ratio of combustion treatment is rising while landfill is going down [1].

However, incineration generates fly ash or air pollution control residue, which contains considerable content of heavy metal and dioxins [2, 3]. The heavy metals are zinc, copper, iron, lead, chromium, and so on, which will be leached out in toxicity characteristic leaching procedure (TCLP) [4], and the concentration will exceed the limitation for normal waste. As a result, it is identified as hazardous waste.

To treat fly ash from municipal solid waste incinerator, there are several methods: melting, chemical stabilization, chemical extraction, cement solidification, sintering or calcining, and recycling.

1.1 Melting

Melting has been developed since the 1990s [5]. It is considered to be a prospective technology for stabilizing MSWI fly ash and bottom ash [5], because the melting process is to heat the fly ash or bottom ash to fusion temperature, normally above 1200 °C, and the residues including fly ash and bottom ash will be transformed to more stable glassy molten slag [5–7]. During the melting process, the organic pollutants decompose and the volume of fly ash and bottom ash can be reduced by 70 %. Moreover, the molten slag can be used in the glass and ceramic industry by some simple pretreatment since the heavy metal has been stabilized in the glassy slag [5–9]. However, the problem of melting fly ash and salts has to be considered and solved.

There are many types of ash melting furnace developed, such as plasma melting furnace, a reflecting surface-melt furnace, a DC electric joule-heating system, etc [10–13]. Figure 7.1 shows the principal of plasma torch and the plasma ash melting system. There are two methods of using plasma torch by a counter-electrode, namely, the transfer and the nontransfer methods [10]. Electricity is used as the heat source for melting. In the actual plant, the electricity generated by the waste incineration is used.

1.2 Chemical Stabilization

Chemical stabilization is to use chemical reagent such as chelate reagent or other chemical compounds to react with the heavy metal in fly ash to form stable compounds such as chelate complex or precipitation [6, 14–18]. The chemical reagents include ethylendiaminetetraacetate (EDTA), diethylenetriaminepentaacetate (DTPA), organic sulfide, thiourea, phosphate, ferrite, sulfide, and so on [14–18]. One of the advantages of chelating agents is that they generally work under moderate pH conditions [18]. EDTA can leach the heavy metals from the ash. NaOH can be used for the leaching of zinc, but the resultant leaching residues should be treated further [16]. Some even added CO_2 to carbonate fly ash to

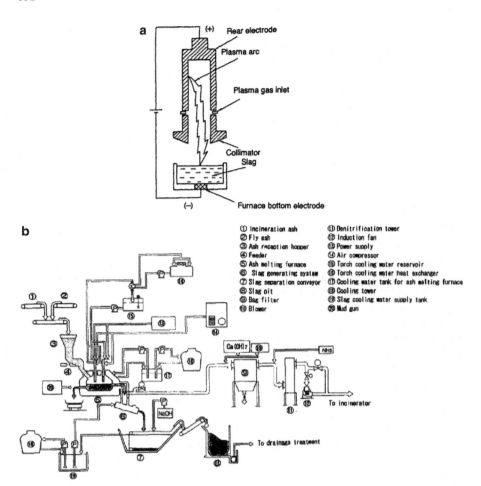

Fig. 7.1 Schematic of a double liner and leachate collection system for a hazardous landfill (Source: [10]). (**a**) Principal of plasma torch (**b**) Plasma ash melting system flow chart

stabilize lead and zinc [19]. As to the phosphate phases, likely to precipitate are extremely numerous, particularly Ca phosphates and heavy metal phosphates [20]. The mechanism of those diverse compounds to stabilize fly ash is associated with the successive precipitation of increasing stable phosphates [21, 22] and microenvironment effects. In media rich in calcium, generally present in the form of calcium carbonates, most of the leachable metals are trapped during the precipitation of calcium phosphates containing traces of metals [20]. Some researchers tend to use polymers to stabilize fly ash; especially thermoplastic polymers which encapsulate the residues in a matrix that coats and disperses them have already been used [17]. Unsaturated polyester (UP) resins are some of the most common thermoset polymers as well as commercial resins [17].

Normally the chemical reagents are liquid and expensive, so as to the fly ash with pH over 12, it had better be neutralized or be adjusted to pH around 10–11 and then use chelate to stabilize. The addition of liquid reagent is around 1–5 % regularly; however, Cu, Ni, Fe, Zn, and other heavy metals will also join in the reaction, so the addition of reagent has to be adjusted.

1.3 Chemical Extraction

Chemical extraction is to use chemical reagent to react with the target element and with the separated deposit or solution to recycle the target elements or stabilize them [6]. To extract the heavy metal in fly ash is very similar to chemical stabilization [23–27]. The difference is that some will use acid, alkaline, or water to extract the heavy metal to get the solution and then use chemical compounds to form deposits [23–26]. However, after the metals were extracted out as solution, hydrometallurgy method is needed such as electrodialytic method. Here we emphasize more on the purpose of recycling [20, 27]. The target metals are normally Cu, Ni, Co, and Zn. There is a typical treatment method for this named "Acid Extraction-Sulfide Stabilization Process (AES Process)," in which water is added to fly ash to convert it to slurry, easily soluble heavy metals are extracted by acidic agent, and sodium sulfide is added to slurry to stabilize the remaining heavy metals, and thus heavy metal leaching from dewatered cake is prevented [28].

Though this method is a little complex and not easy to be carried out into reality, researchers tend to utilize more the sequential extraction of metals in fly ash to study the characteristics of the leaching behavior [29–33]. The sequential chemical extraction (SCE) was first proposed by Tessier [34]. Now it has been widely used to investigate the physicochemical forms of heavy metals in fly ash. To study the characteristics of fly ashes, the sequential chemical extraction sometimes is combined with other methods such as X-ray diffraction (XRD) analysis and X-ray fluorescence (XRF) analysis [30]. The investigated characteristics of fly ash include the chemical and mineralogical characteristics, releasing characteristics of some special elements such as Cr, Cu, Mn, Pb, and Zn [32]. And sometimes the steps of SCE will be simplified according to practical purposes [32].

1.4 Cement Solidification

Cement solidification is a method to treat fly ash by using cement as a combining reagent to stabilize the heavy metal through the hydration process of cement in normal temperature [6, 35–42]. It is a cheap disposal method.

Unconfined compressive strength (UCS) is an important factor to evaluate solidification [35]. When ordinary Portland cement (OPC) is blended with fly ash, with the increases of the fly ash to OPC ratio, the water demand of the mixed cement increased and the strength decreased [35]. Lower strength resulted from the lower Portland cement and much higher waste contents. Generally, a waste/binder ratio of 0.4:0.5 is used with a water/solid ratio of 0.4:0.6. The increase in the binder (cement) content increases the bulk densities because the binders filled the void space of solidified specimen. UCS decreased with increasing crystalline phases [36]. So the initial setting time and hydration process is important in mixing. The standard initial and final setting time reported for OPC should not be less than 45 min and no more than 10 h, respectively [37]. The rate of hydration was reported to be insensitive to temperature over a range of 0–40 °C. It is reported that the lower curing temperatures of a lead-bearing waste/cement matrix decrease the solubility of lead salts formed in the cement, resulting in an increase in gelatinous coatings on grains [38]. Carbonation is another factor to influence the UCS of the final product. Carbonated stabilization/ solidification (S/S) product develops higher strength in comparison to non-carbonated products [39–41]. Carbonation involves reaction with phases like AFt/AFm (hydrated calcium aluminates based on the hydrocalumite-like structure of $4CaO \cdot Al_2O_3 \cdot 13\text{-}19H_2O$), calcium silicate hydrate (CSH) gel, and calcium hydrate (CH). CSH gels are recognized to play an important role in the fixation of toxic species. The fixation will be significantly altered by carbonation [36].

The cement-solidified fly ash will mainly be sent at present to landfill and cannot be used as the blended cement because of the high concentration of chloride, sulfate, and alkali content [42]. As a result, there are still potential environmental risks due to the complex reactions that happened in landfill site [43]. And in fact, fly ash solidified with Portland cement presents some disadvantages, namely, protection against humidity is required to prevent breaking down and leaching of heavy metals [44–47]. Moreover, the volume of the final product is enlarged because of adding cement and water to form a rigid and porous solid [35–43], leading to an increase in the cost of disposal [16]. If we want to recycle fly ash as good ingredient or additives in cement, fly ash should be pretreated [48–58]. Also the fly ash washing pretreatment improved the stabilizing behavior of fly ash-cement mixtures, because the interaction of fly ash with water leads to a rapid formation of hydrate compounds such as syngenite, gypsum ($CaSO_4 \cdot 2H_2O$), ettringite ($3CaO \cdot Al_2O_3 \cdot 3CaSO_4 \cdot 32H_2O$), calcium hydroxyzincate ($CaZn_2(OH)_6 \cdot 2H_2O$), and laurionite ($Pb(OH)Cl$) [49, 50]. One drawback of washing process is that adding the washing water to the mixture increases the final water/solid (w/s) ratio so that it decreases the UCS of the final product. Also cement solidification is not suited for Pb-rich fly ashes, since Pb tends to leach out from the solid phase at high pH, which is caused by cement itself [18].

1.5 Sintering or Calcining

Sintering process or calcining process is similar as melting process, but the operation temperature of both processes is lower than the temperature in melting process [6].

Compared with melting, sintering needs less energy as the heating temperature is lower, while sintering is also effective to stabilize and detoxify fly ash especially to deal with the problem of dioxins [59–61]. Because the operation temperature of both processes is not as high as that in melting process, the volume reduction ratio of the final product is less than that of melting process. Another drawback is that the performance of the sintering process is strongly related to the chemical composition of raw fly ash, and in many cases, this process proves to be ineffective for the conversion of raw fly ash into ceramic materials with good mechanical characteristics [48]. Also, pretreatment had better be adopted [62].

Some researchers studied sintering as the pretreatment of fly ash to reuse it as a concrete aggregate [63]. However, there also needs a washing pretreatment before sintering, because the sintering of MSW fly ashes proved to be ineffective for manufacturing sintered products for reuse as a construction material, and it needs to avoid the adverse chemical characteristics due to sulfate, chloride, and vitrified oxide contents contained in fly ashes [63]. On the other hand, the possibility of using sintered products as concrete aggregates is largely depending on the operating conditions adopted for sintering such as the compaction degree of powders, the sintering temperature and time, as well as the chemical composition of fly ash. Those factors can affect the type and amount of porosity and, consequently, the specific gravity, mechanical strength, and heavy metals' leachability of sintered products, as well as their chemical stability in aqueous solutions [63].

1.6 Recycling

With the concept of 3R (reduce, reuse, recycle) or sustainable development, recycling has been a very popular idea to guide the treatment of waste including the fly ash from municipal solid waste.

One of the most famous technology flows is "3R" technology (3R being the German acronym for Rauchgas-Reinigung mit Ruckstandsbehandlung, which means flue gas purification including residue treatment). This process had been developed in the 1980s by Vehlow [64]. The technology flow is shown in Fig. 7.2. The final product, bottom ash + 3R products will be sent to landfill. The purpose of 3R process is to recover the valuable metals by acid extraction and to stabilize fly ash by returning the residue into the zone of high temperature in the combustion chamber. The 3R technology is suitable for the incinerators equipped with wet scrubber, because the acid which is needed for acid extraction can be obtained from wet scrubber. And the binder such as bentonite is necessary to be added to the treated fly ash, which is sent back to the incinerator, to prevent direct release of additional fly ash, which added the cost of 3R process.

Since the 1990s til now, some researchers began to study the washing process [48–58], which is carried out as the pretreatment to reuse fly ash as the raw material or one of the ingredients in cement [55–58]. Some researchers also suggest adding a heat treatment after the washing process, normally sintering

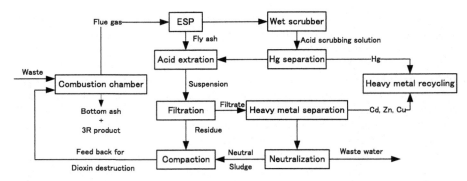

Fig. 7.2 Flow schematic of 3R process [64]

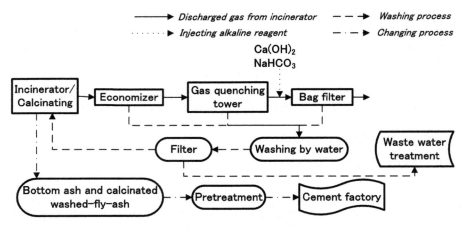

Fig. 7.3 Schematic flow of proposed WCCB system [67]

or calcining [48, 63, 65]. Many aspects have been of concern such as the leaching behavior of heavy metals, intensity of the final product, and so on [48–58, 63, 65]. Also, NaHCO$_3$ was applied to act as a neutralization reagent and some company has successfully used a recycling system for NaHCO$_3$ such as the Solvay Company. It has a patent named NEUTREC®, which is a recycling system for NaHCO$_3$ [66].

There is a new recycling system trying to integrate the above concepts named as WCCB. "W" means washing, the first "C" means calcining, and the "CB" means changing the treated fly ash and bottom ash into raw material in cement industry. In this system two alkaline reagents, both Ca(OH)$_2$ and NaHCO$_3$ are suitable. The schematic flow is shown in Fig. 7.3. Fly ash from economizer, gas quenching tower, and bag filter was firstly washed by water, then dried, and sent back to incinerator for calcining treatment, which can save the energy and cost to dry the washed residue because of a separate new instrument. The treated fly ash together with bottom ash will be used as the raw material in cement industry. In the cement industry, it has been meant to recycle the heavy metal in the treated fly ash and bottom ash by the original heavy metal recycling system combined with rotary kiln or other equipments. Another important aspect is that in the rotary kiln, the temperature is around 1200–1500 °C [68–70], the residence time is long (the gas retention time is approximately 5s), and the turbulence is strong, which can ensure the complete destruction of even the most stable organic compounds such as PCB and dioxins [68–70]. Moreover, heavy metal can be stabilized in the crystal structure of cement product at high temperature.

2 Chlorides in Fly Ash

Many chlorides have been detected in fly ash such as $NaCl$, KCl, $CaCl_2$, $CaClOH$, $CaCl_2 \cdot Ca(OH)_2$, and so on [20, 25, 41, 51–54, 71–74]. Some researchers reported that insoluble chloride in fly ash is Friedel's salt [75]. Besides the chlorides produced in the neutralization process with different alkaline reagents, the incineration itself also forms chlorides and actually some chlorides were detected in raw fly ash from the boiler [74].

There are many chloride sources in municipal solid waste [76–80]. Food, especially the cooked food, is an important source for chlorides such as $NaCl$ and KCl. Plastics is another considerable source for chlorine [76–80]. And the concentration of chlorides is considerable, which are varied from about 6 % to 40 % depending on the alkaline reagent used and the practical condition.

2.1 The Importance of Studying Chlorides in Fly Ash

Chlorides are normal compounds in fly ash with considerable concentration. They will act a negative role in the treatment or recycling process of fly ash. As to melting, at the operation temperature, most of chlorides will surely evaporate and be cooled down on the inner side of the pipes, then finally erode the pipes or block them. As to cement solidification, they will be washed out by rain and then go into the leachate, which will make the leachate more difficult to treat. As to chemical treatment, it is like cement solidification. As to recycling the fly ash for construction material, for example, as the raw material for cement industry, salt is a very critical factor because chlorides in the cement product will erode the embedded steel and high concentration of chlorides in the raw material will finally make the rotary kiln stop working by blocking the pipes at low temperature zone [81–89].

2.2 Chloride Reduction Characteristics by Washing

The three types of fly ashes are fly ash collected in a bag filter with the injection of $Ca(OH)_2$ for acid gas removal (CaFA), the fly ash collected in a bag filter with the injection of $NaHCO_3$ for acid gas removal (NaFA), and raw fly ash collected from the boiler incinerator (RFA). The incinerator is a continuously operated stoker in Japan. The composition of the three types of fly ash was examined by X-ray fluorescence (XRF-1700, Shimadzu Corporation) and inductively coupled plasma-atomic emission spectrometry (ICP-AES, IRIS Intrepid, Optronics Co., Ltd.). The results are shown in Table 7.1.

Ion exchanged water (IEW) was used as the washing solution. The fly ash was mixed with IEW by a vortex mixer (Iuchi Seieido Co., Ltd.) under different liquid (ml) to solid (g) (L/S) ratio and separated by centrifuge (Himac CT4, Hitachi) for 15 min at 3000 rpm after washing. The washed residue was dried in an oven at 105 °C for 24 h. With the literature review and preliminary experiments, it can be identified that washing frequency, L/S ratio, and mixing time are important parameters to influence the results, while washing frequency and L/S ratio act more effective than mixing time [26, 52–54, 56, 67, 90–92].

Tables 7.2, 7.3, and 7.4 shows the washing experiment condition and Figs. 7.4 and 7.5 shows the corresponding results.

Table 7.1 Element content of CaFA, NaFA, and RFA (weight %) [67]

Element	CaFA	NaFA	RFA
O	24.6	13.2	37.3
Si	2.68	1.28	4.97
C	7.55	8.17	3.24
Cl	19.6	37.3	6.36
S	1.81	1.31	3.31
Ca	28.2	10.2	24.7
K	3.38	5.84	2.99
Na	3.46	18.5	2.97
Ti	0.821	0.535	1.51
Mg	0.722	0.341	1.12
P	0.241	0.179	0.666
Al	4.30	5.92	3.99
Fe	1.24	0.713	2.32
Zn	0.390	0.893	0.223
Cu	0.0639	0.147	0.238
Pb	0.212	1.23	0.387
Br	0.269	0.914	0.111

Table 7.2 Experimental design of the single-washing experiments [67]

Code	L/S	Experimental conditions
1#	0	Fly ash
2#	2	By glass rod, 5 min
3#	3	150 rpm, 5 min
4#	4	150 rpm, 5 min
5#	5	150 rpm, 5 min
6#	6	150 rpm, 5 min
7#	8	150 rpm, 5 min
8#	10	150 rpm, 5 min

Table 7.3 Experimental design for the double-washing experiments [67]

Code	L/S	Experimental conditions
9#	3–3	150 rpm, 5 min–150 rpm, 30 min
10#	3–5	
11#	3–7	
12#	3–10	
13#	3–3	150 rpm, 5 min–250 rpm, 10 min
14#	3–5	
15#	3–7	
16#	3–10	
17#	3–3	150 rpm, 5 min–150 rpm, 10 min
18#	3–5	
19#	3–7	
20#	3–10	
21#	3–3	150 rpm, 5 min–250 rpm, 30 min
22#	3–5	
23#	3–7	
24#	3–10	

Table 7.4 Experimental design for the thrice-washing experiment [67]

Code	L/S	Experimental conditions
25#	3–3–3	150 rpm, 5 min–150 rpm, 10 min–150 rpm, 10 min

2.3 Chloride Speciation in Fly Ash

The experimental materials are the same as those in Sect. 2.2. X-ray diffraction is normally used to detect the existence of crystal chlorides in fly ash. And washing experiments are used to identify the percentage of soluble chlorides in fly ash. Many chlorides have been detected in fly ash such as NaCl, KCl, CaCl$_2$, CaClOH, CaCl$_2$·Ca(OH)$_2$, and so on [20, 25, 41, 51–54, 71–74]. Some researchers reported

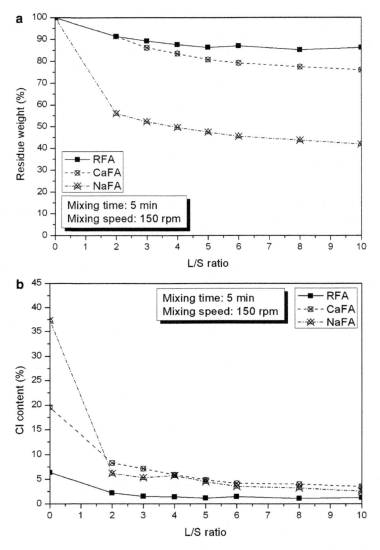

Fig. 7.4 Results of single-washing experiments [67]. (**a**) Weight of residue, (**b**) chlorine content in the residue

that insoluble chloride in fly ash is Friedel's salt [75]. Besides those researches, some researchers have tried to use X-ray absorption near edge structure (XANES) combined with X-ray diffraction (XRD) to do quantitative analysis of chlorides in fly ash [74]. The final results are shown in Fig. 7.6.

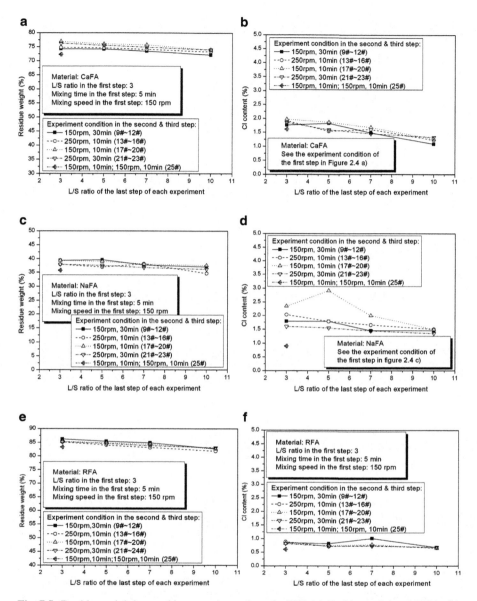

Fig. 7.5 Double- and thrice-washing experimental results [67]. (**a**) Residue weight of CaFA, (**b**) chlorine content in the residue of CaFA, (**c**) residue weight of NaFA, (**d**) chlorine content in the residue of NaFA, (**e**) residue weight of RFA, (**f**) chlorine content in the residue of RFA

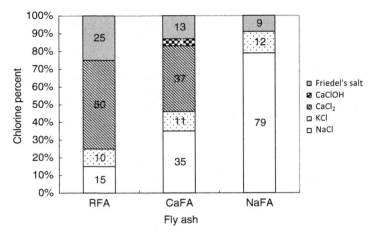

Fig. 7.6 Chloride speciation in fly ashes [74]

2.4 Chloride Behavior in Washing Experiments

Washing is a very popular pretreatment method for fly ash to reduce chlorides. As a result, chloride behavior in washing experiments is an abstractive issue. The experimental materials are the same as those in Sect. 2.2. Washing experiment condition is shown in Tables 7.3 and 7.4. Figure 7.7 shows the chloride behavior in washing experiments.

3 Conclusion, Further Investigations, and Recent Advances

Both the incineration residues of the municipal solid waste (MSW) and the residues of air pollution control facilities are frequently classified as hazardous wastes if they do not pass the toxicity characteristic leaching procedure (TCLP) testing. As the hazardous wastes, they may be either properly disposed of with a cost, or recycled for reuse with a value.

If the above-described residues or fly ashes are mainly composed of calcium and silicon compounds, they have the potential for recycling. In order to utilize the waste residues and reduce their negative environmental impact, the treatment for fly ashes can be generally grouped into three classes: (a) separation processes, (b) solidification or stabilization processes, and (c) thermal processes. The major problem preventing fly ashes from being recycled for reuse is their high content of soluble salts, such as chloride. Water extraction process is a feasible pretreatment of fly ashes for extracting many problematic soluble salts from the ash matrix, which opens the windows for possible resource recovery [95–101].

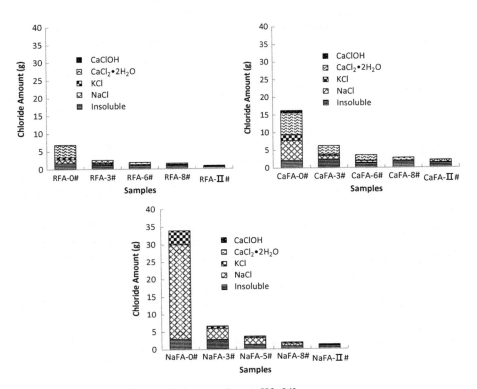

Fig. 7.7 Chloride behavior in washing experiments [93, 94]

Many researchers have reviewed and discussed the fly ash pretreatment technologies, such as solidification/stabilization, thermal process, and separation processes [96, 98, 101]. Lam et al. have (a) studied the chemical properties of the fly ashes, (b) reviewed possible pretreatment prior to utilization of fly ashes, and (c) explored various applications of pretreated fly ashes, such as concrete production, road pavement, glass/ceramic manufacturing, agricultural utilization, stabilization of agent production, adsorption of pollutants, and zeolite production [96]. The practical use of these pretreated waste fly ashes shows a great contribution to waste minimization as well as resource conservation.

Kirkelund et al. have studied the incineration residues (fly ashes) of MSW and other air pollution control residues using an electrodialytic remediation technology [95]. Their results show that the leaching of Cd, Cu, Pb, and Zn can be reduced compared to the initial heavy metal leaching, except when the pH of the residue in suspension is reduced to a level below 8 for the fly ashes. On the other hand, Cr leaching has increased by the electrodialytic treatment. Chloride (Cl^-) leaching from the incineration residues of MSW has been less dependent on experimental conditions and has been significantly reduced in all experiments, compared to the initial levels. Shammas and Wang [101] describe the electrodialytic treatment process in detail.

Although coal ash from power stations has long been used successfully in the cement industry as binders in several Portland formulations, the use of fly ash from MSW has not been successful because the MSW incinerator's fly ash has high concentrations of chloride ranging from 10 to 200 g Cl/kg fly ash (dry weight), exceeding the maximum allowable concentration in most cement mixtures. To reduce chloride content in MSW bottom ash, a laboratory investigation has been carried by Boghetich et al. [99] to study chloride extraction for quality improvement of municipal solid waste incinerator ash for the concrete industry, based on the exhaustive washing in tap water. The influence of operative parameters such as temperature, granulometric properties, and solid/liquid ratio of extraction has been evaluated. In addition to optimization of the operational parameters for full-scale application, their research gives preliminary indications on mechanistic aspects of the washing operation.

Rodella et al. [97] have discussed a new remediation method, based on the use of silica fume, for heavy metal stabilization. The inertization procedure is reported by them and compared with other technologies, involving the use of amorphous silica as stabilizing agent for MSW incinerator fly ash treatment (i.e., colloidal silica and rice husk ash). The produced final materials are characterized in terms of phase analysis and chemical composition. Their reported heavy metal stabilization process appears to be economically and environmentally sustainable.

Ko et al. [100] have reported that the MSW incinerators in Taiwan generate about 300,000 tons of fly ash annually, which is mainly composed of calcium and silicon compounds, and thus have high potential for recycling. They have tried to use a hydrocyclone for reduction of the heavy metals and chloride salts in the MSW incinerator ash in order to make the fly ash nonhazardous.

Their results show that chloride salts can be removed from the fly ash during the hydrocyclone separation process. The presence of a dense medium (quartz sand) is helpful for the removal of chloride salts and separation of the fly ash particles. After the dense-medium hydrocyclone separation process, heavy metals including Pb and Zn have been concentrated in the fine particles so that the rest of the fly ash contain less heavy metal and become both nonhazardous and recyclable [100].

References

1. Municipal Solid Waste in the United States: 2009 Facts and Figures (2011) U.S. Environmental Protection Agency. http://www.epa.gov/wastes/nonhaz/municipal/pubs/msw2009rpt.pdf
2. Lima AT, Ottosen LM, Ribeiro AB (2012) Assessing fly ash treatment: remediation and stabilization of heavy metals. J Environ Manag 95:S110–115
3. Olie K, Vermeulen PL, Hutzinger O (1977) Chlorodibenzo-p-dioxins and chlorodibenzofurans are trace components of fly ash and flue gas of some municipal incinerators in the Netherlands. Chemosphere 6(8):455–459
4. United States Environmental Protection Agency (2011) Toxicity characteristic leaching procedure. http://www.epa.gov/wastes/hazard/testmethods/sw846/pdfs/1311.pdf

5. Li R, Wang L, Yang T, Raninger B (2007) Investigation of MSWI fly ash melting characteristic by DSC-DTA. Waste Manag 27:1383–1392
6. Takeda N, Wang W, He P (2006) Dictionary for special words in waste (Japanese, Chinese, English). Ohmsha (Publisher of Science and Engineering Books), Tokyo
7. Sakai S, Hiraoka M (2000) Municipal solid waste incinerator residue recycling by thermal processes. Waste Manag 20:249–258
8. Lin KL, Chang CT (2006) Leaching characteristics of slag from the melting treatment of municipal solid waste incinerator ash. J Hazard Mater B135:296–302
9. Arvelakis S, Folkedahl B, Frandasen FJ, Hurley J (2008) Studying the melting behaviour of fly ash from the incineration of MSW using viscosity and heated stage XRD data. Fuel 87(10-11):2269–2280
10. Jimbo H (1997) Plasma melting and useful application of molten slag. Waste Manag 16 (5):417–422
11. Yoshiie R, Nishimura M, Moritomi H (2002) Influence of ash composition on heavy metal emissions in ash melting process. Fuel 81:1335–1340
12. Nishigaki M (1997) Reflecting surface-melt furnace and utilization of the slag. Waste Manag 16(5):445–452
13. Nishino J, Umeda J, Suzuki T, Tahara K, Matsuzawa Y, Ueno S, Yoshinari N (2000) DC electric joule-heating system for melting ash produced in municipal waste incinerators. J Jpn Soc Waste Manag Exp 11(3):135–144
14. Mizutani S, van der Sloot HA, Sakai S (2000) Evaluation of treatment of gas cleaning residues from MSWI with chemical agents. Waste Manag 20:233–240
15. Eighmy TT, Crannell BS, Cartledge FK, Emery EF, Oblas D, Krzanowski JE, Shaw EL, Francis CA (1997) Heavy metal stabilization in municipal solid waste combustion dry scrubber residues using soluble phosphate. Environ Sci Technol 31:3330
16. Zhao Y, Song L, Li G (2002) Chemical stabilization of MSW incinerator fly ashes. J Hazard Mater B95:47–63
17. Fuoco R, Ceccarini A, Tassone P, Wei Y, Brongo A, Francesconi S (2005) Innovative stabilization/solidification processes of fly ash from an incinerator plant of urban solid waste. Microchem J 79:29–35
18. Hong K, Tokunaga S, Kajiuchi T (2000) Extraction of heavy metals from MSW incinerator fly ashes by chelating agents. J Hazard Mater B75:57–73
19. Ecke H (2003) Sequestration of metals in carbonated municipal solid waste incineration (MSWI) fly ash. Waste Manag 23:631–640
20. Piantone P, Bodenan F, Derie R, Depelsenaire G (2003) Monitoring the stabilization of municipal solid waste incineration fly ash by phosphation: mineralogical and balance approach. Waste Manag 23:225–243
21. Shyu L, Perez L, Zawacki S, Nancollas G (1983) The solubility of octacalcium phosphate at 37°C in the system Ca(OH)$_2$-H$_3$PO$_4$-KNO$_3$-H$_2$O. J Dent Res 62:398
22. Nancollas GH (1984) The nucleation and growth of phosphate minerals. In: Nriagu JO, Moore P (eds) Phosphate minerals. Springer, Berlin, p 137
23. Kastuura H, Inoue T, Hiraoka M, Sakai S (1996) Full-scale plant study on fly ash treatment by the acid extraction process. Waste Manag 16(5/6):491–499
24. Wilewska-Bien M, Lundberg M, Steenari BM, Theliander H (2007) Treatment process for MSW combustion fly ash laboratory and pilot plant experiments. Waste Manag 27:1213–1224
25. Nagib S, Inoue K (2000) Recovery of lead and zinc from fly ash generated from municipal incineration plants by means of acid and/or alkaline leaching. Hydrometallurgy 56:269–292
26. Zhang F, Itoh H (2006) A novel process utilizing subcritical water and nitrilotriacetic acid to extract hazardous elements from MSW incinerator fly ash. Sci Total Environ 369:273–279
27. Pedersen AJ (2002) Evaluation of assisting agents for electrodialytic removal of Cd, Pb, Zn, Cu and Cr from MSWI fly ash. J Hazard Mater B95:185–198

28. Katsuura H, Inoue T, Hiraoka M, Sakai S (1996) Full-scale plant study on fly ash treatment by the acid extraction process. Waste Manag 16(5/6):491–499
29. Huang S, Chang C, Mui D, Chang F, Lee M, Wang C (2007) Sequential extraction for evaluating the leaching behavior of selected elements in municipal solid waste incineration fly ash. J Hazard Mater 149:180–188
30. Wan X, Wang W, Ye T, Guo Y, Gao X (2006) A study on the chemical and mineralogical characterization of MSWI fly ash using a sequential extraction procedure. J Hazard Mater B134:197–201
31. IAWG (1997) Municipal solid waste incinerator residues. Elsevier Science B.V, Amsterdam
32. Bruder-Hubscher V, Lagarde F, Leroy MJF, Coughanowr C, Enguehard F (2002) Application of a sequential extraction procedure to study the release of elements from municipal solid waste incineration bottom ash. Anal Chim Acta 451:285–295
33. Karlfeldt K, Steenari BM (2007) Assessment of metal mobility in MSW incineration ashes using water as the reagent. Fuel 86:1983–1993
34. Tessier A, Campbell PGC, Blsson M (1979) Sequential extraction procedure for the speciation particulate trace metals. Anal Chem 51(7):844–851
35. Malviya R, Chaudhary R (2006) Factors affecting hazardous waste solidification/stabilization: a review. J Hazard Mater B137:267–276
36. Roy A, Harvill EC, Carteledge FK, Tittlebaum ME (1992) The effect of sodium sulphate on solidification/stabilization of synthetic electroplating sludge in cementitious binders. J Hazard Mater 30:297–316
37. Tay JH (1987) Sludge ash as a filler for Portland cement concrete. J Environ Eng 113:345–351
38. Janusa MA, Heard EG, Bourgeois JC, Kliebert NM, Landry A (2000) Effects of curing temperature on the leachability of lead undergoing solidification/ stabilization with cement. Microchem J 60:193–197
39. Stabilization/solidification of CERCLA and RCRA wastes: physical tests, chemical testing procedures, technology screening and field activities, EPA/625/6-89/022 May 1989
40. Roy A, Carledge FK (1997) Long term behavior of a Portland cement-electroplating sludge waste form in presence of copper nitrate. J Hazard Mater 52:265–286
41. Alba N, Vazquez E, Gasso S, Baldasano JM (2001) Stabilization/ solidification of MSW incineration residues from facilities with different air pollution control systems. Durability of matrices versus carbonation. Waste Manag 21:313–323
42. Lombardi F, Mangialardi T, Piga L, Sirini P (1998) Mechanical and leaching properties of cement solidified hospital solid waste incinerator fly ash. Waste Manag 18:99–106
43. Shimaoka T, Hanashima M (1996) Behavior of stabilized fly ashes in solid waste landfills. Waste Manag 16(5/6):545–554
44. Nzihou A, Sharrock P (2002) Calcium phosphate stabilization of fly ash with chloride extraction. Waste Manag 22:235–239
45. Tsiliyannis CA (1999) Report: comparison of environmental impacts from solid waste treatment and disposal facilities. Waste Manag Res 17:231
46. Andac M, Glasser FP (1998) The effect of test conditions on the leaching of stabilised MSWI-fly ash in Portland cement. Waste Manag 18:309
47. Berardi R, Cioffi R, Santoro L (1997) Matrix stability and leaching behavior in ettringite-based stabilization systems doped with heavy metals. Waste Manag 17(8):535
48. Casa GD, Mangialardi T, Paolini AE, Piga L (2007) Physical-mechanical and environmental properties of sintered municipal incinerator fly ash. Waste Manag 27(2):238–247
49. Mangialardi T, Paolini AE, Polettini A, Sirini P (1999) Optimization of the solidification/ stabilization process of MSW fly ash in cementitious matrices. J Hazard Mater B70:53–70
50. Ubbryaco P, Bruno P, Traini A, Calabrese D (2001) Fly ash reactivity formation of hydrate phase. J Therm Anal Calorim 66:293
51. Mangialardi T (2003) Disposal of MSWI fly ash through a combined washing-immobilization process. J Hazard Mater B98:225–240

52. Wang K, Chiang K, Lin K, Sun C (2001) Effects of a water-extraction process on heavy metal behavior in municipal solid waste incinerator fly ash. Hydrometallurgy 62:73–81
53. Abbas Z, Moghaddam PA, Steenari MB (2003) Release of salts from municipal solid waste combustion residues. Waste Manag 23:291–305
54. Chimenos JM, Fernadez AI, Cervantes A, Miralles L, Fernadez MA, Espiell F (2005) Optimizing the APC residues washing process to minimize the release of chloride and heavy metals. Waste Manag 25:686–693
55. Collivignarelli C, Sorlini S (2002) Reuse of municipal solid wastes incineration fly ashes in concrete mixtures. Waste Manag 22:909–912
56. Bertolini L, Carsana M, Cassago D, Curzio QA, Collepardi M (2004) MSWI ashes as mineral additions in concrete. Cem Concr Res 34:1899–1906
57. Aubert JE, Husson B, Sarramone N (2006) Utilization of municipal solid waste incineration (MSWI) fly ash in blended cement part1: processing and characterization of MSWI fly ash. J Hazard Mater B136:624–631
58. Gao X, Wang W, Ye T, Wang F, Lan Y (2008) Utilization of washed MSWI fly ash as partial cement substitute with addition of dithiocarbamic chelate. J Environ Manag 88:293–299
59. Wunsch P, Greilinger C, Bieniek D, Kettrup A (1996) Investigation of the binding of heavy metals in thermally treated residues from waste incineration. Chemosphere 32 (11):2211–2218
60. Ward DB, Goh YR, Clarkson PJ, Lee PH, Nasserzadeh V, Swithenbank J (2002) A novel energy-efficient process utilizing regenerative burners for the detoxification of fly ash. Process Saf Environ Prot 80(6):315–324
61. Xhrouet C, Nadin C, Pauw ED (2002) Amines compounds as inhibitors of PCDD/Fs de novo formation on sintering process fly ash. Eniron Sci Technol 36:2760–2765
62. Wey M, Liu K, Tsai T, Chou J (2006) Thermal treatment of the fly ash from municipal solid waste incinerator with rotary kiln. J Hazard Mater B137:981–989
63. Mangialardi T (2001) Sintering of MSW fly ash for reuse as a concrete aggregate. J Hazard Mater B87:225–239
64. Vehlow J, Braun H, Horch K, Merz A, Schneider J, Stieglitz L, Vogg H (1990) Semi-technical demonstration of the 3R process. Waste Manag Res 8:461–472
65. Sorensen MA, Mogensen EPB, Lundtorp K, Jensen DL, Christensen TH (2001) High temperature co-treatment of bottom ash and stabilized fly ashes from waste incineration. Waste Manag 21:555–562
66. Introduction of the NEUTREC technology invented by Solvay company. http://www.neutrec.com/library/bysection/result/0,0,-_EN-1000037,00.html
67. Zhu F, Takaoka M, Oshita K, Takeda N (2009) Comparison of two kinds of fly ashes with different alkaline reagent in washing experiments. Waste Manag 29:259–264
68. Ottoboni AP, Souza ID, Menon GJ, Silva RJD (1998) Efficiency of destruction of waste used in the co-incineration in the rotary kilns. Energ Convers Manage 39(16–18):1899–1909
69. Chen G, Lee H, Young KL, Yue PL, Wong A, Tao T, Choi KK (2002) Glass recycling in cement production—an innovative approach. Waste Manag 22:747–753
70. Mujumdar KS, Ranade VV (2006) Simulation of rotary cement kilns using a one-dimensional model. Chem Eng Res Des 84(A3):165–177
71. Kuchar D, Fukuta T, Onyango MS, Matsuda H (2007) Sulfidation treatment of molten incineration fly ashes with Na2S for zinc, lead and copper resource recovery. Chemosphere 67:1518–1525
72. Polettini A, Pomi R, Sirini P, Testa F (2001) Properties of Portland cement- stabilised MSWI fly ashes. J Hazard Mater B88:123–138
73. Bodenan F, Deniard P (2003) Characterization of flue gas cleaning residues from European solid waste incinerators: assessment of various Ca-based sorbent process. Chemosphere 51:335–347
74. Zhu F, Takaoka M, Shiota K, Oshita K, Kitajima Y (2008) Chloride chemical form in various types of fly ash. Environ Sci Technol 42(11):3932–3937

75. Ahn JW, Kim H (2002) Recycling study of Korea's fly ash in municipal solid waste incineration ashes for cement raw material, Feb 17–21 TMS. The Minerals, Metals & Materials Society, Seattle

76. Chiang K, Wang K, Lin F, Chu W (1997) Chloride effects on the speciation and partitioning of heavy metal during the municipal solid waste incineration process. Sci Total Environ 1997 (203):129–140

77. Guo X, Yang X, Li H, Wu C, Chen Y (2001) Release of hydrogen chloride from combustibles in municipal solid waste. Environ Sci Technol 35:2001–2005

78. Cummins EJ, McDonnell KP, Ward SM (2006) Dispersion modelling and measurement of emissions from the co-combustion of meat and bone meal with peat in a fluidised bed. Bioresour Technol 97:903–913

79. Yasuhar A, Katami T, Shibamoto T (2006) Formation of dioxins from combustion of polyvinylidene chloride in a well-controlled incinerator. Chemosphere 62:1899–1906

80. Yasuhara A, Katami T, Shibamoto T (2005) Dioxin formation during combustion of nonchloride plastic, polystyrene and its product. Environ Contam Tox 74:899–903

81. Bolwerk R (2004) Co-processing of waste and energy efficiency by cement plants. In: IPPC conference, Vienna, 21–22 Oct 2004

82. Ahmad S, Al-Kutti WA, Al-Amoudi OSB, Maslehuddin M (2008) Compliance criteria for quality concrete. Constr Build Mater 22:1029–1036

83. Voinitchi D, Julien S, Lorente S (2008) The relation between electrokinetics and chloride transport through cement-based materials. Cem Concr Compos 30:157–166

84. Wang S, Llamazos E, Baxter L, Fonseca F (2008) Durability of biomass fly ash concrete: freezing and thawing and rapid chloride permeability tests. Fuel 87:359–364

85. Zornoza E, Paya J, Garces P (2008) Chloride-induced corrosion of steel embedded in mortars containing fly ash and spent cracking catalyst. Corros Sci 50(6):1567–1575

86. Song H, Lee C, Ann K (2008) Factors influencing chloride transport in concrete structures exposed to marine environments. Cem Concr Compos 30:113–121

87. Glass GK, Buenfeld NR (1997) The presentation of the chloride threshold level for corrosion of steel in concrete. Corros Sci 39:1001–1013

88. Hasson CM, Frolund T, Markussen JB (1985) The effect of chloride type on the corrosion of steel in concrete by chloride salts. Cem Concr Res 15:65–73

89. Farag LM, Abbas M (1995) Practical limits for chlorine cycles in dry process cement plants with precalcining and tertiary air ducting. Zement-Kalk-Gips 48(1):22–26

90. Mangialardi T (2003) Disposal of MSWI fly ash through a combined washing-immobilisation process. J Hazard Mater 98(1–3):225–240

91. Reijnders L (2005) Disposal, uses and treatments of combustion ashes: a review. Resour Conserv Recycl 43(3):313–336

92. Zhang FS, Itoh H (2006) Extraction of metals from municipal solid waste incinerator fly ash by hydrothermal process. J Hazard Mater 136(3):663–670

93. Zhu F, Takaoka M, Oshita K, Kitajima Y, Inada Y, Morisawa S, Tsuno H (2010) Chlorides behavior in raw fly ash washing experiments. J Hazard Mater 2010(178):547–552

94. Zhu F, Takaoka M, Oshita K, Morisawa S, Tsuno H, Kitajima Y (2009) Chloride behavior in washing experiments of two kinds of municipal solid waste incinerator fly ash with different alkaline reagents. J Air Waste Manage Assoc 59:139–147

95. Kirkelund GM, Magro C, Guedes P, Jensen E, Ribeiro AB, Ottosen LM (2015) Electrodialytic removal of heavy metals and chloride from municipal solid waste incineration fly ash and air pollution control residue in suspension: test of a new two compartment experimental cell. Electrochimica Acta, Apr 2015. www.sciencedirect.com

96. Lam CHK, Ip AWM, Barford JP, McKay G (2010) Use of incineration MSW ash: a review. Sustainability 2:1943–1968. doi:10.3390/su2071943

97. Rodella N, Dalipi BR, Zacco A, Borgese L, Depero LE, Bontempi E (2014) Waste silica sources as heavy metal stabilizers for municipal solid waste incineration fly ash. Arab J Chem, King Saud University, Elsevier BV. doi:10.1016/j_arabjc.2014.04.006

98. Zacco A, Borgese L, Gianonelli A, Struis RPWJ, Depero LE, Bontempi E (2014) Review of fly ash inertization treatments and recycling. Environ Chem Lett 12:153–175
99. Boghetich G, Liberti L, Notarnicola M, Palma M, Petruzzelli D (2005) Chloride extraction for quality improvement of municipal solid waste incinerator ash for the concrete industry. Waste Manag Res 23(1):57–61
100. Ko MS, Chen YL, Wei PS (2013) Recycling of municipal solid waste incinerator fly ash by using hydrocyclone separation. Waste Manag 33(3):615–620
101. Shammas NK, Wang LK (2016) Water engineering: hydraulics, distribution and treatment. Wiley, Hoboken, p 805

Chapter 8
Recent Evaluation of Early Radioactive Disposal Practice

Rehab O. Abdel Rahman, Andrey Guskov, Matthew W. Kozak, and Yung-Tse Hung

Contents

1	Introduction	372
2	Marine Disposal	375
	2.1 Early Assessment of Old Practice	377
	2.2 Recent Environmental Assessment of Historical Marine Disposal	379
3	Near-Surface Disposal	384
	3.1 Early Disposal Practices	384
	3.2 Recent Environmental Impact of Old Near-Surface Disposal Practice	385
4	Deep-Well Injection	390
	4.1 Early Assessment of the Old Practice	391
	4.2 Recent Environmental Impact and Assessment of Deep-Well Injection	392
5	Preexisting Cavities and Mine Disposal	393
6	Lessons Learned from Early Disposal Practices	394
References		395

Abstract Developing safe radioactive waste disposal practice is a crucial issue worldwide due to the large amounts of generated wastes of wide chemical, physical, and radiological characteristics that accompany research and application of nuclear

R.O. Abdel Rahman (✉)
Hot Lab. Centre, Atomic Energy Authority of Egypt, Inshas, Cairo 13759, Egypt
e-mail: alaarehab@yahoo.com; karimrehab1@yahoo.com

A. Guskov
Division for Safety of Fuel Cycle Facilities, Scientific and Engineering Centre for Nuclear and Radiation Safety, 11-8 ul. KrasnayaPresnya, Moscow 123242, Russian Federation
e-mail: avguskov@gmail.com

M.W. Kozak
INTERA Inc., 1933 Jadwin Ave, Suite 130, Richland, WA 99354, USA
e-mail: mkozak@intera.com

Y.-T. Hung
Department of Civil and Environmental Engineering, Cleveland State University, 16945 Deerfield Dr, Strongsville, OH 44136-6214, USA
e-mail: yungtsehung@yahoo.com; yungtsehung@gmail.com

© Springer International Publishing Switzerland 2016
L.K. Wang, M.-H.S. Wang, Y.-T. Hung and N.K. Shammas (eds.),
Natural Resources and Control Processes, Handbook of Environmental
Engineering, Volume 17, DOI 10.1007/978-3-319-26800-2_8

sciences and technology in different fields. This work aims to identify learned lessons from early radioactive disposal practice and how these lessons improved the design of disposal facilities. Within this context, early approaches toward safe disposal of radioactive wastes will be summarized. The principals of these approaches and its applications with a special reference to major events in its evolution will be presented. Old regulations and methods for regulating these practices will be addressed, and recent evaluation of these approaches will be given by emphasizing the need for corrective action procedures.

Keywords Radioactive waste disposal • Near surface • Cavities • Marine disposal • Assessment studies

1 Introduction

Throughout the twentieth century, a variety of nuclear activities have been carried out around the world. Initially these activities focused on research to understand radioactivity and the nature of the atom. In the Second World War, activities directed at nuclear weapon production were initiated in Germany, the Soviet Union, the UK, and especially in the USA, where it was known as the Manhattan Project [1]. These activities resulted in a massive increase in the amount of radioactive wastes produced compared to pre-war activities. Following the war, the focus of nuclear activities expanded further to include a large and progressively increasing set of peaceful uses, including nuclear energy and the use of radioactive materials in industry, medicine, and research.

These wide varieties of activities were associated with the generation of radioactive wastes of different chemical, physical, and radiological characteristics which need to be managed in different ways to assure safety of workers, the public, and the environment. Some of these characteristics, which affect the handling requirements, include concentrations and half-lives of radionuclides in the waste, external dose rates, heat generation from decay, and volumes of material to be managed. Many countries have developed national approaches for identifying appropriate management practice for various classes of waste as defined in their national regulations. The International Atomic Energy Agency (IAEA) has issued consolidated guidance on waste classification that represents good practices from national approaches, the most recent of which was issued in 2009 [2]. Within this classification scheme, the wastes are divided into six classes based on their activity and half-life; Fig. 8.1 presented these classes.

In 2007, a study estimated the global inventory of the generated radioactive wastes; it concluded that the estimated volume of the accumulated wastes is $1.9 \times 10^9 \, \text{m}^3$ [3]. Figure 8.2a shows that the majority of the volume of these wastes is mostly generated from mining and milling uranium, whereas the majority of the generated activity is associated with spent nuclear fuel (Fig. 8.2b). These estimates

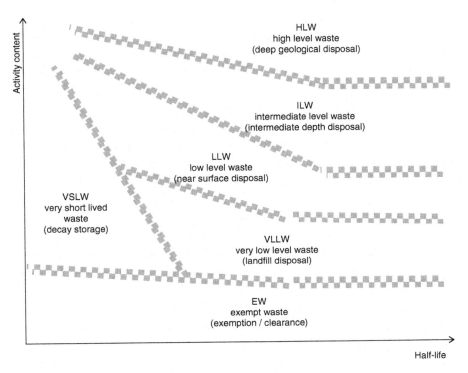

Fig. 8.1 IAEA waste classification scheme (Source: IAEA [2])

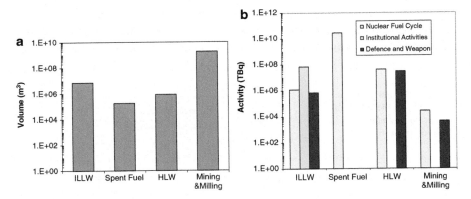

Fig. 8.2 Estimated volumes and activities of radioactive wastes (Derived from [5])

were derived based on the following assumptions: relatively minor amounts of the spent fuel generated by nuclear power plants have been reprocessed, and most of these reprocessed wastes were vitrified, while most of the high-level wastes generated from defense programs were stored as liquid. From that study, it is clearly shown that the estimated volumes and activities of radioactive wastes represent a great challenge for the future of the nuclear industry. This fact was also noted from the examination of the results of the European commission perception on nuclear safety. This perception showed that lack of security to protect nuclear power plant against terrorist attacks and the disposal and management of radioactive waste remain the major threats associated with nuclear energy [4]. Also, it was found that European citizens would like to know more about radioactive waste management and environmental monitoring procedures; and they believe it would be useful to have European legislation regulating nuclear waste management within the European Union and their national territory. That study indicated that if the radioactive waste issues were resolved, the percentage of the public in favor of the nuclear industry will increase with a decrease in the people against this industry (as could be seen from Fig. 8.3).

During early days of the nuclear era, the efforts were focused on the development of nuclear reactor technologies. Long-term management of the associated radioactive waste was not considered a significant problem. Instead, most waste management activities focused their attention on assuring worker safety, with long-term isolation provided either by storage, intended to isolate the wastes, or by disposal methods that were reliant on dilution in the environment to achieve safety. Early disposal practices included the following disposal options for management of radioactive wastes: marine disposal, near-surface disposal, and underground disposal as illustrated in Fig. 8.4.

This work is focused on summarizing early approaches toward the safe disposal of radioactive wastes. Within this context, the principals of early disposal approaches and its applications with a special reference to major events in the evolution of each approach will be summarized. Old regulations and methods for regulating these practices will be addressed, and recent evaluation of these approaches will be given by emphasizing the need for corrective action procedures. Then, concluded summary of the learned lessons will be briefly presented.

Fig. 8.3 Results of public perception on the future of the nuclear industry (Derived from [6])

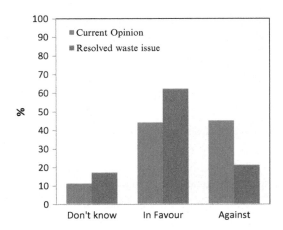

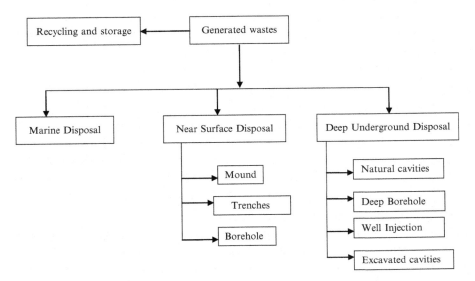

Fig. 8.4 Early disposal options

2 Marine Disposal

Seas have been used to dispose the wastes from human activities. A number of countries used the sea for disposal in the early days of radioactive waste management, in the belief that the sea would rapidly dilute the waste to innocuous levels. The first marine disposal operation took place in 1946 in the USA, about 80 km off the coast of California [5]. The last known dumping operation was in 1993, at the Sea of Japan [6]. Between these two dates, 14 countries have used more than 80 sites to dispose radioactive waste coming from research, medicine, and nuclear industry activities. Three types of radioactive waste were disposed of at seabed, namely, liquid waste, solid waste, and nuclear reactor pressure vessels, with and without fuel. Liquid wastes were disposed either as unpackaged and diluted in surface waters or as contained, but unconditioned, to sea bottom at designated sites. For solid radioactive wastes, two subcategories were disposed at sea; the first is low-level wastes, i.e., paper and textiles from decontamination processes, resins and filters, etc. This subcategory was solidified in cement or bitumen and packaged in metal containers then disposed. The second subcategory included unpackaged large parts of nuclear installations such as steam generators, main circuit pumps, lids of reactor pressure vessels, etc. Finally, reactor vessels containing damaged spent nuclear fuel and reactor vessels without nuclear fuel were disposed at seas. These pressure vessels were usually filled with a polymer-based solidification agent (furfural) to provide an additional protective barrier. In most cases reactor pressure vessels with damaged fuel were further contained in a reactor compartment [6].

The dumping operations were performed under the control of national authorities, and radiological surveys of the sites were carried out from time to time. Samples of seawater, sediments, and deep-sea organisms collected from various disposal sites in the Pacific and North West Atlantic Ocean have rarely shown increase in radionuclide levels above background. However, on several occasions, cesium and plutonium were detected at higher levels in samples taken close to packages at the dumping site [7]. The observed concentrations were considered consistent with safety objectives for marine disposal, but led to increased questions and concerns about the potential dispersion of radionuclides leading to damage to marine resources. This concern had been raised in particular by countries without nuclear energy, which were concerned by the inequity of sea disposal: they could receive the detriment of potentially contaminated seas without receiving the benefits of the nuclear energy. These concerns led to a consensus in 1958, expressed in Article 48 of the Law of the Sea: "every State shall take measures to prevent pollution of the seas from the dumping of radioactive waste, taking into account any standard and regulation which may be formulated by the competent international organizations" [7]. In the 1960s, commercial interest in ocean disposal in USA began to decline and had ended completely by 1970. One of the principal reasons for the decline beside public concern about marine pollution was economics. Ocean disposal was reported to cost $48.75 per 55-gal (200 L) drum compared to $5.15 per drum for burial on land [7]. Table 8.1 illustrates the chronological sequence of major events that lead to the complete ending of sea disposal in 1994 [6, 8].

IAEA has developed an inventory database for the radioactive material disposed in marine environment [6]. This database considers five practices leading to potential dispersion of radionuclides in the sea: seabed and deep seabed disposal, accidents and losses at sea involving radioactive materials, controlled coastal discharges of low-level radioactive liquid effluents, releases from nuclear weapon testing, and accidental releases from land-based nuclear installations [6]. Figure 8.5a-c represents the total activity of radioactive wastes disposed in the oceans and country contribution of these activities, whereas the used method to assess doses is illustrated in Fig. 8.5d [6–11]. Table 8.2 shows the waste types that were disposed in the ocean.

Table 8.1 Chronological sequence of major marine disposal events

Year	Event
1946	First sea dumping operation
1958	First United Nations Conference on the Law of Sea (UNCLOS I)
1972	Adoption of the convention on the Prevention of Marine Pollution by Dumping of Wastes and Other Matter
1985	Resolution calling of a voluntary moratorium on radioactive waste dumping
1993	Resolution on Sea Disposal of Radioactive wastes and other radioactive matter
1994	Total prohibition on radioactive waste disposal at sea came into force

2.1   Early Assessment of Old Practice

The London Convention (LC) prohibited disposal of high-level radioactive wastes in seas and establishes IAEA responsibility to identify which wastes will be prohibited from being dumped. Based on Group of Experts on the Scientific Aspects of Marine Pollution (GESAMP), a modeling methodology was proposed to assess the radiological impact of marine disposal practices [12]. This methodology divides the disposal system into two subsystems, namely, near field (the region in the vicinity of the release, where the concentration is significantly greater than the ocean average) and far field (the rest of the ocean). The source includes the waste form and the package, which consist of the canister and lining. The main processes that lead to the release were identified to include canister corrosion, degradation of the lining and cap, and finally release of radionuclides from the waste form. The potential media of importance were identified as bottom sediment, benthic boundary layer, and open oceans. Table 8.3 lists some recommended models to assess the performance of the marine disposal [12]. The GESAMP methodology led to the recommended constraints on marine disposal shown in Table 8.4.

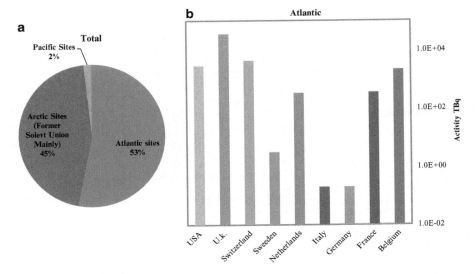

Fig. 8.5 (**a**) percentage distribution of the radioactive wastes disposed in marine sites, (**b**) activity distribution of the disposed wastes in the Atlantic sites, and (**c**) activity of radioactive wastes dumped in Pacific sites, (**d**) Source–pathway–receptor analysis for marine disposal option

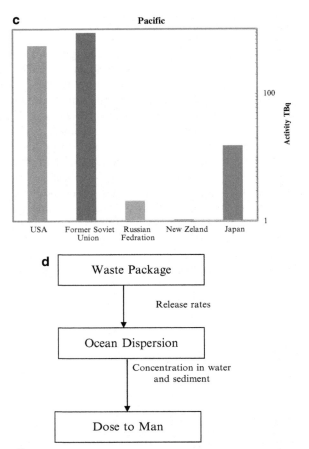

Fig. 8.5 (continued)

Table 8.2 Activity of the dumped wastes in marine disposal (TBq)

Waste type	Atlantic	Pacific	Arctic	% Activity
Reactor with spent fuel	–	–	36,867	43.34
Reactor without spent fuel	1221	166	143	1.80
Low-level solid wastes	44,042.5	820.9	585.4	53.42
Low-level liquid wastes	<0.001	458.5	764.7	1.44
Total	45,263.5	1445.4	38,369.1	

Table 8.3 Recommended GESAMP models to assess the radiological impact of marine disposal

Near field	Far field
Simple finite ocean diffusion model	Contaminant with a long residence time is modeled using well-mixed box
Modified version of the model to account for source size and scavenging	One-dimensional scavenging model
Plume solution if the size of near field exceeds the scale within which diffusion dominates	Simple three-dimensional diffusive model with scavenging
	Medium resolution box model
	Two- and three-dimensional finite difference models

Table 8.4 Summary of the GESAMP-derived limits

Near field	Far field
Dose limits	1 millisievert per year[a]
Dumping period	1000 year[b]
Concentration limit	The concentration limits and limit on mass-dumping rate were set as a cap on the total amount of activity that could ever be dumped per year in a single ocean basin. The mass-dumping rate of 108 kg/year
Dumping mass	1000 tons[c]
Source upper bound	No specific value for a dose upper bound was selected

[a]This limit was consistent with the International Commission on Radiological Protection (ICRP) recommendation at that time
[b]Consistent with the time periods over which the use of nuclear power may be used
[c]Derived based on the average mass dumped in 1978

2.2   Recent Environmental Assessment of Historical Marine Disposal

A number of studies were conducted to estimate the consequences of old disposal practice in oceans. These studies were conducted on international, regional, and national scales. This section will summarize these efforts and their most important findings.

A coordinated research project was initiated within the International Atomic Energy Agency (IAEA). This program was conducted to estimate the average concentration of some radionuclides in surface waters of the Pacific and Indian Oceans [13]. The assessment was conducted by dividing the oceans into 17 regions, which were chosen according to ocean circulation, global fallout patterns, and the location of nuclear weapon test sites. Present levels and time trends in radionuclide concentrations in surface water for each region were studied, and the corresponding "effective half-lives" were estimated. These effective half-lives include both the radiological half-life and the rate of removal of radionuclides by natural transport

Table 8.5 Surface water concentration and effective half-life in the Indian and Pacific Oceans

Element	^{90}Sr	^{137}Cs	$^{293, 240}Pu$
Surface water (µBq/L)	100-150	100-280	0.1-5.2
Effective half-life			
North Pacific (year)	12 ± 1		7 ± 1
South Pacific (year)	20 ± 1		12 ± 4
Equatorial Pacific (year)	21 ± 2		10 ± 2
Indian	N.A.	21 ± 2 years	9 ± 1

Source: Povinec et al. [13]

and chemical processes. The estimated surface water concentrations of ^{90}Sr, ^{137}Cs, and $^{239,240}Pu$ in latitudinal belts of the Pacific and Indian Oceans for the year 2000 were suggested to be used as baseline levels, against which any new contribution from nuclear facilities, nuclear weapon tests, radioactive waste dumping, or possible nuclear accidents can be evaluated. Table 8.5 lists the average values of the surface water concentration and the effective half-life for these radionuclides in the studied oceans.

2.2.1 Arctic Ocean

The joint Russian–Norwegian expeditions, in 1992–1994, visited four principal radioactive waste dumping sites in Kara Sea in the Arctic. Seawater, sediment, and biota samples were collected for activity analysis. The results of these expeditions showed that the influence of the dumped radioactive waste on the general levels of radioactive contamination in Kara Sea was insignificant [14], but the sediment samples taken in the immediate vicinity of waste containers showed elevated levels of Co, Sr, Cs, and Pu. In 2012, other joint expeditions lunched to update the investigations at nuclear submarine K-27 and solid radioactive waste dumps. In- and ex-radiological measurements revealed that no leak indication in the vicinity of the reactor unit K-27 and the activity concentration in seawater, sediment, and biota are lower than those reported in 1990s. The study concluded that despite the current environmental levels of radioactivity are not of concern, there should be continuous monitoring for the sites [15]. The former Soviet Union also disposed of radioactive waste in the Far Eastern Seas, although, unlike in the Arctic, no reactors containing fuel were dumped there. The joint Japanese–Korean–Russian expeditions carried out during 1994 and 1995 took samples of seawater, seabed sediments, and biota from dump sites and from reference sites. The results show that the concentrations of ^{90}Sr, ^{137}Cs, and $^{238,239,240}Pu$ in the Far Eastern Seas were low and were predominantly due to global fallout [16–19].

The Arctic Nuclear Waste Assessment Program was implemented at the end of the last century. This program aimed to assess the impact of the disposed nuclear waste in the Arctic Ocean. Within this project, the major sources of nuclear contamination to the Arctic were identified. These sources include global fallout

Main Arctic process
  1) Ice uptake & movement of radionuclides and sediment;
        • Density-driven currents on Arctic shelves;
        • Sediment dynamics in the Kara Sea;
        • Interactions between colloids and radionuclides in the Arctic river systems;
        • Corrosion and impairment of disposal barrier materials:
  2) Identification of sentinel organisms for the monitoring and evaluation of Arctic radionuclide contamination;
  3) Radionuclide levels, bio-concentration factors, and food chain interaction in Arctic animals;
  4) Deposition of radionuclides due to interactions with phytoplankton; and
  5) Sublethal biological effects from radionuclide contamination

The ANWAP risk assessment addressed the following Russian wastes, media, and receptors:
        • Dumped nuclear submarines and icebreaker in Kara Sea: marine pathways;
        • Solid reactor parts in Sea of Japan and Pacific Ocean: marine pathways;
        • Thermoelectric generator in Sea of Okhotsk: marine pathways;
        • Current known aqueous wastes in Mayak reservoirs and Asanov Marshes: riverline to marine pathways; and
        • Alaska as receptor

For these wastes and source terms addressed, other pathways, such as atmospheric transport, could be considered under future-funded research efforts for impacts to Alaska. The ANWAP risk assessment did not address the following wastes, media, and receptors:
        • Radioactive sources in Alaska (except to add perspective for Russian source terms);
        • Radioactive wastes associated with Russian naval military operations and decommissioning;
        • Russian production reactor and spent-fuel reprocessing facilities nonaqueous source terms;
        • Atmospheric, terrestrial and nonaqueous pathways; and
        • Dose calculations for any circumpolar locality other than Alaska

Fig. 8.6 Main processes and waste sources addressed in the Arctic sea project

arising from anthropogenic sources, ocean dumped or discharged wastes, and disposal and discharge in open fields, pits, landfills, wetlands, and reservoirs [20–23]. The program comprised 80 different research projects covering field surveys, laboratory experiments, modeling studies, and archival data analysis. The main processes and waste sources studied in the project are illustrated in Fig. 8.6. The results of the program are summarized as follows:

- Chernaya Bay on southwestern end of Novaya Zemlya was found to contain localized high-concentration zones [24–28].
- Russian rivers were not introducing radionuclides to the Arctic Ocean in any great quantity [29–31].
- Elevated concentration levels for Cs-137 were found near the mouth of the Yenisey River, but most of the radioactivity is trapped in bottom sediments of the lower river estuaries [32, 33].
- Ob and Yenisey watersheds were found to have considerable capacity to retain any releases with possible exception of radionuclides such as strontium-90 that are closely associated with the aqueous phase [34, 35].
- Discovering ^{137}Cs in sediment-laden sea ice close to Alaska in Chukchi Sea suggested that ice formation processes in Kara Sea have the potential to entrain Cs-rich fine-grained sediments and indicate that some contamination could be transported by ice into the Canadian Basin of the Arctic Ocean or initiate from Chernobyl or similar accidents [36, 37].

- Concerning waters closer to Alaska, anthropogenic radionuclide levels were not due to Soviet-era dumping of nuclear waste but due to atmospheric testing of nuclear weapons [38–40].

A risk assessment study was conducted to evaluate the risk to the biota and humans; in this study [23, 41], the source–pathway–receptor analysis was conducted through which potential sources of release and contributors were identified, the radionuclide transport and deposition were modeled, and the uptake into Arctic fishes and marine mammals was estimated. The assessment identified the sources of radionuclides to include former Soviet Union sources of Kara Sea and the Northwest Pacific and potential sources through river transport from Russian watersheds to the Arctic Ocean. Results of the risk assessment can be summarized as follows:

(a) The identified sources were compared to the already existing fallout levels of key radionuclides, wastes from the Chernobyl incident, releases from the European fuel-reprocessing facilities at Sellafield (UK) and La Hague (France), and naturally occurring radioactivity. Except for localized instances in the Kara Sea near dumped reactors and nuclear testing sites, the existing fallout levels and the Sellafield reprocessing source terms were found to dominate in the Arctic.

(b) Over 95 % of the potential human and ecological risks in Kara Sea, Northwest Pacific, and inland is from ^{137}Cs, ^{239}Pu, ^{241}Am, and ^{90}Sr.

(c) The primary potential risk from the submarine reactor cores in Kara Sea were found to arise from ^{137}Cs, and the primary potential risks from the land-based sources arise from ^{90}Sr.

(d) The estimated maximum total release of radionuclides under the worst-case scenario is summarized in Table 8.6.

Table 8.6 Results of the worst-case release scenario for the Arctic Ocean disposed wastes

Site	Scenario	Duration of max. release (year)	Max. total release (GBq/year)
Kara Sea	Breaching of containment occurs and all of the materials are released instantaneously	At year 2050	>1300
Sea of Japan Pacific Ocean East Coast of Kamchatka	Reactor solid objects are subject to direct corrosion, at a rate of 0.05 mm/year	After 1000 years	≈1 >0.01
Sea of Okhotsk	Radioisotopes in thermoelectric generator will decay before they are released and not be a source of concern		
West Siberian basin	Mayak reservoirs releasing radioactivity to near-surface groundwater		1,400,000 for only 1 year

(e) Dose from worst-case release scenarios was used to assess the potential for radiological effects to marine organisms, including potential detrimental effects on reproductive success in sensitive Alaskan marine species. The predicted concentrations of radionuclides from former Soviet sources are not expected to affect the survival of reproducing populations of marine mammals, fish, and other biota of human dietary importance in Alaska coastal waters. The predicted dose rates were found to be too low to cause any loss of endangered species or any significant ecological impacts.

(f) A worst-case scenario assessment of risk to humans was performed for people in north and northwestern coastal Alaska whose subsistence diet includes fish and marine mammals from the Arctic Ocean. It was found that the largest doses occurring in the Alaskan coastal communities who subside on seafoods came from naturally occurring ^{210}Po, ^{137}Cs, and ^{90}Sr from global fallout. The estimated doses were found to be below background levels and global fallout.

(g) A newly published research modeled the transport of iodine (^{129}I) from Sellafield and La Hague processing plants during 1966–2012 and estimated the values ^{129}I that introduced to the Arctic Ocean to be 5.1 and 16.6 TBq [42].

Elevated cesium concentrations were found in fine-grained sediments entrained in multiyear sea ice floes grounded in Resolute Bay near the center of Northwest Passage through the Canadian Arctic Archipelago. These high-specific activities (1800–2000 Bq kg^{-1} dry weight) are about two orders of magnitude higher than average specific activities detected in previous studies of sea ice-rafted sediments from the Arctic Ocean [43]. The study suggested that the sediments were probably from different sources and were likely mixed during sea ice transport. In 2007, the radionuclide levels were determined for underwater disposal sites in Kara Sea and Oga, Tsivolky, Stepovoy, and Abrosimov Bays. The measurements were carried out in zones both near to and remote from buried solid radioactive waste in the outer and inner parts of the bays. It was found that at the repository of the solid radioactive waste containers in the inner part of the Stepovoy Bay and Abrosimov Bay, the concentration of Cs is higher than background [44].

2.2.2 Pacific Ocean

Between 1946 and 1970, approximately 47,800 large containers of low-level radioactive waste were dumped in the Pacific Ocean west of San Francisco. These containers, mostly 55-gal (200 L) drums, were dumped at three designated sites in the Gulf of Farallones, but many were not dropped on target, probably because of inclement weather and navigational uncertainties. The drums are spread over 1400 km^2 area of seafloor. In 1990, the US Geological Survey (USGS) and the Gulf of Farallones National Marine Sanctuary began a cooperative survey for 200 km^2 of the waste dump using side-scan sonar technique [45]. Radiological surveys of the Northeast Pacific and North West Atlantic Ocean sites are carried out from time to time by the US Environmental Protection Agency and US National

Oceanic and Atmospheric Administration. So far, samples of seawater, sediments, and deep-sea organisms collected near various sites have not shown any excess in the level of radionuclides above background, except in certain instances where isotopes of cesium and plutonium were detected at elevated levels in sediment samples taken close to disposed packages [46].

The long-term benthic infaunal monitoring for the dredged disposal material in northern California was conducted [47]. At this work 135 benthic infaunal samples were collected from San Francisco Deep-Ocean Disposal Site (SF-DODS) over a period from January 1996 to September 2004. The monitoring of the Eastern Pacific deep sea showed that no regional impact or degradation of benthic fauna was detected due to dredged material disposal. Within SF-DODS species, richness and diversity were found to be reduced. The study demonstrated that dredged material disposal at SF-DODS has not caused regional degradation outside of the disposal site nor even at the boundaries of the site. The data clearly indicated that benthic communities in the vicinity of SF-DODS are highly resilient and capable of reworking small amounts of dredged material and recovering rapidly from larger deposits.

2.2.3 Atlantic Ocean

IAEA carried out a project to understand the distribution of radionuclides in the Atlantic [48]. In this project, the high concentrations of ^{137}Cs, ^{99}Tc, and ^{129}I radionuclides were attributed to the discharge from the reprocessing facilities and Chernobyl.

3 Near-Surface Disposal

Near-surface disposal of radioactive waste has been started more than 70 years ago [49]. The first disposal facility, which was in the USA, dates back to the mid-1940s; land repositories followed in many other countries in the 1950s and 1960s (in the UK, India, the former Soviet Union Republics, the Czech Republic, Hungary, Poland, Bulgaria, Norway, South Africa, and others). These disposal facilities were constructed using at-surface designs (mounds) or shallow trenches and then vaults and boreholes.

3.1 Early Disposal Practices

In these practices, safety assessments were not used to derive systematic site selection criteria or design requirements intended to contain and isolate the wastes. Instead, on a site-specific basis, some sites carried out ad hoc studies to show that the disposal would be adequate, while in others rather less evaluation was conducted. In the science of the time, there was a general belief in the capacity of geological systems to provide sufficient retention and dilution to assure safety. This attitude was accentuated for many of the early sites by their remote locations, far from potentially exposed people. Waste safety in the early days often accentuated worker safety far more than concern about potential releases into the environment.

Early trench disposal generally involved clearing and grading the land surface and excavating shallow unlined trenches, generally less than 15 m deep, to be used for waste disposal. The waste was generally placed into trenches on a first-come, first-served basis. In many of the early trenches, disposal resembled municipal landfill disposal, and waste was simply dumped in with minimal packaging or structure and without detailed information about the contents. Trenches were then backfilled using material removed during trench excavation, compacted, and graded to create an earthen mound cap to prevent rainwater ponding and to promote runoff. The principle for this disposal option was the assumption that the nature and rates of natural processes acting on the earthen trench system would be sufficient to slow the movement of radionuclides from the disposal trenches to allow dilution and dispersion and until the wastes decayed to acceptable background levels found in nature [50].

This philosophy gradually began to change, and a perception grew that the use of engineered barriers and more careful disposal practices is a better technical solution for waste disposal. The motivation for some of these changes was practical, to solve operational difficulties with early trenches, and for some changes, the motivation was an increased perception of the waste hazard. Still, radioactive waste often remained unconditioned, and there were sometimes no specific packaging requirements for the waste. It was often packaged in a variety of container types and randomly dumped or stacked into the trenches and vaults, with different approaches taken by different national organizations and sites. The waste was generally placed into the repository on a first-come, first-served basis.

A further evolution of these old facilities came with the introduction of waste conditioning methods. Solid waste would be cemented or encased in bitumen to reduce its leachability, improve its structural strength, and/or minimize surface dose rate. These changes were instituted in different countries at different times, but were generally driven by improvements in technology and worker safety rather than environmental concerns.

3.2 Recent Environmental Impact of Old Near-Surface Disposal Practice

Historical near-surface repositories reflected the understanding of safety at the time they were constructed. Subsequently, ideas about safety evolved, and current regulatory structures and approaches differ from the historical norms. International and national safety requirements were developed based on the practical experience, lessons learned, and scientific and technological progress. Now historical facilities can be assessed using modern concepts in site suitability studies, assessment methodologies and tools, quality assurance systems, and strategies for building confidence. In the mid 1990s, several countries had started to develop formal methodologies for assessing the safety of near-surface disposal facilities for low- and intermediate-level radioactive waste [51]. Internationally, IAEA initiated a series of four projects between the 1990s and 2015 to improve confidence in the results of safety assessment approaches for near-surface disposal facilities [52–57]. IAEA safety assessment methodology for near-surface disposal used in these programs was based on prior national experience developed over several decades [58–60]. The methodology comprises a series of interrelated steps, leading to improved understanding of the system and its uncertainties, namely; assessment context, disposal system description, development and justification of scenarios, formulation and implementation of models, and analysis of the assessment results and building confidence.

Recognizing the problem of legacy waste in some countries initiated different programs to evaluate or upgrade old near-surface disposal facilities. These programs in general aimed to achieve compliance with modified regulatory requirements and are often focused on specific decisions such as repairing existing unsafe conditions, preventing unsafe conditions from developing in the future, permitting continuing operation, applying new technological developments, restarting operations after suspension, and responding to public and stakeholder demands. These upgrade programs may include one or more of the following actions: adopting new waste acceptance criteria and container specifications, building additional engineered barriers, installing hydrogeological cutoff walls, and improving cover systems or partial or complete waste retrieval. These programs include the identification, assessment, and selection of remedial alternatives, as needed, development of action plans, and identification and implementation of appropriate technologies to be used.

IAEA had identified the key steps to perform an upgrade process for corrective action, which includes: definition of the initiating event, identification and assessment of potential corrective actions, planning and implementation of preferred actions, and confirmation of effectiveness [61]. The initiating events are defined as the circumstances at a specific disposal that lead to the need for corrective actions. These events may be categorized as follows:

(a) Changes in regulatory and standards requirement
(b) Detection or prediction of releases that exceed safe limits
(c) Stakeholder concerns

The identification of root causes might be a simple process of evaluating the assembly and analysis of existing information. If the existing information is not sufficient to identify the causes, there will be a need to identify gaps in information, including:

(a) Baseline site characterization
(b) Changes over time to site or repository conditions
(c) Records on the types and amounts of waste emplaced in the repository
(d) Knowledge of the physical and chemical characteristics of waste forms and related degradation
(e) Knowledge of the performance of engineered barriers utilized in the repository
(f) Extent of water ingress and egress
(g) Environmental monitoring data for all relevant media

A wide range of corrective actions may be applied to a specific initiating event, and selection of a specific action is a complex process that involves many factors. An example of a decision-making process is illustrated in Fig. 8.7 [61]. Figures 8.8 and 8.9 illustrate the initiating events related to the change in the regulatory requirements and detection of unsafe releases, along with their corresponding possible actions. Table 8.7 lists the typical methods to retrieve the radioactive wastes from a near-surface repository found to need corrective actions.

Widespread types of old disposal facilities, some of which have been found to need corrective actions, are the Radon-type facilities. These facilities were historically used in the former Soviet Union and some eastern European countries for near-surface disposal of institutional waste of low- and intermediate-activity level. Established in the early 1960s, the Radon system included 35 specialized facilities

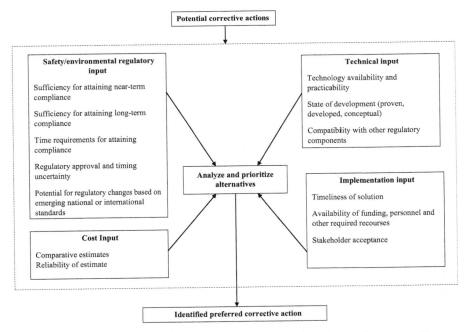

Fig. 8.7 Example of a decision-making approach for selection of corrective actions (Source: IAEA [59])

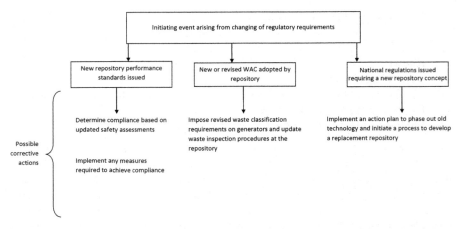

Fig. 8.8 Initiating events and corrective action options for changes in the regulatory requirements

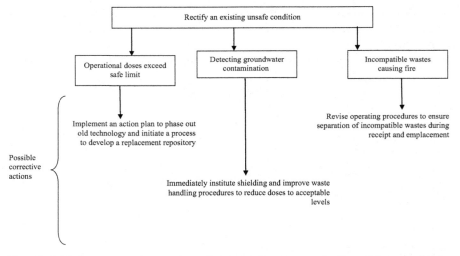

Fig. 8.9 Initiating events and corrective action options for releases of radionuclides exceeded the safe limits

in the Soviet Union with 16 Radon facilities in the territory of Russian Federation, including the two largest facilities – the MosNPO RADON and Leningrad Special Combine. Currently, 14 out of the original 16 Russian facilities are still in operation and have about 10 % of their repositories available for future waste storage.[1] Fifty years' experience in low- and intermediate-level radioactive waste isolation in

[1] Beginning in 2000, new regulatory requirements in the Russian Federation specify that Radon facilities are licensed to "store" rather than "dispose" the waste.

Table 8.7 Method to retrieve the wastes during upgrade of old shallow land disposal

Waste category	Technique	Equipment	Procedure
Loose LLW, low dose rate	Manual	Bucket Small crane	Initial segregation and characterization Waste is packed in containers awaiting for further actions[a]
Waste in intact containers	Manual	Crane Forklift truck	Depending on the condition of the container, it may be over packed awaiting for further action
High dose rate	Remote	Custom designed robotics, remote grapple, shielded tasks	The retrieved wastes placed immediately in shielded casks
In situ conditioned wastes	Manual or remote depending on the dose rate	Cutting equipments such as diamond saws Crane	Measure should be taken to minimize the risks of cutting the waste and minimize exposure to airborne contaminant during cutting and transport
Sand, soil, and gravel backfill		Small diggers, vacuum equipment	Measure should be taken to detect contamination
Liquids		Suitable pump, suitable tanks with shield if necessary	Collected water should await for further actions

[a]Further actions are taken based on the dose rate of the retrieved wastes and may include one or more of the following actions, sending the wastes to treatment plant, conditioning of the wastes, and transferring the waste to storage or disposal

typical near-surface repositories at Russian Radon facilities has shown that a lot of operational and natural factors can influence the natural and engineered barriers of the disposal system and may lead to releases from the facility or other real or perceived difficulties in operation. Examples of issues seen at Radon facilities include biological intrusion, perched water in combination with freeze–thaw cycles, potential erosion, and flooding [62]. Based on the lessons learned, two new types of near-surface facilities were recently constructed and are in operation at the site of MosNPO RADON. Both of them were designed and constructed as storage facilities that can be transformed into final disposal, if the regulatory environment allows such a transition. The first one is the vault constructed above the ground level, whereas the second one is a large-diameter borehole (LDB). Two LDB-type storage facilities with diameter of 1.5 m and depth of 38 m are filled with cemented low- and intermediate-level RAW in retrievable form for about 7 years under continuous monitoring to define future perspectives of their wide implementation. Some corrective actions were performed for historical Radon-type facilities, including re-cementation of unconditioned waste and construction of multilayer cap above the facility (MosNPO RADON), recovering protective properties of the natural barrier in near field, retrieval, conditioning, packaging, and emplacement into existing repositories, those packages that are acceptable for near-surface localization.

4 Deep-Well Injection

Deep-well injection of liquid radioactive waste was based on the practice of deep injection of nonradioactive waste widely used in the middle of the 1950s in the USA and was used for significant volumes of liquid radioactive waste generated associated with the nuclear arms race. Attempts to establish facilities for deep injection of liquid radioactive waste in the USA (Idaho, Oak Ridge, New Mexico) failed, and studies suggested that the geologies of these locations were not favorable for deep injection. Another concept (hydro-fracture grouting) was developed to suit the geology of these areas, which relies on pumping premixed cementitious waste form into underground shale layer. The pumped grout pressure will cause fractures in the shale allowing the cementitious waste to penetrate along the horizontal bedding planes of the shale in layers. The operations of Oak Ridge disposal continue from 1964 to 1984 to dispose low- and intermediate-level radioactive wastes composed of mixture of all kinds of generated liquid wastes including those generated from hot cell, pilot plant, and reactor operations besides organic reagents and solvents [63].

In Soviet Union, a governmental decision was taken in the mid-1950s to start geological survey at four sites to study and establish deep-injection facilities. The principle of this option relies on confining liquid wastes in deep geological formations by injecting them in reservoir horizons [64]. It was proposed that this option will:

1. Obviate the need for surface construction of additional liquid radioactive waste and industrial waste storage sites
2. Reduce environmental contamination resulting from discharging these wastes into lakes and rivers
3. Lead to significant cost savings
4. Allow time for the natural decay of radionuclides and isolate the waste by geochemical reactions of the waste with host rocks

Table 8.8 lists the deep-well injection practice in Russia and its evolution.

Table 8.8 Examples for deep-well injection in Russia

Place	Injection depth (m)	Type of reservoir	Beginning at	Waste volume (10^6 *m^3)
Siberian Chemical Combine	270–320	Sand, sandstone, freshwater	1963	43.5
	314–386			
Krasnoyarsk-26	180–280	Sand, freshwater	1967	6.1 LLW
	355–500			2.25 ILW&HLW
Research Institute of Nuclear Reactors	1130–1410	Limestones, brines	1966	2.5
	1440–1550			

4.1   Early Assessment of the Old Practice

4.1.1   Hydro-Fracture Groutting in Oak Ridge

The performance of the practice was evaluated by constructing experimental injection wells (HF-1 and HF-2) and 24 observation and monitoring wells. Cement grout doped with radioactive tracer was injected into the Pumpkin Valley Shale, and the tracing results indicated the acceptability of the grout performance to isolate radio-contaminate hazards [63]. Two hydro-fracture facilities were installed, and the stabilized radioactive wastes that resulted from the operation of the old facility formed grout sheet at 240–300 m below ground which is estimated to be up to 480–980 m wide and 6×10^{-3} m thick. The new facility produced a stabilized grout sheet about 300–340 m underground which is estimated to be 1200 m in diameter [64].

4.1.2   Deep Liquid Injection in Experimental–Industrial Test Site (Former Soviet Union)

In 1958, the decision was made to undertake research investigations by creating a disposal system to inject drainage water, wash water, and decontamination water [65]. The site was named the Experimental–Industrial Test Site (EITS). Two sand reservoir horizons were used at depths of 270–320 m and 314–386 m. Preliminary geological investigations were conducted near the Siberian Chemical Combine, and these studies confirmed predictions and data from earlier geological investigations. These positive results led to the creation of a deep-well injection facility (at industrial scale) for low-level wastes and intermediate-level wastes. Additional deep-well injection facilities were developed at RIAR at Dimitrovgrad, commissioned in 1967, in the Ulyanoskaya Region, commissioned in 1966–1973, and at the Mining and Chemical Combine (Krasnoyarsk-26) in 1967–1969.

EITS occupies approximately 6.5 km^2, surrounded by an exclusion zone of 52 km^2. It is situated within an ancient erosional depression filled with sand–clay strata reaching 550 m below the ground surface. Interspersed are three aquifers of quartz-feldspar, gravelites, sands, and sandstones. The lower two aquifers, Horizons I and II, at 355–500 m and 180–280 m below ground, are used for injection of intermediate-level waste (ILW)/high-level waste (HLW) and low-level waste (LLW), respectively [65]. The injection system for HLW and ILW consists of eight injection wells, eight relief wells, and 54 monitoring and observation wells. HLWs were injected one to two times per year in batches of 1000–2000 m^3. ILWs were regularly injected from spring to fall at rates of up to 300 m^3 per day. The increase of pressure in Horizon I due to injection operations is relieved by relief wells located approximately 1 km to the south of the injection array. The LLW system consists of four injection wells, four relief wells, and 37 monitoring wells. Low-level wastes were injected from spring to fall at rates of up to 600 m^3 per day.

However, the site continues to hold a mining license allowing disposal of all classes of wastes (LLW, ILW, and HLW). The license is renewed every 5 years.

4.2 Recent Environmental Impact and Assessment of Deep-Well Injection

4.2.1 Hydro-Fracture Grouting in Oak Ridge

A recent hazard assessment document concluded that no credible accident scenario could release the stabilized material inventory due to the physical properties of the grout and the lack of credible energy resource that could impact the material under soil, rock, and shale [63, 66].

4.2.2 Deep Liquid Injection in Experimental–Industrial Test Site (Former Soviet Union)

Russian institutes, such as All-Russian Design and Research Institute of Production Engineering (VNIPIPT), the Institute of Geology of Ore Deposits, Petrography, Mineralogy, and Geochemistry of the Russian Academy of Science (IGEM), and the International Institute for Applied Systems Analysis (IIASA), have assessed the concept and implementation of deep-well injection. These three groups made very different assumptions, and the degree of confidence they have in their results also differs. Despite these differences, there is a remarkable convergence of the results from the three studies, indicating that the existing system of deep-well injection at Krasnoyarsk is functioning as designed. Under the current best understanding of site conditions, there is very little likelihood that the injected wastes would reach the earth's surface at concentrations above standards set for drinking water [67].

Results from mathematical modeling of the deep-well injection of radioactive and nonradioactive wastes, carried out at of SSC RF–NIIAR and Chepetsk mechanical plants, have confirmed the feasibility and safety of deep-injection disposal of toxic and radioactive wastes into permeable aquifers. These results have been used to justify continuing such disposal of wastes and the development of measures for monitoring this method [68]. Another study that was directed to model the distribution of the injected low-level radioactive organic waste was performed [69]. It was found that the distribution of waste does not exceed tens of meters from the injection well and is associated with the transfer of the organic phase into an immobile condition. The presence of the radiation field in the zone of maximum radionuclide accumulation on the rock causes radiological decomposition of the organic contaminants. When the aqueous radioactive wastes are injected, they are more easily transported from the injection zone. Hence, repeated injection of waste will not increase the concentration of organic material within the injection zone. The presence of the waste does not markedly increase the temperature in the injection zone. The maximum temperature is significantly less than boiling point under formation conditions. Thus, this injection technology is regarded by the operators as a radiolysis reactor for treating liquid-organic radioactive waste. The effectiveness and safety of the low-level radioactive organic waste decomposition processes are defined by the radioactivity accumulated in the rock and the waste injection cycle length.

5   Preexisting Cavities and Mine Disposal

Mines and excavated tunnels were used to dispose LLW since the 1940s in several countries. During the conversion of some mines for disposal purposes, it was necessary to reinforce some parts with concrete and to construct drainage systems for any water entering the mine. Limestone, salt, and uranium mines were extensively used in the 1960s; some have since been closed, but others remain in operation [70]. Both solid and liquid LILW were disposed in caverns. Various waste conditioning and closure strategies were followed in different countries. Table 8.9 summarizes this disposal practice worldwide.

As indicated in Table 8.9, in Germany salt mines were selected to host a repository for LLW and ILW radioactive wastes. LLW drums were stacked in old mining chambers by loading vehicles or simply emplaced dumped (tip disposal).

Table 8.9 Summary of some underground disposal

Place	Depth (m)	Type of reservoir	Begin of the practice	Waste volume
Czechoslovakia				
Hostim	30	Limestone mine	At 1940 end 1997	400 m^3
Richard	70–80	Limestone mine waste	1964 in operation	2700 m^3
Bratrstvi	–	Uranium mine	1974 in operation	700 drum
Germany				
Asse	725–750	Salt mine	1967–1978 chamber will be backfilled till 2013	47,000 m^3
Morsleben	400–600	Potash and salt mine	1978 in operation	36,752 m^3
Swedish final repository	50 below Baltic Sea	Metamorphic bedrock	1988	60,000 m^3
Finland				
Olkiluoto	60–100	Crystalline bedrock	1992	8400 m^3
Loviisa	70–100	–	1997	4000 m^3
USA WIPP	655	Rock salt formation	1999	–

Generally, the remaining voids were backfilled by crushed salt or brown coal filter ash. ILW was lowered into inaccessible chambers through a borehole from a loading station above using a remote control. Thirty years ago, the feasibility of both borehole and drift disposal concepts were studied and proved in the Asse mine [71]. The emplacement of HLW has been investigated since 1980, and the investigations included several full-scale in situ tests that were conducted to simulate borehole emplacement of vitrified HLW canisters and the drift emplacement of spent fuel in Pollux casks. Quasi-closed system (QCS) approach was used to study LLW disposal in a German salt mine with a focus on disposed waste forms, geo-engineered barriers, and backfill strategies [72]. The study focused on geochemical tools and a thermodynamic database for modeling highly concentrated salt systems. It was shown that QCS approach provides essential data to study the long-term geochemistry and related radionuclide concentrations to be used in performance assessment and safety analysis.

Richard repository, in the Czech Republic, currently contains about 10^{15} Bq [73] of waste. In 2003, the Czech Radioactive Waste Repository Authority (RAWRA) launched a project that aims to reduce the burden from past practices during the first phase of Richard repository operation and at improving its overall long-term safety. Reviews of the preliminary closure concept and its related safety assessment indicated that the existing concept was deficient in regard to postclosure performance. A decision was made to develop a new concept for the closure of individual waste chambers. The main technological element of this concept is the installation of an additional engineered barrier called a "hydraulic cage" around the waste chambers. The hydraulic cage was designed to decrease the hydraulic gradient and minimize advective flow through the repository.

6   Lessons Learned from Early Disposal Practices

Many of the early approaches for disposing the radioactive wastes were simple, conducted without engineered barriers or with simple engineered barriers. These practices have the following common characteristics:

(a) They were developed before modern national laws, international guidance, or conventions were in place.
(b) They were developed before current regulatory requirements took effect, or they are inconsistent with modern site suitability guidance, technological advances, safety assessment methodologies, or quality assurance systems.
(c) In many cases, unpackaged bulk waste was disposed.
(d) Waste packages, if existed, were often emplaced nonuniformly or through simple tip disposal.
(e) A heterogeneous mixture of waste packages, waste types, waste forms, or waste classes was often disposed in the same facility. Waste items may also

have unexpectedly high-dose rates, or other unexpected features such a pyrophoricity, owing to the lack of standardized waste acceptance criteria.

(f) Poor documentation of the wastes, their radiological and chemical content, their characteristics, the waste forms used, and the location of the wastes in the repository is normal for these facilities.

(g) Unknown or poorly documented information on non-radiological hazards (e.g., asbestos, organic solvents, pathological agents, and toxic chemicals).

Some past practices need action to meet modern regulatory requirements, correct an existing unsafe condition, prevent any unsafe condition from occurring in the future, and respond to societal demands. However, the difficulties in working with such facilities means that worker risks associated with undertaking such activities may be high, and it may be necessary to balance potential risks in the far future to hypothetical members of the public with real risks to workers undertaking remedial actions. Based on these lessons, recent disposal practices have been developed with greater consideration for record keeping, waste isolation, and facility control, all with a view to minimize the potential need to modify repositories in the future as the result of future changes in understanding and regulatory philosophy.

Lessons learned from the operation of early disposal facilities have resulted in development of new designs for radioactive waste disposal facilities that aims to provide adequate isolation of these wastes using containment and confinement strategy. These designs rely on the multi-barrier concept to achieve isolation of the disposed waste for appropriate time taking into account the waste and site characteristics and safety requirements [74–79]. This concept helps in avoiding overreliance on one component of the disposal system (i.e., natural barriers) to provide the necessary safety and allow for certain component to fail without compromising the overall safety of the disposal system [78, 79]. These designs are developed in accordance of holistic and graded approaches, and the process of design development is having iterative nature that allows the designer to modify the design to achieve optimum performance consistence with good engineering practice and meet regulatory requirements [80].

References

1. Rhodes R (1995) The making of the atomic bomb. Simon & Schuster, New York
2. IAEA (2009) Classification of radioactive waste: safety guide, GSG-1.IAEA safety standards series. International Atomic Energy Agency, Vienna
3. IAEA (2007) Estimation of global inventories of radioactive waste and other radioactive materials, IAEA-TECDOC-1591. International Atomic Energy Agency, Vienna
4. Europeans and Nuclear Safety Report (2010) TNS opinion & social. Special eurobarometer 324, Mar 2010, Avenue Herrmann Debroux, 40, 1160 Brussels
5. Calmet DP (1989) Ocean disposal of radioactive waste: status report. A number of studies are being done to more fully assess the impact of sea disposal. IAEA Bullet 4/1989:47–50
6. IAEA (1999) Inventory of radioactive waste disposals at sea, IAEA-TECDOC-1105. International Atomic Energy Agency, Vienna

7. Ryan MT, Lee MP, Larson HJ (1999) History and framework of commercial low-level radioactive waste management in the United States, NUREG-1853. U.S. Nuclear Regulatory Commission/Advisory Committee on Nuclear Waste, Washington, DC

8. Hamblin JD (2006) Hallowed lords of the sea: scientific authority and radioactive waste in the United States, Britain, and France. Osiris 21:209–228

9. Yablokov AV (2001) Radioactive waste disposal in seas adjacent to the territory of the Russian Federation. Mar Pollut Bull 43(1–6):8–18

10. U.S. EPA (1980) Fact sheet on ocean dumping of radioactive waste materials. U.S. Environmental Protection Agency, Washington, DC

11. Dumping of radioactive waste at sea, DSC-Env1, Issue 3.0 – 21 Aug 2006 http://www.mod.uk/ NR/rdonlyres/76CA4EF7-EED5-491E-9519-DC3058BC7C26/0/seadumpradwastebrf.doc

12. Hagen A, Ru B (1986) Deep-sea disposal: scientific bases to control pollution: a status report on the technical work of the IAEA and NEA. IAEA Bullet 28:29–32

13. Povinec PP, Aarkrog A, Buesseler KO, Delfanti R, Hirose K, Hong GH (2005) 90Sr, 137Cs and 239,240Pu concentration surface water time series in the Pacific and Indian Oceans – WOMARS results. J Environ Radioact 81(1):63–87

14. Hirose K, Amano H, Baxter MS, Chaykovskaya E, Chumichev VB, Hong GH, Isogai K, Kim CK, Kim SH, Miyao T, Morimoto T, Nikitin A, Od K, Pettersson HBL, Povinec PP, Seto Y, Tkalin A, Togawa O, Veletova NK (1999) Anthropogenic radionuclides in seawater in the East Sea/Japan Sea: results of the first-stage Japanese-Korean-Russian expedition. J Environ Radioact 43:1–13

15. Gwynn JP, Nikitin A, Shershakov V, Heldal HE, Lind B, Teien HC, Lind OC, Sidhu RS, Bakke G, Kazennov A, Grishin D, Fedorova A, Blinova O, Sværen I, Liebig PL, Salbu B, Wendell CC, Strålberg E, Valetova N, Petrenko G, Katrich I, Logoyda I, Osvath I, Levy I, Bartocci J, Pham MK, Sam A, Nies H, Rudjord AL (2015) Main results of the 2012 joint Norwegian Russian expedition to the dumping sites of the nuclear submarine K-27 and solid radioactive waste in Stepovogo Fjord, Novaya Zemlya. J Environ Radioact 151:1–10. doi:10. 1016/j.jenvrad.2015.02.003

16. Noshkin VE, Wong KM (1980) Plutonium in the North Equatorial Pacific, processes determining the input behavior and fate of radionuclides and trace elements in continental shelf environments. (Abstract workshop, Gaithersburg, 1979), Conf-790382, 11 (SPENCER D, Convener), U.S. Department Energy, Washington, DC

17. Hirose K (1997) Complexation scavenging of plutonium in the ocean, Radionuclides in the Oceans Part 1, Inventories, behaviour and processes. In: Radioprotection-colloques, vol 32 C2 (Germain, P., et al., Eds), p 225

18. Noshkin VE, Wong KM, Jokela TA, Brunk JL, Eagle RJ (1983) Comparative behavior of plutonium and americium in the equatorial Pacific. Lawrence Livermore National Lab, Livermore, UCRL-88812

19. Robison WL, Noshkin VE (1999) Radionuclide characterization and associated dose from long-lived radionuclides in close-in fallout delivered to the marine environment at Bikini and Enewetak Atolls. Sci Total Environ 237/238:311–328

20. Landers DH (ed) (1995) Ecological effects of arctic airborne contaminants. Sci Total Environ 160/161 (xvii–xviii):870

21. Kang DJ, Chung CS, Kim SH, Kim KR, Hong GH (1998) Distribution of 137Cs and 239, 240Pu in the surface waters of the East Sea (Sea of Japan). Mar Pollut Bull 35:305–312

22. Yablokov AV, Karasev VK, Rumyantsev BM, Kokeev ME, Petrov OI, Lystov VN, Yemelyanenkov AF (1993) Facts and problems related to radioactive waste disposal in seas adjacent to the territory of the Russian Federation. Office of the President of the Russian Federation, Moscow

23. Champ MA, Makeyevt VV, Brooks JM, Delaca TE, Van Der Horst KM, Englett MV (1998) Assessment of the impact of nuclear wastes in the Russian arctic. Mar Pollut Bull 35 (7–12):200–221

24. ANWAP (Arctic Nuclear Waste Assessment Program) (1995) Annual report (FY 1993–1994). Office of Naval Research (ONR). ONR 322-95-5. Code 322. Room 207, BCT 3, 8011 N. Quincy Street, Arlington, 22217–5660. 247 pp, plus appendices

25. Hamilton TF, Ballestra S, Baxter MS, Gastaud J, Osvath I, Parsi P, Provinec PP, Scott EM (1994) Radiometric investigations of Kara Sea sediments and preliminary radiological assessment related to dumping of radioactive wastes in the Arctic Seas. J Environ Radioact 25 (1–2):113–134

26. Strand P, Nikitin A, Rudjord AL, Salbu B, Christensen G, Foyn L, Krysbev II, Chumichcv VB, Dahlgaard H, Holm E (1994) Survey of artificial radionuclides in the Barents Sea and the Kara Sea. J Environ Radioact 25(1–2):99–112

27. Forman SL, Polyak L, Smith J, Ellis K, Matishov G, Bordikov Y, lvanov G (1995) Radionuclides in the Barents and Kara Sea bottom sediments, distribution, sources, and dispersal pathways. In: Morgan J, Codispoti L (eds) Department of Defense Nuclear Waste Assessment Program (ANWAP), FY's 1993–94, ONR 322-95-5, Office of Naval Research, pp 67–72

28. Timms SJ, Lynn NM, Mount ME, Sivintsev Y (1995) Modeling the release to the environment in the Kara Sea from radioactive waste in the dumped reactor compartment of the icebreaker Lenin. In: Preller RH, Edson R (eds) Proceedings of the ONR/NRL workshop on modeling the dispersion of nuclear contaminants in the Arctic Seas, Rep NRIJMR/7322-95-7584, Naval Research Laboratory, pp 268–293

29. Trapeznikov A, Aarkrog A, Pozolotina V, Nielsen SP, Polikarpov G, Molchanova I, Karavaeva E, Yushkov P, Trapcznikova V, Kulikov N (1994) Radioactive pollution of the Ob river system from urals nuclear enterprise 'MAJAK'. J Environ Radioact 25(1–2):85–98

30. Baskaran M, Asbill S, Santschi P, Davis T, Brooks JM, Champ MA, Makleyev V, Khlebovich V (1995) Distribution of 239, 241 and 238Pu concentrations in sediments from the Ob and Yenisey Rivers and the Kara Sea. Appl Radiat Isot 46(11):1109–1119

31. Baskaran M, Asbill S, Santschi P, Brooks JM, Champ MA, Adkinson D, Colmer MR, Makeyev VV (1996) Pu, 137Cs, and excess 210Pb in the Russian Arctic sediments. Earth Sci Letter 140:243–257

32. Brooks JM, Champ MA (1995) An overview of GERG's 1993 and 1994 Russian Arctic ANWAP cruises. Arctic Nuclear Waste Assessment Workshop, May 1tt95, Woods ltole Oceanographic Institution

33. Champ MA, Brooks JM, Makeyev VV, Wade TL, Kennicutt MC II, Baskaran M (1995) Preliminary results of studies of industrial and nuclear contaminants in the Ob and Yenisey Rivers and the Kara Sea to assess the environmental and human health risks in the Russian Arctic. In: Kirk CD (ed) Ocean pollution in the Arctic North and tile Russian Far East. American Association for the Advancement of Science, Washington, DC, pp 28–65

34. Paluszkiewicz T, Hibler LF, Richmond MC, Becker P (1995) An assessment of the flux of radionuclide contamination through the Ob and Yenisey Rivers and estuaries to the Kara Sea. Arctic Nuclear Waste Assessment Program (ANWAP) Workshop, Woods Hole Oceanographic Institution, May 1995

35. Beasley TM, Cooper LW, Grebmeier JM (1995) Gamma ray spectroscopy, transuranic radionuclides, iodine-129 and technitium-99. In: Morgan J, Codispoti L (eds) Department of Defense Nuclear Waste Assessment Program, FY's 1993–94, Arlington, AV. ONR 322-95-5, Office of Nawd Research Department, p 2–13

36. Meese D, Tucker WB (1995) Radionuclide contaminants in Central Arctic Sea ice. In: Preller RH, Edson R (eds) Proceedings of the ONR/NRL workshop on modeling the dispersion of nuclear contaminants in the Arctic Seas, Rep NRL/MR/7322-95-7584, Naval Research Laboratory, pp 53–59

37. Cooper LW, Grebmeier JM, Larsen IL, Ravina K, Beasley TM (1995) Radionuclide contamination of the Arctic Basin. In: Morgan J, Codispoti L (eds) Department of Defense Nuclear Waste Assessment Program, FY's 1993–94, ONR 322-95-5, Office of Naval Research, pp 42–51

38. Beasley TM, Cooper LW, Grebmeier JM, Orlandini K and Kelley JM (1995) Fuel reprocessing plutonium in the Canadian Basin. Arctic Nuclear Waste Assessment Workshop, Woods Hole Oceanographic Institution, May 1995
39. Alexander C, Windhom H (1995) Particle bound radionuclide transport to the Bering Sea in the Anadyr River. In: Morgan J, Codispoti L (eds) Department of Defense Nuclear Waste Assessment Program, FY's 1993–94, ONR 322-95-5, Office of Naval Research Department, pp 112-115
40. Radvanyi J (1995)1 Workshop results: Japan-Russia United States study group on dumped nuclear waste in the Sea of Japan, Sea of Okhotsk and the North Pacific Ocean. Arctic Nuclear Waste Assessment Program (ANWAP) workshop, Woods Hole Oceanographic Institution, May 1995
41. Layton D, Edson R, Varela (Engle) M, Napier B (1997) Radionuclides in the Arctic Seas from the former SovietUnion: potential health and ecological risks. Arctic Nuclear Waste Assessment Program (ANWAP), Office of Naval Research
42. Villa M, López-Gutiérrez JM, Suh KS, Min BI, Periáñez R (2015) The behaviour of 129I released from nuclear fuel reprocessing factories in the North Atlantic Ocean and transport to the Arctic assessed from numerical modeling. Mar Pollut Bull 90:15–24
43. Cota GF, Cooper LW, Darby DA, Larsen IL (2006) Unexpectedly high radioactivity burdens in ice-rafted sediments from the Canadian Arctic Archipelago. Sci Total Environ 366 (1):253–261
44. Stepanets O, Borisov A, Ligaev A, Solovjeva G, Travkina A (2007) Radioecological investigations in shallow bays of the Novaya Zemlya Archipelago in 2002–2005. J Environ Radioact 96(1–3):130–137
45. U.S. DOI (2001) Search for containers of radioactive waste on the sea floor beyond the Golden Gate— oceanography, geology, biology, and environmental issues in the Gulf of the Farallones, U.S. Department of the Interior U.S. Geological Survey
46. Ikeuchi Y, Amano H, Aoyama M, Berezhnov VI, Chaykovskaya E, Chumichev VB, Chung CS, Gastaud J, Hirose K, Hong GH, Kim CK, Kim SH, Miyao T, Morimoto T, Nikitin A, Oda K, Pettersson HB, Povinec PP, Tkalin A, Togawa O, Veletova NK (1999) Anthropogenic radionuclides in seawater of the Far Eastern Seas. Sci Total Environ 237/238:203–212
47. Blake JA, Maciolek NJ, Ota AY, Williams IP (2009) Long-term benthic infaunal monitoring at a deep-ocean dredged material disposal site off Northern California. Deep-Sea Res 56:1775–1803
48. IAEA (2005) Worldwide marine radioactivity studies (WOMARS) radionuclide levels in oceans and seas final report of a coordinated research project, IAEA-TECDOC-1429. International Atomic Energy Agency, Vienna
49. IAEA (2005) Upgrading of near surface repositories for radioactive waste. Technical reports series, No 433. International Atomic Energy Agency, Vienna
50. IAEA (2007) Retrieval and conditioning of solid radioactive waste from old facilities, Technical reports series, No. 456. International Atomic Energy Agency, Vienna
51. IAEA (1995) Model inter-comparison using simple hypothetical data (test case 1), First report of NSARS, IAEA-TECDOC-846. International Atomic Energy Agency, Vienna
52. IAEA (2004) Safety assessment methodologies for near surface disposal facilities, results of a coordinated research project, volume 1: review and enhancement of safety assessment approaches and tools, IAEA-ISAM. International Atomic Energy Agency, Vienna
53. IAEA (2004) Safety assessment methodologies for near surface disposal facilities, results of a coordinated research project, volume 2: test cases, IAEA-ISAM. International Atomic Energy Agency, Vienna
54. IAEA (2002) Safety assessment methodologies (ASAM), coordinated research project on application of safety assessment methodologies for near-surface radioactive waste disposal facilities. International Atomic Energy Agency, Vienna
55. Hossain S, Grimwood PD (1997) Results of the IAEA co-ordinated research programme on the safety assessment of near surface radioactive waste disposal facilities. In: Proceedings of an

international symposium on experience in the planning and operation of low level waste disposal facilities, International Atomic Energy Agency, Vienna, 17–21 June 1996, pp 453–467

56. ASAM News No 1, IAEA, Vienna, 2003
57. PRISM: practical illustration and use of the safety case concept in the management of near-surface disposal. http://www-ns.iaea.org/projects/prism/default.asp?s=8&l=67
58. USNRC (2000) A performance assessment methodology for low-level radioactive waste disposal facilities, NUREG-1573. U.S. Nuclear Regulatory Commission, Washington, DC
59. NCRP (2005) Performance assessment of low-level waste disposal facilities, Report of the National Council on Radiation Protection and Measurements (NCRP), report 152. NCRP, Bethesda
60. Kozak MW, Chu MSY, Mattingly PA (1990) A performance assessment methodology for low-level waste facilities. U.S. Nuclear Regulatory Commission, Washington, DC, Sandia National Laboratories
61. IAEA (2005) Upgrading of near surface repositories for radioactive waste, Technical reports series, No 433. International Atomic Energy Agency, Vienna
62. Prozorov L, Guskov A, Tkachenko A, Litinsky Y (2005) The loading and monitoring system in large diameter boreholes: the first experience of borehole application as a LILW repository at the MosNPO "Radon". In: Proceedings – 10th international conference on environmental remediation and radioactive waste management, ICEM'05, pp 1264–1268
63. Abdel Rahman RO, Rakhimov RZ, Rakhimova NR, Ojovan MI (2014) Cementitious materials for nuclear waste immobilisation. Wiley, New York, pp 201–219
64. Gault G (2007) Safety evaluation report for the old and new hydrofracture facilities, hazards assessment document, SER-HAD-GROUT-SBT-03-59, REV 2
65. Compton KL, Novikov V, Parker FL (2000) Deep well injection of liquid radioactive waste at Krasnoyarsk-26. International Institute for Applied Systems Analysis, Laxenburg
66. U.S. DOE (2008) Oak ridge environmental management program, Melton valley remediation completed. DOE, Oak Ridge
67. Rybalchenko AI, Pimenov MK, Kurochkin VM, Kamnev EN, Korotkevich VM, Zubkov AA, Khafizov RR (2005) Deep injection disposal of liquid radioactive waste in Russia, 1963–2002: results and consequences. Dev Water Sci 52:13–19
68. Baydariko EA, Rybalchenko AI, Zinin AI, Zinina GA, Ulyushkin AM, Zagvozkin AL (2005) Deep-well injection modeling of radioactive and nonradioactive wastes from russian nuclear industry activities, with examples from the injection disposal sites of SSC RF–NIIAR and Chepetsk mechanical plants, deep injection disposal of liquid radioactive waste in Russia, 1963–2002: results and consequences. Dev Water Sci 52:501–509
69. Balakhonov BG, Zubkov AA, Matyukha VA, Noskov MD, Istomin AD, Zhiganov AN, Egorov GF (2005) Safety assessment of deep liquid-organic radioactive waste disposal, deep injection disposal of liquid radioactive waste in Russia, 1963–2002: results and consequences. Dev Water Sci 52:481–485
70. Rempe NT (2007) Permanent underground repositories for radioactive waste. Prog Nucl Energy 49:365–374
71. Brewitz W, Rothfuchs T (2007) Concepts and technologies for radioactive waste disposal in rock salt. Acta Montan Slovaca 12(1 Special Issue):67–74
72. Kienzler B, Metz V, Lützenkirchen J, Korthaus E, Fanghänel T (2007) Geochemically based safety assessment. J Nucl Sci Technol 44(3):470–476
73. Haverkamp B, Biurrun E (2006) Closure of the Richard underground repository for low and intermediate level radioactive waste – application of the hydraulic cage concept. ATW – InternationaleZeitschrift fur Kernenergie 51(12):797–802
74. Abdel Rahman RO, El Kamash AM, Zaki AA, El Sourougy MR (2005) Disposal: a last step towards an integrated waste management system in Egypt. International conference on the safety of radioactive waste disposal, IAEA-CN-135/81, IAEA, Tokyo, 3–7 Oct 2005, pp 317–324

75. Abdel Rahman RO, El-Kamash AM, Zaki AA, Abdel- Raouf MW (2005) Planning closure safety assessment for the Egyptian near surface disposal facility. International conference on the safety of radioactive waste disposal, IAEA-CN-135/09, IAEA, Tokyo, 3–7 Oct 2005, pp 35–38
76. Abdel Rahman RO, Zaki AA, El-Kamash AM (2007) Modeling the long-term leaching behavior of (137)Cs, (60)Co, and (152,154) Eu radionuclides from cement clay matrices. J Hazard Mater 145(3):372–380
77. Abdel Rahman RO (2012) Planning and implementing of radioactive waste management system. In: Abdel Rahman RO (ed) Radioactive waste. InTech, Rijeka, pp 3–18. doi:10. 5772/39056, http://www.intechopen.com/books/radioactive-waste/planning-and-implementa tion-of-radioactive-waste-management-system. ISBN 978-953-51-0551-0
78. Abdel Rahman RO, Ibrahim HA, Abdel Monem NM (2009) Long-term performance of zeolite Na A-X blend as backfill material in near surface disposal vault. Chem Eng J 149:143–152
79. Abdel Rahman RO, Kozak MW, Hung YT (2014) Radioactive pollution and control. In: Hung Y-T, Wang LK, Shammas NK (eds) Handbook of environment and waste management. World Scientific Publishing, Singapore, pp 949–1027, http://dx.doi.org/10.1142/9789814449175_0016
80. Abdel Rahman RO (2011) Preliminary evaluation of the technical feasibility of using different soils in waste disposal cover system. Environ Prog Sustainable Energy 30(1):19–28

Chapter 9
Recent Trends in the Evaluation of Cementitious Material in Radioactive Waste Disposal

Rehab O. Abdel Rahman and Michael I. Ojovan

Contents

1   Introduction 404
2   Hydration of Cement 407
    2.1   Hydration Mechanisms 409
    2.2   Hydration Kinetics 410
    2.3   Effect of the Hydration Phases on the Hardened Cement Properties 411
3   Trends in Assessing the Long-Term Behavior of Cementitious Material 412
    3.1   Experimental Assessment 413
    3.2   Predictive Assessment of the Cementitious Material Performance 414
    3.3   International and National Assessment Programs 423
4   Cementitious Material Radioactive waste disposal 425
    4.1   International and National Trends in Regulation and Guidance 425
    4.2   Performance of Cementitious Waste Form 425
    4.3   Container Materials 436
    4.4   Evaluation of Concrete Performance 437
    4.5   Cement-Based Backfill/Buffer 439
5   Summary and Recommendations 441
References 442

Abstract Cementitious materials are essential parts in any radioactive waste disposal facility (either shallow or deep underground facilities). Despite these materials having been extensively used and studied, there is still a need to investigate and understand their long-term behavior due to the fact that disposal is a passive system and regulatory requirements for the safe disposal range from 300 to few thousands of years, so there is a need to ensure that the system will work as

R.O. Abdel Rahman (✉)
Hot Lab. Management Centre, Atomic Energy Authority of Egypt,
Inshas, Cairo 13759, Egypt
e-mail: alaarehab@yahoo.com; karimrehab1@yahoo.com

M.I. Ojovan
Immobilization Science Laboratory, Department of Material Science and Engineering,
University of Sheffield, Mappin Street, Sheffield S1 3JD, UK
e-mail: m.ojovan@sheffield.ac.uk

© Springer International Publishing Switzerland 2016
L.K. Wang, M.-H.S. Wang, Y.-T. Hung and N.K. Shammas (eds.),
Natural Resources and Control Processes, Handbook of Environmental
Engineering, Volume 17, DOI 10.1007/978-3-319-26800-2_9

intended for this period. An extensive array of researches have been devoted to study the feasibility of using cement or cement-based materials in immobilizing and solidifying different radioactive wastes and the feasibility of its potential use as engineering barriers. In this work, the current understanding of cement hydration mechanisms and kinetics will be presented. Experimental and predictive trends to assess the long-term behavior of cementitious material will be reviewed, and recent researches on the utilization of cements or cement-based materials as engineering barriers in radioactive waste disposal will be summarized. Finally this work will examine how these researches contributed to the current knowledge of long-term behavior of cementitious materials in disposal conditions and analyze if this knowledge has provided required inputs to regulation.

Keywords: Radioactive wastes • Disposal • Cement • Engineered barrier • Assessment

Nomenclature

a, b	Tortuosity constant (dimensionless)
A_1	Effective area of concrete surrounding the reinforcement bar (cm^2)
A_i	Radioactivity in the waste form (Bq)
a_i	Specific radioactivity of the solution (Bq/L)
As_1	Area of one reinforcement bar (cm^2)
B	Linear strain caused by one mole of sulfate reduced in 1 m^3 (m^3/mole)
b_f	Fracture aperture (cm)
C	Concentration (mg/L)
Cl	Chloride concentration in bulk solution (mg/L)
D	Diffusion coefficient (cm^2/s)
Dr	Reinforcement bar diameter (cm)
E	Young modulus (20 GPa)
g	Gravitational Acceleration (980 cm/s^2)
h_2	Distance between neutral axis and the lower face (cm)
I	Source/sink term for contaminant (dimensionless)
K	Hydraulic conductivity (cm/s)
k	Reaction rate constant (s^{-1})
K_d	Concentration factor of radionuclide (mL/g)
K_{mp}	Permeability of cement matrix (cm^2)
m	Quantity of sulfate reacted with cement (mole/kg anhydrous cement)
NR_i	Normalized leaching rate of nonradioactive nuclides (g/cm^2day)
q_i	Specific activity of a given radionuclide in a waste form (Bq/g)
R_d	Retardation coefficient (dimensionless)
s	Crack spacing (cm/crack)
S	Surface area of the waste form in contact with water (cm^2)

t	Time (s)
tb	Bottom of the concrete cover over reinforcement (cm)
U_o	Maximum network dissolution velocity (cm/d)
v	Poisson's ratio (dimensionless)
V	Solution volume (L)
W	Moisture source/sink term (dimensionless)
WC	Water-to-cement weight ratio (dimensionless)
w_i	Mass of nuclide in the waste form (g)
w_o	Mass of the waste form (g)
X	Depth of carbonation penetration (cm)
x	Spatial coordinate in the direction of flow (cm)
x_c	Concrete thickness over steel reinforcement (cm)
x_d	Depth of deterioration (cm)
y	Spatial coordinate normal to the direction of flow (cm)

Symbol

ζ	Soil moisture capacity $\zeta = \delta\theta/\delta\psi$
λ	Decay constant (s^{-1})
ρ	Density of water (g/cm^3)
γ	Fracture energy of concrete (J/m^2)
ψ	Water head Pressure (cm)
α	Roughness factor for fracture pain (dimensionless)
ε	Strain in steel reinforcement (dimensionless)
μ	Viscosity (g/cm s)
θ	Volumetric water content (cm^3/cm^3)
ρ_c	Bulk density (g/mL)
θ_c	Total porosity (dimensionless)
α_1, α_t	Longitudinal and transverse dispersivity (cm)

Suffix

Alum	Aluminate in unhydrated cement
c	Contaminant
cc	Critical chloride for initiation of corrosion
ch	Chloride in surrounding material
E	Reacted sulfate as ettringite
F	Final state
i	Intrinsic
ic	Total inorganic carbon in water
ini	Initial critical chloride in concrete pore water
k	Sulfate in kinetic experiment

Mg	Magnesium in bulk solution
o	Initial state
port	Bulk portlandite in concrete solid
s	Sulfate in bulk solution
sat	Saturation in aqueous solution
spall	Characteristic for reaction

1 Introduction

Radioactive waste generation accompanied the application of nuclear technologies in peaceful and defense fields. The development of safe practices for managing these wastes is one of the critical issues for the nuclear industry future. Handling, transportation, treatment, conditioning, storage, and disposal are main elements of the waste management practice, which should be complementary and compatible with each other in order to achieve the overall safety goal [1, 2]. The last step in any radioactive waste management practice is disposal which is designed to ensure radioactive waste isolation under controlled conditions. That should allow the radioactivity to either decay naturally or to slowly disperse to an acceptable level for long period of time. There are different disposal options that range from deep geological disposal to near-surface disposal. Table 9.1 lists the feature and limitation of different disposal options [2]. The choice of a disposal option depends on the waste type and on local conditions, including geological and hydro-geological conditions, radiological performance requirements, and considerations of sociopolitical acceptance [3]. Regardless which option is selected to dispose the radioactive wastes, the design of disposal system should be based on passive safety concept at which there is no reliance on surveillance and maintenance. The isolation of radioactive waste in a disposal facility is achieved by using the multibarrier concept to restrict the release of radionuclides into the environment. These restricting barriers can be either natural or engineered. Figure 9.1 illustrates the multibarrier concept for different disposal options, where the function of these barriers and possible materials that could be used in the design of each barrier are presented in Table 9.2 [4–8].

Cementitious materials are most often incorporated in the design of radioactive waste disposal systems as structural components and engineered barriers to impede water flow and delay radionuclide transport. Hydraulic cements represent important classes of materials that are used in radioactive waste disposal. They are inorganic materials that have the ability to react with water at ambient conditions to form a hardened and water-resistant product. Cementation of radioactive waste has been practiced for many years basically for immobilization of low- and intermediate-level wastes to produce stable waste forms. The prominent advantages of immobilization by cementation are [2]:

- Inexpensive and readily available material
- Simple and low-cost processing at ambient temperature

Table 9.1 Features and limitation of different disposal options

Option	Features	Limitations
Near surface without engineered barrier systems	Excavated trenches covered with a layer of soil. Simple and inexpensive	Suitable only for short-lived and low-level wastes. Erosion, intrusion, and percolation of rainwater may affect the performance
Near surface with engineered barrier system	Multibarrier approach to enhance the safety of disposal. Suitable for most low- and intermediate-level wastes. Long experience of operation	Limited amount of long-lived waste. Erosion, intrusion, and percolation of rainwater may affect the performance
Borehole and cavities at intermediate depth	The depth is adequate to eliminate the risk of erosion, intrusion, and rainwater percolation. Possibility to use existing disused cavities and mines. Simple and not expensive (boreholes)	Geological barriers are site dependent
Geological disposal (including borehole)	Suitable for all waste categories. Enhanced confinement	Site-dependent geological formations. High cost. Complex technology involved. Extensive safety and performance analyses

- Acting as diffusion barrier and providing sorption and reaction sites
- Suitable for sludge, liquors, emulsified organic liquids, and dry solids
- Good thermal, chemical, and physical stability of waste form
- Alkaline chemistry which ensures low solubility for many key radionuclides
- Nonflammability of waste form
- Good self-shielding
- Good waste form compressive strength which facilitates handling
- Easily processed remotely
- Flexible, can be modified for particular waste form

The most common utilized cements are those based on calcium silicates, such as the ordinary portland cements (OPCs). Several OPC-based mixtures were tested and are currently used to improve the characteristics of waste forms and overcome the incompatibility problems associated with the chemical composition of certain types of radioactive waste. These mixtures offer cost reduction, energy saving, and potentially superior long-term performance. Also, cementitious materials have been tested and applied in the design of container, backfill, disposal vaults, and lining cavities. Although these materials have been extensively used and studied, there is still a need to investigate and understand their long-term behavior due to the long

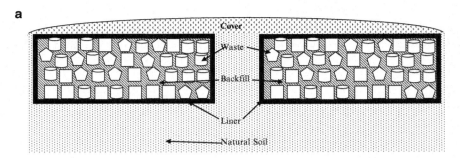

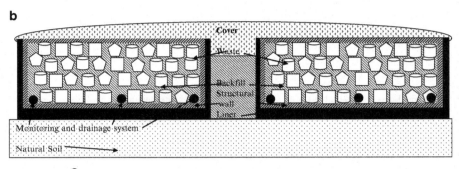

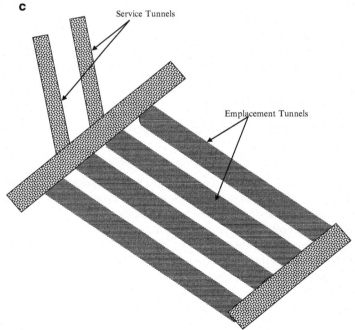

Fig. 9.1 Multibarrier concept for different disposal options. (**a**) Simple engineered barrier in trench facility. (**b**) Multibarrier concept for vault disposal facility. (**c**) Service and emplacement tunnels in geological disposal concept

Table 9.2 Function of each engineered barrier

Barrier	Function	Material
Container	Mechanical strength Limit water ingress Retain radionuclides	Concrete Metal
Waste form	Mechanical strength Limit water ingress Retain radionuclides	Cement Bitumen Polymer
Backfill	Void filling Limit water infiltration Radionuclide sorption Gas control	Cement based Clay based
Structural materials	Physical stability containment barrier	Concrete Steel
Cover	Limit water infiltration Control of gas release Erosion barrier Intrusion barrier	Clay Gravel/cobble Geotextile

Source: IAEA [38]

service lifetime for the disposal systems that extend to hundreds of years in case of short-lived radionuclides and thousands to hundreds of thousands of years for long-lived radionuclides. Moreover the history of cementitious material is relatively short compared to these service lifetimes.

This work will review the current understanding of cement hydration mechanisms and kinetics and the effect of the presence of certain hydration phases on the final hardened cement. Recent researches that have been devoted to study the feasibility of using cements or cement-based materials for immobilization of different radioactive waste types and for designing other engineered barriers will be summarized. Finally this work will examine how these researches contributed to the current knowledge of long-term behavior of cementitious materials in disposal conditions, analyze if this knowledge has provided required inputs to regulation, and highlight areas that still need more investigation to produce more realized assessment for the overall disposal system.

2   Hydration of Cement

Cementitious materials are commonly present in both shallow ground and deep underground radioactive waste disposal facilities. Most of these materials are based on the utilization of OPC because of its commercial availability and low cost. OPC is a hydraulic cement produced by pulverizing clinker and calcium sulfate as an inter-ground addition. It comprises mainly from lime (60–65 wt% CaO), silica (21–24 wt% SiO_2), alumina (3–8 wt% Al_2O_3), ferric oxide (3–8 wt% Fe_2O_3),

Table 9.3 Unhydrated and hydrated phases of portland cement

Phase	Name	Abbreviation	Composition
(a) Unhydrated phases of portland cement			
Major unhydrated phases			
Alite	Tricalcium silicate	C_3S	$3CaO.SiO_2$
Belite	Dicalcium silicate	C_2S	$2CaO.SiO_2$
Aluminate	Tricalcium aluminate	C_3A	$3CaO.Al_2O_3$
Ferrite	Tetracalcium aluminoferrite	C_4AF	$4CaO.Al_2O_3.Fe_2O_3$
Minor unhydrated phases			
Periclase			MgO
Free lime			CaO
Anhydrite			$CaSO_4$
Gypsum	Calcium sulfate dihydrate		$CaSO_4 \cdot H_2O$
(b) Hydrated phases of portland cement			
	Calcium silicate hydrate	C–S–H	Ca/Si molar ratio $1.7 + \_0.1$
		Tobermorite	$Ca_4(Si_3O_9H)2]Ca{-}_8H_2O$
		Jennite	$[Ca_8(Si_3O_9H)_2(OH)_8]Ca{-}_6H_2O$
		Afwillite	$3CaO \cdot 2SiO_2 \cdot 3H_2O$, $Ca_3Si_2O_4(OH)_6$
Ettringite	Calcium trisulfoaluminate hydrate	Aft	$3CaO.Al_2O_3.3CaSO_4.32H_2O$
Monosulfate	Calcium monosulfoaluminate hydrate	AFm	$3CaO.Al_2O_3.CaSO_4.12H_2O$
	Tetracalcium aluminate hydrate	C_4AH_{13}	$3CaO.Al_2O_3.Ca(OH)_2.12H_2O$
Portlandite	Calcium hydroxide	CH	$Ca(OH)_2$
Hydrogarnet		$C_3AH_6{-}C_3ASH_4$	$Ca_3Al_2(OH)_{12}{-}Ca_3Al_2Si(OH)_8$
	Dicalcium aluminate hydrate	C_2AH_8	Not precisely determined but may $Ca_2Al(OH)_6Al(OH)_4ânH_2O$ or $Ca_2Al(OH)_6Al(OH)_3(H_2O)_3{-}OH$
Brucite	Magnesium hydroxide		$Mg(OH)_2$

Sources: Ojovan and Lee [2], Abdel Rahman et al. [3], and Mindess et al. [11] (original table based on these references)

minor amounts of magnesia (0–2 wt % MgO), and sulfur trioxide (1 – 4 wt % SO_3) and other oxides introduced as impurities from the raw materials used in its manufacture. The major unhydrated phases of OPC are alite, belite, aluminate, and ferrite. Other phases such as periclase, free lime, anhydrite, or gypsum in addition to additives that slow OPC setting and alkali sulfates (Na_2SO_4 and K_2SO_4) might exist usually less than 1 mass %. Table 9.3a shows the compositions and abbreviations of unhydrated OPC [2, 9–12].

2.1 Hydration Mechanisms

Mixing OPC powder with water initiates series of chemical reactions that lead to hardening. Each phase of the major cement unhydrated phases (see Table 9.3a) hydrates in the presence of water to produce new hydrated phases (see Table 9.3b). Alite and belite hydrate to form amorphous calcium silicate hydrate (C—S—H) gel and portlandite. Although aluminate and ferrite phases comprise less than 20 % of the bulk of cement, their reactions with water are very important and dramatically affect the hydration of the calcium silicate phases because the hydration of aluminate is very fast compared to alite. In the absence of additives, aluminate reacts with water to form two intermediate phases, namely, C_2AH_8 and C_4AH_{13} that spontaneously transform into the fully hydrated, thermodynamically stable hydrogarnet phase [9]. In hydrogarnet, aluminum exists as highly symmetrical octahedral Al $(OH)_6$ units; if the very rapid and exothermic hydration of aluminate is allowed to proceed in unhindered cement, then the setting occurs too quickly, and the cement does not develop strength. In the presence of gypsum, aluminate and ferrite form ettringite phase. Ferrite hydration is much slower than that of aluminate, and water is observed to bead up on the surface of ferrite particles. This may be due to the fact that iron is not as free to migrate in the pastes as aluminum, which may cause the formation of a less permeable iron-rich layer at the surface of the ferrite particles and isolated regions of iron hydroxide. In cement, if there is insufficient gypsum to convert all of the ferrite to ettringite, then an iron-rich gel will form at the surface of the silicate particles which is proposed to slow down their hydration. Ettringite formation is more favorable due to the presence of higher ratio of SO_4^{2-} than Al $(OH)_4^-$; by the end of the hydration process, the release of SO_4^{2-} is almost completed where the continued supply of $Al(OH)_4^-$ will reduce the ratio between the two ions leading to the dissolution of ettringite to monosulfate [11]. The abovementioned hydration reactions can be summarized as follows [13–16]:

$$2(C_3S) + 6H = 3C - S - H + 3C - H + 114\,KJ/mole \quad (9.1)$$

$$2(C_2S)^+ + 4H = 3C - S - H + C - H + 43\,KJ/mole \quad (9.2)$$

$$C_3A + 30H + 3CS_2H = AFt + 200\,KJ/mole \quad (9.3)$$

$$C_4AF + 10H + 2C - H = C_3F_6H + C_3A_6H + 100\,KJ/mole \quad (9.4)$$

Each of the produced phases formed during cement hydration influences the structure and the properties of the final hardened cement. The structure of the C—S—H phase consists of nanoscale solid particles (tobermorite and calcium hydroxide or a fine-scale mixture of tobermorite, jennite, and afwillite) that are packed, or agglomerated, into randomly oriented structures containing internal gel pores that forms in a standard cement paste [12, 17]. The setting and hardening behavior, strength, and dimensional stability of the hardened cement depend primarily on this phase. Development of the microstructure of hydrated cement occurs after the paste has set and continues for months (and even years) after placement.

OPC is typically 95–98 % hydrated after 12 months and comprises an aqueous phase, which is largely confined to filling pores less than 1 mm in radius (pore water) and a heterogeneous paste matrix.

2.2 Hydration Kinetics

The hydration of cement is more complex than the sum of the hydration reactions of its individual phases [11, 17, 18]. The typical depiction of a cement grain involves larger silicate particles surrounded by the much smaller aluminate and ferrite particles. The more reactive aluminate and ferrite phases react first, and these reactions dramatically affect the silicate-phase hydration. Cement hydration can be broken down into three distinct periods, namely, dormant, setting, and hardening. In the dormant period, the first few minutes of hydration, the aluminate and ferrite phases react with gypsum to form an amorphous gel at the surface of the cement grains, and short rods of ettringite grow (Fig. 9.2b). In the setting period that occurs over hours, the buildup in the ettringite and portlandite is continued, and C—S—H started to form. During the period from 3 to 24 h, about 30 % of cement reacts to form portlandite and C—S—H. The development of C—S—H in this period occurs in two phases. In the first, within 10 h, alite reacts to produce outer C—S—H

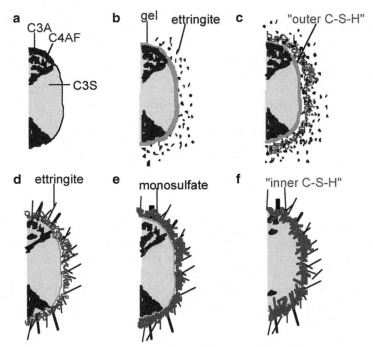

Fig. 9.2 Kinetics of the hydration reaction (*Source*: Barron [15])

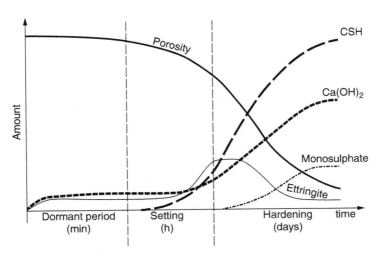

Fig. 9.3 Timescale for different hydration periods (*Source*: Ojovan and Lee [2])

which grows out from the ettringite rods rather than directly out from the surface of the alite particles (Fig. 9.2c). Therefore, in the initial phase of the reaction, the silicate ions must migrate through the aluminum and iron-rich phase to form the C—S—H. In the second, within 18 h of hydration, aluminate continues to react with gypsum, forming longer ettringite rods (Fig. 9.2d). This network of ettringite and C—S—H appears to form hydrating shell about 1 μm from the surface of anhydrous alite. A small amount of inner C—S—H forms inside this shell. Finally, in the hardening period, reactions slow down and the deceleratory period begins (Fig. 9.2e). Aluminate reacts with ettringite to form some monosulfate. Inner C—S—H continues to grow near the alite surface, narrowing the 1 μm gap between the hydrating shell and anhydrous alite. The rate of hydration is likely dependent on the diffusion rate of water or ions to the anhydrous surface. After 2 weeks (Fig. 9.2f), the gap between the hydrating shell and the grain is completely filled with C—S—H, and the original outer C—S—H becomes more fibrous [16]. The timescale for each period and the development of hydrated phases in each time period are dependent on the hydration conditions, cement composition, and the presence of additives; an example for hydrated phase development as a function of time is illustrated in Fig. 9.3 [2].

2.3 Effect of the Hydration Phases on the Hardened Cement Properties

Cement composition and fineness play a major role in controlling final hydrated cement properties. The average finenesses of cement range from 3000 to 5000 cm^2/g. Greater fineness increases the surface available for hydration, causing greater early

Table 9.4 ASTM specification for the mineral composition of different cement types

Type/designation according to C150	Special characteristics/intended use	Composition			
		C_3S	C_2S	C_3A	C_4AF
I	Normal, general purpose	50	24	11	8
II	Moderate sulfate resistance/moderate heat of hydration	42	33	5	13
III	High early strength	60	13	9	8
IV	Low heat of hydration	26	50	5	12
V	High-sulfate resistance	40	40	4	9

Source: Abdel Rahman et al. [9]

strength and more rapid heat generation. Coarse cement produces pastes with higher porosity than those produced by finer cement. The microstructure of the cement hydrates determines the mechanical behavior and durability of the resulting concrete.

A hardened cement paste is a heterogeneous multiphase system. At room temperature, a fully hydrated portland cement paste may consist of 50–60 % C—S—H, 20–25 % portlandite, 15–20 % ettringite, and monosulfate by volume. The minor hydration products form in small quantities depending on the composition of the cementing material and hydration conditions. Alite and belite have the most influence on long-term structure development. Aluminates contribute to the formation of ettringite, which can lead to fracture in concrete. Cements high in alite hydrate more rapidly and develop higher early strength. However, the hydration products formed make it more difficult for late hydration to proceed resulting in a lower ultimate strength. Cements high in belite hydrate much more slowly, at 20 °C taking approximately 1 year to reach a good ultimate strength, leading to a denser ultimate structure and higher long-term strength [2].

Early hydration of cement is principally controlled by the amount and activity of alite, balanced by the amount and type of sulfate inter-ground with the cement. Alite hydrates rapidly and influences the early bonding characteristics. Abnormal hydration of alite and poor control of its hydration by sulfate can lead to problems such as flash set, slump loss, and cement–admixture incompatibility. Based on this information, a number of cements were designed with different disabilities or high early strengths. The five recognized portland cements are produced by adjusting the proportions of their minerals and finenesses as shown in Table 9.4; [2, 9].

3   Trends in Assessing the Long-Term Behavior of Cementitious Material

Various investigations were carried out to study the microscale to the macro (structure) scale of cementitious materials from science and engineering points of view. Research efforts in different countries were focused on both practical and predictive aspects of using cementitious barriers for radioactive waste disposal.

3.1   Experimental Assessment

The experimental research efforts could be categorized to laboratory, field, and natural analogue studies. Laboratory tests are performed under controlled conditions according to internationally approved methods and standards. These tests are limited to well-defined, controlled conditions (small sample sizes, ambient laboratory conditions, prescribed leachate replacement schedules, etc.) that do not represent actual repository conditions. Usually these tests are designed to provide conservative estimates of contaminant release. The main advantage of standardized laboratory tests is that variability between tests conducted at different institutions is controlled and, therefore, experimental uncertainties are small. This allows the intercomparison of results.

Field or in-situ tests are being carried out as a complement to the laboratory tests. Field tests generally provide data that represents more closely the actual or anticipated repository conditions, including interaction with the host media. Field testing can also be used to validate assumptions concerning the degree of conservatism in laboratory tests and provide data for more realistic estimates of contaminant release. Field tests are generally tests conducted with samples that have real dimensions and under field conditions. The time frame for field tests is on the order of few years or even few tens of years. Parameters measured in these tests are expected to provide data that are more realistic for source-term model validation. A potential shortcoming of the field tests is that there are no standard procedures because, by definition, they are site specific. Field tests also have limitations, perhaps the most serious being the duration of the testing period and the high costs, which limit the number of tests. Therefore, they cannot provide an absolute level of confidence.

Archaeological and natural analogues can be used to enhance the level of confidence in field and laboratory data, by providing some long-term validation for important processes affecting cementitious material behavior. Natural and archaeological analogue studies provide information on long-term stability and durability. Analogue studies are improving the impact of chemical process understanding (e.g., sorption, solubility, pH, Eh, etc.) on contaminant transport. Due to the limited knowledge of conditions, data from natural analogue studies are often difficult to use in mathematical models of performance.

However, experimental data often provide confirmatory evidence and provide an independent method of demonstrating the long-term performance. The performance of the cementitious barrier is dependent on the functional requirements of this barrier. Generally these barriers are designed to retain radionuclides, reduce water ingress, and increase the structural stability of the overall disposal system. The parameters that are used to evaluate the performance of the barrier are listed in Table 9.5 [19].

Table 9.5 Parameter that is used to assess the performance of cementitious barriers

Type of tests	Properties	Characteristics and parameters
Laboratory tests	Durability	Free water content Compressive strength Tensile strength Radiation stability Thermal stability Porosity Permeability Corrosion rate Biodegradation

	Radionuclide release – liquid phase	Release rate parameters
		  Leaching rate
		  Diffusion coefficient
		  Dissolution rate
		Chemical parameters
		  Sorption coefficient
		  Solubility
		  Solution chemistry (pH, Eh, major ions, etc.)

	Radionuclide release – gas phase	Release rate mechanisms Radiolysis Biodegradation Corrosion
Field tests	Durability	Biodegradation Corrosion
	Radionuclide release and transport	Leaching rate Diffusion coefficient Sorption coefficient
Archaeological and natural analogues	Durability	Thermal stability Corrosion rate
	Radionuclide release and transport	Leaching rate Diffusion coefficient Sorption coefficient Colloidal formation Solubility

Source: IAEA [19]

3.2 Predictive Assessment of the Cementitious Material Performance

Radiological safety implications of disposed radioactive wastes are usually assessed using predictive models. The modeling efforts can be divided to process modeling and to assessment modeling.

3.2.1   Process Models

Process models are used in conjunction with laboratory, field, and analogue experiments to understand, quantify, and rank which are the most important processes that take place in a disposal barrier, i.e., waste matrix, engineering barrier, geosphere, or biosphere. The experimental results could be extrapolated by using semiempirical equations or by using the fundamental mechanistic equation that govern the performance of the barrier [20].

3.2.1.1   Degradation Process

Carbonation and sulfate, magnesium, and chloride attacks are main chemical processes that affect the performance of the cementitious barrier. Carbonation is the process where carbon dioxide enters cement and reacts with calcium hydroxide to form calcium carbonate. Under dry unsaturated conditions, there is insufficient water to drive the carbonation reaction, while under saturated conditions the reaction is slowed by the reduced rate of diffusion of carbon dioxide into the concrete. Carbonation increases the strength of cement, except for high-sulfate cement and, in the case of portland cement pastes, reduces the permeability and increases the hardness [21]. The shrinking caused by carbonation may cause cracking or joint separation, and the reduced pH may enhance the mobility of some radionuclides. Carbonation can also depassivate steel reinforcement allowing corrosion. The depth of penetration of carbonation X (cm) could be given by

$$X = \left(2D \frac{C_{ic}}{C_{port}} t \right)^{0.5} \tag{9.5}$$

where:

D = diffusion coefficient of Ca in concrete (cm^2/s)
C_{port} = bulk concentration of Ca(OH)$_2$ in concrete solid (mol/cm^3)
C_{ic} = total inorganic carbon concentration in groundwater or soil moisture (mol/cm^3)
t = time (s)

Sulfate and magnesium ions present in groundwater; they can migrate into cementitious barrier and react with it. The resulting reaction products displace more space in the cement than the reactants, causing a physical disruption of the structure of cement. Sulfate reacts with aluminum phases to form calcium aluminum sulfates such as ettringite and higher sulfate concentrations gypsum, while magnesium ions can migrate into concrete and react to form brucite [21]. Empirical models were developed to represent the depth of deterioration due to these reactions. These models should be used with caution, because if they are extrapolated beyond the range of experimental conditions on which they are based, unreliable and nonconservative results may be obtained. The following empirical model could be used to calculate the depth of deterioration due to sulfate and manganese attack:

$$x_d = 1.86 * 10^6 C_{\text{alum}} (C_{\text{Mg}} + C_s) D_i t \qquad (9.6)$$

where:

x_d = depth of deterioration (cm)
C_{alum} = weight percent of aluminate in unhydrated cement (%)
C_{Mg} = concentration of magnesium ions in bulk solution (mole/L)
C_s = sulfate concentration in bulk solution (mole/L)
t = time (s)
D_i = intrinsic diffusion coefficient in cement (cm^2/s)

Mechanistic models may provide more defensible results; Atkinson and Hearne described the kinetics of the sulfate attack as follows [22]:

$$
\begin{aligned}
m &= m_c \log \left(\frac{t_{\text{spall}} C_s}{t_r C_k} \right) \quad m \prec m_c \\
m &= m_c \qquad m \succ m_c \\
C_E &= m \left(\frac{\text{mass}}{\text{volume}} \frac{\text{cement}}{\text{concrete}} \right) \\
X_{\text{spall}} &= \frac{2\alpha\gamma(1-v)}{E(BC_E)^2} \qquad t_{\text{spall}} = \frac{X^2 C_E}{2 D_i C_s} \\
R &= \frac{EB^2 C_E C_s D_i}{\alpha\gamma(1-v)}
\end{aligned}
\qquad (9.7)
$$

where:

C_k = sulfate concentration in kinetic experiment (mole/m^3)
C_s = sulfate concentration in bulk solution (mole/m^3)
C_E = concentration of reacted sulfate as ettringite (mole/m^3)
E = Young modulus (20 GPa)
m = quantity of sulfate reacted with cement (mole/kg anhydrous cement)
m_c = value of m at the end of reaction (mole/kg anhydrous cement)
t_{spall} = characteristic time for reaction(s)
R = degradation rate (m/s)
α = roughness factor for fracture pain (dimensionless)
B = linear strain caused by one mole of sulfate reduced in 1 m^3 (m^3/mole)
γ = fracture energy of concrete (J/m^2)
v = Poisson's ratio (dimensionless)
D_i = intrinsic diffusion coefficient in cement (cm^2/s),

The initial alkaline environment of intact concrete protects steel reinforcement from corrosion. As concrete ages, however, corrosive agents such as chloride and oxygen may penetrate the concrete and reach steel reinforcement. Steel expands as it corrodes, causing the surrounding concrete to crack. Continued corrosion of the steel may lead to structural instability. The time to the initiation of chloride attack of steel reinforcement has been estimated from an empirical model of the form [21]

$$t = \frac{129 x_c^{1.22}}{\text{WCCl}^{0.42}} \tag{9.8}$$

where:

x_c = concrete thickness over the steel reinforcement in (cm)
WC = water-to-cement weight ratio (dimensionless)
Cl = chloride concentration in the bulk solution (mg/L)

When the critical chloride level is reached at the depth of the reinforcement, breakup of the passive layer is assumed to initate. So by solving the mass transport equation for the arrival time of chloride, the time to the initiation of chloride attack could be estimated as follows [22]:

$$\frac{C_{cc} - C_{ini}}{C_{ch} - C_{ini}} = \text{erfc} \left(\frac{X}{\left(\frac{4Dt}{R_d} \right)^{0.5}} \right) \tag{9.9}$$

where:

C_{cc} = critical chloride concentration for initiation on corrosion (mg/L)
C_{ini} = initial critical chloride concentration in concrete pore water (mg/L)
C_{ch} = chloride concentration in surrounding material (mg/L)
X = depth of cover over reinforcement (cm)
erfc = complementary error function
D = diffusion coefficient (cm^2/s)
R_d = retardation coefficient (dimensionless)

This equation is solved iteratively to obtain the time when the critical threshold is reached at the level of reinforcement. Then the rate of corrosion and the percent remaining could be calculated using

$$R_{corrosion} \cong 9.4 D_i \left(\frac{C_{ch}}{\Delta X} \right)$$
$$\% remaining = 100 \left(1 - \frac{4 sr R_{corrosion}}{\pi d_r^2} \right) \tag{9.10}$$

D_r = reinforcement bar diameter (cm)
sr = space between reinforcement (cm)

3.2.1.2 Hydraulic Process

The permeability of cementitious barrier is strongly affected by the formation of cracks, particularly microcracks. Microcracks are induced due to physical loading, drying shrinkage, and freeze–thaw cycles. The permeability of cracked slab (evenly spaced cracks of constant aperture) is given by [22]

$$K = \frac{\rho g b_f^2}{24 \mu s} + K_{mp} \tag{9.11}$$

where:

$K =$ hydraulic conductivity (cm/s)
$\rho =$ density of water (g/cm^3)
$g =$ Gravitational acceleration (980 cm/s^2)
$b_f =$ fracture aperture (cm)
$\mu =$ viscosity of water (0.01 g/cm s)
$K_{mp} =$ permeability of cement matrix (cm^2)
$s =$ the crack spacing (cm/crack)

$$\frac{s}{D_r} = \left(25.7 \left(\frac{t_b}{h_2} \right)^{4.5} + 1.66 \left(\frac{A_1}{A_{S1}} \right)^{0.333} \right) \left(\frac{\varepsilon^2}{0.236 \times 10^{-6}} \right)$$

$t_b =$ bottom of concrete cover over reinforcement (cm)
$h_2 =$ distance between neutral axis and the lower face (cm)
$A_1 =$ effective area of concrete surrounding the reinforcement bar (cm^2)
$A_{S1} =$ area of one reinforcement bar (cm^2)
$\varepsilon =$ strain in steel reinforcement (dimensionless)
$D_r =$ reinforcement bar diameter (cm)

3.2.1.3 Radionuclide Leaching from Cementitious Waste Form

The mechanisms, those that instigate the leaching phenomena of the cementitious waste matrices, could be evaluated by fitting the experimental data to mechanistic mass transport models. If the experimental data could not be adequately represented by a single mechanism, the transport may be resulted from a combination of these mechanisms [23]. There are various mechanistic mass transport models that could be used to calculate the radionuclide leaching from cementitious waste form [23–26].

First-order reaction model is used if the radionuclide leaching is controlled by the exchange kinetics between the surface of the waste matrix and leaching solution. The surface exchange rate could be governed by first-order reaction rate, so it will be proportioned to the readily available amount of radionuclides in the waste matrix (Q, mg/g) as follows [25]:

$$\frac{dQ}{dt} = kQ \tag{9.12}$$

where:

k = reaction rate constant (s^{-1})

Diffusion model is used if the radionuclide transport is controlled by diffusion; the solution of Fick's second law in semi-infinite medium and Fick's first law could be used to obtain the flux of the diffusing materials $J(t)$ through the immobilized waste matrix as follow:

$$J(t) = -D\frac{\partial C}{\partial x}\bigg|_{x=0} = -C_o\sqrt{\frac{D}{\pi t}} \tag{9.13}$$

where:

C_o = initial concentration in the waste matrix (mg/cm³)
D = diffusion coefficient (cm²/s)

Dissolution model is used if the leaching species are structurally a major component of the waste form; its release into leaching solution leads to a structural breakdown of the matrix, a process which is referred to as dissolution. The kinetics of dissolution can be represented by a network dissolution velocity U, defined as the volume of solid material being dissolved per unit time and per unit surface area of solid exposed [25]:

$$U(t) = U_o\left(1 - \frac{C^w(t)}{C_{sat}^w}\right) \tag{9.14}$$

where:

U_o = maximum network dissolution velocity (cm/s)
C_{sat}^w = saturation concentration in the aqueous solution (mg/L)

3.2.1.4 Radionuclide Transport Through Cementitious Barrier

Depending on the nature of water migration through the cementitious barrier, the barrier could be classified into saturated or unsaturated barrier. Saturated cementitious barrier represents the most conservative case, where all the pores are filled with water. In this case, the transport of radionuclides through the barrier could be described by the following mass transport equation, if the release is controlled by diffusion mechanism [27–33]:

$$\frac{\partial C_c}{\partial t} = \frac{D_c}{R_d}\left[\frac{\partial^2 C_c}{\partial x^2} + \frac{\partial^2 C_c}{\partial y^2}\right] - \lambda C_c \qquad (9.15)$$

where:

x = spatial coordinate in the direction of flow (cm)
y = spatial coordinate normal to the direction of flow (cm)
t = time (s)
C_c = contaminant concentration (mg/L)
D_c = diffusion coefficient of contaminant (cm^2/s)
R_d = retardation coefficient $(1 + K_d\rho_c/\theta_c)$ (dimensionless)
K_d = concentration factor of radionuclide (L/g)
ρ_c = bulk density of barrier (g/L)
θ_c = total porosity of barrier (dimensionless)
λ = decay constant (s^{-1})

In more realistic case, unsaturated cementitious barriers are found under normal evolution scenario for the disposal facilities, where the pores are partially filled with water, and the performance of the barrier is assessed by solving multiphase problem of air, water, and solute transport. The coupled problem of groundwater flow and contaminant transport through the unsaturated cement could be adequately used to predict the performance of this barrier [34]. The water pressure head at the barrier could be estimated using Richard's equation that can be written in two dimensions as follows:

$$\xi\frac{\partial \psi}{\partial t} = \frac{\partial}{\partial x}\left[K(\psi)\frac{\partial \psi}{\partial x}\right] + \frac{\partial}{\partial z}\left[K(\psi)\frac{\partial \psi}{\partial z}\right] - \frac{\partial}{\partial z}K(\psi) + W \qquad (9.16)$$

where:

ξ = soil moisture capacity, $\xi = \delta\theta/\delta\psi$ (dimensionless)
θ = volumetric water content (cm^3/cm^3)
ψ = pressure water head (cm)
K = hydraulic conductivity (cm/s)
W = moisture source/sink term (dimensionless)

The solution of the above equation requires the knowledge of the water retention and hydraulic conductivity curves. A suitable empirical model could be used to describe those curves.

The two-dimensional solute advective dispersive tarnsport could be described by

$$\frac{\partial \theta R_d C_c}{\partial t} = \frac{\partial}{\partial x}\left[\theta\left(D_{xx}\frac{\partial C_c}{\partial x} + D_{xz}\frac{\partial C_c}{\partial z} - C_c V_x\right)\right]$$
$$+ \frac{\partial}{\partial z}\left[\theta\left(D_{xz}\frac{\partial C_c}{\partial x} + D_{zz}\frac{\partial C_c}{\partial z} - C_c V_z\right)\right] + I \qquad (9.17)$$

where:

R_d = retardation coefficient (dimensionless)
C_c = contaminant concentration (mg/L)
I = source/sink term for contaminant (dimensionless)

The velocities and the dispersion coefficient component are given by

$$V_x = -K(\psi)\frac{\partial \psi}{\partial x} \qquad V_z = -K(\psi)\left[\frac{\partial \psi}{\partial z} + 1\right]$$

$$D_{xx} = \alpha_l\frac{V_x^2}{V} + \alpha_t\frac{V_z^2}{V} + D_o \text{ae}^{b\theta} \qquad D_{zz} = \alpha_t\frac{V_x^2}{V} + \alpha_l\frac{V_z^2}{V} + D_o \text{ae}^{b\theta}$$

$$D_{xz} = (\alpha_l - \alpha_t)\frac{V_z V_x}{V}$$

where:

D_o = diffusion coefficient of solute molecules in free water (cm^2/s)
a, b = tortuosity constant (dimensionless)
α_l, α_t = longitudinal and transverse dispersivity (cm)

3.2.1.5 Mechanical Performance of Deformed Cementitious Barrier

Change in the mechanical and hydraulic performance of the cementitious barrier might be attributed to the changes in cementitious material and operating forces due to additional deformation of the surrounding materials. In disposal conditions, the deformation of the cementitious materials could be attributed to the swelling pressure of surrounding materials and/or the rock pressure. This behavior could be expressed by using a nonlinear elasticity model. The rigidity and strength of cementitious material are reduced due to calcium leaching progress. This process could be modeled using Mohr–Coulomb's failure criterion and an iterative calculation scheme for stress redistribution. These mechanical and hydraulic changes could be calculated based on the input data which represent chemical transition in the materials. The kinetics of deformation and stress state could be obtained by solving the balance of operating forces considering the mechanical property changes, where the distribution of hydraulic conductivity is calculated based on the chemical property changes as well as the obtained deformation. The concept of mechanical behavior analysis for the cement is illustrated in Fig. 9.4 [35].

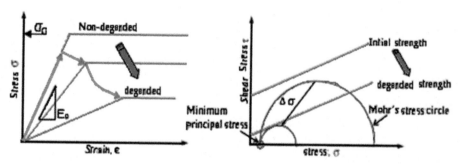

Fig. 9.4 Conceptual model for the assessment of cementitious barrier mechanical behavior (*Source*: Sahara et al. [35])

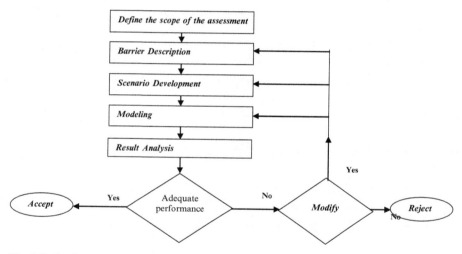

Fig. 9.5 Barrier assessment procedure

3.2.2 Assessment Models

In assessment modeling, the important processes are linked together to predict the overall performance of the disposal system. The assessment of the engineering barriers is an iterative procedure, at which the confidence in the assessment is progressively increased. Figure 9.5 illustrates the major steps in the assessment procedure. The interpretation of the assessment results confirms or not if the barrier has the potential to meet design and regulatory requirements [36]. The iterative nature of this procedure allows the designer to modify the barrier design to achieve the optimum performance that meets regulatory requirements and is consistence with good engineering practice, operational needs, and cost constrains. The complexity of the assessment studies varies according to the level of iteration and the assessment context. To model the engineering barrier performance, various waste

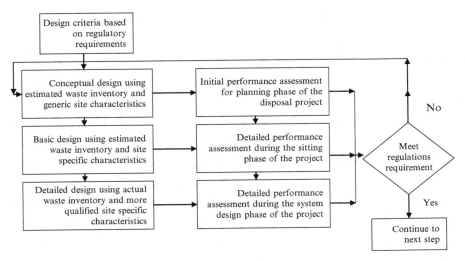

Fig. 9.6 Relationships between different low-level radioactive waste disposal design phases and quality of information (*Source*: Abdel Rahman [37])

forms, engineering barrier materials, and site characteristic information are required. The quality of the required information is largely dependent on the barrier design phase. Figure 9.6 illustrates the relationships between different low-level radioactive waste disposal design phases and the quality of the required information [37]. The main objective of the conceptual design phase is to select the disposal option; this phase uses estimated radioactive waste inventory and characteristics and generic site characteristics in developing the performance assessment. At the basic design phase, the designer has to confirm that the disposal option could become a licensable operational option [38]. During this stage, the design is expanded, and site-specific information are incorporated, technical feasibility of materials are carried out, and the data gaps are identified.

3.3   International and National Assessment Programs

IAEA has established a cooperative research program (CRP) that aims to investigate the behavior and performance of cementitious materials used in radioactive waste disposal. Several research objectives were defined that include the following: (i) cementitious materials for radioactive waste packaging, including radioactive waste immobilization into waste form, waste backfilling, and containers; (ii) emerging and alternative cementitious systems; (iii) physical–chemical processes occurring at the production of cement compounds (cement hydration) and their influence on the cement compound quality; (iv) methods of production of cementitious materials for immobilization into waste form, backfills, and containers; (v) conditions envisaged for packages (physical and chemical conditions, temperature variations, groundwater, radiation fields); (vi) testing and non-destructive monitoring techniques for cementitious materials; and (vii) waste acceptance criteria for waste packages, waste forms, and backfills, transport, long-term storage, and disposal requirements[20, 39, 40]. The results of this program showed that interactions between a cement system and the waste stream are complex, and more research in this area is needed to fully understand these systems. Initial research has highlighted that by controlling the internal chemistry, microstructure of the matrix and hydration products formed, through additives, curing temperature and water/cement ratio; a system may be developed to immobilize a specific waste ion. More research though is required on the physical and chemical effects of waste ions on the cement structure during solidification, the formation of further hydration products or species of waste ion, and the stability of these phases. These areas must be understood because a slight change in the matrix chemistry could result in significant change in immobilization capacity and the release of radioactive material into the biosphere. Interactions and durability especially of phases containing waste ions are the focus of much current research.

Several countries had initiated research programs to assess the long-term behavior of cementitious material. France has a major research program on long-term performance of cementitious waste forms. This program is tied directly to the French Legal Reference Case for management of nuclear wastes [41]. Canadian research program has made substantial progress in developing mechanistic models for predicting service life of concrete structures. Mechanistic degradation analyses and durability prediction approaches that have been applied and tested on conventional concrete applications provide an excellent starting point for evaluating, designing, and improving containment structures and entombments for radioactive waste disposal.

Cooperative research projects have also been established in this respect; a research project was initiated between the Energy Research Centre of the Netherlands, DHI Water & Environment (Denmark), and Vanderbilt University (USA). This effort has made extensive progress in characterizing leaching phenomena and developing associated mechanistic release models for engineered cementitious materials (primarily waste forms) and wastes. Additional notable international research programs related to engineered cementitious materials used in waste applications are being carried out in the United Kingdom and Switzerland. Areas of advancement include cement–waste interactions and the chemistry of contaminant transport [32–44].

4 Cementitious Material in Radioactive Waste Disposal

4.1 International and National Trends in Regulation and Guidance

National regulations of radioactive waste disposal differ from country to country, but usually they are derived based on ICRP recommendations and IAEA safety standard series. ICRP had recommended the long-term safety constraint for radioactive waste disposal to include dose and risk of 0.3 mSv/year or 10^{-5} years from quantitative prediction in the range 1000–10,000 year for final disposal. Behind these timescales dose and risk should be used as reference values [45]. Task group 80 prepared a publication that describes and clarifies the application of ICRP 101 and 103 recommendation for the protection against occupational and public exposures that may result from the geological disposal of long-lived solid radioactive waste [46]. IAEA is prepared new safety standard publications that are directed to provide safety requirements and performance measure related to radioactive waste disposal [47–49]. Also, IAEA had initiated some projects that address performance and safety assessment approaches for radioactive waste disposal facilities [20]. Recently, the safety case concept was introduced to demonstrate the long-term safety of the geological disposal; then this concept was extended to include all radioactive waste disposal options [48, 50]. So the assessment of the dose and/or risk criteria is performed within a border process that reflects the increasing role of individual equity, safety culture, and stakeholder involvement into the decision-making process. An essential part of the safety case is the quantitative predication that is conducted by assessing the performance of the individual barrier of the disposal.

In order to facilitate the implementation of national regulations, national regulatory bodies had developed guidance documents. The aim from producing this kind of documents is to enhance its staff capability to review and evaluate license application and/or to summarize the knowledge of the disposal system behavior. These guidance may include information that must be addressed to demonstrate that the disposal practice meets the regulatory requirements. The performance measure of the cement-based waste form was addressed in some countries in details as discussed in Sect. 4.2, where there were no specific requirements or recommendations regarding performance measure of the other cementitious barriers.

4.2 Performance of Cementitious Waste Form

Immobilization techniques consist of entrapping the radionuclides within a solid matrix, i.e., cement, cement-based material, glass, or ceramic. Despite the existence of several disadvantages in the utilization of cement immobilization technique such as its low-volume reduction and relatively higher leachability, the choice of this technique has been worldwide employed for the immobilization of low- and intermediate-level radioactive wastes because of its compatibility with aqueous waste streams and capability of activated several chemical and physical immobilization mechanisms for a wide range of inorganic waste species [2, 51–55]. Also, cement immobilization possesses good mechanical characteristics, radiation and thermal stability, simple operational conditions, availability, and low cost. During immobilization two simultaneous processes occurred, namely, stabilization and solidification. The radioactive waste is stabilized by converting the radionuclides to less mobile form as a result of chemical change, where solidifying these wastes aims to improve their mechanical performance [56].

The immobilization of radioactive waste in cements includes chemical fixation of radionuclides due to chemical interactions between the hydration products of the cement and the radiocontaminants, physical adsorption of the radiocontaminants on the surface of hydration products of the cements, and physical encapsulation of radioactive waste due to the low permeability of the hardened pastes [6]. The first two mechanisms depend on the nature of the hydration products and the contaminants, and the third aspect relates to both the nature of the hydration products and the density and physical structure of the paste. If the leaching and mechanical performance of the produced waste forms meet appropriate regulatory criteria and waste acceptance criteria at the disposal facility, the cement-based waste forms will be disposed. Waste acceptance criteria for any disposal facility include specifications that guarantee the long-term safety of the overall disposal facility. For the immobilized waste from these specifications included factors that affect the mechanical strength of the waste form and its radionuclide retention.

The long-term behavior of the immobilized waste form under disposal conditions is very hard to be predicted and simulated in laboratory. So their performance is evaluated using a combination of several physical and chemical tests. Each test is used to provide partial insights into the behavior of these wastes. These tests are conducted to assess whether the immobilized waste form meets the regulatory and waste acceptance criteria or is used to understand failure mechanisms that could be categorized as mechanical and radionuclide retention performance tests.

The reliability of the immobilization process is characterized by the rate at which radionuclides can be released from the waste form during long-term storage and/or disposal. As the most plausible path for reintroduction of radioactivity into the biosphere is via water, the most important parameters that characterize the ability of waste form to hold onto the active species are the leach rates. The leaching behavior of waste forms containing different amounts of waste radionuclides is compared using the normalized leaching rates NR for each i-th nuclide expressed in $g/cm^2 day$, and the normalized mass losses NL, expressed in g/cm^2. These are determined measuring concentrations c_i (g/L) of inactive constituents or activities a_i (Bq/L) of radionuclides in the water solution in contact with the waste form after a time interval Δt expressed in days. The mass fraction of a given nuclide i in a waste form is defined as [2]

$$f_i = \frac{w_i}{w_o} \qquad (9.18)$$

where:

w_i = mass of nuclide in the waste form (g)
w_o = mass of waste form (g)

The specific activity of a given radionuclide in a waste form q_i (Bq/g) is defined as

$$q_i = \frac{A_i}{w_o} \qquad (9.19)$$

where A_i is the radioactivity of radionuclides in the waste form (Bq). The normalized leaching rate of nonradioactive nuclides NR_i (g/cm^2day) is calculated using the expression.

$$NR_i = \frac{C_i V}{f_i S \Delta t} \qquad (9.20)$$

where:

S = surface area of waste form in contact with water (cm^2)
V = solution volume (L)
Δt = test duration (d)

The normalized leach rate of radioactive nuclides NR_i (g/cm^2day) is calculated using the expression

$$NR_i = \frac{a_i V}{q_i S \Delta t} \qquad (9.21)$$

where:

a_i = specific radioactivity of solution which is in contact with waste form (Bq/L)

A set of standard tests to determine the durability of vitrified waste and other waste forms was developed at the Materials Characterization Centre (MCC) of the Pacific Northwest National Laboratory, USA. These MCC tests are now the internationally approved standards used worldwide. The most important tests are given in Table 9.6 [2, 57]. The results of these tests can be expressed in the form of the incremental fractional release (IFR) or cumulative fractional release (CFR). The IFR is the amount of contaminant released normalized to the initial contaminant inventory while CFR is the ratio between the amount released and the initial inventory in the waste form. The CFR is the sum of all IFR and is a measure to retain radionuclides. CFR and IFR can be translated into NR_i using the radionuclide-specific inventory and surface area of the waste form.

Table 9.6 Standard tests on immobilization reliability

Test	Conditions	Use
ISO 6961, MCC-1	Deionized water. Static. Monolithic specimen. Sample surface to water volume (S/V) usually 10 m^{-1}. Open to atmosphere. Temperature 298 K (25 °C) for ISO test and 313 K (40 °C), 343 K (70 °C), and 363 K (90 °C) for MCC-1 test	For comparison of waste forms
MCC-2	Deionized water. Temperature 363 K (90 °C). Closed	Same as MCC-1 but at high temperatures
PCT (MCC-3)	Product consistency test. Deionized water stirred with glass powder. Various temperatures. Closed	For durable waste forms to accelerate leaching
SPFT (MCC-4)	Single-pass flow-through test. Deionized water. Open to atmosphere	The most informative test
VHT	Vapor-phase hydration. Monolithic specimen. Closed. High temperatures	Accelerates alteration product formation

Sources: Ojovan and Lee [2]

The fractional leaching rate is the IFR divided by the time between measurements. The rate is measured by standard leaching tests, such as ANS 16.1 or ISO 6961. ISO 6961 test is the most frequently used to quantify the leach resistance of cementitious waste forms. It determines how well the radionuclides of concern are retained within the waste form in a wet environment. There are two mechanisms that influence leaching behavior. One is the creation of a physical barrier between the radionuclide and the environment, which is how most polymer and in part bitumen systems work. The other mechanism involves chemical reaction between the radionuclides and the matrix, which occurs most often in cement-based systems. This behavior is waste form and radionuclide specific and can be altered by the waste chemistry, the formulation of the immobilization matrix, and the leaching water chemistry. Most transuranic elements are retained well by common cement phases owing to the high-pH (basic) conditions and the chemical reactions that occur in the matrix. The choice of cement type and cementation technology depends on a number of factors, although waste acceptance criteria are among the most important. Waste acceptance criteria (e.g., Table 9.8) specify required characteristics of matrix materials and waste packages [2]. The normalized leaching rate for the main waste radionuclide ^{137}Cs is as low as $1.3 \cdot 10^{-6}$ g/cm^{2}day, which is at the level of some glasses and ceramics (see Fig. 9.7) [2, 58, 60]. The leaching rates for ^{60}Co are low at the level of $(4.2–7.3)10^{-5}$ cm/day [61]. The results from an old opened experimental disposal [40-year-old disposal] had demonstrated limited degradation with time and high retention of radionuclides in disposal conditions. Figure 9.7 shows blocks of cemented waste in an opened experimental repository after 40 years of burial [62, 63]. The leachability index is a material parameter of the leachability of diffusing species, which is used to catalogue the efficiency of a matrix material to solidify a waste and is given by [54]

Table 9.7 Acceptance criteria for cemented radioactive waste in Russian Federation

Criterion	Permitted limits
Specific radioactivity Of β- and γ-nuclides Of α-nuclides	$<1 \cdot 10^{-3}$ Ci/kg $<1 \cdot 10^{-6}$ Ci/kg
Leaching rate of ^{137}Cs and ^{90}Sr	$<10^{-3}$ g/cm^2·day
Mechanical durability (compressive strength)	Depending on transport requirements, interim storage, and disposal, but not less than 5 MPa
Radiation durability	Mechanical durability not less than permitted (at least 5 MPa) at irradiation dose 10^8 rad
Thermal cycle durability	The same after 30 freeze–thaw cycles ($-40\ ^\circ$C, $+40\ ^\circ$C)
Durability to long-term water immersion	Mechanical durability not less than permitted (at least 5 MPa) after 90 days of immersion in water

Source: Ojovan and Lee [2]

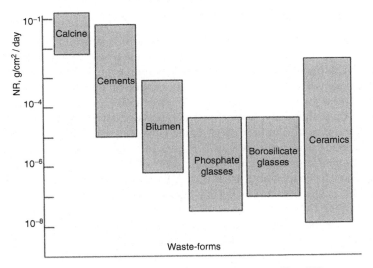

Fig. 9.7 Water leaching of various waste forms (*Source*: Ojovan and Lee [2])

$$L = -\log(D) \tag{9.22}$$

The value of 6 is the threshold to accept a given matrix as adequate for the immobilization of radioactive wastes.

The ability of the immobilized waste form to resist mechanical stresses could be measured by studying the unconfined compressive strength (UCS) before and after immersion in water. There are numerous standard methods for its determination, all of which involve vertical loading of a monolithic waste form to failure. Standard

Fig. 9.8 Blocks of
cemented waste in an
opened experimental
repository after 40 years of
burial (*Source*: Sobolev
et al. [63])

methods vary mainly with regard to the waste form shape and size. Since these variables have an effect on the test result, they must be clearly reported. Measurement of strength after immersion, as well as before, is important to ensure that the waste form has set and hardened chemically rather than merely dried and to ensure that deleterious swelling reactions do not occur in the presence of excess water. Because of its simplicity, unconfined compressive stress measurement is also suitable for use as a compliance test. In the USA, cement-solidified class B and C wastes include an average UCS of 3.5 MPa (500 psi) using ASTM C39/C39M [64, 65]. In the Netherlands and France, a UCS of 1 MPa is suggested for disposal [10]. In Russian Federation a UCS of 5 MPa is used as acceptance criterion for disposal. It has also been suggested that the UCS with immersion should not be less than 80 % of the UCS without immersion [66–69]. Given that UCS is an indicator of the progress or inhibition of cement hydration reactions, it may be appropriate to also consider this more indirect aspect in setting performance criteria.

Long-term field tests of cemented aqueous radioactive wastes in an experimental mound-type surface repository were carried out at Moscow Scientific and Industrial Association "Radon" from 1965 to 2004. Aqueous radioactive wastes of different compositions containing short-lived radionuclides including ^{90}Sr and ^{137}Cs at concentrations from 0.34 to 1.8 MBq/L were immobilized using cementation technology. Water-to-cement ratio was 0.66, grout mixing time was 10–15 min, and cement paste hardening time was 7 days. Seventy-three cement blocks with a volume of 0.027 m^3 were disposed of for long-term tests in a simple mound-type surface repository. The atmospheric precipitates, which contacted radioactive cement blocks, were collected and analyzed for the content of radionuclides. In August 2004 the experimental repository was opened, cemented blocks; underlying and covering materials were retrieved for analyses. X-ray diffraction (XRD) analyses showed that along with amorphous tobermorite gel, the main crystal phases in cements are calcite and portlandite. Both visual inspection and radiometric analyses demonstrate that cemented blocks are in good condition and that the cement paste has retained the immobilized radionuclides. Thus after 39 years of storage in the mound-type repository, the cemented aqueous wastes are reliably immobilized [70].

In India, radioactive chemical sludge has been fixed in cement before disposal in near-surface facilities. Experiments were carried out to optimize the ratio of sludge to cement and to study the effect of additives like vermiculite and bentonite on the leaching behavior. Results indicated that the blocks with the sludge-to-cement ratio 1.0:1.5 by weight showed good compressive strength but poor leaching characteristics in tap water. It was found that adding bentonite or vermiculite (5 % of weight of cement) reduced the compressive strength but improved the leaching behavior. The study recommended the addition of vermiculite to cement to minimize the leaching [71]. The corrective action at the Centralised Waste Management Facility (CWMF) in India was done to immobilize 160 m^3 of radioactive chemical sludge generated from treatment of several batches of radioactive liquid wastes by chemical precipitation method. The process consisted of diluting the sludge with low active effluents/water for homogenization and facilitating the transfer of sludge, dewatering of the slurry utilizing decanter centrifuge, and fixation of dewatered concentrate in OPC with vermiculite as an additive using in-drum mixing method [72].

To improve the performance of cementitious waste form in retarding important radionuclides, chemical sludge obtained from the radioactive aqueous treatment plant in Egypt was immobilized with OPC, and OPC blended with natural local clays known with their high sorptivity. The results indicate that adding 10 % clays to OPC at water-to-cement ratio of 0.4 reduces the leach pattern as OPC–bentonite < OPC–red clay < OPC for matrices containing ^{137}Cs, ^{60}Co, and $^{152-154}$Eu. The calculated apparent diffusion coefficients and leachability indices indicated that the performance of the prepared matrices was acceptable [54].

Another study investigates the mechanical properties of cement-based radioactive waste form containing two kinds of sludge in different contents. The first kind is obtained by treatment of liquid radioactive effluents using a PA-type anionic polyelectrolyte, and the other one simulates the sludge obtained by decontamination of contaminated surfaces using a pNaAc-type hydrogel. The influence of the content of sludge on the setting time of the mortar paste, as well as on the compressive and bending strength of the cement-based radioactive waste forms, was studied [73].

Alkali-activated cements consist of an alkaline activator and cementing components, such as blast furnace slag, coal fly ash, phosphorus slag, steel slag, metakaolin, etc., or a combination of two or more of them. Properly designed alkali-activated cements can exhibit both higher early and later strengths than conventional portland cement. The main hydration product of alkali-activated cements is C—S—H with low Ca/Si ratios at room temperature. Under hydrothermal conditions, C—S—H, tobermorite, xonotlite, and/or zeolites are the main hydration product. The metastable crystalline compounds such as CH and AFT and AFM did not exist. Alkali-activated cements also exhibit excellent resistance to corrosive environments. The leachability of contaminants from alkali-activated cement-stabilized hazardous and radioactive wastes is lower than that from hardened portland cement-stabilized wastes [74]. The feasibility of immobilizing phosphate effluent produced from reprocessing plant using ground-granulated blast furnace slag (GGBS) or Super-Cement was studied. Results indicated that to achieve a stable waste form, it was necessary to pretreat the phosphate effluent with lime before encapsulation, the compressive strengths of grouted waste exceeded 10 MPa after 28 days of curing, and a waste loading exceeding 13 wt% was achieved [75]. Another investigation was directed to assess the suitability of GGBS for the encapsulation of medium-level radioactive waste. The grout was prepared using 90 % GGBS to 10 % OPC portland cement to reduce the heat generated within the grout and minimize thermal cracking and diffusion of chemicals. The grade GGBS grout has a controlled particle size distribution that gives the grout high flow and extended working life with low bleed, which are vital for encapsulation process. The study indicates that GGBS can be used in concrete requiring high-durability performance in aggressive ground conditions and is also suitable in designated chemical classes [76]. The suitability of using fly ash- and blast furnace slag-blended cements for encapsulation of Cs–Ionsiv in a monolithic waste form was also investigated. A small fraction (≤ 1.6 wt%) of the Cs inventory was released from the encapsulated Ionsiv during leaching experiments carried out on hydrated samples. Furthermore, it was found that slag-based cements showed lower Cs release than the fly ash-based cements [77].

A 25 year mortar and concrete testing project in Serbia is conducted to evaluate the factors that can influence the design of the engineered trench system for a future central radioactive waste disposal center. Within this project, the immobilization of radioactive evaporator concentrate containing ^{137}Cs, ^{60}Co, and ^{85}Sr was tested, and the influence of natural additives (bentonite and zeolite) on the leaching behavior was assessed [78, 79]. The retardation factors and distribution coefficients were determined using a simplified mathematical model for analyzing the leaching behavior [80, 81]. It was found that increasing the amount of bentonite additive causes a significant reduction in the leaching rate, because of its good sorption characteristics and ion selectivity [82].

The aqueous durability of Materials Testing Reactor (MTR) cementitious waste form was studied by performing a series of medium-term (up to 92 months) durability tests, without leachate replacement. The waste form was made from cement, GGBS, and simulated waste liquor. The mechanical behavior of the solidified waste form was acceptable. The inhomogeneous waste form mixture contained calcite, C—S—H phase, hydrotalcite, and unreacted slag particles. After leaching for 92 months, the crystallinity of the C—S—H phase increased, and the steady-state conditions prevailed after 3–6 months of leaching. The highest releases of matrix elements were for Na (37 %), K (40 %), and S (16 %). By the end of the experiment, it was found that in an alteration layer about 80 μm deep, calcium has been depleted. Na, K, and Sr showed signs of diffusion toward the outer part of the cement samples [83].

The effect of temperature on the leaching kinetics of Cs ions from cement paste solids was studied at two temperatures (30 °C and 70 °C) and some intermediate temperatures using two different specimens' thicknesses. The leaching results indicated that the 20–30 % of Cs tends to be immobilized in the matrix, while elution of the readily leachable portion follows Fick's law reasonably well. The long-term leaching experiments (up to 8 years) revealed the acceleration of the elution process (not detectable in short-term tests), attributable to increase in cement porosity due to matrix erosion. Sorption experiments of Cs ions by cement granules indicated that adsorption on cement pore surfaces is not significant. On the other hand, the leaching tests at two different temperatures or with intermediate changes in temperature between 30 °C and 70 °C yielded activation energies that indicated a more complicated kinetic behavior [84].

Wet chemistry experiments and X-ray absorption fine structure (XAFS) measurements were carried out to investigate the immobilization of nonradioactive Sr and ^{85}Sr in calcite-free and calcite-containing portland cement. The partitioning of pristine Sr between hardened cement paste (HCP) and pore solution and the uptake of ^{85}Sr and nonradioactive Sr were investigated in batch-type sorption/desorption experiments. Sr uptake by HCP was found to be fast and nearly linear for both cements, indicating that differences in the compositions of the two cements have no influence on Sr binding. The partitioning of pristine Sr bound in the cement matrix and ^{85}Sr between HCP and pore solution could be modeled in terms of a reversible sorption process using similar Kd values. Wet chemistry and spectroscopic data further indicate that Sr binding to C—S—H phases is likely to be the controlling uptake mechanism in the cement matrix, which allows Sr uptake by HCP to be predicted based on a Ca–Sr ion exchange model previously developed for Sr binding to C—S—H phases. The latter finding suggests that long-term predictions of Sr immobilization in the cementitious near field of repositories for radioactive waste can be based on a simplified sorption model with C—S—H phases [85].

The possibility of solidifying exhausted synthetic zeolite A, loaded with ^{137}Cs and/or ^{90}Sr radionuclides, in OPC was investigated. The obtained results showed that the presence of zeolite A in the final cemented wastes improves the mechanical characteristics of the solidified cement matrix (mechanical strength and setting times) toward the safety requirements and reduces considerably the radionuclide's leach rates [86].

The conditioning of spent radioactive sources containing ^{60}Co and ^{137}Cs radionuclides at Cekmece Waste Processing and Storage Facility (CWPSF) in Turkey was performed. Reinforced metal drums (200 L) and cement matrix were used for conditioning of these sources. Maximum dose rates at the surface of the conditioned waste package were found to fulfill the transportation, storage, and disposal requirements (<2 mSv h^{-1}) [87].

The feasibility of immobilizing the ash produced from the incineration of solid radioactive wastes was assessed. OPC was mixed with 6, 12, 20, and 30 wt% of ashes at different water-to-cement (w/c)–ash ratio. The results shows that the extent

of solidification of the studied waste matrices is acceptable based on a comparison
with the compressive strength threshold of 35 kg/cm^2 and that ^{137}Cs leaching
resulted from first-order reaction between the surface of the waste matrix and the
leaching solution followed by diffusion through the studied matrices. The leaching
of ^{60}Co was found to be as a result of four subsequent mechanisms that include
release of loosely bound ^{60}Co followed by first-order reaction the diffusion and
finally dissolution. It was found that the studied immobilized waste matrices have
acceptable mechanical performance. The values of the leachability indices indicate
that the performance of the studied matrices in ^{137}Cs stabilization is not
acceptable [23].

Different samples of simulated radioactive borate waste have been prepared and
solidified after mixing with cement–water extended polyester composite. The
polymer–cement composite samples were prepared from recycled polyethylene
terephthalate waste and cement paste (water/cement ratio of 40 %). The samples
aged in their molds for 7 years, and then IAEA leaching test was used to evaluate
the performance of these samples. The results indicate that the studied samples have
adequate chemical stability required for the long-term disposal process [88].

The optimization of the best source-term conceptual model that represents Cs
release from 14 different cementitious waste matrices was conducted [26]. The
investigated matrices included three categories; the first are simulated intermediate-
level waste (ILW) immobilized in OPC, BFS, and a mixture of both of them. The
second category includes cementitious OPC–ILW matrices with different inorganic
resin additives, and the last contains OPC–ILW–polymer-immobilized matrices
with different organic exhausted resins. The experimental data of the cumulative
leach fraction from long-term static test were linearly and nonlinearly regressed to
different developed source-term conceptual models. The regression results indicated that Cs releases are best described by first-order reaction/diffusion model. The
utilization of this method was found to give more realistic prediction for the
cumulative leach fraction for the first and second waste categories, where IAEA
recommended method produces more realistic estimation for the third waste
category.

Recently, an extensive array of leaching studies has been addressed to reduce the
leachability of different radionuclides from immobilized waste matrices by mixing
the cement with different materials having significant sorption capacity such as
bentonite, fly ash, silica fume, ilmenite, blast furnace slag, kaolin, rice husk, and
zeolites [12, 26, 89–97]. Leaching test and mechanical strength of the prepared
waste forms were conducted for different water-to-cement ratio and different waste
loading ratios. Table 9.8 summarizes the research conducted to assess the cementitious waste form [12, 23, 26, 27, 54–104]. Recently, the binding of Sr and Cs metal
ions in cementitious waste forms with different additive contents (bentonite of
0–15 wt%) was investigated [12, 104]. The studies indicated that mechanical
performance of the matrices was reduced by increasing the percentage of bentonite.
But, generally, it was acceptable, and the enhanced compressive strength was
attributed to the progression in the formation of cement hydrated phases and the
pozzolanic reaction between bentonite and lime in cement–bentonite matrices. The

Table 9.8 Summary of the recent studies conducted to assess the performance of cement-based waste form

Waste type	Studied waste form	Performance criteria	References
Radioactive aqueous waste	OPC	Radiometric analyses	[70]
		Microbiological activities	[98]
Radioactive chemical sludge	OPC OPC+ bentonite OPC+ red clay	Leaching	[54]
Radioactive chemical sludge	OPC OPC +vermiculite OPC + bentonite	Mechanical using compressive strength Leaching	[71, 72]
Sludge from PA-type anionic polyelectrolyte Simulated the sludge pNaAc-type hydrogel	OPC	Mechanical using compressive and bending strength	[73]
	Alkali-activated cements	Characterization of hydration phases Leaching	[74]
Phosphate effluent produced from reprocessing plant	Ground-granulated blast furnace slag (GGBS)	Mechanical using compressive strength	[75]
Medium-level radioactive waste	GGBS + CEM1	Durability	[76]
Cs–Ionsiv	Fly ash- and blast furnace slag-blended cements	Leaching	[77]
Evaporator concentrate	OPC + bentonite OPC + zeolite	Leaching	[78–83]
Simulated waste liquor	GGBS	Durability	[84]
Cs ions	OPC	Long-term leaching	[85]
Sr ions	Calcite-free OPC calcite–OPC	Batch-type sorption/ desorption	[85]
Exhausted synthetic zeolite	OPC	Mechanical using compressive strength Leaching	[86]
Spent radioactive sources	OPC	Dose rates	[87]
Incineration ash	OPC	Mechanical using compressive strength Leaching	[23]
Simulated borate radioactive liquid waste	OPC – polyethylene terephthalate	Leaching	[88]
Borate radioactive liquid waste	Blast furnace slag cement	Leaching	[89]
Cs	Alkaline-activated fly ash	Leaching	[90]
	OPC + kaolin clay	Leaching	[91]
Low-level waste	OPC OPC + kaolinite	Leaching	[92]

(continued)

Table 9.8 (continued)

Waste type	Studied waste form	Performance criteria	References
Liquid radioactive waste	OPC + rice husk ash	Leaching	[93]
Co and Cs	Treated zeolites	Leaching	[94]
	Zeolite/cement	Leaching	[95]
Low-level radioactive resins	ASC zeolite blends		[96]
Radioactive liquid waste	Zeolite		[97]
NORM scale	OPC	Compressive strength	[99]
Nitrate-containing liquid waste	OPC	Leaching Compressive strength Microbiological	[100]
Al waste	OPC	Mechanical	[101, 102]
Liquid scintillator	Cement–natural clay	Leaching	[103]

results indicated that the presence of the contaminants and bentonite did not lead to formation of new hydrated phases. The investigation of $SrCl_2$ and major phase speciation formed evidences that strontium containment in cement is due to Ca substitution in ettringite structure and exchange with cations on the negatively charged montmorillonite lattice [12]. The intermolecular channels in ettringite and CSH structures were found to possibly contribute to the physical entrapment of Cs and Sr ions within the solidified matrices. The addition of 15 % bentonite to the matrices was found to slightly reduce the interstitial pore fluid in the hydrated phases and affect Ca and Al solubility [104]. The mathematical analysis of the long-term leaching results indicated that strontium leaching resulted from a combination of first-order reaction and diffusion mechanisms, where the leaching of Cs and Sr metal ions from their binary matrices was found to be best represented by a combination of first-order, diffusion, dissolution, and instantaneous release of contaminant mechanisms [12, 104].

4.3 Container Materials

The waste container can range from a simple carbon steel container to high-density polyethylene (HDPE) container with a concrete overpack. The function of the container is to provide safety during transportation of the waste to the disposal site and during its emplacement in the disposal facility. With regard to the functional requirements during the post-closure phase, the waste container should provide both physical and chemical containment of waste form. The container serves as the second barrier in the multibarrier system of the disposal. Estimation of container lifetime is necessary to establish how much credit should be assigned to this barrier. The performance and lifetime of the container are function of the container material and design, degradation mechanism and rate, environmental conditions, and groundwater chemistry. For concrete containers, degradation of concrete material needs to be considered to estimate the container lifetime.

To demonstrate long-term behavior of container type CE-2a used in El Cabril disposal site, a full-scale test cell was constructed. The test cell includes a full-scale concrete container, gravel backfill, and upper concrete slab. The cell was then waterproofed to maintain the thermal insulation and waterproofing of the real disposal cell. The cell had a lower drainage outlet connecting it to the outside, similar to those connecting the real cells. During the manufacturing of the container, a series of sensors were incorporated prior to concreting and to mortar injection, located both on the inner surface of the structural walls of the container and embedded in the mortar of the block. It was found that container reinforcement and the steel of the drums, located in the container, are perfectly passive. The concrete is in process of drying and densification and the oxygen content inside it is high. The study concluded that the temperature has a great influence in all of the parameters as a consequence of seasonal cycles, resistivity, deformation, and oxygen availability [105].

A comparison of package performance was conducted by comparing drum and cask package of solidified pressurized water reactor waste in cement. It was found that the drum can decrease over 50 % final volume of the waste. Furthermore, drum manufacture and transportation costs are cheaper than that of the cask, but an additional shielding may be needed for the higher-level wastes. More waste can be contained if an appropriate in-drum mixer is used, while secondary waste will be unavoidable if the out-drum mixing is adopted. A carriage can make the decontamination of the equipment surface and the floor easier; furthermore the manufacture and maintenance of the carriage are more economical than that of roller conveyor. That study recommended the optimized cementation recipe [106].

Cylindrical supercontainer is considered to be the most promising Belgian design to enclose the vitrified high-level radioactive waste and the spent fuel assemblies. This type of containers is based on the use of an integrated waste package composed of a carbon steel overpack surrounded by an OPC buffer. A study was conducted to reduce the early-age cracking of the buffer to retard the transport through the container [107]. Finite element simulations are performed for the first production step of the supercontainer, where the concrete buffer is cast into the outer steel liner, in order to investigate the problem of early-age cracking. Long-term creep test and shrinkage test are performed to predict the behavior of the concrete buffer. These mechanical properties are implemented into the material database of the simulation tool HEAT. From the simulations, the study concluded that if the environmental temperature does not exceed a value of 20 °C, no early-age cracking is expected in stage 1. The axial, radial, and tangential stresses remain, at all times, smaller than $0.7f_{ct}$. The insertion of a heat-emitting source inside the buffer also does not create additional cracking.

4.4 Evaluation of Concrete Performance

Factors that can affect the performance of the concrete structure for El Cabril disposal project were identified and studied. The design requirement for these structures is to have a design life longer than 300 years. The potential aggressive conditions for cement-based materials include carbonation, water permeation (leaching), and reinforcement corrosion. The study indicates that despite chlorides are not in the environment, they are inside the drums as part of analytical wastes. Also, it was found that disposal vaults and waste containers are made of a very similar concrete composition, while mortar was designed to be pumpable, with low hydration heat and low shrinkage and permeability [108].

The long-term integrity of geological disposal for radioactive waste was assessed by simulating the geochemical interactions between a concrete engineered barrier and a mudrock on time periods of up to 106 years with the reactive transport code Hytec. The results suggested that illite and quartz destabilization rates are key parameters governing the geochemical evolution of the degraded interface and that coupling between pH dependence of mineral stability and feedback of mineral precipitation on pH sharpens the cementation front [109].

In the field of radioactive waste management, concrete structures are expected to undergo significant heating due to the waste thermal power and significant drying (in the French design, the temperature is not expected to exceed 80 °C). The durability assessment of such structures thus requires the knowledge of the evolution of the water vapor sorption properties versus temperature. The latter can be easily estimated using the Clausius–Clapeyron (CC) equation: this approach requires the knowledge of the isosteric heat of adsorption and one desorption isotherm (at ambient or any other temperature). It was concluded that by using this equation and the desorption isotherms at two different temperatures, it is possible to estimate accurately the desorption isotherm at any other temperature [110]. The interaction between concrete/cement and swelling clay (bentonite) has been modeled in another study. The geochemical transformations observed in laboratory diffusion experiments at 60 °C and 90 °C between bentonite and different high-pH solutions (K—Na—OH and $Ca(OH)_2$ saturated) were reconciled with the reactive transport code CrunchFlow. For K—Na—OH solutions (pH = 13.5 at 25 °C) partial dissolution of montmorillonite and precipitation of Mg silicates (talc-like), hydrotalcite, and brucite at the interface are predicted at 60 °C, while at 90 °C the alteration is wider. Alkaline cations diffused beyond the mineralogical alteration zone by means of exchange with Mg^{2+} in the interlayer region of montmorillonite. Very slow reactivity and minor alteration of the clay are predicted in the $Ca(OH)_2-$ bentonite system. The model is a reasonable description of the experiments but also demonstrates the difficulties in modeling processes operating at a small scale under a diffusive regime [111].

The biological attack of the concrete barrier was studied by examining the fungal deterioration of the barrier in microcosms simulating a heterogeneous environment with an external source of nutrients for the fungi. Fungi successfully colonized a concrete barrier, generally avoiding granite aggregates, and biochemically (by excretion of protons and ligands) and biomechanically deteriorated the concrete. Fungi dissolved the cement matrix structural elements and accumulating them within the fungal biofilm and associated microenvironment. Oxalate-excreting Aspergillus niger formed abundant calcium oxalate crystals on the concrete and encrusting fungal hyphae [112].

Cement-based grout plays a significant role in the design and performance of nuclear waste repositories: if used correctly, it can enhance their safety. However, the high water-to-binder ratios, which are required to meet the desired workability and injection ability at early age, lead to high porosity that may affect the durability of this material and undermine its long-term geochemical performance. A methodology was presented to compromise between these two conflicting requirements. It involves the combined use of the computer programs CEMHYD3D for generating digital-image-based microstructures and CrunchFlow for reactive transport calculations. This methodology was applied to two grout types, standard mix 5/5, used in the upper parts of the structure, and the "low-pH" P308B, to be injected at higher depths. It was found that diffusion of solutes in the pore solution was the dominant transport process. A single scenario was studied for both mix designs and their performances were compared. The reactive transport model adequately reproduces the process of decalcification of the C—S—H and the precipitation of calcite, which is corroborated by empirical observations. It was found that the evolution of the deterioration process is sensitive to the chemical composition of groundwater, its effects being more severe when grout is set under continuous exposure to poorly mineralized groundwater. Results obtained appear to indicate that a correct conceptualization of the problem was accomplished and support the assumption that, in the absence of more reliable empirical data, it might constitute a useful tool to estimate the durability of cement-based structure [113].

To modify the concrete properties such as the pore fluid composition and the microstructure of the hydrated products, OPC with high contents of silica fume and/or fly ashes was prepared, and the resistance to long-term groundwater aggression was evaluated. The results show that the use of OPC cement binders with high silica content produces low-pH pore waters and the microstructure of these cement pastes is different from the conventional OPC ones, generating C—S—H gels with lower CaO/SiO_2 ratios that possibly bind alkali ions. Leaching tests show a good resistance of low-pH concretes against groundwater aggression although an altered front can be observed [114].

To assess the long-term evolution of physical and chemical properties of concrete, a project was developed in France to determine the variation of transfer properties (diffusivity, permeation) with hydrolysis/decalcification [115]. To simulate the two time phases of the structure life according to the external conditions, transfer properties were examined through two tests: a dynamic (temperature between 20 °C and 80 °C, under a hydraulic pressure drop from 2 to 10 MPa) and a static test (NH_4/NO_3 attack).

4.5 Cement-Based Backfill/Buffer

The UK disposal concept for intermediate-level and low-level radioactive waste (ILW/LLW) utilizes a cementitious material for backfilling vaults containing cementitious waste packages. The backfill is porous and is designed to promote an alkaline environment in which the mobility of radionuclides is reduced. Initial investigation into the use of multidimensional reactive transport geochemical models to examine the spatial and temporal effects that may occur as cementitious backfill and waste packages interact with groundwater was carried out [116]. Models have been developed, verified against experimental data, and used to examine the mineralogical reactions in the cement materials and their effect on pH buffering and porosity.

In Japan, the multibarrier concept consisting of cement-based backfill, structures and support materials, and a bentonite-based buffer material has been studied for transuranic (TRU) waste disposal design. Concerns regarding the utilization of bentonite-based materials in that disposal environment were based on the knowledge of long-term alteration under hyper-alkaline conditions due to the presence of cementitious materials. Experimental tests were designed to simulate this interaction between bentonite and cement. The results showed that formation of secondary minerals due to alteration reactions under the conditions expected for geological disposal of TRU waste (equilibrated water with cement at low liquid/solid ratio) has not been observed, although alteration was observed under extremely hyper-alkaline conditions with high temperatures. This was attributed to the difficulties in detecting C—S—H gel formed at the interface as a secondary mineral, because of its low crystallinity and low content [117]. Within the Japanese disposal project, the gas production event initiated by anaerobic corrosion of metals and microbial degradation of disposed organic materials is included in the failure scenario. This scenario deals with gas accumulation leading to the existence of pressure level that will subsequently affect the performance of the engineered barrier. To face these concerns, large-scale gas migration test (GMT) was initiated in 1997 at Nagra's Grimsel Test Site (GTS) in Switzerland under the leadership of the Radioactive Waste Management Funding and Research Centre (RWMC) in Japan. The field tests were completed in December 2004; interpretation of the GMT field data, incorporating the supporting laboratory tests and using five different modeling approaches, was performed [118].

The loess terrains near Kozloduy Nuclear Power Plant (NPP) in Bulgaria are among the prospective areas for the disposal of low- and intermediate-level wastes. The analysis of the loess properties has shown two main problems: a loess collapsibility and water permeability. Using a soil–cement cushion under the repository foundation and a soil–cement backfill is a possibility to avoid these disadvantages. In this connection loess–cement mixtures with bentonite and clinoptilolite additives have been investigated. The function of the proposed mixtures is to improve the impermeability and sorption properties against radionuclide migration [119]. This mixture was categorized into two kinds; the first are compacted at the optimum moisture content until the maximum dry density, and the second are compacted at higher moisture content equal to the liquid limit of loess. For the first type of mixtures, the unconfined compressive strength varies from 2 to 6 MPa depending on the cement and additives percents. Permeability measurements have shown satisfactory results, while for the second type, the unconfined compressive strength is less than the first type, but is sufficient for a backfill between the waste containers.

The study concluded that the loess–cement mixtures, especially these with clinoptilolite additive, are prospective as barriers of a low- and intermediate-level radioactive waste repository.

Highly permeable mortar is foreseen to be used as backfill for the engineered barrier of the Swiss repository for LILW. The backfill is considered to be a chemical environment with some potential for colloid generation and, due to its high porosity, for colloid mobility. Colloid concentration measurements were carried out using an in-situ liquid particle counting system. This system was tested using suspensions prepared from certified size standards. The concentrations of colloids with size range 50–1000 nm were measured in cement pore water, which was collected from a column filled with a highly permeable backfill mortar. Colloid concentrations in the backfill pore water were found to be typically lower than 0.1 ppm. The specific surface areas of the colloid populations were in the range 240–770 m^2g^{-1}. The low colloid inventories observed in this study can be explained by the high ionic strength and Ca concentrations of the cement pore water. The study concluded that these conditions are favorable for colloid–colloid and colloid–backfill interactions and unfavorable for colloid-enhanced nuclide transport [120].

Bentonite and concrete are essential components in construction of a geological high-level waste repository. OPC used for concretes gives a pore water leachate with a pH as high as 13.5 in contact with groundwater. This alkaline plume of leaching waters might perturb the engineered barrier system, which might include bentonite buffer, backfill material, or the near-field host rock. The accepted solution to maintain the bentonite stability, which is controlled by the pH, is to develop cementitious materials with pore water pH around 11. Four lixiviation experiments representative of long-term interaction of solids and pore fluids at the concrete/bentonite interface were performed with two types of cement paste, portland and calcium aluminate cement, before and after being carbonated under supercritical conditions, with granite at 80 °C. The evolution of the pH indicates that the supercritical carbonation reduced the alkalinity of the cement pastes and calcite likely controls the equilibrium of Ca at the end of the experiments. The bentonite helps to buffer the alkalinity of concrete leachate through several reactions such as dissolution of montmorillonite and precipitation of secondary products such as trioctahedral smectite, zeolites (gismondine), and presumably Mg hydroxides and amorphous gels. Carbonation may reduce propagation of the alkaline plume and enhance the barrier performance [121].

5 Summary and Recommendations

The main design requirements for utilizing cementitious barriers in radioactive waste disposal are to limit the water flux, retard the radiocontaminant, provide physical stability, and deter intrusion and physical disruption. An intensive array of research efforts were devoted to study the hydration mechanisms and assess the behavior of OPC with additive and other cement types in retarding the radionu-clides and provide acceptable mechanical performance. Other efforts were directed to assess the durability of the cementitious barrier under disposal conditions and its interactions with the host environment. The following conclusions could be drawn:

1. Except the waste form, there is a lack of specific requirements or recommenda-tions regarding performance measure of cementitious barriers.
2. Most often either simulated wastes or simply aqueous solutions of most impor-tant radionuclides are used to perform the laboratory experiments. Actual wastes, especially intermediate- and high-level waste, are very difficult to be used due to the radiation hazard.
3. The experimental laboratory assessment of the cementitious waste form is typically done based on tests developed for the bathtub scenario, which is a very conservative approach. There are no experimental assessments and stan-dards based on a more realistic scenario.
4. The hydrological performance of unsaturated cementitious barrier has to be addressed.
5. Despite hydrated phases of the cement can affect the performance of the barrier in retarding the radioactive contaminants, there is not a clear understanding on the effect of additives and/or radioactive wastes on the formation of the hydrated phase of cement.
6. There is a lack of procedures and standards that determine the mixing conditions for the cementitious paste and the measures that should be taken to guarantee its homogeneity during the operation.
7. The relationships between microstructure of the hardened barrier and flow and radiocontaminant transport parameters have not yet addressed adequately.
8. The effect of the microstructure on the aging or exposure to certain environ-mental conditions has not been adequately explored.
9. There is an acknowledged need to have a comprehensive database for the studied barrier materials and the mechanisms that control the hydraulic, retarding, and mechanical performance of these materials.

Acknowledgment This research has been conducted within the IAEA-coordinated research project (CRP) "Behaviors of Cementitious Materials in Long Term Storage and Disposal."

References

1. Abdel Rahman RO, El-Kamash AM, Zaki AA, El Sourougy MR (2005) Disposal: a last step towards an integrated waste management system in Egypt. In: Proceedings of the Interna-tional Conference on the Safety of Radioactive Waste Disposal, Tokyo, pp. 317–324 (IAEA-CN-135/81)
2. Ojovan MI, Lee WE (2005) An introduction to nuclear waste immobilisation. Elsevier Science, Amsterdam. ISBN: 0-080-44462-8.
3. Abdel Rahman RO, Kozak MW, Hung YT (2014) Radioactive pollution and control. In: Hung YT, Wang LK, Shammas NK (eds) Handbook of environmental and waste

management. Vol. 2: land and groundwater pollution control. World Scientific, Singapore, pp 950–1028

4. International Atomic Energy Agency (2001) Performance of engineered barrier materials in near surface disposal facilities for radioactive waste. IAEA, Vienna (TECDOC-1255)

5. International Atomic Energy Agency (1984) Design, construction, operation, shutdown and surveillance of repositories for solid radioactive wastes in shallow ground. Safety series no. 63. IAEA, Vienna

6. Gupta MP, Mondal NK, Bodke SB, Bansal NK (1997) Indian experience in near surface disposal of low level radioactive solid wastes, planning and operation of low level waste disposal facilities. In Proceedings of the international symposium, Vienna. IAEA, Vienna, pp 275–284

7. Sakabe Y (1997) Design concept and its development for the Rokkasho low level radioactive waste disposal centre, planning and operation of low level waste disposal facilities. In Proceedings of the symposium, Vienna. IAEA, Vienna, pp 123–132

8. International Atomic Energy Agency (2003) Safety considerations in the disposal of disused sealed radioactive sources in borehole facilities. IAEA, Vienna (IAEA-TECDOC-1368)

9. Abdel Rahman RO, Rakhimov RZ, Rakhimova NR, Ojovan MI (2014) Cementitious materials for nuclear waste immobilisation. Wiley, New York

10. Spence RD, Caijun Shi (2005) Stabilization and solidification of hazardous, radioactive, and mixed wastes. CRC Press, Boca Raton

11. Mindess S, Young JF, David D (2003) Concrete, 2nd edn. Prentice Hall, Pearson Education, New York

12. Abdel Rahman RO, Zein DH, Abo SH (2013) Assessment of Strontium immobilization in cement-bentonite matrices. Chem Eng J 228:772–780

13. Odler I (2006) Setting and hardening of Portland cement. In: Hewlett PC (ed) Lea's chemistry of cement and concrete. Elsevier Science, Amsterdam, pp 241–297

14. Bishop M, Bott SG, Barron AR (2003) A new mechanism for cement hydration inhibition: solid-state chemistry of calcium nitrilotris (methylene) triphosphonate. Chem Mater 15:3074–3088

15. Barron AR (2010) Hydration of Portland cement. Connexions website: http://cnx.org/content/m16447/1.11/content_info. Last accessed 21 Aug 2014

16. Taylor HF (1997) Cement chemistry. Thomas Telford, London

17. Thomas JJ, Jennings HM, Allen AJ (2010) Relationships between composition and density of tobermorite, jennite, and nanoscale CaO-SiO_2-H_2O. J Phys Chem C 114:7594–7601

18. Schindler AK (2004) Prediction of concrete setting. In: Weiss J, Kovler K, Marchand J, Mindess S (eds) Proceedings of the RILEM international symposium on advances in concrete through science and engineering. RILEM Publications SARL, Evanston. http://www.eng.auburn.edu/users/antons/Publications/RILEM-Schindler-Concrete-Setting-Dec-03.pdf

19. International Atomic Energy Agency (2004) Long term behavior of low and intermediate level waste packages under repository conditions. IAEA, Vienna (IAEA-TECDOC-1397)

20. Drace Z, Mele I, Ojovan MI, Abdel Rahman RO (2012) An overview of research activities on cementitious materials for radioactive waste management. Mater Res Soc Symp Proc 1475:253–264

21. International Atomic Energy Agency (2004) Safety assessment methodologies for near surface disposal facilities, vol 1. IAEA, Vienna

22. Walton JC, Plansky LE, Smith RW (1990) Models for estimation of service life of concrete barriers in low-level radioactive waste disposal. U.S. Nuclear Regulatory Commission, Washington (NUREG/CR-5542 EGG-2597)

23. Abdel Rahman RO, Zaki AA (2009) Assessment of the leaching characteristics of incineration ashes in cement matrix. Chem Eng J 155:698–708

24. Suzuki K, Ono Y (2008) Leaching characteristics of stabilized/solidified fly ash generated from ash-melting plant. Chemosphere 71:922–932

25. Geankoplis CJ (1993) Principles of unsteady-state and convective mass transfer. In: Betty S (ed) Transport processes and unit operations. Prentice Hall, Englewood Cliffs
26. Abdel Rahman RO, Zaki AA (2011) Comparative study of leaching conceptual models: Cs leaching from different ILW cement based matrices. Chem Eng J 173:722–736
27. El-Kamash AM, Mohamed RO, Nagy ME, Khalill MY (2002) Modeling and validation of radionuclides releases from an engineered disposal facility. Int J Waste Manage Environ Restoration 22(4):373–393
28. Johnston HH, Wilinot DJ (1992) Sorption and diffusion studies in cementitious grouts. Waste Manage 12(2–3):289–297
29. Mohamed RO, El-Kamash AM, El-Sourougy MR (2000) Performance assessment for backfill materials in waste repository site. In: International Conference on the Safety of Radioactive Waste Management, Spain, 13–17 March (IAEA-CN-78/69)
30. Plecas IB, Dimovic SD (2009) Mathematical modelling of immobilization of radionuclides 137Cs and 60Co in concrete matrix. J Open Waste Manage 2:43–46
31. Marsavina L, Audenaert K, De Schutter G, Faur N, Marsavina D (2009) Experimental and numerical determination of the chloride penetration in cracked concrete. Construct Build Mater 23:264–274
32. Garbalińska H, Kowalski SJ, Staszak M (2010) Linear and non-linear analysis of desorption processes in cement mortar. J Cem Concr Res 40:752–762
33. Abdel Rahman RO, Ibrahim HA, Abdel Monem NM (2009) Long-term performance of zeolite Na A-X blend as backfill material in near surface disposal vault. Chem Eng J 149:143–152
34. Abdel Rahman RO (2005) Performance assessment of unsaturated zone as a part of waste disposal site. PhD thesis, Faculty of Engineering, Alexandria University, Alexandria
35. Sahara F, Murakami T, Kobayashi I, Mihara M, Ohi T (2008) Modelling for the long-term mechanical and hydraulic behavior of betonite and cement-based materials considering chemical transitions. Phys Chem Earth 33:531–537
36. Abdel Rahman RO (1999) Performance assessment of engineered barrier in waste disposal site. MSc thesis, Faculty of Engineering, Alexandria University, Alexandria
37. Abdel Rahman RO (2011) Preliminary evaluation of the technical feasibility of using different soils in waste disposal cover system. Environ Prog Sustainable Energy 30(1):19–28
38. International Atomic Energy Agency (2001) Procedures and techniques for closure of near surface disposal facilities for radioactive waste. IAEA, Vienna (TECDOC-1260)
39. Drace Z, Ojovan MI (2009) The behaviors of cementitious materials in long term storage and disposal: an overview of results of the IAEA coordinated research program. Mater Res Soc Symp Proc 1193:663–672
40. Ojovan MI, Drace Z (2013) Processing and disposal of radioactive waste: selection of technical solutions. Mater Res Soc Symp Proc 1518:203–209 (mrsf12-1518-ll09-02). doi:10.1557/opl.2012.1569
41. Dauzeres A, Le Bescop P, Sardini P, Dit Coumes CC (2010) Physico-chemical investigation of clayey/cement-based materials interaction in the context of geological waste disposal: experimental approach and results. J Cem Concr Res 40(8):1327–1340
42. NIREX (2008) The longevity of intermediate-level radioactive waste packages for geological disposal: a review. Environment Agency, Rio House, Bristol (NWAT/Nirex/06/003)
43. Berner U (2003) Project opalinus clay: radionuclide concentration limits in the cementitious near-field of an ILW repository, Nuclear Energy and Safety Research Department Laboratory for Waste Management. PSI Bericht 02–26. PSI, Villigen. http://www.iaea.org/inis/collection/NCLCollectionStore/_Public/34/065/34065552.pdf
44. Langton CA, Kosson DS, Garrabrants A (2007) Engineered cementitious barriers for low-activity radioactive waste disposal. Workshop summary and recommendations for DOE Office of Environmental Management (WSRC-TR-2006-00226)
45. ICRP (2007) Recommendation of the international commission on radiological protection. ICRP publication 103. Ann ICRP 37:2–4

46. ICRP (2013) Task group 80 geological disposal of long-lived solid radioactive waste. Ann ICRP 42(3):1–57
47. IAEA (2013) Near surface disposal of radioactive waste. Draft safety guide no. DS356. IAEA, Vienna
48. IAEA (2013) The safety case and safety assessment for radioactive waste disposal. Draft safety guide no. DS 355. IAEA, Vienna
49. IAEA (2013) Disposal of radioactive waste. Draft safety guide no. DS354. IAEA, Vienna
50. Joint CNRA/CRPPH/RWMC Workshop (1997) Regulating the long-term safety of radioactive waste disposal. In: Proceedings of an NEA international workshop held in Córdoba, Spain, 20–23 Jan1997. OECD, Paris
51. Glasser FP (1992) Progress in the immobilization of radioactive wastes in cement. J Cem Concr Res 22:201–209
52. Burns RH (1971) Solidification of low and intermediate level waste. At Energy Rev 9:547–552
53. Plecas I, Dimovic S (2003) Immobilization of Cs and Co in concert matrix. Ann Nucl Energy 30:1899–1903
54. Abdel Rahman RO, Zaki AA, El-Kamash AM (2007) Modeling the long-term leaching behavior of 137Cs, 60Co, and 152,154Eu radionuclides from cement-clay matrices. J Hazard Mater 145(3):372–380
55. Ojovan MI, Varlackova GA, Golubeva ZI, Burlaka ON (2011) Long-term field and laboratory leaching tests of cemented radioactive wastes. J Hazard Mater 187:96–302
56. Batchelor B (2006) Overview of waste stabilization with cement. Waste Manage 26:689–698
57. Strachan DM (2001) Glass dissolution: testing and modelling for long-term behaviour. J Nucl Mater 298:69–77
58. Lee WE, Ojovan MI, Stennett MC, Hyatt NC (2006) Immobilisation of radioactive waste in glasses, glass composite materials and ceramics. Adv Appl Ceram 105(1):3–12
59. Ojovan MI, Ojovan NV, Startceva IV, Barinov AS (2002) Some trends in radioactive waste form behaviour revealed in long term field tests. In: Proceedings of the WM'02, Tucson, Arizona, 24–28 Feb 2002, 6p (CD-ROM; 385.pdf)
60. Dmitriev SA, Barinov AS, Ozhovan NV, Startseva IV, Golubeva ZI, Sobolev IA, Ozhovan MI (2005) Results of 16 years of testing of vitrified intermediate-level wastes from the Kursk nuclear power plant. At Energy 98(3):196–201
61. Plecas I, Pavlovic R, Pavlovic S (2004) Leaching behaviour of ^{60}Co and ^{137}Cs from spent ion exchange resins in cement-bentonite clay matrix. J Nucl Mater 327:171–174
62. Varlakova GA, Golubeva ZI, Barinov AS, Sobolev IA, Ojovan MI (2009) Properties and composition of cemented radioactive wastes extracted from the mound-type repository. In: Material Research Society symposium proceedings, Warrendale (1124-Q05-07)
63. Sobolev IA, Dmitriev SA, Barinov AS, Varlackova GA, Golubeva ZI, Ojovan MI (2005) Reliability of cementation technology proved via long-term field tests. In: Proceedings of the ICEM'05, Glasgow, 4–8 September. ASME, Fairfield (ICEM05-1216)
64. NRC US (1991) Technical position on waste form. Low-Level Waste Management Branch, Division of Low-Level Waste Management and Decommissioning, U.S. Nuclear Regulatory Commission, Office of Nuclear Material Safety and Safeguards, Washington
65. ASTM. Standard test method for compressive strength of cylindrical concrete specimens. ASTM International, West Conshohocken (C39/C39M – 09a). http://www.astm.org/Standards/C39.htm
66. Environment Agency (2001) Guidance on the disposal of contaminated soils, version 3. Environment Agency, Bristol
67. The Dutch List (1994) Intervention values and target values: soil quality standards. Ministry of Housing, Spatial Planning and Environment, The Hague
68. Stegemann JA, Côté PL (1996) A proposed protocol for evaluation of solidified wastes. Sci Total Environ 178(1–3):103–110
69. Sherwood PT (1993) Soil stabilization with cement and lime. HMSO, London

70. Sobolev IA, Dmitriev SA, Barinov AS, Varlakova GA, Golubeva ZI, Startceva IV, Ojovan MI (2006) 39-years performance of cemented radioactive waste in a mound type repository. Mater Res Soc Symp Proc 932:721–726
71. Sinha PK, Shanmugamani AG, Renganathan K, Muthiah R (2009) Fixation of radioactive chemical sludge in a matrix containing cement and additives. Ann Nucl Energy 36:620–625
72. Anji Reddy D, Khandelwal SK, Muthiah R, Shanmugamani AG, Paul B, Rao SVS, Sinha PK (2010) Conditioning of sludge produced through chemical treatment of radioactive liquid waste; operating experiences. Ann Nucl Energy 37(7):934–941
73. Dogaru D, Jinescu C, Dogaru G (2009) Influence of the sludge content on the mechanical properties of the cemented based radioactive waste form. Rev Chim 60(8):826–829
74. Shi C, Fernández-Jiménez A (2006) Stabilization/solidification of hazardous and radioactive wastes with alkali-activated cements. J Hazard Mater 137(3):1656–1663
75. Sugaya A, Horiguchi K, Tanaka K, Akutsu S (2009) Cement based encapsulation experiments of low radioactive phosphate effluent. Mater Res Soc Symp Proc 1124:373–378
76. Connell M (2008) Sellafield chooses GGBS due to long-term durability and immobilisation benefits. Concrete 42(6):26–28
77. Jenni A, Hyatt NC (2010) Encapsulation of caesium-loaded Ionsiv in cement. Cem Concr Res 40(8):1271–1277
78. Plecas I, Dimovic S (2006) Immobilization of 137Cs and 60Co in concrete matrix. J Porous Media 9(2):181–184
79. Plecas I (2005) Leaching study in process of solidification of radionuclide ^{60}Co in concrete. Pol J Environ Stud 14(5):699–701
80. Plecas I, Dimovic S (2006) Immobilization of ^{137}Cs and ^{60}Co in a concrete matrix. Part two: mathematical modeling of transport phenomena. Porous Media 9(5):483–489
81. Plećaš I, Dimovic S (2006) Curing time effect on compressive strength and incremental leaching rates of ^{137}Cs and ^{60}Co in cement immobilized sludge. Prog Nucl Energy 48 (7):629–633
82. Plecas I, Dimovic S, Smiciklas I (2006) Utilization of bentonite and zeolite in cementation of dry radioactive evaporator concentrate. Prog Nucl Energy 48(6):495–503
83. McGlinn PJ, Brew DRM, Aldridge LP, Payne TE, Olufson KP, Prince KE, Kelly IJ (2008) Durability of a cementitious waste form for intermediate level waste. Mater Res Soc Symp Proc 1107:101–108
84. Papadokostaki KG, Savidou A (2009) Study of leaching mechanisms of caesium ions incorporated in Ordinary Portland Cement. J Hazard Mater 171(1–3):1024–1031
85. Wieland E, Tits J, Kunz D, Dähn R (2008) Strontium uptake by cementitious materials. Environ Sci Technol 42(2):403–409
86. El-Kamash AM, El- Nagga r MR, El-Dessouky MI (2006) Immobilization of cesium and strontium radionuclides in zeolite-cement blends. J Hazard Mater 136(2):310–316
87. Osmanlioglu AE (2006) Management of spent sealed radioactive sources in Turkey. Health Phys 91(3):258–262
88. Guerrero A, Goni S (2002) Efficiency of blast furnace slag cement for immobilize simulated borate radioactive liquid waste. Waste Manage 22:831–836
89. Fernandez-Jimenz A, Macphee DE, Lachowsk EE, Dalomo A (2005) Immobilization of Cs in alkaline activated fly ash matrix. J Nucl Mater 346:185–193
90. Osmanlioglu AE (2002) Immobilization of radioactive waste by cementation with purified kaolin clay. Waste Manage 22:481–483
91. El Kamash AM, El Dakroury AM, Aly HF (2002) Leaching kinetic of Cs and Co radionuclide fixed in cement and cement based materials. J Cem Concr Res 32:1797–1803
92. Sakr K, Sayed MS, Hafez N (1997) Comparison studies between cement and cement-kaolinite properties for incorporation of low level radioactive wastes. J Cem Concr Res 27:1919–1926
93. El-Dakroury AMS (2008) Gasser, Rice husk ash (RHA) as cement admixture for immobilization of liquid radioactive waste at different temperatures. J Nucl Mater 381(3):271–277

94. Foldsova M, Luckac P (1996) Leachability of Co and Cs from natural and chemically treated zeolites. J Radioanal Nucl Chem 214:479–487

95. Dyer A, Las T, Zubair M (2001) The use of natural zeolites for radioactive waste treatment studies on leaching from zeolite/cement composites. J Radioanal Nucl Chem 243:839–841

96. Junfeng L, Gong ZW, Jianlong W (2005) Solidification of low level-radioactive resins in ASC zeolite blends. Nucl Eng Des 235:817–821

97. Osmanlioglu AE (2006) Treatment of radioactive liquid waste by sorption on natural zeolite in Turkey. J Hazard Mater 136:310–316

98. Varlakova GA, Dyakonova AT, Netrusov AI, Ojovan MI (2011) Microbiological activities on cementitious waste forms in a shallow-ground repository. J Mater Sci Eng B1:591–596

99. Hussein OH-O, Ojovan M, Kinoshita H (2011) Immobilisation of BaSO4: phases and microstructure of OPC-BaSO4 system cured at an elevated temperature. In: Proceedings of the WM'11 conference, Phoenix, 27 Feb–3 Mar 2011, p 11 (WM – 11012)

100. Varlackova GA, Golubeva ZI, Barinov AS, Roschagina SV, Dmitriev SA, Sobolev IA, Ozhovan MI (2009) Evaluation of the cemented radioactive waste with prolonged tests in mound type repository. At Energy 107(1):32–38

101. Spasova LM, Ojovan MI (2008) Characterisation of Al corrosion and its impact on the mechanical performance of composite cement wasteforms by the acoustic emission technique. J Nucl Mater 375:347–358

102. Spasova LM, Ojovan MI, Scales CR (2007) Acoustic emission technique applied for monitoring and inspection of cementitious structures encapsulating aluminium. J Acoust Emission 22:51–68

103. Eskander SB, Bayoumi TA, Saleh HM (2013) Leaching behavior of cement-natural clay composite incorporating real spent radioactive liquid scintillator. Prog Nucl Energy 67:1–6

104. Abdel Rahman RO, Zein DH, Abo Shadi H (2014) Cesium binding and leaching from single and binary contaminant cement-bentonite matrices. Chem Eng J 245:276–287

105. Navarro Santos M (2004) Behaviour of concrete container CE-2 a under disposal conditions. IAEA, Vienna (Tec Doc 1397)

106. Chen L, Li J-H (2009) Analysis of cementation technology for liquid radioactive-waste in PWR NPPs. Nucl Power Eng 30(2):113–116

107. Craeyea B, De Schutter G, Van Humbeeck H, Van Cotthem A (2009) Early age behaviour of concrete supercontainers for radioactive waste disposal. Nucl Eng Des 239:23–35

108. Andrade C, Martínez I, Castellote M, Zuloaga P (2006) Some principles of service life calculation of reinforcements and in situ corrosion monitoring by sensors in the radioactive waste containers of El Cabril disposal (Spain). J Nucl Mater 358(2–3):82–95

109. Trotignon L, Devallois V, Peycelon H, Tiffreau C, Bourbon X (2007) Predicting the long term durability of concrete engineered barriers in a geological repository for radioactive waste. Phys Chem Earth 32(1–7):259–274

110. Poyet S, Charles S (2009) Temperature dependence of the sorption isotherms of cement-based materials: heat of sorption and Clausius-Clapeyron formula. Cem Concr Res 39 (11):1060–1067

111. Fernández R, Cuevas J, Mäder UK (2010) Modeling experimental results of diffusion of alkaline solutions through a compacted bentonite barrier. Cem Concr Res 40(8):1255–1264

112. Fomina M, Podgorsky VS, Olishevska SV, Kadoshnikov VM, Pisanska IR, Hillier S, Gadd GM (2010) Fungal deterioration of barrier concrete used in nuclear waste disposal. Geomicrobiol J 24(7–8):643–653

113. Galíndez JM, Molinero J (2010) Assessment of the long-term stability of cementitious barriers of radioactive waste repositories by using digital-image-based microstructure generation and reactive transport modeling. Cem Concr Res 40(8):1278–1289

114. García Calvo JL, Hidalgo A, Alonso C, Fernández Luco L (2010) Development of low-pH cementitious materials for HLRW repositories. Resistance against ground waters aggression. Cem Concr Res 40(8):1290–1297

115. Perlot C, Bourbon X, Carcasses M, Ballivy G (2007) The adaptation of an experimental protocol to the durability of cement engineered barriers for nuclear waste storage. Mag Concr Res 59(5):311–322. ISSN: 0024-9831; E-ISSN: 1751-763X.
116. Small JS, Thompson OR (2009) Modelling the spatial and temporal evolution of pH in the cementitious backfill of geological disposal facility. Mater Res Soc Symp Proc 1124:327–332
117. Sakamoto H, Shibata M, Owada H, Kaneko M, Kuno Y, Asano H (2007) Development of an analytical technique for the detection of alteration minerals formed in bentonite by reaction with alkaline solutions. Phys Chem Earth 32(1–7):311–319
118. Shimura T, Fujiwara A, Vomvoris S, Marschall P, Lanyon GW, Ando K, Yamamoto S (2006) Large-scale gas migration test at grimsel test site. In: Proceedings of the 11th International High-level Radioactive Waste Management Conference, IHLRWM, pp 784–791
119. Antonv D (2003) Soil based barriers for a low and intermediate level radioactive waste disposal. In: Proceedings of the International Conference on Radioactive Waste Management and Environmental Remediation, ICEM 1, pp 571–573
120. Wieland E, Spieler P (2001) Colloids in the mortar backfill of a cementitious repository for radioactive waste. Waste Manage 21(6):511–523
121. Huertas FJ, Hidalgo A, Rozalén ML, Pellicione S, Domingo C, García-González CA, Andrade C, Alonso C (2009) Interaction of bentonite with supercritically carbonated concrete. Appl Clay Sci 42(3–4):488–496

Chapter 10
Extensive Monitoring System of Sediment Transport for Reservoir Sediment Management

Chih-Ping Lin, Chih-Chung Chung, I-Ling Wu, Po-Lin Wu, Chun-Hung Lin, and Ching-Hsien Wu

Contents

1 Introduction 451
2 Review of Surrogate Techniques for Suspended-Sediment Monitoring 454
  2.1 Optical Turbidity 454
  2.2 Acoustic (Ultrasonic) 459
  2.3 Other SSC Surrogate Techniques 461
3 Time-Domain Reflectometry Method 462
  3.1 Principles of TDR SSC Measurement 463
  3.2 New TDR Probe Design and Data Reduction 465
  3.3 Experimental Verification of TDR SSC Measurements 468
4 Case Study: Monitoring Program in the Shihmen Reservoir 471
  4.1 Problem Background of the Shihmen Reservoir 471
  4.2 Planning of SSC Monitoring Program 475
  4.3 Results and Discussion of the First Full-Event Monitoring 477
  4.4 Characteristics of Sediment Transport and Sluicing Operation 482
References 490

C.-P. Lin (✉)
Department of Civil Engineering, National Chiao Tung University, 1001 Ta-Hsueh Rd, Hsinchu 300, Taiwan
e-mail: cplin@mail.nctu.edu.tw

C.-C. Chung
Disaster Prevention and Water Environment Research Center, National Chiao Tung University, Hsinchu, Taiwan, ROC
e-mail: chung.chih.chung@gmail.com

I.-L. Wu • P.-L. Wu • C.-H. Lin
Disaster Prevention and Water Environment Research Center, National Chiao Tung University, 1001 Ta-Hsueh Rd, Hsinchu 300, Taiwan
e-mail: testqk@gmail.com; plwu@mail.nctu.edu.tw; chlin.ce@gmail.com

C.-H. Wu
Water Resources Planning Institute, Water Resources Agency, 1340 Chung-Cheng Rd., Wu-Fong, Taichung, Taichung County, Taiwan
e-mail: wcs@wrap.gov.tw

© Springer International Publishing Switzerland 2016
L.K. Wang, M.-H.S. Wang, Y.-T. Hung and N.K. Shammas (eds.),
Natural Resources and Control Processes, Handbook of Environmental Engineering, Volume 17, DOI 10.1007/978-3-319-26800-2_10

Abstract Sedimentation is a serious threat to long-term water resource management worldwide. In particular, reservoir sedimentation is becoming more serious in Taiwan due to geological weathering and climate change in watersheds. Large amount of sediments transport to reservoirs during storm events at hyperpycnal concentration. Full-event monitoring of sediment transport in a reservoir plays an important role in sustainable reservoir management. This chapter begins by reviewing existing surrogate techniques in need for monitoring suspended-sediment transport in reservoirs with high concentration range and wide spatial coverage. More commercially available techniques suffer from particle-size dependency and limited measurement range. This chapter introduces a relatively new technique based on time-domain reflectometry. It possesses several advantages, including particle-size independence, high measurement range, durability, and cost-effective multiplexing. This chapter describes a modified TDR technique for better field applicability and demonstrates its application in an extensive SSC monitoring program for reservoir management through a case study in Shihmen Reservoir, Taiwan. Monitoring stations were installed at the major inflow river mouth and outlet works with fixed protective structures to provide inflow and outflow sediment-discharge records. To capture the characteristics of density currents, a multi-depth monitoring station was designed and deployed on floating platforms in the reservoir. Some of the data collected during typhoons are presented as an example to demonstrate the effectiveness and benefits of the TDR-based monitoring program.

Keywords Reservoir sedimentation • Density current • Sediment monitoring • Sediment transport • Time-domain reflectometry (TDR)

Nomenclature

$c =$ Speed of light (2.998×10^8 m s^{-1}) [m s^{-1}]
$dt =$ Sampling interval [s]
$G_s =$ Specific gravity of suspended sediment [—]
$L =$ Length of the sensing waveguide[m]
$ppm =$ Parts per million (or milligram per liter [mg L^{-1}])
$SS =$ Suspended-sediment concentration in terms of volume fraction [—]
$T =$ Measured temperature in degree Celsius [°C]
$\Delta t =$ Round-trip travel time of the EM pulse in the sensing waveguide [s]
$\varepsilon =$ Dielectric constant [—]
$\varepsilon_{ss} =$ Dielectric constant of suspended sediment [—]
$\varepsilon_w =$ Dielectric constant of water [—]

1   Introduction

Sedimentation is a serious threat to long-term water resource management world-wide. In Taiwan, due to geological weathering and climate change, sediment yields from soil erosion and landslides are experiencing unexpected increase in many catchments in the last decade. The 1999 Mw 7.6 Chi-Chi earthquake has triggered at least 20,000 landslides in the Western Foothills fold-and-thrust belt of the Taiwan orogen. The coseismic and postseismic landslides have resulted in elevated postseismic erosion rates and a cascade of sediment from hillslopes to channels [1]. On top of that, records of extreme precipitation have been broken constantly. Torrential rains carried by typhoons often causes extensive landslides in upland areas. As an example, Fig. 10.1 revealed many storm-triggered landslides in the Shihmen Reservoir watershed during Typhoon Aere in 2004. Large amount of sediments transport to reservoirs during typhoon events at hyperpycnal concentration. The photograph in Fig. 10.2 vividly shows such a phenomenon at the river mouth flowing into Wushe reservoir during Typhoon Morakot in 2009.

The annual sediment yield being deposited into existing reservoirs varies worldwide over a range from 20 to 5000 m^3/km^2, with a world average of about 100 m^3/km^2. The total storage loss and annual sedimentation rate were estimated to be about 11.8 % and 0.52 %, respectively, according to Sumi and Hirose in 2002 [2]. Recently, Taiwan has experienced dramatic increase in sedimentation rate. As of 2013, the total storage loss of the 21 main reservoirs with original design storage greater than 10×10^6 m^3 has now reached 37 %, much higher than the world

Fig. 10.1 Photo showing extensive landslides in the catchment of the Shihmen Reservoir during Typhoon Aere in 2004

Fig. 10.2 Photo showing large amount of sediments transports to the Wu-She reservoir at hyperpycnal concentrations during Typhoon Morakot in 2009 (Photo taken by Mr. Ke Wun-Tao, Courtesy of Prof. H Capart, Morphohydraulics Research Group, National Taiwan University, Taiwan)

average and threatening the functions of dams. In particular, three of the four major large dams, Zengwen, Shihmen, and Nanhua, have lost 36.8 %, 33 %, and 36.6 % of total storage, respectively.

Most existing dams were planned and designed to work for a finite period of time (i.e., the design life), frequently as short as 100 years, which will eventually be terminated by sedimentation loss. Little thoughts have been given to reservoir replacement when the excessive effective storage is consumed by sedimentation or to measures to maintain reservoir services and sustain long-term use despite continued sediment inflow. Reservoirs' construction requires sites with particular hydrologic, geologic, topographic, and geographic characteristics, and existing reservoirs generally occupy the best available sites. Potential sites for new dams are very limited and may not be feasible from the economic, social, political, or environmental standpoints. In addition, the costs of safely decommissioning a dam at the end of its useful life can be quite substantial. Therefore, it is imperative for water resource management to focus on how to sustain the function of existing dams as they are increasingly affected by sedimentation [3].

Because of high cost and multiple problems associated with sediment removal and disposal on a massive scale, the sedimentation of large reservoirs is to a large extent an irreversible process. Sustainable reservoir management involves a combination of watershed management, the construction of low-level outlets or bypass works, temporary removal of dam service for sediment removal activities, release of increased volumes of water downstream for sediment discharge, and dredging. Characterization of sediment yield and sediment transport in a reservoir is crucial in many of the aforementioned measures. Several large remediation projects have been embarked in Taiwan to either renovate existing outlet structures or construct new sluicing tunnels to slow down sedimentation. Real-time monitoring of sediment transport plays an important role in planning and operation of these facilities when completed.

The difficulty in quantifying sediment volumes transporting through natural streams has always impeded understandings of catchment hydrology and impacts on land management. Streams carry most of the total sediment transports during flood events, which often occur at night and are hard to predict. Although suspended-sediment concentration (SSC) measurement by sediment sampling is the most direct approach, there is a considerable difficulty and expense for a full runoff event monitoring due to the large spatial and temporal variability associated with the suspended-sediment transportation. Apparently, an automated surrogate measurement system is inevitable to estimate the discrete storm event loads. In general, the relation between instantaneous measurements of water discharge and suspended-sediment concentration varies dramatically for such a purpose. Serious over- or underestimating of loads using sediment rating curves has been observed particularly for short time frames [4]. Although bias-correction procedures can be applied, the substantial scatter evidenced by most rating relationships and complexities associated with hysteresis and exhaustion effects are considered to preclude any major improvements under the reliability of rating curves [5]. Methods are required to obtain more accurate load estimates for discrete storm events at reasonable costs.

In addition to the sediment load of streams and catchment sediment yield, characterization of sediment transport in a reservoir is crucial to planning and operation of low-level outlets or bypass tunnels. The transport and deposition of the sediments in narrow alpine reservoirs is often caused by turbidity currents. They follow the thalweg of the lake to the deepest area near the dam, where the sediments can affect the operation of bottom outlet and intake structures. An ideal monitoring system of reservoir sediment transport should provide in real time the following information: (a) sediment inflow into the reservoir, (b) discharge of sediments through outlets, (c) characteristics of turbidity currents (including the occurrence time, corresponding SSC, thickness, and moving speed), and (d) sediment lofting and settling of stagnant muddy water. Such an extensive monitoring program is challenging considering the lack of surrogate techniques for SSC monitoring that can be deployed extensively and robustly in the field with sufficient temporal and spatial resolution.

This chapter introduces an innovative SSC monitoring technique based upon time-domain reflectometry (TDR) and demonstrates its application to an extensive SSC monitoring program for reservoir management through a case study in Shihmen Reservoir, Taiwan. Monitoring stations were installed at the major inflow river mouth and outlet works with fixed protective structures to provide inflow and outflow sediment-discharge records. To capture the characteristics of turbidity currents, a multi-depth monitoring station was designed and deployed on floating platforms in the reservoir. Some of the data collected during typhoon events are presented as an example to demonstrate the effectiveness and benefits of the TDR-based monitoring program.

2 Review of Surrogate Techniques for Suspended-Sediment Monitoring

Surrogate techniques for SSC measurement have been reported, including optical backscatter, acoustic, focused beam reflectance, laser diffraction, nuclear, optical transmission, spectral reflectance, vibrating tube, differential pressure, etc. [6–8]. The operating principles, advantages, and disadvantages of various techniques are summarized in Table 10.1.

More commercial instruments operating on bulk optic (turbidity), laser optic, pressure difference, and acoustic backscatter have been or are the foci of field or laboratory tests by the US Geological Survey (USGS) and other organizations in riverine settings [7]. These instruments were evaluated based on cost, reliability, robustness, accuracy, sample volume, susceptibility to biological fouling, and suitable range of mass concentration and particle-size distribution. They show varying degrees of promise toward supplanting traditional data collection methods based on routine collection of physical samples and subsequent laboratory analyses. However, these potential instruments are subjected to at least one of the following limitations: (a) small measurement range, (b) strong particle-size dependency [6, 9, 10], and (c) too expensive and fragile instruments in torrent regimes. Suspended-sediment concentrations of density currents in reservoirs can be in excess of 10 g/L (or 10,000 ppm) or even 100 g/L (or 100,000 ppm), as increasingly encountered during typhoon events in Taiwan since 2004. When SSC exceeds the range of the automated measurement device, information is lost during such critical periods. The continuous monitoring techniques most readily available as adequate commercial products are turbidity probes based on optical backscatter. Nonetheless, these probes are mostly suitable for low SSC measurements [11]. Optical and acoustic probes also exhibit strong particle-size dependency. Site-specific calibrations aimed to account for particle-size dependency can be extremely difficult because the particle size of suspended sediments may vary drastically with water depth and flow velocity. Moreover, the main sensing components of existing instruments are packaged inside the probe to be submerged in water. These instruments are prone to damage during flood events. They are also often too expensive to deploy in wide spatial coverage. Several techniques of particular interest for sediment transport monitoring in reservoir are introduced in more detail in the following sections.

2.1 Optical Turbidity

Among the potential SSC surrogate technologies, turbidity is the most common of such surrogate technology and the first to be sanctioned by the US Geological Survey for use in producing data used in concert with water discharge data to compute sediment concentrations and fluxes for storage in the National Water Information System [8]. The technology is well-established and inexpensive with

Table 10.1 Summary of selected attributes of suspended-sediment surrogate technologies

Technology	Operating principle	Advantages	Limitations
Acoustic backscatter[a] (e.g., ADCP)	Sound backscattered from sediment is used to determine size distribution and concentration	Good spatial and temporal resolution, measures over wide vertical range, nonintrusive	Backscattered acoustic signal is difficult to translate, signal attenuation at high particle concentration
Acoustic attenuation	Acoustic attenuation through sediment is used to determine concentration	Simple, good temporal resolution, allows remote deployment and data logging, wide concentration measuring range, allows remote deployment and data logging	Strong particle-size dependency, flow intrusive, at-a-point measurement
Focused beam reflectance[a]	Time of reflection of laser incident on sediment particles is measured	No particle-size dependency, wide particle size, and concentration measuring range[b]	Expensive, flow intrusive, at-a-point measurement
Laser diffraction[a]	Refraction angle of laser incident on sediment particles is measured	No particle-size dependency	Unreliable, expensive, flow intrusive, at-a-point measurement, limited particle-size range
Nuclear[a]	Backscatter or transmission of gamma or X-rays through water-sediment samples is measured	Low power consumption, wide particle size, and concentration measuring range	Low sensitivity, radioactive source decay, regulations, flow intrusive, at-a-point measurement
Optical[a]	Backscatter, nephelometer, or transmission of visible or infrared light through water-sediment sample is measured	Simple, good temporal resolution, allows remote deployment and data logging, relatively inexpensive	Exhibits strong particle-size dependency, flow intrusive, at-a-point measurement, instrument fouling, limited concentration range except nephelometer
Remote spectral reflectance[a]	Light reflected and scattered from body of water is remotely measured	Able to measure over broad areas	Poor resolution, poor applicability in fluvial environment, particle-size dependency
Pressure difference[c]	Simultaneous measurements from two exceptionally sensitive pressure transducer sensors for reading difference in pressure that is converted to the water density and sediment concentration	The pressure difference technology's inference of SSC in a single vertical is an improvement over at-a-point measurements, relatively robust	Low concentration, data unreliable; veracity of higher concentrations unresolved, particle-size dependency

(continued)

Table 10.1 (continued)

Technology	Operating principle	Advantages	Limitations
Vibrating tube	The vibrational period of the tube is monitored for converting to the water density and sediment concentration	Simple, maybe allows remote deployment and data logging	Flow intrusive, at-a-point measurement, particle-size dependency unknown, sediment settlement as low flow velocity, expensive
Capacitance	The output current of the capacitance is measured to convert to the sediment concentration	Simple, maybe allows remote deployment and data logging, wide concentration measuring range	Particle-size dependency, low sensitivity at low concentration, at-a-point measurement, signal attenuation at high salinity

[a]Modified after Wren et al. [6]
[b]Wide concentration measuring range means measurement range is greater than 10 g/L
[c]Modified after Gray and Gartner [7]

respect to other suspended-sediment surrogate technologies. Turbidity is an expression of the optical properties of a sample that causes light rays to be scattered and absorbed rather than transmitted in straight lines through the sample [12]. All turbidity measurements detect the amount of light either transmitted through or scattered by the particles in a sample of water.

The detection angle between the centerline of the incident light beam and the centerline of the detector's receiving angle has a significant impact on the detection of particles from a size perspective and on the turbidity range of the instrument. Depending on the detection angle, turbidimeters operate on three different modes: 90° nephelometry, 180° transmissometry, and backscatter modes. The 90° detection angle, also known as nephelometric detection angle, is the default technology for most regulatory applications because of its sensitivity to a broad range of particle sizes. Nephelometers are calibrated using standards containing formazin to give NTU (nephelometric turbidity unit) readings. It is very susceptible to color interference and is best used at low turbidity levels. Turbidity can indicate the presence of sediment in waters. However, the relationship between NTU and SSC can vary widely in utility. Transmissometers employ a light source beamed directly at a light detector. The instrument measures the fraction of visible light from a collimated light source that reaches the detector. The fraction of light reaching the detector is converted to a beam attenuation coefficient, which is related to SSC. The transmissometry mode has the greatest susceptibility to absorbance and color interferences. And the measurement sensitivity is highly wavelength dependent. The optical backscatter (OBS) mode measures the light that is scattered in the direction of the incident light beam. The backscatter angle (typically between 0° and 45° from the incident light) is sensitive to light scatter that is reflected back in the direction of the incident light source, which is characteristic with extremely high turbidity samples. This is not an appropriate technique for low-level turbidity monitoring because it has inherent poor sensitivity at these levels. Over the past several decades, instrument technology has advanced dramatically, and many turbidity measurement techniques have resulted. Some turbidity measurement technology involves the use of signal ratios of two or more detectors to determine the turbidity value. One detector is at 90° from the incident source and the other detectors can be at any of several different angles. This ratioing technique can help compensate for color interference and in optical changes such as light source degradation. Transmissometers and nephelometers are more sensitive at low SSC, but OBSs have superior linearity in more turbid water. OBSs are normally used in regions where reservoir sedimentation is a problem.

All turbidity measurements detect the amount of light either transmitted through or scattered by the particles in a sample of water. It is important to remember that turbidity is not a measure of the quantity of suspended solids in a sample but, instead, an aggregate measure of the combined scattering effect of the water sample's suspended particles on an incident light source. The most important of the factors associated with the use of turbidimeters is particle-size distribution, which can vary in a factor of 1000 in the environment. To a less extent, particle shape and composition can also affect the turbidity under constant SSC. Other drawbacks associated with the use of turbidimeters include nonlinear responses of sensors to SSC and instrument saturation at instrument-specific turbidity values [7, 13–15]. The maximum SSC limits for turbidity instruments also depend on particle-size distribution. The OBS tends to reach the instrumentation saturation at lower SSC as particle gets finer. Sensor saturation that may occur at the higher flows that are most influential in sediment transport can limit the usefulness of the derived turbidity data.

Prior to developing the new TDR-based SSC technique, we tested and evaluated a few more commercially available optical and acoustic instruments for sediment transport monitoring in reservoirs. As examples to illustrate the difficulties encountered in SSC surrogate techniques, Fig. 10.3 shows two instruments (one dual-beam

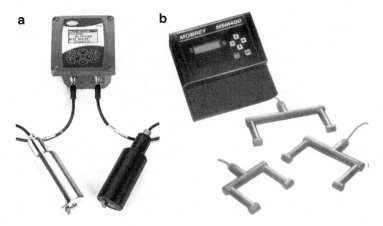

Fig. 10.3 Photographs showing two example instruments tested: (**a**) Hach Solitax sc optical sensor; (**b**) Solartron Mobrey 433 ultrasonic sensor

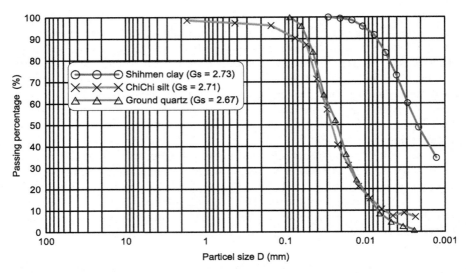

Fig. 10.4 Particle-size distribution and specific gravity Gs of sediments used for testing, including Shihmen clay, Chi-Chi silt, and ground quartz

OBS sensor and one ultrasonic acoustic transmissometer) tested to reveal the relation between surrogate sensor response and SSC. Three types of sediments were used for experiments, including a clayey sediment (Gs = 2.73) from the Shihmen Reservoir in Northern Taiwan, a sandy silt (Gs = 2.71) from the Chi-Chi weir in central Taiwan, and a man-made ground quartz (Gs = 2.67) grinded from glass materials. The particle-size distributions (PSDs) of these three sediments are presented in Fig. 10.4. The particle size of the ground quartz was chosen such that its average particle size is close to that of Chi-Chi silt, but more uniform. In addition, the ground quartz is mainly composed of silica, while the mineral composition of Chi-Chi silt is diverse. Because of the high SSC over 100 g/L encountered in turbidity density current in the Shihmen Reservoir, experiments were performed for SSC up to 200 g/L.

The experimental results for the optical sensor, Hach Solitax sc dual-beam sensor (with nephelometric and backscatter photoreceptors), are shown in Fig. 10.5. The instrument has two modes: turbidity mode for NTU measurements from 0.001 to 4000 NTU and suspended solid (SS) mode for SSC measurements from 0.001 mg/L to 50 g/L. Because of the high SSC range tested, the SS mode was used. As can be expected, the sensor has a much higher response to SSC in clayey sediments than in silty and sandy sediments. The OBS gain in Shihmen sediments is twice that of Chi-Chi sediments at the same SSC and becomes nonlinear with SSC at lower concentration. Earlier study showed OBS gain is minimally affected by changes in PSDs in the range of 200–400 μm but is greatly affected by changes if particles are smaller than about 44 μm [16]. Although the above numbers may depend on the wavelength of the light source, it might explain why the relation between OBS and SSC for Chi-Chi sediments is similar to that for ground quartz in

Fig. 10.5 The experimental relation between the sensor output (using the default conversion from OBS gain to SSC) and SSC for the three sediments tested

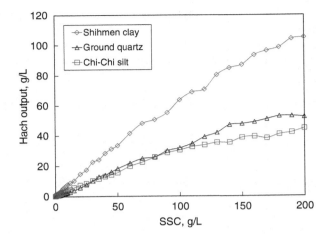

the linear range. Because of the relation between OBS gain and the particle-size distribution, an OBS is best suited for application at sites with relatively stable PSDs. However, the sediments in reservoirs are rich in fine particles smaller than 44 μm, and the PSD can vary significantly with depth and flow velocity during each storm event. Site-specific calibrations aimed to account for particle-size dependency can be extremely difficult since it is time dependent.

2.2   Acoustic (Ultrasonic)

In the context of measuring suspended solids, acoustic measurements are similar to optical measurements due to the same nature of wave propagation. Acoustic measurements can also operate in backscatter or transmissometry mode. In recent years, there is a lot of attempts to characterize suspended sediments from acoustic backscatter (ABS) instruments. Short bursts of high-frequency sound (1–5 MHz) emitted from a transducer are directed toward the measurement volume. Sediment in suspension will direct a portion of this sound back to the transducer [17]. The backscattered strength is dependent on particle size as well as concentration. The ABS technology can be broadly classified into two approaches. The first approach uses specially designed acoustic instrumentation using multiple frequencies to compute SSCs and grain sizes over relatively short ranges (1–2 m). The size-dependent response to different frequencies is used to determine particle size. And the concentration is then calculated using particle-size information. The water column is sampled in discrete increments based on the return time of the echo. This approach has primarily been applied using fixed deployments to study near-bed sediment transport processes in the marine environment. Although there is an ample amount of literature on the development and application of this approach, it is not yet widely available at a competitive cost, and the system calibration and procedure involved to convert backscatter data into sediment concentration and size distribution is quite complex. A good overview of the technique is referred to Thorne and Hanes in 2002 [18]. The second approach uses commercially available in situ acoustic Doppler current profiles (ADCPs), which are primarily single-frequency instruments at present. It is impossible to differentiate between a change in mass concentration and a change in PSD (without sufficient calibrations) when using a single-frequency instrument, as changes in both SSC and PSD can result in a change in the backscatter signal strength. In addition, there is an appropriate or optimum acoustic frequency for a given PSD. Multiple single-frequency ADCPs can be combined to establish a multi-instrument, multifrequency system at a higher cost for segregating size fractions [7, 8, 19].

Multifrequency acoustic backscatter shows merit for use in measuring suspended-sediment concentration. Its ability to measure sediment concentration in a larger sampling volume than point measurement techniques, while estimating the PSD, would make it a good choice for many applications. However, further refinement and research are necessary before widespread use of this instrumentation is possible. The acoustic backscatter system is in principle an indirect method of measurement; an inversion algorithm is required for determining sediment concentration with measured backscattered signal strength. The acoustic backscatter equations provide the basis for the development of such an algorithm [20]. In situ calibrations are required to convert the ABS measurements to SSC. Complex post-processing requires compensations for physical properties of ambient water such as temperature, salinity, and pressure, and, in some cases, suspended materials. Additional compensations are needed for instrument characteristics such as frequency, power, and transducer design. Furthermore, the method appears appropriate for use in SSC only up to several g/L. Quantification of higher SSC (>20 g/L) may be problematic, especially when using higher acoustic frequencies that are more prone to attenuation by suspended sediments.

Another type of acoustic measurement for higher SSC (>1 up to 150 g/L) monitoring uses the transmission attenuation of ultrasound to measure the concentration. In the attenuation regime, ultrasonic signal attenuation is proportional to the applied frequency and to the suspended solid concentration. Measurement techniques based on attenuation spectrometry have been developed for slurry characterization. When ultrasound passes through slurry, the signal strength is reduced by the interaction of the ultrasound with the particles within the slurry. This ultrasonic signal attenuation can be analyzed to provide real-time in situ measurement of slurry concentration and particle-size distribution [21–24]. The response of acoustic transmission attenuation to particle-size distribution can be quite complex. Commercially available instruments based on acoustic attenuation use a single transmitter-receiver pair operating in a single frequency or two switchable frequencies. Figure 10.3b shows one such instrument that we evaluated using the same materials for testing the optical device. The experimental results are shown in Fig. 10.6. For each SSC, both frequencies (1 and 3.3 MHz) were measured. The acoustic attenuation is higher for higher frequency, but both frequencies show similar trend in the relation between acoustic attenuation and SSC. The acoustic

Fig. 10.6 The experimental relation between the acoustic attenuation of Solartron Mobrey 433 sensor and SSC for the three sediments tested

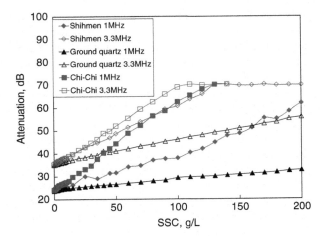

response (attenuation)-SSC relation shows much higher dependency on particle-size distribution than the optical response-SSC relation. Furthermore, although the ground quartz and Chi-Chi silt possess similar mean diameter (see Fig. 10.4), their attenuation-SSC relations are dramatically different. The small fraction of coarser and finer particles in the Chi-Chi sediment drastically changes the acoustic attenuation response. The particle-size distribution could be so diverse that it cannot be simply represented by a mean diameter. Until a robust multifrequency technique is available for a wide range of PSD and SSC, it is a merit to have an SSC measurement technique that is independent of PSD.

2.3   Other SSC Surrogate Techniques

Surrogate technologies to continuously monitor suspended sediment are much hoped for to supplant traditional data collection methods requiring costly collection and analysis of water samples. In addition to the optical and acoustic techniques, various other efforts have been put forward in finding a good SSC surrogate technique. Lewis and Rasmussen in 1996 [25] proposed a differential pressure approach to measure the water column mass density changes due to various sediment volumes in the water. The pressure difference technology is designed for monitoring SSCs at a single vertical as opposed to at-a-point measurements. This technology may be unique in that the measurement accuracy theoretically improves with concentrations increasing above 10–20 g/L, addressing a unique monitoring niche for measurements in highly concentrated or hyperconcentrated flows [7, 8]. The theoretical underpinnings of this technology are straightforward. However, it is easily affected by the water temperature, water velocity, turbulence, and dissolved solid in water. Tollner and Rasmussen in 2005 [26] further indicated that although the laboratorial accuracy of differential pressure approach was 10–1000 mg/L, produced error in the field was up to 65,000 mg/L due to suspended objects in the water, indicating that the apparatus is not yet feasible under in-stream condition. Its performance has been marginal in other field applications [7, 8].

A new technique based on time-domain reflectometry (TDR) was recently introduced, taking advantages of TDR's unique features, e.g., wider measurement ranges, insensitiveness to particle-size distribution, simple calibration, robustness, easy maintainability and cost-effective multiplexing [27]. Most of the existing technologies are based on scattering of waves (either acoustic or electromagnetic); particle sizes relative to the wavelength present attenuation efforts that potentially affect the reliability of their performance. TDR, which measures spatial average of the dielectric property, provides a way to overcome the limitations of existing technologies. Being a well-established method for measuring soil moisture content, TDR is suitable for monitoring medium-to-high concentrations. Chung and Lin in 2011 [27] reported that SSC resolution of about 3 g/L and measurement accuracy half the resolution can be achieved by the system they used. This new technique will be the center of this chapter. The modified TDR technique for better resolution and field applicability will be introduced. It will also be demonstrated how the TDR technique can be deployed as a full monitoring program for reservoir management through a case study in Shihmen Reservoir, Taiwan. Monitoring stations at upstream riverbank and outflow channels were installed with fixed protective structures to provide inflow and outflow sediment-discharge records. To capture the characteristics of density currents, multi-depth monitoring stations were designed and deployed on floats in the reservoir. Some of the data collected during typhoons were presented as an example to demonstrate the effectiveness and benefits of the TDR-based monitoring program.

3   Time-Domain Reflectometry Method

Time-domain reflectometry (TDR) is based on transmitting an electromagnetic pulse via a coaxial cable connected to a sensing waveguide and watching for reflections of the transmission due to changes in characteristic impedance along the sensing waveguide. Depending on the design of the waveguide and the analysis method, the reflected signal can be used to measure various engineering parameters, such as soil moisture content, electrical conductivity, displacement, and water level [28–32]. Dissimilar to other techniques having a transducer with a built-in electronic sensor, TDR-sensing waveguides are simple and durable mechanical device without any electronic components. When connected to a TDR pulser above water for measurement, the submerged TDR-sensing waveguide is rugged and can be replaced economically if damaged. Multiple TDR-sensing waveguides can be connected to a TDR pulser through a multiplexer and automated, hence increasing both temporal and spatial resolutions. In light of several advantages of TDR monitoring technique, it was aimed to develop a TDR-based SSC measurement technique that features high measurement range, easy calibration, robustness, good maintainability, as well as cost-effectiveness for multiplexing.

3.1 Principles of TDR SSC Measurement

A TDR measurement installation is composed of a TDR device and a transmission line system. A TDR device generally consists of a pulse generator, a sampler, and an optional oscilloscope; the transmission line encompasses a leading coaxial cable and a sensing waveguide, as shown in Fig. 10.7. The pulse generator delivers an electromagnetic (EM) pulse along a transmission line, and the sampler is used to record returning reflections from the sensing waveguide. Reflections occur at impedance discontinuities along the transmission line; the reflected waveform depends on the impedance mismatches and electrical properties of the insulating materials in the transmission line. The TDR SSC measurement methodology is much similar to the TDR soil-water content measurement. As illustrated in Fig. 10.7, the step pulse is reflected at the beginning and end of a sensing waveguide. The travel time analysis of the two reflections can determine the round-trip travel time (Δt) of the EM pulse in the sensing waveguide of length (L). Propagation velocity of the EM pulse depends on dielectric permittivity of the material surrounding the conductors. The round-trip travel time of EM wave, Δt, along the waveguide is related to the dielectric constant of the medium ε as

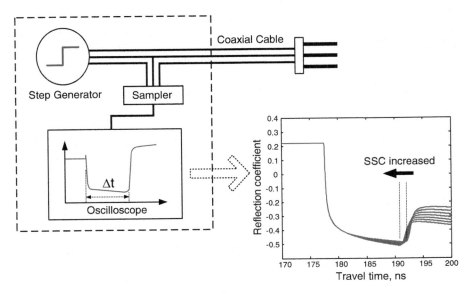

Fig. 10.7 TDR components and illustration of the SSC measurement principle

$$\Delta t = \frac{2L}{c}\sqrt{\varepsilon} \qquad (10.1)$$

where L is the length of the waveguide and c is the speed of light (2.998×10^8 m/s). That is, the EM wave velocity in a medium is equal to the speed of light divided by the square root of its dielectric constant. EM wave velocity is one ninth of the light speed in the water, much lower than that in soil minerals with velocity ranging from a quarter to one third of the light speed. Therefore, the average velocity of a sediment suspension depends largely on the SSC. The round-trip travel time along the sensing waveguide in the sediment suspension would decrease as the SSC increases, as illustrated in Fig. 10.7. TDR can also be used to measure electrical conductivity (EC) from the long-time steady-state voltage [29]. But besides sediment concentration, the EC of sediment suspension is highly dependent on water salinity. Therefore, the dielectric-based method is adopted for SSC measurement.

A sediment suspension is mainly composed of water and soil solid. The dielectric constant soil solid is temperature independent and narrowly ranges from 3 to 9, depending on its mineral composition [33]. On the contrary, dielectric constant of water ε_w is much higher and temperature dependent as indicated in [34]

$$\varepsilon_w(T) = 78.54 \cdot \left(1 - 4.58 \cdot 10^{-3}(T - 25) + 1.19 \cdot 10^{-5}(T - 25)^2 - 2.8 \cdot 10^{-8}(T - 25)^3\right)$$

$$(10.2)$$

where T is the measured temperature in degree Celsius. It has been well documented that the dielectric constant is also a function of water salinity [35]. But it is neglected in Eq. 10.2 because the effect of salinity on the dielectric constant of water is insignificant in our targeted freshwater environment where EC of water is lower than 1000 μs cm^{-1}. The bulk dielectric permittivity of sediment suspension can be expressed as a function of SSC by the volumetric mixing model [36] as

$$\sqrt{\varepsilon} = (1 - SS)\sqrt{\varepsilon_w(T)} + SS\sqrt{\varepsilon_{ss}} \qquad (10.3)$$

where ε is the dielectric constant of the sediment suspension; SS is the SSC in terms of volume fraction, which ranges from zero to one; and ε_{ss} is the dielectric constant of the suspended-sediment solid. The assumption of two-phase medium is made in Eq. 10.3 and throughout the following derivation. Other liquid or solid mixtures and entrapped air are considered as sources of uncontrollable error. Substituting Eqs. 10.3 into 10.1, the measured travel time is related to the SS by the following equation:

$$\Delta t = \frac{2L}{c}\left[(1 - SS)\sqrt{\varepsilon_w(T)} + SS\sqrt{\varepsilon_{ss}}\right] \qquad (10.4)$$

Water temperature is simultaneously measured to provide necessary compensation of temperature effect on dielectric constant of water. Once the $\varepsilon_w(T)$ and ε_{ss} are known, the volume fraction SS can be determined from the measured travel time Δt in the sediment suspension as

$$SS = \frac{\Delta t - \frac{2L}{c}\sqrt{\varepsilon_w(T^oC)}}{\frac{2L}{c}\left(\sqrt{\varepsilon_{ss}} - \sqrt{\varepsilon_w(T^oC)}\right)} \tag{10.5}$$

The volume fraction SS can be converted into ppm (or milligram per liter [mg L^{-1}]) unit, commonly used in hydraulic engineering, as:

$$ppm(mgl^{-1}) = \frac{SS \cdot G_S}{1 - SS}10^6 \tag{10.6}$$

in which the G_s is the specific gravity of suspended sediment, typically ranging from 2.6 to 2.8. The TDR-sensing waveguide for SSC measurement and the corresponding data reduction for precise travel time determination will be introduced in the following section.

3.2 New TDR Probe Design and Data Reduction

Since SSC measurement requires the highest possible accuracy than that of water content measurement in soil, special attention was introduced while designing the TDR SSC probes. A metallic shielding head was utilized to prevent leakage of electromagnetic waves. Chung and Lin in 2011 [27] tested both the balanced and unbalanced configurations of the conductors and concluded that a balanced configuration is required to achieve stable measurements for SSC. The original version of TDR SSC waveguide proposed by Chung and Lin in 2011 [27] is a trifilar (three-rod) type. In practice, this type of waveguide would suffer from interference of foreign objects, such as woods, leafs, and plants, in water caught by the multiple rods. In addition, the fouling, both organic and inorganic, would cause measurement bias if the TDR waveguide is submerged for a long time. To obtain more stable measurements, this chapter introduces a new TDR SSC waveguide to resolve aforementioned problems. The new TDR waveguide is a 25 cm long coaxial-type probe, as shown in Fig. 10.8. The outer tube and inner rod form the required conductors for the sensing waveguide. The outer tube kept interfering objects out and allowed water flowing into the sensing range near the central conductor via open holes and slots. Both outer and inner conductors are coated with Teflon® to reduce the speed of fouling and make routine maintenance much easier.

To precisely determine the round-trip travel time of the EM wave in the sensing waveguide, it requires a consistent, accurate approach for locating the reflection

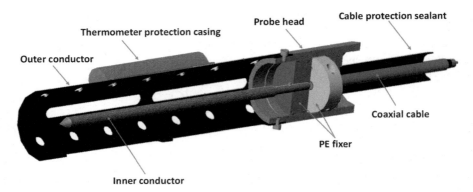

Fig. 10.8 Illustration of the new coaxial TDR SSC probe

points. However, the precise locations of actual reflections may be difficult to define in the time domain. Following the recommendation of Heimovaara in 1993 [37] made for water content measurement, Chung and Lin in 2011 [27] used an electrical marker, constructed by connecting a splice connector whose impedance is apparently less than the cable impedance, to clearly define a beginning characteristic point in the waveform. The relationship between the measured travel time from the characteristic point and the actual travel time is a time offset to be calibrated. Any ambiguity in the determining the time arrival of end reflection in the time domain is also absorbed in this offset term. The time offset and probe length should be precisely calibrated by making two measurements in two media of known dielectric constants (e.g., water and air). In the modified probe design, shown in Fig. 10.8, impedance is matched in the probe head so that no significant reflection occurs before the EM wave reaches the sensing medium. A new waveform analysis is proposed to eliminate the time offset term and allow simpler probe length calibration using only clear water to benchmark 0 SSC.

A typical TDR waveform of a step pulse input is shown in Fig. 10.9a. It is often easier to determine the travel time in the impulse waveform, which is obtained by taking the derivative of the step pulse waveform, as shown in Fig. 10.9b. The impedance match yields clearly two distinct reflections, one from the beginning of the sensing section and the other from the end. The dielectric permittivity and phase velocity of a multiphase medium is in general a function of frequency due to interfacial polarization. The apparent travel time deduced from first arrivals by tangent line methods normally corresponds to the phase velocity at higher frequency of the TDR bandwidth. In the context of soil moisture measurement, Lin in 2003 [38] showed that the interfacial polarization is negligible beyond 200 MHz, and water dipolar polarization is minimal below a couple of GHz, as illustrated in Fig. 10.10. It is within this optimal frequency range that the dielectric constant of a soil depends only on water content and is practically independent of soil type or particle-size distribution. A sediment suspension has much higher water content than a soil, making it even less dispersive within the optimal frequency range. The

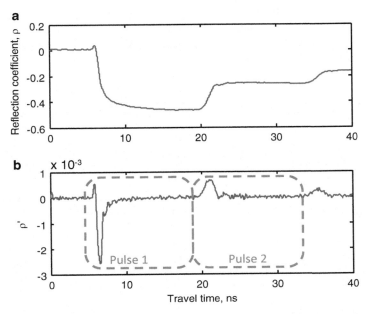

Fig. 10.9 (a) Typical step pulse waveform of the new coaxial TDR SSC probe and (b) the corresponding derivative of the waveform

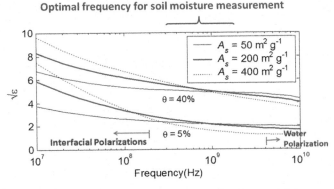

Fig. 10.10 An illustration of the square root of soil dielectric permittivity as a function of frequency, in which θ is the volumetric water content and A_S is the effective specific surface of a soil

signal shown in Fig. 10.9 can be passed to a band-pass filter within the aforementioned optimal frequency range, and the travel time can be numerically computed using cross correlation. By doing so, a noise control can be achieved simultaneously by the band-pass filter, and the effective frequency of the travel time analysis is ensured to be within the optimal frequency range. Furthermore, there is no time offset time involved in the calibration. Only the effective length of sensing section, L in Eqs. 10.4 and 10.5, should be calibrated by one single TDR measurement in clear water (i.e., $SS = 0$). The effect of conductor coatings is also taken into account during this calibration process.

3.3 Experimental Verification of TDR SSC Measurements

The TDR probe is ready for SSC measurements once the zero calibration (i.e., effective probe length for 0 SSC) is completed. In a TDR SSC measurement, water temperature and TDR waveform are acquired at the same time. The TDR waveform is reduced to the round-trip travel time in the sensing waveguide, while the water temperature determines the dielectric constant of water using Eq. 10.2. However, the dielectric constant of suspended sediment, ε_{ss}, is required to ultimately estimate the SSC value using Eq. 10.5. The sensitivity of TDR SSC estimation on ε_{ss} was experimentally evaluated. TDR measurements were conducted on sediment suspensions of known SSC varying from 0 to 150 g/L for the three sediments shown in Fig. 10.4. Due to limited samples, the highest SSC for Chi-Chi silt only reached 50 g/L. The dielectric constant of each of the three sediments was back-calculated by regression analysis. Linear regression yielded sediment dielectric constant $\varepsilon_{ss} = 7.5$, 6.9 and 3.6 for Shihmen clay, Chi-Chi silt, and ground quartz, respectively, which all fell within the reasonable range of dielectric permittivity of soil minerals. The ground quartz has a significant different ε_{ss} value from Shihmen clay and Chi-Chi silt due to the apparently different mineralogy of silica from natural suspended sediments. The obtained dielectric constant of ground quartz is quite close to that of quartz in the literature [33]. To illustrate the impact of ε_{ss} variability on SSC estimation, $\varepsilon_{ss} = 7.0$ is fixed in Eq. 10.5 to compute estimated SSCs for all three sediments. The results are compared with the actual SSCs in Fig. 10.11. Although the dielectric constant of Chi-Chi silt differs from that of Shihmen clay by 8 % (i.e., $\varepsilon_{ss} = 7.5$ vs. 6.9), the estimated SSCs still fall closely on the 1:1 line, regardless of their significant difference in PSD. As the ε_{ss} is reduced by half in ground quartz (i.e., $\varepsilon_{ss} = 3.6$ vs. 7.0), the estimated SSC deviates from the 1:1 line by 14 %. Although this 14 % discrepancy due to mineralogy may appear obvious in Fig. 10.11, it is insignificant compared to the effect of particle size on optical and acoustic instruments, in which 100 % and 800 % difference was, respectively, observed for measurements among the three sediments. While sediment particle size can vary significantly during a full runoff event, it is believed that the mineralogy of the natural sediments does not vary significantly with time in the same region. Therefore, the sediment dielectric constant can be easily calibrated with a few direct SSC measurements by sampling. Even if the mineralogy does vary, the dielectric constants of different minerals fall within a small range (from 3 to 9). The SSC error due to mineralogy will be bounded within 15 %. A major advantage of TDR SSC method over optical and acoustic methods is its invariance to the particle size.

Fig. 10.11 The estimated SSCs in comparison with the actual SSCs for Shihmen clay, Chi-Chi silt, and ground quartz, using a fixed $\varepsilon_{ss} = 7.0$ in Eq. 10.5

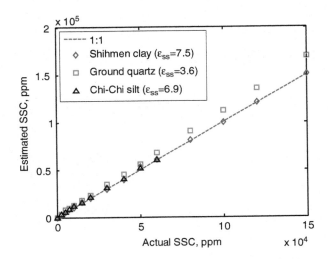

SSC measurements require much higher resolution and accuracy than those of soil-water content measurements. Sensitivity is first defined for resolution analysis to theoretically examine effects of acquisition and probe parameters as well as the limitation of TDR SSC measurements. The estimation of SSC by the TDR method relies on the measurement of the EM wave travel time in the TDR probe. Thus, the measurement sensitivity can be defined as the change of travel time due to a unit change of volumetric sediment content SS:

$$\text{Measurement Sensitivity} = \frac{\partial \Delta t}{\partial SS} = \frac{2L}{c}\left(\sqrt{\varepsilon_{ss}} - \sqrt{\varepsilon_w(T)}\right) \qquad (10.7)$$

The measurement sensitivity of SSC is a function of dielectric constants of water and suspended sediment and more importantly the probe length L. It increases linearly with probe lengths. The resolution of TDR SSC measurement can then be defined as the relative SS change in response to a unit travel time change (i.e., sampling interval dt). From Eq. 10.7, the resolution of TDR SSC measurement can be written as

$$\text{Resolution} = \frac{dt}{\frac{2L}{c}\left(\sqrt{\varepsilon_w(T)} - \sqrt{\varepsilon_{ss}}\right)} \qquad (10.8)$$

The unit of the TDR SSC measurement resolution in Eq. 10.8 is volume fraction SS. It can be transferred into milligram per liter or ppm by Eq. 10.6. The measurement resolution is proportional to the sampling interval dt and inversely proportional to the probe length L. The sampling interval is limited by the TDR sampling device. For the above laboratory experiments, a TDR device with 12.5 ps sampling interval was used. A TDR device with better sampling resolution of 5 ps was later used in

Fig. 10.12 Field
verifications of the new
TDR SSC coaxial probe by
other sampling methods

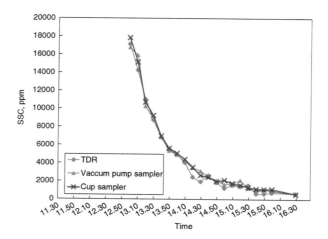

field applications. The TDR SSC probe currently used is 25 cm long. Longer probe
length can further improve the SSC measurement resolution, but the spatial resolution, signal attenuation due to electrical conductivity, and how to obtain average
water temperature within the entire sensing section should also be considered. For
the TDR probe and data acquisition setting used in the laboratory, the theoretical
instrument resolution is about 3000 ppm (3 g/L). However the measurement errors
were mostly within 1500 ppm, which is half of the instrument resolution. This is
attributed to the travel time analysis algorithm being able to obtain travel time with
accuracy twice the sampling interval.

A field verification test was performed in an outlet open channel of Chi-Chi Weir
in central Taiwan. The 6 m wide, 6 m high open channel is made of reinforced
concrete and located near several sediment sluiceways. The outlet intake is shut off
during sluicing operation and provides regular opportunities for field testing when it
is reopened and the water is temporarily high in SSC. A TDR SSC probe is installed
with a hollow steel pipe that is attached to the side wall of the open channel. This
setup kept the TDR probe 1.5 m below normal water level and are easily lifted up
for routine maintenance. During the field verification test, a submersible pump and
a cup sampler were used for comparisons. Comparison of all measurements based
on different methods is shown in Fig. 10.12. All three measurements showed
similar trends with discrepancies less than 1500 ppm. Both the laboratory and
field tests validated the feasibility and practicality of the new coaxial TDR SSC
probe for its subsequent applications in the reservoir. The independence of TDR
method to sediment particle makes it a unique and attractive surrogate technique for
SSC monitoring. Current TDR measurement accuracy is about 1500 ppm (or better
with higher resolution TDR device and temperature sensor) with theoretically
unlimited measurement range, making it suitable for applications with medium-
to-high concentrations.

4 Case Study: Monitoring Program in the Shihmen Reservoir

This section demonstrates the application of the TDR SSC monitoring technique in an extensive SSC monitoring program for reservoir management through a case study in Shihmen Reservoir, Taiwan. Due to geological weathering and climate change, sediment yields from soil erosion and landslides are experiencing unexpected increase in the Shihmen watershed in the last decade. The major strategy to slow down the sedimentation involves renovating an existing low-level outlet structure to increase its outlet capacity and planning new sluicing tunnels for sediment routing. Real-time monitoring of sediment transport plays an important role in planning and operation of these facilities when completed. Monitoring stations were installed at the riverbank of the major inflow river mouth and outflow channels with fixed protective structures to provide inflow and outflow sediment-discharge records. To capture the characteristics of turbidity currents, a multi-depth monitoring station was designed and deployed on a floating platform in the reservoir. Some of the data collected during typhoon events are presented as an example to demonstrate the effectiveness and benefits of the TDR-based monitoring program.

4.1 Problem Background of the Shihmen Reservoir

The Shihmen Reservoir watershed is located in the Tahan River watershed in Northern Taiwan (Fig. 10.13). The Shihmen Dam, completed in 1964, is located in Shihmen Valley in the midstream of the Dahan River. The area of the watershed is 760.2 km^2, and the elevation ranges from 158 m (Shihmen Reservoir dam site) to 3524 m. Except for a small number of igneous rock facies, the majority of the rocks in this area are low-grade metamorphic and sedimentary. Most of the soil is stony, with riverbanks and valleys composed of alluvium. The average annual rainfall is approximately 2580 mm and is highly concentrated between May and September due to heavy rainfall from typhoons. Rainfall also occurs during thunderstorms triggered by southwest air currents and intermittent tropical depressions. The Shihmen Reservoir serves as an important hydraulic facility with multiple functions of public water supply, irrigation water supply, hydraulic power generation, flooding prevention, and tourism. However, the watershed of the Shihmen Reservoir has deep valleys and vulnerable geology characteristics, particularly after the Chi-Chi earthquake in 1999 [39]. More soils from landslides inrush into the upstream of the reservoir [40] and cause reservoir sedimentation during subsequent torrential rainfalls. The storage loss of the Shihmen Reservoir due to sedimentation over time is shown in Fig. 10.14. During the period from 23 to 26 August 2004, Typhoon Aere brought a total precipitation of almost 1600 mm in the watershed and flushed 27.88 × 10^6 m^3 sediments into the reservoir. The effective storage volume

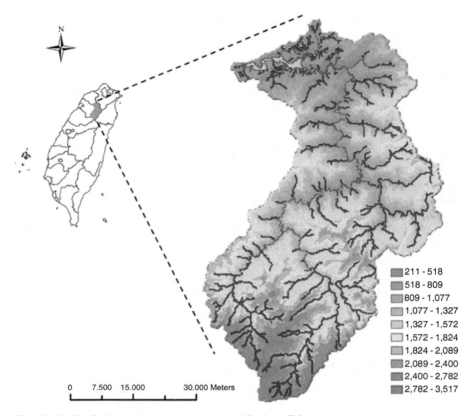

Fig. 10.13 The Shihmen Reservoir watershed in Northern Taiwan

of the reservoir is decreased by about 10 % subsequent to this event. Failure of the upstream Baling Sabo Dam during 2007 Typhoon Wipha further aggravated the sedimentation problem in the reservoir [41]. Although several conservation approaches in watershed were adopted in the early days after 1964 Typhoon Gloria [42], recent sediment yield from landslides in any specific typhoon event can be significantly larger than the average sediment yield of past several decades. The total deposited sediment volume in 2013 is 91.98×10^6 m³, which is equal to 29.76 % of the total storage capacity.

Figure 10.15 shows all outlet works in the Shihmen Reservoir and their corresponding intake elevations and maximum discharge capacities. During typhoon events, large amount of sediments often transport to reservoir at hyperpycnal concentrations and form density currents. Current sediment discharge during flood events in the Shihmen Reservoir is limited due to the lack of high-capacity outlet at low elevations. Large remediation projects have been embarked to renovate existing outlet works and construct new sluicing tunnels to slowdown dam sedimentation. The first immediate countermeasure against sedimentation was

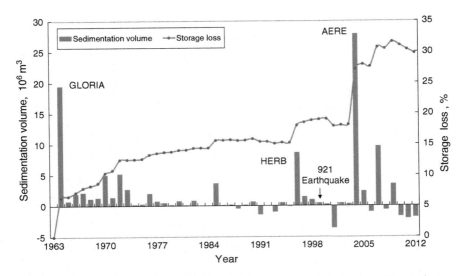

Fig. 10.14 Annual sedimentation volume and total storage loss of the Shihmen Reservoir over time due to sedimentation

Fig. 10.15 Outlet works of the Shihmen Reservoir, with intake elevation and maximum discharge capacity indicated in the box

an ingenious modification of the power plant penstocks. As illustrated in Fig. 10.16, two penstocks were originally constructed for the two power generators. The maximum discharge capacity for each penstock was 68 cms. The penstock #2 was cut off and diverted to a jet flow gate. The construction was completed in 2012. The maximum discharge capacity was increased from 68 to 300 cms.

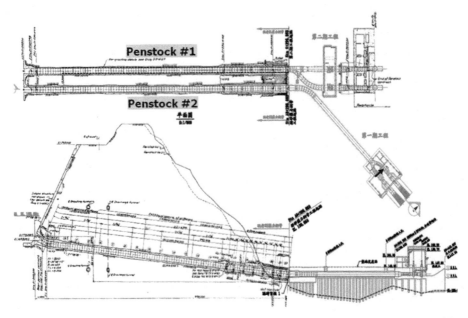

Fig. 10.16 Illustration of modification of the power plant penstocks (*Top*: plan view; *Bottom*: side view)

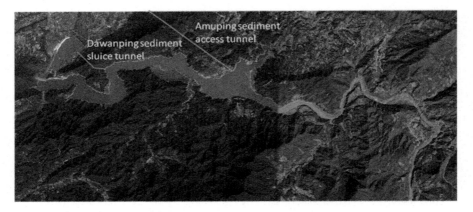

Fig. 10.17 Illustration of the two sediment bypass tunnels under planning

The penstock #1 is under reconstruction to be cut off and split to supply the two power generators. Further countermeasure against sedimentation calls for constructing sediment bypass tunnels. Four plans of bypass tunnels were investigated. Feasibility studies lead to two bypass tunnel projects (Amuping sediment access tunnel and Dawanping sediment sluice tunnel) as illustrated in Fig. 10.17. Understanding the behavior of sediment transport in a reservoir is crucial in

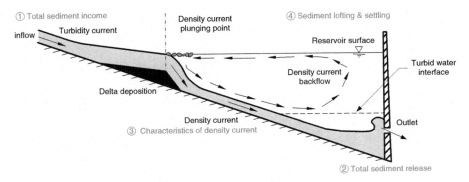

Fig. 10.18 Main questions defined for the SSC monitoring program in a reservoir

planning and designing the bypass tunnels. Real-time monitoring of sediment transport plays an important role in characterizing sediment transport as well as in the operation of desilting facilities during flood events.

4.2   Planning of SSC Monitoring Program

The primary questions to be answered by sediment transport monitoring scheme in a reservoir are first defined and illustrated in Fig. 10.18. These include (a) sediment inflow into the reservoir, (b) discharge of sediments through outlets, (c) characteristics of turbidity currents (including the occurrence time, corresponding SSC, thickness, and moving speed), and (d) sediment lofting and settling of stagnant muddy water. As a result, a sediment monitoring program was initiated in 2008. The program began with manual SSC samplings at the upstream reservoir boundary of the main tributary and all outlets. The location of the upstream boundary was selected at Luofu (at Section 32 in Fig. 10.19). The outlets monitored included spillway, tunnel spillway, power plant outlet, permanent river outlet (PRO), and the Shihmen canal outlet. Measurements of SSC at Luofu and the discharge open channel of Shihmen canal outlet were first automated using the TDR technique. TDR probes were installed at desired elevations with fixed protective structure at the upstream riverbank and the Shihmen canal outlet channel. Measurements of SSC at the rest of outlets were later subsequently automated by a TDR probe in a monitoring tank in which an electromagnetic valve periodically opens and takes water from the outlet by tubing.

To capture the characteristics of density currents, a multi-depth monitoring station was designed and deployed on a floating platform at Section 24 of the reservoir in 2008. Section 24 is located downstream not far from the expected plunging point of density currents. A plan view of the distribution of current monitoring stations is shown in Fig. 10.19. More automated monitoring stations on floats were added subsequently following the success of the first float station on

Fig. 10.19 Plan view of the distribution of sediment monitoring stations

Section 24. These include two stations at Section 20 (close to the intake of Amuping sediment access tunnel), a station at Section 15, two stations at Section 12 (close to the intake of Dawanping sediment sluice tunnel), a station at Section 7 (close to the intake of tunnel spillway), and a station at Section 4 near the dam.

Implementation of monitoring station on a float is a challenging work. It was considered too challenging to install a winch for profiling measurements during storm events. Instead, multiple TDR probes are attached to a hanging wire and lowered down to the reservoir with the lowest probe located 1 m above the reservoir bed. Since the water level does not change significantly during wet season, such a design should serve the purpose for monitoring the density current. Figure 10.20 displays the layout of the TDR SSC system on a floating platform. An automated TDR measurement system with solar power supply is responsible for the TDR waveform acquisition and SSC data reduction. Multiple TDR-sensing waveguides are connected to a TDR pulser through a multiplexer and automated, hence improving both temporal and spatial resolutions. One 8-channel multiplexer was installed on the platform to connect 8 SSC waveguides (or probes) and provide SSC data at different depths. More probes can be added simply by adding multiplexers. Each probe is coupled with a thermometer to obtain water temperature for compensation of temperature effect on dielectric constant of water. An embedded system was adopted to control TDR devices, measure temperatures, and process TDR waveforms in real time to yield SSC data. This automated system reduced the amount of data for efficient wireless communication via GPRS. A data center was also implemented to facilitate management of multiple monitoring stations [43]. A schematic diagram of the connection between the TDR monitoring system controlled by a field embedded system and data center is shown in Fig. 10.21. In the diagram, it encompasses three major components, i.e., the front-end module (clients), the TDR monitoring system, and the back-end services (data center) including database servers. The front-end module handles user interfaces via browsers. It establishes user sessions to provide services via the Internet. The TDR monitoring

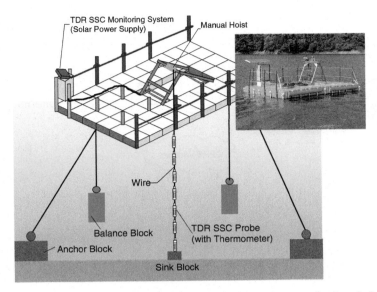

Fig. 10.20 Schematic and photo of the TDR SSC monitoring system on a floating platform

system is shown at the bottom of the diagram. The back-end module facilitates services and database management. The architecture supports communication and connectivity via the Internet as well as general packet radio service (GPRS) mechanism.

4.3   Results and Discussion of the First Full-Event Monitoring

This study utilized TDR as the core technique to provide real-time sediment hydrograph during typhoon events. One of the automated TDR SSC monitoring stations was installed at the Shihmen canal outlet (as shown in Fig. 10.19). At this station, a steel bar was installed on the side wall with an attached TDR SSC waveguide at EL. 193.5 m (near the bottom of the channel) as shown in Fig. 10.22a. The fixed bar was designed so that it can be lifted by a small winch during routine maintenance of TDR waveguide. Direct SSC measurements by manual cup samplers were carried out to evaluate the in situ performance of TDR SSC measurements during typhoon events. As an example, Fig. 10.22b shows comparison of the results for TDR measurements and manual samplings during Typhoon Fung-Wong in 2008. Both sediment hydrographs from TDR and manual sampling demonstrated similar pattern. The peak value of the manual sampling is slightly lower than that of the TDR SSC measurement. It should be noted that the manual cup sampling took only the surface water in rush flow and may

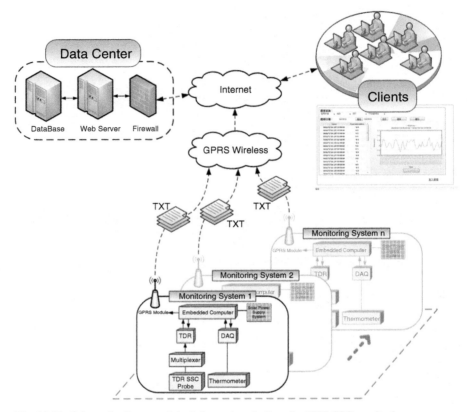

Fig. 10.21 Schematic diagram of the information platform for TDR SSC monitoring program

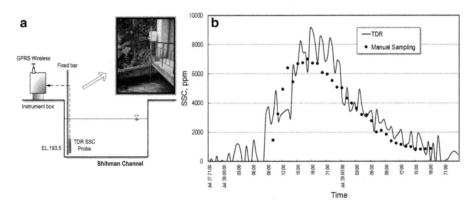

Fig. 10.22 (**a**) Field configuration and (**b**) comparison of TDR SSC with manual sampling measurements at Shihmen canal outlet

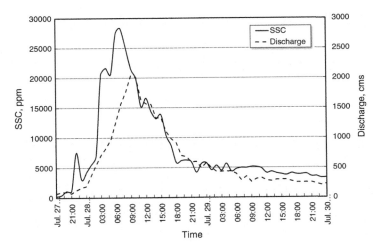

Fig. 10.23 Sediment hydrograph at Luofu station (Section 32) during Typhoon Fung-Wong

underestimate the actual SSC. TDR measurements showed greater fluctuation partly due to the level of stability of TDR pulse sampler when it was first deployed in field monitoring. The resolution and stability of the TDR device were further improved and used for the subsequent monitoring stations. Even with the level of scatter shown in Fig. 10.22b, the automated TDR measurements provided a satisfactory estimation for sediment outflow.

The inflow and outflow of sediment obtained during a full single event are valuable for the assessment of watershed management and reservoir desilting operation. Empirical model can be drawn from accumulated data collected during different types of storm events. The first full-event observation was 2008 Typhoon Fung-Wong, which is used as an example to demonstrate what can be obtained by this monitoring program. Figure 10.23 presents the inflow sediment hydrograph obtained by automated TDR measurements at Luofu riverbank (at Section 32) during Typhoon Fung-Wong. The peak discharge was 2040 cms, not particularly high compared to historical data. The highest SSC reached almost 30,000 mg/L before the discharge peak and subsequently induced a density current, which will be presented later.

A sediment discharge rating curve was previously established at an upstream hydrological station (Siayun gauging station) not far from Section 32. The SSC data at that station was collected by manual sampling about 30 times a year, mostly during low inflow condition (<1000 cms). The suspended-sediment rating curve utilizing the data from 1963 to 2005 is plotted in Fig. 10.24. Significant amount of data was collected in a single event under the full-event monitoring program. The suspended-sediment and discharge data during Typhoon Fung-Wong was shown as data points in Fig. 10.24. Although sediment discharge shows highly scattered relation to water discharge, it is apparent that the data points are significantly higher than the previous rating curve. Estimation of suspended-sediment load based on

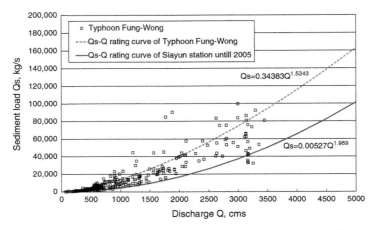

Fig. 10.24 Comparison between the sediment discharge data collected during Typhoon Fung-Wong and the sediment discharge (Qs-Q) rating curve from historical data at Siayun gauging station

Table 10.2 Inflow sediment load and outflow sediment discharge of each outlet during Typhoon Fung-Wong

Reservoir inflow (10^3 m³)	124,470
Reservoir outflow (10^3 m³)	120,300
Total sediment income (10^3 t)	1794.2
Total sediment released (10^3 t)	231.8
Total sluicing ratio	13 %
(Contribution from each outlet)	
Power plant outlet (EL. 173 m)	9 %
Spillway (EL. 235 m)	1.2 %
Spillway tunnel (EL. 220 m)	2.6 %
Shihmen canal outlet (EL. 193.5 m)	0.3 %

historical data may result in significant error as the behavior of sediment discharge may be greatly altered by the Chi-Chi earthquake in 1999. By monitoring each single event, the sediment inflow can be accurately estimated, and the sediment yield characteristics of the watershed can be studied dynamically. Table 10.2 summarizes the sediment inflow and sediment outflow of each outlet during Typhoon Fung-Wong. About 13 % of sediment inflow is discharged by the reservoir operation during Typhoon Fung-Wong. This amount may depend on the inflow condition and the operation of the outlets at different elevations. Analyzing the contribution of each outlet, it is clear that the power plant outlet, which at that time was the only low-elevation outlet with significant discharge capacity, plays an important role in sluicing out sediments during flood events.

The first TDR SSC monitoring system installed on a float platform is located downstream of but not far from the expected plunging point of density current. With multiple TDR-sensing points, the variation of SSC with depth and time can be gathered. The results obtained during Typhoon Fung-Wong are shown in

Fig. 10.25 Variation of SSC profile with time during Typhoon Fung-Wong

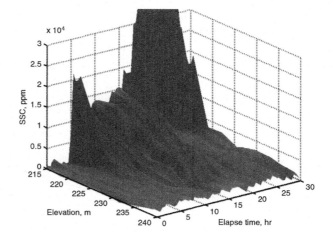

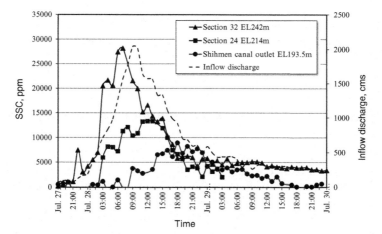

Fig. 10.26 Inflow discharge and suspended-sediment hydrographs at different locations during Typhoon Fung-Wong

Fig. 10.25, from which the formation of density current was observed. The SSC was about 12,000 ppm at EL. 215 and dropped to 6000 ppm at EL. 228. The thickness of the density current was estimated to be about 15 m. The density current lasted for about 15 h and recessed. The 3D data presented in Fig. 10.25 is quite valuable. And it can be readily realized using the TDR technique.

The sediment hydrographs at different locations from upstream to downstream are combined in Fig. 10.26, illustrating the moving fronts of high sediment concentration flow at different sections. A water quality monitoring float at Section 15 also picked up the arrival of high turbidity. By analyzing the moving fronts of high-concentration flow, the velocity of sediment transport can be estimated, as demonstrated in Fig. 10.27. In particular, from the deep water at Section 24 to the dam, the

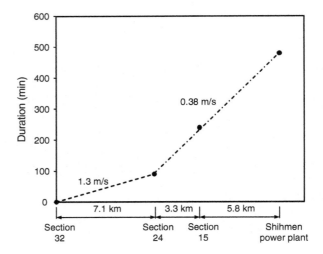

Fig. 10.27 Estimation of sediment transport velocity during Typhoon Fung-Wong

density current traveled at an average speed of 0.38 m/s. Real-time monitoring at the floating platform provides early warning for contingencies of reservoir operation.

4.4 Characteristics of Sediment Transport and Sluicing Operation

The sediment monitoring program in the Shihmen Reservoir was initiated in 2008. Since then, full runoff measurements were conducted for every typhoons event. SSC profile monitoring in the reservoir began with a single float station on Section 24. It was expanded subsequently and now has seven float stations in the deep water. A clearer and more refined picture of the density current movement can be captured by the extensive monitoring program. Figure 10.28 shows the SSC profile with time at several float stations in a recent event during Typhoon Soulik in 2013. Section 20 near the planned intake of Amuping sediment access tunnel is wide and instrumented with two float stations. Both stations, located not far from the expected plunging point, clearly show density current taking shape. Some subtle differences can be observed since the left station is located in the deep channel, and the right station is toward the right bank closer to the intake location and the junction of a small creek. Before the density took place, the increase in SSC at shallow depth is more pronounced at the station near the deep channel due to dispersion of incoming turbid water. When the density current occurred, the main density current spread almost the entire cross section and is picked up simultaneously by both monitoring stations with a peak SSC about 30 g/L. It is worth noting that the peak rainfall intensity of the Shihmen watershed broke the record during Typhoon Soulik. The peak hourly precipitation was more than twice the

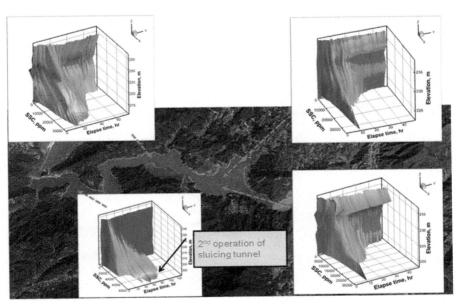

Fig. 10.28 SSC profile with time at several float stations in a recent event during Typhoon Soulik in 2013

precious record high. Fortunately it did not last long and the density current dissipated after about 10 h. The density current migrated downstream and was observed subsequently by the float stations downstream. The high SSC profile dissipated more slowly in the middle (Section 15) and lower region than the upper region at Section 20. In the lower region, the deep turbid water brought in by density current was accumulated and settled. During settling, a distinct interface was observed between the mass of settling solids in the sludge zone and the clear supernatant above formed from the water of separation. The zone settling of the stagnant muddy lake took place quite slowly due to the high SSC. The monitoring program in the reservoir realized by TDR technique provided detailed field observation of density current that was unprecedented. The spatial distribution of SSC observed by the monitoring program provides valuable information for sluicing operation.

The monitoring station at Section 32 is responsible for estimating the sediment load delivered to the Shihmen Reservoir for each single storm event. During Typhoon Soulik, the suspended-sediment hydrographs at Section 32 and float stations' lowest points are shown in Fig. 10.29. With the monitoring points in series from the upstream to downstream, it clearly depicts the migration of high-concentration flow. It also shows that the peak SSC in the density current is at least 50 % less than that in the shallow water region at Section 32, indicating a significant proportion of larger particles settles down along the way in the shallow water region.

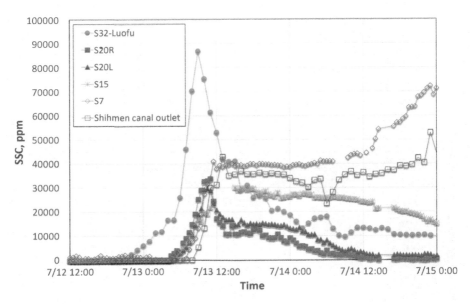

Fig. 10.29 Suspended-sediment hydrographs at Luofu (Section 32) and float stations' lowest points during Typhoon Soulik

Similar to Fig. 10.24, the sediment load vs. discharge was plotted in Fig. 10.30 using all full-event data obtained in the past 5 years at Section 32 and compared to the sediment discharge rating curve previously established at a nearby gauging station using sparse data collected from 1963 to 2005 mostly under low-discharge conditions. In particular, the data points collected during Typhoon Soulik is separately shown. The effect of record-breaking rainfall intensity is clearly seen when comparing the data from Typhoon Soulik with the rest of the data. A subgroup of data after the peak hourly precipitation shows apparent higher trend in the sediment load discharge relation. The study of dynamics of sediment transport, sediment yield characteristics, and watershed management will certainly benefit from the full-event monitoring program.

As a first measure to increase the low-level outlet capacity, the penstocks of Shihmen power plant were to be reconstructed as illustrated in Fig. 10.16. The first reconstruction phase involving the Penstock #2 was completed in 2012. The maximum discharge capacity was increased from 68 to 300 cms. The Penstock #2 was later renamed as the sluice tunnel and officially operated for the first time during Typhoon Soulik in 2013. The operation records of all outlets in terms of sediment hydrograph were shown in Fig. 10.31 together with the inflow sediment and discharge hydrograph. The low-level outlets, including permanent river outlet (EL. 169.5 m), power plant (Penstock #1, EL. 173 m), sluice tunnel (Penstock #2, EL. 173 m), and the Shihmen canal outlet (EL. 193.5 m), for a long period of time discharged muddy water with SSC much higher than the initial SSC of the density

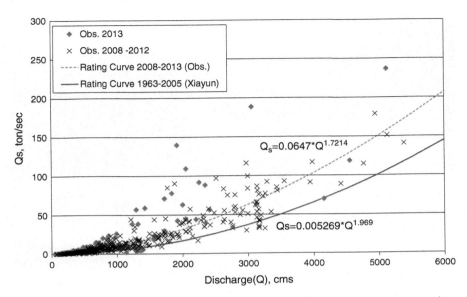

Fig. 10.30 Sediment load (Qs) vs. discharge (Q) data observed by the monitoring program since 2008 in comparison with the sediment-discharge rating curve from historical data at Siayun gauging station

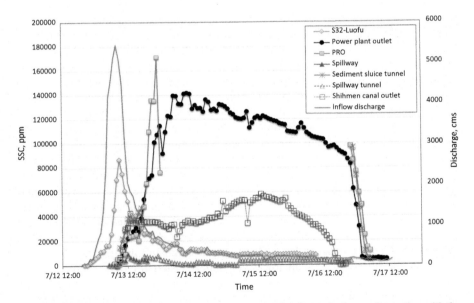

Fig. 10.31 The operation records of all outlets in terms of sediment hydrograph together with the inflow sediment (from Luofu station) and discharge hydrograph

Fig. 10.32 Photo showing muddy water sluiced out from the sluice tunnel as the density current reached the dam

current due to hindered settling and compression settling. The bathymetry revealed that the reservoir bed near the dam is at EL. 185 m. Dredging and a buttress wall kept the intakes of the low-level outlets from being buried by the sedimentation. Based on the monitoring results, the sluice tunnel was fully opened as the density current reached the dam. It effectively sluiced out the muddy water brought in by the density current, as vividly shown in Fig. 10.32. After about 10 h of operation, the sluice tunnel was shut down because the released muddy water caused the turbidity to rise to an unacceptable level at the intake of a water treatment plant downstream. Other low-level outlets with lower discharge continued to release muddy water, which was diluted by more clear water from spillway to keep the water supply within acceptable turbidity level. After the sluice tunnel was shut down, the SSC in the rest of low-level outlets continued to rise and reached over 140 g/L in the power plant outlet and PRO. Typhoon Aere, 2004, brought in significant amount of large woody debris. Some of the driftwood sink into the reservoir and blocked the trash rack of low-level outlet intakes. Cleaning and repair work were carried out by divers, but some remained buried in the mud. Since then, low-level outlets, especially the PRO, often experienced sudden increase and dropdown in SSC during sediment sluicing, as also shown in Fig. 10.31. After Typhoon Soulik, the zone-settling velocity of the stagnant muddy reservoir was so slow that the ultrahigh SSC in low-level outlets lasted for over 3 days. It was decided to open the sluice tunnel again during off-peak hours of water supply of the water treatment plant downstream. The SSC in all outlets drop dramatically, and the water cleared out after 4 h of sluicing operation. This can also be seen in Fig. 10.28, in which the spatial distribution of the muddy water was revealed by the float monitoring stations. The second operation of the sluice tunnel quickly drained out the stagnant muddy water in the lower region.

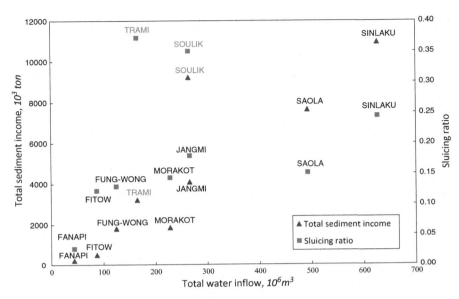

Fig. 10.33 Total sediment income and sluicing ratio in relation with total water inflow

Since 2008, full-event monitoring was carried out for every typhoon that affected the Shihmen Reservoir. By the end of 2013, a total of nine typhoon events were recorded. For each typhoon event, the reservoir inflow, total sediment income from Shihmen watershed, reservoir outflow, and total sediment released were inventoried. It was found that the total sediment income has a good linear correlation with reservoir inflow except for the Typhoon Soulik, during which the peak rainfall intensity exceeded two times the previous record high, as shown in Fig. 10.33. The ratio of reservoir outflow to inflow is mostly about 1:0 by the flood control operation in the Shihmen Reservoir during typhoons, which means the Shihmen Reservoir has abundant water for sluicing operation during typhoons. The sluicing ratio is defined here as the ratio of sediment released to sediment income. Before the sluice tunnel was in service, it was also found that the sluicing ratio had a linear correlation with reservoir inflow under the operation practice at that time. In 2013, the sluicing tunnel was operated during two typhoon events, Typhoon Soulik and Trami. These two data points soar above the previous trend in the relation between sluicing ratio and reservoir inflow, as revealed in Fig. 10.33. The addition of the sluice tunnel in the reservoir operation significantly increased the efficiency of sediment sluicing. At present, the operation of the sluice tunnel is constrained by the condition that the turbidity at the water intake of the water treatment plant downstream should not exceed the purification capability. The Jhongjhuang bankside reservoir project is under construction that when completed will supply clean water to the water treatment plant during typhoon and release the constraint on sluicing operation in the Shihmen Reservoir. It is believed that the extensive sediment monitoring program will provide needed detailed data to develop an empirical model that will optimize future sluicing operation.

Glossary

Acoustic backscatter Acoustic backscatter (ABS) measurement is a technique that emits and receives the acoustic (or echoes) energy for the monitoring of suspended-sediment particles in the water column.

Acoustic Doppler current profiler An acoustic Doppler current profiler (ADCP or ADP) is a hydro-acoustic current meter similar to a sonar, attempting to measure water current velocities over a depth range using the Doppler effect of sound waves scattered back from particles within the water column.

Dam A dam is a barrier that impounds water or underground streams. Dams generally serve the primary purpose of retaining water.

Density currents A highly turbid, relatively dense current carrying large quantities of clay, silt, and sand in suspension which flows down a submarine slope through less dense water.

Dielectric constant The dielectric constant is the ratio of the dielectric permittivity of a substance to the dielectric permittivity of free space.

Dielectric permittivity Dielectric permittivity is the measure of the resistance that is encountered when forming an electric field in a medium. In other words, permittivity is a measure of how an electric field affects, and is affected by, a dielectric medium. It is an expression of the extent to which a material concentrates electric flux.

Flow discharge The amount of water passing a cross section of a stream per unit time, often presented in cubic meter per second (cms).

Hydrograph A hydrograph is a plot of the variation of discharge with respect to time (it can also be the variation of stage or other water property with respect to time).

Hydrology Hydrology is the study of the movement, distribution, and quality of water on Earth and other planets, including the hydrologic cycle, water resources, and environmental watershed sustainability.

Hyperpycnal flow A denser inflow that occurs when a sediment-laden fluid flows down the side of a basin and along the bottom as a turbidity current.

Nephelometer A nephelometer measures suspended particulates by employing a light beam (source beam) and a light detector set to one side (often 90°) of the source beam.

Nephelometric turbidity units The units of turbidity from a calibrated nephelometer are called nephelometric turbidity units (NTUs).

Optical backscatter Optical backscatter measures the light that is scattered in the direction of the incident light beam.

Outlet works A combination of structures and equipment required for the safe operation and control of water released from a reservoir to serve various purposes.

Particle-size distribution The particle-size distribution (PSD) of a granular material is a list of values or a mathematical function that defines the relative amount, typically by mass, of particles present according to size. PSD is also known as grain size distribution.

Penstock A penstock is a sluice or gate or intake structure that controls water flow or an enclosed pipe that delivers water to hydroturbines and sewerage systems.

Polarization Polarization is a phenomenon that as a charged body is placed close to a nonconducting substance, the molecule orientation of the substance gets migrated

Plunging point The plunge point is the main mixing point between river and reservoir water, especially used for the density current transport behavior.

Precipitation Precipitation is a major component of the water cycle and is responsible for depositing most of the fresh water on the planet.

Reservoir A reservoir is a natural or artificial lake, storage pond, or impoundment from a dam which is used to store water.

Sediment load The amount of soil solid that is transported by a natural agent, especially by a stream.

Sediment Rating Curves The relationship between the sediment load and flow discharge.

Sediment yield The amount of sediment per unit area removed from a watershed by flowing water during a specified period of time.

Sedimentation The process of deposition of sediment, especially in a reservoir.

Specific gravity Specific gravity is the ratio of the density of a substance to the density (mass of the same unit volume) of a reference substance, such as water.

Spillway A spillway is a structure used to provide the controlled release of flows from a dam or levee into a downstream area, typically being the river that was dammed.

Storage of a reservoir Volume or storage of a reservoir is usually divided into distinguishable areas. Dead or inactive storage refers to water in a reservoir that cannot be drained by gravity through a dam's outlet works, spillway, or power plant intake and can only be pumped out. Active or live storage is the portion of the reservoir that can be utilized for flood control, power production, navigation, and downstream releases.

Suspended-sediment concentration Suspended-sediment concentration (SSC) is generally transported within and at the same velocity as the surrounding fluid. The stronger the flow and/or the finer the sediment, the greater the amount or concentration of sediment that can be suspended by turbulence.

Time-domain reflectometry Time-domain reflectometry (TDR) is a measurement technique used to determine the characteristics of electrical lines by observing reflected waveforms. It uses a time-domain reflectometer (TDR) as an electromagnetic (EM) wave radar to characterize and locate faults in metallic cables. TDR is also used to determine moisture content or suspended-sediment concentration in soil-water mixture.

Transmissometry A transmissometry measures suspended particulates by employing a light beam (source beam) and a light detector set to opposite side (often 180°) of the source beam.

Turbidity Turbidity is an expression of the optical properties of a sample that causes light rays to be scattered and absorbed rather than transmitted in straight lines through the sample.

References

1. Dadson SJ, Hovius N, Chen H, Dade WB, Lin J, Hsu M, Lin C, Horng M, Chen T, Milliman J, Stark CP (2004) Earthquake-triggered increase in sediment delivery from an active mountain belt. Geology 32(8):733–736. doi:10.1130/G20639.1
2. Sumi T, Hirose T (2002) Accumulation of sediments in reservoirs, EOLSS – Encyclopedia of Life Support Systems
3. Garcia MH (2008) Sedimentation engineering, process, measurements, modeling, and practice, ASCE manuals and reports on engineering practice no. 110. American Society of Civil Engineers, Reston
4. Walling DE, Webb BW (1981) Reliability of suspended sediment load data. In: Erosion and sediment transport measurement, IAHS-AISH publication no. 133. IAHS Press, Wallingford, pp 177–194
5. Walling DE, Webb BW (1988) The reliability of rating curve estimates of suspended sediment yield: some further comments. In: M. P. Bordas and D. E. Walling (eds.) Sediment budgets, IAHS publication no. 174. IAHS Press, Wallingford, pp 337–350
6. Wren DG, Barkdoll BD, Kuhnle RA, Derrow RW (2000) Field techniques for suspended-sediment measurement. J Hydraul Eng, ASCE 126(2):97–104
7. Gray JR, Gartner JW (2009) Technological advances in suspended-sediment surrogate monitoring. Water Resour Res 45(4). doi:10.1029/2008WR007063
8. Gray JR, Gartner JW (2010) Surrogate technologies for monitoring suspended-sediment transport in rivers. In: Poleto C, Charlesworth S (eds) Sedimentology of aqueous systems. Wiley-Blackwell, Chichester, West Sussex, pp 3–45
9. Sutherland TF, Lane PM, Amos CL, Downing J (2000) The calibration of optical backscatter sensors for suspended sediment of varying darkness levels. Mar Geol 162(2–4):587–597
10. Gentile F, Bisantino T, Corbino R, Milillo F, Romano G, Trisorio LG (2010) Monitoring and analysis of suspended sediment transport dynamics in the Carapelle torrent (southern Italy). Catena 80(1):1–8
11. Campbell CG, Laycak DT, Hoppes W, Tran NT, Shi FG (2005) High concentration suspended sediment measurements using a continuous fiber optic in-stream transmissometer. J Hydrol 311(1–4):244–253
12. Anderson CA (2005) Turbidity, in National Field Manual for the Collection of Water-Quality Data, U.S. Geological Survey Techniques of Water Resources Investigations., Book 9, U.S. Geological Survey, Reston. Available at http://water.usgs.gov/owq/FieldManual/Chapter6/6.7_contents.html
13. Downing JP (1996) Suspended sediment and turbidity measurements in streams: what they do and do not mean. Paper presented at Automatic Water Quality Monitoring Workshop, B. C. Water Quality Monitoring Agreement Coordinating Committee, Richmond
14. Rasmussen PP, Gray JR, Glysson GD, Ziegler AC (2009) Guidelines and procedures for computing time-series suspended-sediment concentrations and loads from in-stream turbidity-sensor and streamflow data. US Geological Survey Techniques and Methods report 3 C4. http://pubs.usgs.gov/tm/tm3c4/

15. Downing J (2006) Twenty-five years with OBS sensors: the good, the bad, and the ugly. Cont Shelf Res 26:2299–2318
16. Conner CS, DeVisser AM (1992) A laboratory investigation of particle size effects on optical backscatterance sensor. Mar Geol 108:151–159. doi:10.1016/0025-3227(92)90169-I
17. Thorne PD, Vincent CE, Harcastle PJ, Rehman S, Pearson N (1991) Measuring suspended sediment concentrations using acoustic backscatter devices. Mar Geol 98:7–16. doi:10.1016/0025-3227(91)90031-X
18. Thorne PD, Hanes DM (2002) A review of acoustic measurements of small-scale sediment processes. Cont Shelf Res 22:603–632
19. Topping DJ, Wright SA, Melis TS, Rubin DM (2007) High resolution measurement of suspended-sediment concentrations and grain size in the Colorado River in Grand Canyon using a multi-frequency acoustic system. Paper presented at 10th International Symposium on River Sedimentation, World Association for Sediment and Erosion Research, Moscow
20. Thorne PD, Meral R (2008) Formulations for the scattering properties of suspended sandy sediments for use in the application of acoustics to sediment transport. Cont Shelf Res 28 (2):309–317
21. Bamberger JA, Kytomaa HK, Greenwood MS (1998) Slurry ultrasonic particle size and concentration characterization. In: Schulz WW, Lombardo NJ (eds) Science and technology for disposal of radioactive tank wastes. Plenum Press, New York, pp 485–495
22. Stolojanu V, Prakash A (2001) Characterization of slurry systems by ultrasonic techniques. Chem Eng J 84:215–222
23. Hauptmann P, Hoppe N, Puttmer N (2002) Application of ultrasonic sensors in the process industry. Meas Sci Technol 13:73–83. doi:10.1088/0957-0233/13/8/201
24. Bamberger JA, Greenwood MS (2004) Using ultrasonic attenuation to monitor slurry mixing in real time. Ultrasonics 42:145–148
25. Lewis AJ, Rasmussen TC (1996) A new, passive technique for the in situ measurement of total suspended solids concentrations in surface water. Technical completion report for project no. 14-08-001-G-2013 (07). U.S. Department of the Interior, U.S. Geological Survey, Reston
26. Tollner EW, Rasmussen TC (2005) Simulated moving bed form effects on real-time in-stream sediment concentration measurement with densitometry. J Hydraul Eng, ASCE 131 (12):1141–1144
27. Chung C-C, Lin C-P (2011) High concentration suspended sediment measurements using time domain reflectometry. J Hydrol 401(1–2):134–144
28. Topp GC, Davis JL, Annan AP (1980) Electromagnetic determination of soil water content and electrical conductivity measurement using time domain reflectometry. Water Resour Res 16:574–582
29. Lin C-P, Chung C-C, Tang S-H (2007) Accurate TDR measurement of electrical conductivity accounting for cable resistance and recording time. Soil Sci Soc Am J 71(4):1278–1287
30. O'Connor KM, Dowding CH (1999) Geomeasurements by pulsing TDR and probes. CRC Press, Boca Raton
31. Lin C-P, Chung C-C, Tang S-H, Lin C-H (2007) Some innovative developments of TDR technology for geotechnical monitoring. In: 7th International Symposium on Field Measurement in Geomechanics, Boston, 24–27 Sept 2007
32. Chung C-C, Lin CP, Wu I-L, Chen P-H, Tsay T-K (2013) New TDR waveguides and data reduction method for monitoring of stream and drainage stage. J Hydrol 505:346–351
33. Robinson DA (2004) Measurement of the solid dielectric permittivity of clay minerals and granular samples using a time domain reflectometry immersion method. Vadose Zone J 3:705–713
34. Pepin S, Livingston NJ, Hook WR (1995) Temperature-dependent measurement errors in time domain reflectometry determinations of soil water. Soil Sci Soc Am J 59:38–43
35. Klein LA, Swift CT (1977) An improved model for the dielectric constant of sea water at microwave frequencies. IEEE Trans Antennas Propag 25:104–111

36. Dobson MC, Ulaby FT, Hallikainen MT, EL-Rayes MA (1985) Microwave dielectric behavior of wet soil – part II: dielectric mixing models. IEEE Trans Geosci Remote Sens GE-23:35–46
37. Heimovaara TJ (1993) Design of triple-wire time domain reflectometry probes in practice and theory. Soil Sci Soc Am J 57:1410–1417
38. Lin C-P (2003) Analysis of a non-uniform and dispersive TDR measurement system with application to dielectric spectroscopy of soils. Water Resour Res 39(1), 1012
39. Liu H-C, You C-F, Chung C-H, Huang K-F, Liu Z-F (2011) Source variability of sediments in the Shihmen Reservoir, Northern Taiwan: Sr isotopic evidence. J Asian Earth Sci 41 (3):297–306
40. Tsai Z-X, You J-Y, Lee H-Y, Chiu Y-J (2013) Modeling the sediment yield from landslides in the Shihmen Reservoir watershed, Taiwan. Earth Surf Process Landf 38(7):661–674
41. Tseng W-H, Shieh C-L, Lee S-P, Tsang Y-C (2009) A study on the effect of a broken large sabo dam on the sediment transportation in channel – an example of Baling-sabo-dam. Paper presented at EGU General Assembly, Vienna
42. Lin Y-J, Chang Y-H, Tan Y-C, Lee H-Y, Chiu Y-J (2011) National policy of watershed management and flood mitigation after the 921 Chi-Chi earthquake in Taiwan. Nat Hazards 56 (3):709–731
43. Hsieh S-L, Hung S-H, Chung C-C, Lin C-P (2012) Development and implementation of geo-nerve monitoring system using time domain reflectometry. Int Rev Comput Softw 7 (6):3238–3244

Chapter 11
Glossary of Land and Energy Resources Engineering

Mu-Hao Sung Wang and Lawrence K. Wang

Contents

Glossary of Land and Energy Resources Engineering 493
References 622

Abstract Technical and legal terms commonly used by land pollution control engineers and energy engineers are introduced. This chapter covers mainly the glossary terms used in the following two books:

1. Natural Resources and Control Processes
2. Environmental and Natural Resources Engineering

The above two books form a miniseries in the field of natural resources management.

Keywords Natural resources engineering • Environmental engineering • Energy engineering • Glossary • Hydraulic fracturing • Radioactive waste management • Land pollution control • Landfill • Land treatment • Solid waste disposal • Electric power generation • Natural processes

Glossary of Land and Energy Resources Engineering

Abandoned vehicles It refers to any motor vehicles and trailers left on public or private property for an extended period of time and usually inoperable or in hazardous condition and with only scrap value (Washington, DC).

M.-H.S. Wang (✉) • L.K. Wang
Lenox Institute of Water Technology, Newtonville, NY 12128-0405, USA

Rutgers University, New Brunswick, NJ, USA
e-mail: lenox.institute@gmail.com; lawrencekwang@gmail.com

© Springer International Publishing Switzerland 2016
L.K. Wang, M.-H.S. Wang, Y.-T. Hung and N.K. Shammas (eds.),
Natural Resources and Control Processes, Handbook of Environmental
Engineering, Volume 17, DOI 10.1007/978-3-319-26800-2_11

Abandoned well It refers to a well that is no longer in use, whether dry, inoperable, or no longer productive.

Aboveground disposal It refers to the disposal of toxic or hazardous waste, such as low-level radioactive waste in engineered structures built such that all wastes are placed above final grade with no natural material cover. Aboveground disposal includes, but is not limited to, the use of aboveground vaults.

Absolute humidity The ratio of the mass of water vapor to the volume occupied by a mixture of water vapor and dry air.

Absorbed dose The amount of energy absorbed per unit mass in any kind of matter from any kind of ionizing radiation. Absorbed dose is measured in rads or grays.

Absorbent A material that extracts one or more substances from a fluid (gas or liquid) medium on contact and which changes physically and/or chemically in the process.

Absorber The component of a solar thermal collector that absorbs solar radiation and converts it to heat or, as in a solar photovoltaic device, the material that readily absorbs photons to generate charge carriers (free electrons or holes).

Absorption The passing of a substance or force into the body of another substance.

Absorption chiller A type of air-cooling device that uses absorption cooling to cool interior spaces.

Absorption coefficient In reference to a solar energy conversion devices, the degree to which a substance will absorb solar energy. In a solar photovoltaic device, the factor by which photons are absorbed as they travel a unit distance through a material.

Absorption cooling A process in which cooling of an interior space is accomplished by the evaporation of a volatile fluid, which is then absorbed in a strong solution, then desorbed under pressure by a heat source, and then recondensed at a temperature high enough that the heat of condensation can be rejected to a exterior space.

Absorption refrigeration A system in which a secondary fluid absorbs the refrigerant, releasing heat, then releases the refrigerant and reabsorbs the heat. Ammonia or water is used as the vapor in commercial absorption cycle systems, and water or lithium bromide is the absorber.

Absorptivity In a solar thermal system, the ratio of solar energy striking and absorbed by the absorber to the solar energy striking a black body (perfect absorber) at the same temperature. The absorptivity of a material is numerically equal to its emissivity.

Accent lighting Draws attention to special features or enhances the aesthetic qualities of an indoor or outdoor environment.

Accumulator A component of a heat pump that stores liquid and keeps it from flooding the compressor. The accumulator takes the strain off the compressor and improves the reliability of the system.

Accuracy The degree of agreement between a measurement and its true value. The accuracy of a data set is assessed by evaluating results from standards or spikes containing known quantities of an analyte.

Acid rain A term used to describe precipitation that has become acidic (low pH) due to the emission of sulfur oxides from fossil fuel burning power plants.

Action plan An action plan addresses assessment findings and root causes that have been identified in an audit or an assessment report. It is intended to set forth specific actions that the site will undertake to remedy deficiencies. The plan includes a timetable and funding requirements for implementation of the planned activities.

Active agricultural use It refers to lands used for agricultural purposes no less than two of the five calendar years.

Active cooling The use of mechanical heat pipes or pumps to transport heat by circulating heat-transfer fluids.

Active maintenance It refers to any significant activity needed during the period of institutional control to maintain a reasonable assurance that the performance objectives of the related state or federal environmental laws or regulations are met. Such active maintenance includes ongoing activities such as the pumping and treatment of water from a disposal unit or one-time measures such as repair or replacement of all or part of a disposal unit. Active maintenance does not include custodial activities such as fence repair, replacement or repair of monitoring equipment, revegetation, minor additions to soil cover, minor repair of disposal unit covers, and general disposal site upkeep such as mowing grass.

Active power The power (in watts) used by a device to produce useful work. Also called input power.

Active solar heater A solar water or space-heating system that uses pumps or fans to circulate the fluid (water or heat-transfer fluid like diluted antifreeze) from the solar collectors to a storage tank subsystem.

ACToR It is the USEPA's online warehouse of all publicly available chemical toxicity data, which can be used to find all publicly available data about potential chemical risks to human health and the environment. ACToR aggregates data from over 500 public sources on over 500,000 environmental chemicals searchable by chemical name, other identifiers, and chemical structure.

Adiabatic Without loss or gain of heat to a system. An adiabatic change is a change in volume and pressure of a parcel of gas without an exchange of heat between the parcel and its surroundings. In reference to a steam turbine, the adiabatic efficiency is the ratio of the work done per unit weight of steam, to the heat energy released, and theoretically capable of transformation into mechanical work during the adiabatic expansion of a unit weight of steam.

Adjustable speed drive An electronic device that controls the rotational speed of motor-driven equipment such as fans, pumps, and compressors. Speed control is achieved by adjusting the frequency of the voltage applied to the motor.

Adobe A building material made from clay, straw, and water, formed into blocks, and dried; used traditionally in the southwestern USA.

Advisory committee It refers to the advisory committee on siting and disposal method selection for permanent disposal facilities established pursuant to the appropriate state or federal environmental conservation laws or regulations.

Aerobic It refers to the life or processes that require, or are not destroyed by, the presence of oxygen.

Aerobic bacteria Microorganisms that require free oxygen, or air, to live and that which contribute to the decomposition of organic material in soil or composting systems.

Agricultural land It refers to a land on which agricultural animals, foods, feeds, fruits, Christmas trees, rubber trees, fiber crops, etc., are grown. This includes range land or land used as pasture.

Agronomic rate A rate of sludge application that is designed to (1) provide the amount of nitrogen needed by a crop or vegetation grown on the land and (2) minimize the amount of nitrogen in the sewage sludge that passes below the root zone of the crop or vegetation grown on the land to the groundwater.

Air A mixture of gases that surrounds the Earth and forms its atmosphere, composed of, by volume, 20.95 % oxygen, 78.09 % nitrogen, 0.93 % argon, 0.039 % carbon dioxide, and small amounts of other gases. Air also contains a varible amount of water vapor, 0.4–1.0 % over entire atmosphere.

Air change A measure of the rate at which the air in an interior space is replaced by outside (or conditioned) air by ventilation and infiltration; usually measured in cubic feet per time interval (hour), divided by the volume of air in the room.

Air collector In solar heating systems, a type of solar collector in which air is heated in the collector.

Air conditioner A device for conditioning air in an interior space. A room air conditioner is a unit designed for installation in the wall or window of a room to deliver conditioned air without ducts. A unitary air conditioner is composed of one or more assemblies that usually include an evaporator or cooling coil, a compressor and condenser combination, and possibly a heating apparatus. A central air conditioner is designed to provide conditioned air from a central unit to a whole house with fans and ducts.

Air conditioning The control of the quality, quantity, and temperature humidity of the air in an interior space.

Air diffuser An air distribution outlet, typically located in the ceiling, which mixes conditioned air with room air.

Air infiltration measurement A building energy auditing technique used to determine and/or locate air leaks in a building shell or envelope.

Air pollution The presence of contaminants in the air in concentrations that prevent the normal dispersive ability of the air and that interfere with biological processes and human economics.

Air pollution control The use of devices to limit or prevent the release of pollution into the atmosphere.

Air quality standards The prescribed level of pollutants allowed in outside or indoor air as established by legislation.

Air register   The component of a combustion device that regulates the amount of air entering the combustion chamber.

Air retarder/barrier   A material or structural element that inhibits air flow into and out of a building's envelope or shell. This is a continuous sheet composed of polyethylene, polypropylene, or extruded polystyrene. The sheet is wrapped around the outside of a house during construction to reduce air infiltration and exfiltration, yet allow water to easily diffuse through it.

Air space   The area between the layers of glazing (panes) of a window.

Airlock entry   A building architectural element (vestibule) with two airtight doors that reduce the amount of air infiltration and exfiltration when the exterior most door is opened.

Air-source heat pump   A type of heat pump that transfers heat from outdoor air to indoor air during the heating season and works in reverse during the cooling season.

Airtight drywall approach (ADA)   A building construction technique used to create a continuous air retarder that uses the drywall, gaskets, and caulking. Gaskets are used rather than caulking to seal the drywall at the top and bottom. Although it is an effective energy-saving technique, it was designed to keep airborne moisture from damaging insulation and building materials within the wall cavity.

Air-to-air heat pump   See air-source heat pump.

Air-to-water heat pump   A type of heat pump that transfers heat in outdoor air to water for space or water heating.

Albedo   The ratio of light reflected by a surface to the light falling on it.

Alcohol   A group of organic compounds composed of carbon, hydrogen, and oxygen; a series of molecules composed of a hydrocarbon plus a hydroxyl group; includes methanol, ethanol, isopropyl alcohol, and others.

Algae   Primitive plants, usually aquatic, capable of synthesizing their own food by photosynthesis.

Alluvial fan   A cone-shaped deposit of alluvium made by a stream where it runs out onto a level plain.

Alluvium   Sedimentary material deposited by flowing water such as a river.

Alpha radiation   The least penetrating type of radiation. Alpha radiation can be stopped by a sheet of paper or the outer dead layer of skin.

Alternating current   A type of electrical current, the direction of which is reversed at regular intervals or cycles; in the USA, the standard is 120 reversals or 60 cycles per second; typically abbreviated as AC.

Alternative fuels   A popular term for "nonconventional" transportation fuels derived from natural gas (propane, compressed natural gas, methanol, etc.) or biomass materials (ethanol, methanol).

Alternator   A generator producing alternating current by the rotation of its rotor and which is powered by a primary mover.

Ambient lighting   Provides general illumination indoors for daily activities, and outdoors for safety and security.

Ambient air The air external to a building or device.

Ambient temperature The temperature of a medium, such as gas or liquid, which comes into contact with or surrounds an apparatus or building element.

Ammonia A colorless, pungent, gas (NH_3) that is extremely soluble in water may be used as a refrigerant; a fixed nitrogen form suitable as fertilizer.

Amorphous semiconductor A noncrystalline semiconductor material that has no long-range order.

Ampere A unit of measure for an electrical current; the amount of current that flows in a circuit at an electromotive force of one volt and at a resistance of one ohm. Abbreviated as amp.

Amp-hours A measure of the flow of current (in amperes) over one hour.

Anaerobic bacteria Microorganisms that live in oxygen-deprived environments.

Anaerobic digester A device for optimizing the anaerobic digestion of biomass and/or animal manure and possibly to recover biogas for energy production. Digester types include batch, complete mix, continuous flow (horizontal or plug flow, multiple tank, and vertical tank), and covered lagoon.

Anaerobic digestion (1) A complex process by which organic matter is decomposed by anaerobic bacteria. The decomposition process produces a gaseous by-product often called "biogas" primarily composed of methane, carbon dioxide, and hydrogen sulfide. (2) An aerobic biological process involving the fermentation of organic sludge waste by anaerobic hydrolytic microorganisms and the production of fatty acids, carbon dioxide (CO_2), and hydrogen (H_2). Short fatty acids are then converted into acetic acid (CH_3COOH), H_2, CO_2.

Anaerobic It refers to the life or process that occurs in, or is not destroyed by, the absence of oxygen.

Anaerobic lagoon A holding pond for livestock manure that is designed to anaerobically stabilize manure and may be designed to capture biogas, with the use of an impermeable, floating cover.

Analyte It refers to a substance or chemical constituent being analyzed and 3019 of the US Resource Conservation and Recovery Act (RCRA) and the regulations promulgated pursuant to these sections, including the US 40 CFR Parts 260 through 272.

Anemometer An instrument for measuring the force or velocity of wind; a wind gauge.

Angle of incidence In reference to solar energy systems, the angle at which direct sunlight strikes a surface, the angle between the direction of the sun and perpendicular to the surface. Sunlight with an incident angle of 90° tends to be absorbed, while lower angles tend to be reflected.

Angle of inclination In reference to solar energy systems, the angle that a solar collector is positioned above horizontal.

Angstrom unit A unit of length named for A. J. Angstome, a Swedish spectroscopist, used in measuring electromagnetic radiation equal to 0.000,000,01 cm, or 10^{-10} m.

Anhydrous ethanol One hundred percent alcohol, neat ethanol.

Annual fuel utilization efficiency (AFUE) The measure of seasonal or annual efficiency of a residential heating furnace or boiler. It takes into account the cyclic on/off operation and associated energy losses of the heating unit as it responds to changes in the load, which in turn is affected by changes in weather and occupant controls.

Annual load fraction That fraction of annual energy demand supplied by a solar system.

Annual pollutant loading rate (APLR) The maximum amount of a pollutant that can be applied to a unit area of land during a 365-day period. This term describes pollutant limits for sewage sludge that is given away or sold in a bag or other container for application to the land.

Annual solar savings The annual solar savings of a solar building is the energy savings attributable to a solar feature relative to the energy requirements of a nonsolar building.

Annual whole sludge application rate The maximum amount of sewage sludge on a dry weight basis that can be applied to a land application site during a 365-day (1 year) period.

Anode The positive pole or electrode of an electrolytic cell, vacuum tube, etc. (see also sacrificial anode).

Anthracite (Coal) A hard, dense type of coal, which is hard to break, clean to handle, difficult to ignite, and burns with an intense flame and with the virtual absence of smoke because it contains a high percentage of fixed carbon and a low percentage of volatile matter.

Anthropogenic Referring to alterations in the environment due to the presence or activities of humans.

Antifreeze solution A fluid, such as methanol or ethylene glycol, added to vehicle engine coolant or used in solar heating system heat-transfer fluids, to protect the systems from freezing.

Antireflection coating A thin coating of a material applied to a photovoltaic cell surface that reduces the light reflection and increases light transmission.

Aperture An opening; in solar collectors, the area through which solar radiation is admitted and directed to the absorber.

Apparent day A solar day; an interval between successive transits of the sun's center across an observer's meridian; the time thus measured is not equal to clock time.

Apparent power (kVA) This is the voltage-ampere requirement of a device designed to convert electric energy to a nonelectrical form.

Appliance (1) A device for converting one form of energy or fuel into useful energy or work. (2) An instrument or apparatus.

Appliance energy efficiency ratings The ratings under which specified appliances convert energy sources into useful energy, as determined by procedures established by the US Department of Energy.

Aquiclude It refers to an impermeable body of rock that may absorb water slowly, but does not transmit it.

Aquifer A water-bearing unit of permeable rock or soil that will yield water in usable quantities to wells. *Confined aquifers* are bounded above and below by less permeable layers. Groundwater in a confined aquifer is under a pressure greater than the atmospheric pressure. *Unconfined aquifers* are bounded below by less permeable material but are not bounded above. The pressure on the groundwater at the surface of an unconfined aquifer is equal to that of the atmosphere.

Aquifer It refers to an underground geological formation, or group of formations, containing water, which is a source of groundwater for wells and springs.

Aquitard It is a geological formation that may contain groundwater but is not capable of transmitting significant quantities of it under normal hydraulic gradients.

Area of cropland An area of cropland that is not a single field and has been subdivided into several strips, and each strip represents an individual field unit.

Argon A colorless, odorless inert gas sometimes used in the spaces between the panes in energy-efficient windows. This gas is used because it will transfer less heat than air. Therefore, it provides additional protection against conduction and convection of heat over conventional double-pane windows.

Array (solar) Any number of solar photovoltaic modules or solar thermal collectors or reflectors connected together to provide electrical or thermal energy.

As low as reasonably achievable (ALARA) It takes into account the state of technology and the economics of improvements in relation to (1) benefits to the environment and public health and safety, (2) other societal and socioeconomic considerations, (3) the utilization of radioactive materials in the public interest, and (4) an approach to radiation protection that advocates controlling or managing exposures (both individual and collective) to the work force and the general public and releases of radioactive material to the environment as low as social, technical, economic, practical, and public policy considerations permit. As used in the US Department of Energy, ALARA is not a dose limit but, rather, a process that has as its objective the attainment of dose levels as far below the applicable limits of the order as practicable.

Ashes (1) Noncombustible residue from the burning of wood, coal, coke, and other combustible materials in homes, stores, institutions, and industrial establishments for the purpose of heating, cooking, and disposing of waste combustible material and, unless otherwise specified, does not include tin cans, scrap metal, and glass (Baltimore County, MD). (2) The residue from the burning of wood, coal, coke, or other combustible materials (Washington, DC). (3) The residue of the combustion of solid fuels (Bettendorf, IA). (4) They include noncombustible residue from the burning of wood, coal, and other combustible materials in homes, stores, institutions, and industrial establishments for the purpose of heating, cooking, and disposing of waste combustible material (Prince George's County, MD).

ASHRAE Abbreviation for the American Society of Heating, Refrigeration, and Air-Conditioning Engineers.

Assay It refers to a test for a specific chemical, microbe, or effect.

Assets It refers to all existing and all probable future economic benefits obtained or controlled by a particular entity.

ASTM Abbreviation for the American Society for Testing and Materials, which is responsible for the issue of many standard methods used in the energy industry.

Asynchronous generator A type of electric generator that produces alternating current that matches an existing power source.

Atmospheric pressure The pressure of the air at sea level; one standard atmosphere at zero degrees centigrade is equal to 14.695 lb per square inch (1.033 kg per square centimeter).

Atrium An interior court to which rooms open.

Attic The usually unfinished space above a ceiling and below a roof.

Attic fan A fan mounted on an attic wall used to exhaust warm attic air to the outside.

Attic vent A passive or mechanical device used to ventilate an attic space, primarily to reduce heat buildup and moisture condensation.

Audit (energy) The process of determining energy consumption, by various techniques, of a building or facility.

Automatic (or remote) meter reading system A system that records the consumption of electricity, gas, water, etc., and sends the data to a central data accumulation device.

Automatic damper A device that cuts off the flow of hot or cold air to or from a room as controlled by a thermostat.

Auxiliary energy or system Energy required to operate mechanical components of an energy system, or a source of energy or energy supply system to back up another.

Availability Describes the reliability of power plants. It refers to the number of hours that a power plant is available to produce power divided by the total hours in a set time period, usually a year.

Available heat The amount of heat energy that may be converted into useful energy from a fuel.

Average demand The demand on, or the power output of, an electrical system or any of its parts over an interval of time, as determined by the total number of kilowatt-hours divided by the units of time in the interval.

Average master file A method of calculating the average raw wastewater concentration for each pollutant of interest in a subcategory. The average master file was calculated using all available data collected in the landfills industry study.

Average wind speed (or velocity) The mean wind speed over a specified period of time.

Avoided cost The incremental cost to an electric power producer to generate or purchase a unit of electricity or capacity or both.

AWG The abbreviation for American wire gauge; the standard for gauging the size of wires (electrical conductors).

Awning An architectural element for shading windows and wall surfaces placed on the exterior of a building; can be fixed or movable.

Axial fans Fans in which the direction of the flow of the air from inlet to outlet remains unchanged; includes propeller, tubaxial, and vaneaxial type fans.

Axial flow compressor A type of air compressor in which air is compressed in a series of stages as it flows axially through a decreasing tubular area.

Axial flow turbine A turbine in which the flow of a steam or gas is essentially parallel to the rotor axis.

Azimuth (solar) The angle between true south and the point on the horizon directly below the sun.

Backdrafting The flow of air down a flue/chimney and into a house caused by low indoor air pressure that can occur when using several fans or fireplaces and/or if the house is very tight.

Background radiation Natural and man-made radiation such as cosmic radiation and radiation from naturally radioactive elements and from commercial sources and medical procedures.

Backup energy system A reserve appliance; for example, a standby generator for a home or commercial building.

Bacteria Single-celled organisms, free-living or parasitic, that break down the wastes and bodies of dead organisms, making their components available for reuse by other organisms.

Baffle A device, such as a steel plate, used to check, retard, or divert a flow of a material.

Bagasse The fibrous material remaining after the extraction of juice from sugarcane; often burned by sugar mills as a source of energy.

Bagged sewage sludge Sewage sludge that is sold or given away in a bag or other container (i.e., either an open or closed receptacle containing 1 metric ton or less of sewage sludge).

Baghouse An air pollution control device used to filter particulates from waste combustion gases; a chamber containing a bag filter.

Balance point An outdoor temperature, usually 20–45 °F, at which a heat pump's output equals the heating demand. Below the balance point, supplementary heat is needed.

Balance of system In a renewable energy system, refers to all components other than the mechanism used to harvest the resource (such as photovoltaic panels or a wind turbine). Balance-of-system costs can include design, land, site preparation, system installation, support structures, power conditioning, operation and maintenance, and storage.

Baling A means of reducing the volume of a material by compaction into a bale.

Ballast A device used to control the voltage in a fluorescent lamp.

Ballast efficacy factor The measure of the efficiency of fluorescent lamp ballasts. It is the relative light output divided by the power input.

Ballast factor The ratio of light output of a fluorescent lamp operated on a ballast to the light output of a lamp operated on a standard or reference ballast.

Band gap In a semiconductor, the energy difference between the highest valence band and the lowest conduction band.

Band gap energy The amount of energy (in electron volts) required to free an outer shell electron from its orbit about the nucleus to a free state and thus promote it from the valence to the conduction level.

Barrel (petroleum) 42 US gallons (306 lb of oil, or 5.78 million Btu).

Basal metabolism The amount of heat given off by a person at rest in a comfortable environment; approximately 50 Btu per hour (Btu/h).

Base power Power generated by a power generator that operates at a very high capacity factor.

Baseboard radiator A type of radiant heating system where the radiator is located along an exterior wall where the wall meets the floor.

Baseline flow Estimated air, water, or wastewater discharge flow rate for a selected facility in specific time.

Baseload capacity The power output of a power plant that can be continuously produced.

Baseload demand The minimum demand experienced by a power plant.

Baseload power plant A power plant that is normally operated to generate a base load and that usually operates at a constant load; examples include coal-fired and nuclear-fueled power plants.

Basement The conditioned or unconditioned space below the main living area or primary floor of a building.

BAT (best available technology) See best available technology (BAT).

Batch heater This simple passive solar hot water system consists of one or more storage tanks placed in an insulated box that has a glazed side facing the sun. A batch heater is mounted on the ground or on the roof (make sure your roof structure is strong enough to support it). Some batch heaters use "selective" surfaces on the tank(s). These surfaces absorb sun well but inhibit radiative loss. Also known as bread box systems or integral collector storage systems.

Batch process A process for carrying out a reaction in which the reactants are fed in discrete and successive charges.

Batt/blanket A flexible roll or strip of insulating material in widths suited to standard spacings of building structural members (studs and joists). They are made from glass or rock wool fibers. Blankets are continuous rolls. Batts are precut to four or eight foot lengths.

Battery An energy storage device composed of one or more electrolyte cells.

Battery energy storage Energy storage using electrochemical batteries. The three main applications for battery energy storage systems include spinning reserve at generating stations, load leveling at substations, and peak shaving on the customer side of the meter.

BCT (best conventional pollutant control technology) See best conventional pollutant control technology (BCT).

Beadwall A form of movable insulation that uses tiny polystyrene beads blown into the space between two window panes.

Beam radiation Solar radiation that is not scattered by dust or water droplets.

Bearing wall A wall that carries ceiling rafters or roof trusses.

Becquerel (Bq) A unit of radioactivity equal to one nuclear transformation per second.

Belowground disposal It refers to the disposal of toxic or hazardous waste, such as low-level radioactive waste, such that all wastes are placed totally below final grade in engineered structures which are located within the upper 30 m of the Earth's surface and are covered with natural material. Belowground disposal methods include, but are not limited to, buried vaults, lined augered holes, and earth-mounded bunkers.

Benefits charge The addition of a per unit tax on sales of electricity, with the revenue generated used for or to encourage investments in energy efficiency measures and/or renewable energy projects.

Best available technology (BAT) In the USA, it means the best available technology economically achievable, applicable to effluent limitations to be achieved by July 1, 1984, for industrial discharges to surface waters, as defined by Sec. 304(b)(2)(B) of the Clean Water Act (CWA).

Best conventional pollutant control technology (BCT) In the USA, it is applicable to discharges of conventional pollutants from existing industrial point sources, as defined by Sec. 304(b)(4) of the Clean Water Act (CWA).

Best management practice (BMP) A method that has been determined to be the most effective, practical means of preventing or reducing pollution from non-point and point sources.

Best practicable control technology currently available (BPT) In the USA, it is applicable to effluent limitations to be achieved by July 1, 1977, for industrial discharges to surface waters, as defined by Sec. 304(b)(l) of the Clean Water Act (CWA).

Beta radiation Electrons emitted from a nucleus during fission and nuclear decay. Beta radiation can be stopped by an inch of wood or a thin sheet of aluminum.

Bimetal Two metals of different coefficients of expansion welded together so that the piece will bend in one direction when heated, and in the other when cooled, and can be used to open or close electrical circuits, as in thermostats.

Bin method A method of predicting heating and/or cooling loads using instantaneous load calculation at different outdoor dry-bulb temperatures and multiplying the result by the number of hours of occurrence of each temperature.

Binary cycle Combination of two power plant turbine cycles utilizing two different working fluids for power production. The waste heat from the first turbine cycle provides the heat energy for the operation of the second turbine, thus providing higher overall system efficiencies.

Binary cycle geothermal plants Binary cycle systems can be used with liquids at temperatures less than 350°F (177°C). In these systems, the hot geothermal liquid vaporizes a secondary working fluid, which then drives a turbine.

Biochemical oxygen demand, five-day (BOD$_5$) The weight of oxygen taken up mainly as a result of the oxidation of the constituents of a sample of water by biological action; expressed as the number of parts per million of oxygen taken up by the sample from water originally saturated with air, usually over a period of five days at 20° centigrade. A standard means of estimating the degree of contamination of water. See five-day biochemical oxygen demand (BOD_5).

Biocide It refers to any substance that kills or retards the growth of microorganisms.

Bioconversion The conversion of one form of energy into another by the action of plants or microorganisms. The conversion of biomass to ethanol, methanol, or methane.

Biodegradation It refers to the chemical breakdown of materials under natural conditions.

Bioenergy The conversion of the complex carbohydrates in organic material into energy.

Biogas (1) A combustible gas created by anaerobic decomposition of organic material, composed primarily of methane, carbon dioxide, and hydrogen sulfide and (2) a product from anaerobic digestion containing gases such as methane (CH_4), CO_2, and trace elements. Biogas can be used as a source of energy.

Biogasification or biomethanization The process of decomposing biomass with anaerobic bacteria to produce biogas.

Biomass As defined by the Energy Security Act (PL 96-294) of 1980, "any organic matter which is available on a renewable basis, including agricultural crops and agricultural wastes and residues, wood and wood wastes and residues, animal wastes, municipal wastes, and aquatic plants."

Biomass energy Energy produced by the conversion of biomass directly to heat or to a liquid or gas that can be converted to energy.

Biomass fuel Biomass converted directly to energy or converted to liquid or gaseous fuels such as ethanol, methanol, methane, and hydrogen.

Biomass gasification The conversion of biomass into a gas, by biogasification (see above) or thermal gasification, in which hydrogen is produced from high-temperature gasifying and low-temperature pyrolysis of biomass.

Biophotolysis The action of light on a biological system that results in the dissociation of a substrate, usually water, to produce hydrogen.

Biosolids Biosolids are solids, semisolids, or liquid materials, resulting from biological treatment of domestic sewage that has been sufficiently processed to permit these materials to be safely land applied. The term biosolids was introduced by the wastewater treatment industry in the early 1990s and has been recently adopted by the United States Environmental Protection Agency (US EPA) to distinguish high-quality, treated sewage sludge from raw sewage sludge and from sewage sludge containing large amounts of pollutants.

Blackbody An ideal substance that absorbs all radiation falling on it and reflecting nothing.

Blower The device in an air conditioner that distributes the filtered air from the return duct over the cooling coil/heat exchanger. This circulated air is cooled/

heated and then sent through the supply duct, past dampers, and through supply diffusers to the living/working space.

Blower door A device used by energy auditors to pressurize a building to locate places of air leakage and energy loss.

Blown in insulation (see also loose fill) An insulation product composed of loose fibers or fiber pellets that are blown into building cavities or attics using special pneumatic equipment.

Boiler A vessel or tank where heat produced from the combustion of fuels such as natural gas, fuel oil, or coal is used to generate hot water or steam for applications ranging from building space heating to electric power production or industrial process heat.

Boiler feedwater The water that is forced into a boiler to take the place of that which is evaporated in the generation of steam.

Boiler horsepower A unit of rate of water evaporation equal to the evaporation per hour of 34.5 lb (15.66 kg) of water at a temperature of 212°F (100°C) into steam at 212°F (100°C).

Boiler pressure The pressure of the steam or water in a boiler as measured; usually expressed in pounds per square inch gauge (psig).

Boiler rating The heating capacity of a steam boiler; expressed in Btu per hour (Btu/h), or horsepower, or pounds of steam per hour.

Bone (oven) dry In reference to solid biomass fuels, such as wood, having zero moisture content.

Bone dry unit A quantity of (solid) biomass fuel equal to 2400 lb bone dry.

Booster pump A pump for circulating the heat-transfer fluid in a hydronic heating system.

Boot In heating and cooling system distribution ductwork, the transformation pieces connecting horizontal round leaders to vertical rectangular stacks.

Boron The chemical element commonly used as the dopant in solar photovoltaic device or cell material.

Bottled gas A generic term for liquefied and pressurized gas, ordinarily butane, propane, or a mixture of the two, contained in a cylinder for domestic use.

Bottoming-cycle A means to increase the thermal efficiency of a steam electric generating system by converting some waste heat from the condenser into electricity. The heat engine in a bottoming cycle would be a condensing turbine similar in principle to a steam turbine but operating with a different working fluid at a much lower temperature and pressure.

BPT (best practicable control technology currently available) See the best practicable control technology currently available (BPT).

Brayton cycle A thermodynamic cycle using constant pressure, heat addition, and rejection, representing the idealized behavior of the working fluid in a gas turbine-type heat engine.

Bread box system This simple passive solar hot water system consists of one or more storage tanks placed in an insulated box that has a glazed side facing the sun. A bread box system is mounted on the ground or on the roof (make sure your roof structure is strong enough to support it). Some systems use "selective" surfaces on the tank(s). These surfaces absorb sun well but inhibit radiative loss. Also known as batch heaters or integral collector storage systems.

Brine Water saturated or strongly impregnated with salt.

British thermal unit (Btu) The amount of heat required to raise the temperature of one pound of water one degree Fahrenheit; equal to 252 cal.

Buffer zone It refers to a portion of the disposal site that is controlled by the licensee and that lies under the disposal units and between the disposal units and the boundary of the site.

Building debris or waste It includes any refuse or residue resulting from minor noncommercial repairs to a private dwelling made by the owner or occupant thereof (Bettendorf, IA), and (2) it includes any and all refuse or residue resulting directly from building construction, reconstruction, repair or demolition; from grading, shrubbing, or other incidental work in connection with any premises; or from replacement of building equipment or appliances (Bettendorf, IA).

Building energy ratio The space-conditioning load of a building.

Building envelope The structural elements (walls, roof, floor, foundation) of a building that encloses conditioned space; the building shell.

Building heat-loss factor A measure of the heating requirements of a building expressed in Btu per degree day.

Building orientation The relationship of a building to true south, as specified by the direction of its longest axis.

Building overall energy loss coefficient-area product The factor, when multiplied by the monthly degree days, that yields the monthly space-heating load.

Building overall heat-loss rate The overall rate of heat loss from a building by means of transmission plus infiltration, expressed in Btu per hour, per degree temperature difference between the inside and outside.

Bulb The transparent or opaque sphere in an electric light that the electric light transmits through.

Bulb turbine A type of hydroturbine in which the entire generator is mounted inside the water passageway as an integral unit with the turbine. These installations can offer significant reductions in the size of the powerhouse.

Bulk density The weight of a material per unit of volume compared to the weight of the same volume of water.

Bulk refuse or waste (1) Any waste articles more than eight feet (2.44 m) in length or more than fifty pounds (22.7 kg) in weight, tree trimmings and hedge trimmings over five inches (12.7 cm) in diameter or over four feet (1.22 m) in length, and tree roots or stumps larger than can be contained in a bushel basket (Bettendorf, IA) and (2) any large items of solid waste such as appliances, furniture, large auto parts, tree and branches, stumps, flotage, and the like (Washington, DC).

Bulk sewage sludge Sewage sludge that is not sold or given away in a bag or other container for application to the land.

Burner capacity The maximum heat output (in Btu per hour) released by a burner with a stable flame and satisfactory combustion.

Burning point The temperature at which a material ignites.

Bus (electrical) An electrical conductor that serves as a common connection for two or more electrical circuits; may be in the form of rigid bars or stranded conductors or cables.

Busbar The power conduit of an electric power plant; the starting point of the electric transmission system.

Busbar cost The cost of producing electricity up to the point of the power plant busbar.

Bypass An alternative path. In a heating duct or pipe, an alternative path for the flow of the heat-transfer fluid from one point to another, as determined by the opening or closing of control valves both in the primary line and the bypass line.

Cage The component of an electric motor composed of solid bars (of usually copper or aluminum) arranged in a circle and connected to continuous rings at each end. This cage fits inside the stator in an induction motor in channels between laminations, thin flat disks of steel in a ring configuration.

Calorie The amount of heat required to raise the temperature of a unit of water, at or near the temperature of maximum density, one degree Celsius (or centigrade [C]); expressed as a "small calorie" (the amount of heat required to raise the temperature of 1 g of water one degree C) or as a "large calorie" or "kilogram calorie" (the amount of heat required to raise one kilogram [1,000 g] of water one degree C); capitalization of the word calorie indicates a kilogram calorie.

Calorific value The heat liberated by the combustion of a unit quantity of a fuel under specific conditions; measured in calories.

Candela The luminous intensity, in a given direction, of a source that emits monochromatic radiation of frequency $540 \times 1,012$ Hz and that has a radiant intensity in that direction of 1/683 W per steradian.

Candle power The illuminating power of a standard candle employed as a unit for determining the illuminating quality of an illuminant.

Capability The maximum load that a generating unit, power plant, or other electrical apparatus can carry under specified conditions for a given period of time, without exceeding its approved limits of temperature and stress.

Capability margin The difference between net electrical system capability and system maximum load requirements (peak load); the margin of capability available to provide for scheduled maintenance, emergency outages, system operating requirements, and unforeseen loads.

Capacitance A measure of the electrical charge of a capacitor consisting of two plates separated by an insulating material.

Capacitor An electrical device that adjusts the leading current of an applied alternating current to balance the lag of the circuit to provide a high power factor.

Capacity The load that a power generation unit or other electrical apparatus or heating unit is rated by the manufacture to be able to meet or supply.

Capacity (condensing unit) The refrigerating effect in Btu/h produced by the difference in total enthalpy between a refrigerant liquid leaving the unit and the total enthalpy of the refrigerant vapor entering it. Generally measured in tons or Btu/h.

Capacity (effective, of a motor) The maximum load that a motor is capable of supplying.

Capacity (heating, of a material) The amount of heat energy needed to raise the temperature of a given mass of a substance by one degree Celsius. The heat required to raise the temperature of 1 kg of water by 1 °C is 4186 J.

Capacity factor The ratio of the average load on (or power output of) a generating unit or system to the capacity rating of the unit or system over a specified period of time.

CAPDET It refers to the computer-assisted procedure for the design and evaluation of wastewater treatment systems, which was developed by the US Army Corp. of Engineers (CAPDET), is intended to provide planning level cost estimates to analyze alternate design technologies for wastewater treatment systems.

Capital costs The amount of money needed to purchase equipment, buildings, tools, and other manufactured goods that can be used in production.

Captive A term used to describe a facility that only accepts wastes generated on site and/or by the landfill owner/operator at the facility.

Carbon dioxide A colorless, odorless noncombustible gas with the formula CO_2 that is present in the atmosphere. It is formed by the combustion of carbon and carbon compounds (such as fossil fuels and biomass), by respiration, which is a slow combustion in animals and plants, and by the gradual oxidation of organic matter in the soil.

Carbon monoxide A colorless, odorless but poisonous combustible gas with the formula CO. Carbon monoxide is produced in the incomplete combustion of carbon and carbon compounds such as fossil fuels (i.e., coal, petroleum) and their products (e.g., liquefied petroleum gas, gasoline) and biomass.

Carbon zinc cell battery A cell produces electric energy by the galvanic oxidation of carbon; commonly used in household appliances.

Carnot cycle An ideal heat engine (conceived by Sadi Carnot) in which the sequence of operations forming the working cycle consists of isothermal expansion, adiabatic expansion, isothermal compression, and adiabatic compression back to its initial state.

Casing It refers to the pipe cemented in the well to seal off formation fluids and to keep the hole from caving in.

Catalytic converter An air pollution control device that removes organic contaminants by oxidizing them into carbon dioxide and water through a chemical reaction using a catalysis, which is a substance that increases (or decreases) the rate of a chemical reaction without being changed itself; required in all automobiles sold in the United States and used in some types of heating appliances.

Categorical exclusion A proposed action that normally does not require an environmental assessment or an environmental impact statement and that the responsible environmental agency has determined does not individually or cumulatively have a significant effect on the human environment.

Cathedral ceiling/roof A type of ceiling and roof assembly that has no attic.

Cathode The negative pole or electrode of an electrolytic cell, vacuum tube, etc., where electrons enter (current leaves) the system; the opposite of an anode.

Cathode disconnect ballast An electromagnetic ballast that disconnects a lamp's electrode heating circuit once it has started; often called "low-frequency electronic" ballasts.

Cathodic protection A method of preventing oxidation of the exposed metal in structures by imposing between the structure and the ground a small electrical voltage.

Caulking A material used to seal areas of potential air leakage into or out of a building envelope.

Ceiling The downward facing structural element that is directly opposite the floor.

Ceiling concentration limits (CCL) The *ceiling concentration limits* are the maximum concentrations of the nine trace elements allowed in biosolids to be land applied. Sewage sludge exceeding the ceiling concentration limit for even one of the regulated pollutants is not classified as biosolids and, hence, cannot be land applied.

Ceiling fan A mechanical device used for air circulation and to provide cooling.

Cell (landfill) An area of a landfill that is separated from other areas by an impervious structure. Each cell has a separate leachate collection system or would require a separate leachate collection system if one were installed. Individual leachate collection systems that are combined at the surface are considered separate systems by this definition.

Cell (electrical) A component of an electrochemical battery. A "primary" cell consists of two dissimilar elements, known as "electrodes," immersed in a liquid or paste known as the "electrolyte." A direct current of 1–1.5 V will be produced by this cell. A "secondary" cell or accumulator is a similar design but is made useful by passing a direct current of correct strength through it in a certain direction. Each of these cells will produce 2 V; a 12 V car battery contains six cells.

Cellulase An enzyme complex, produced by fungi and bacteria, capable of decomposing cellulose into small fragments, primarily glucose.

Cellulose The fundamental constituent of all vegetative tissue; the most abundant material in the world.

Cellulose insulation A type of insulation composed of waste newspaper, cardboard, or other forms of waste paper.

Central heating system A system where heat is supplied to areas of a building from a single appliance through a network of ducts or pipes.

Central power plant A large power plant that generates power for distribution to multiple customers.

Central receiver solar power plants Also known as "power towers," these use fields of two-axis tracking mirrors known as heliostats. Each heliostat is individually positioned by a computer control system to reflect the sun's rays to a tower-mounted thermal receiver. The effect of many heliostats reflecting to a common point creates the combined energy of thousands of suns, which produces high-temperature thermal energy. In the receiver, molten nitrate salts absorb the heat energy. The hot salt is then used to boil water to steam, which is sent to a conventional steam turbine generator to produce electricity.

Certification It refers to a decision issued by the state government pursuant to the State Environmental Conservation Law to the effect that one or two proposed disposal sites and the disposal method or methods proposed for use at such site or sites are in conformance with the applicable provisions of the state or federal environmental laws or regulations.

Cetane number A measure of a fuel's (liquid) ease of self-ignition (CFR 122.2).

Char A by-product of low-temperature carbonization of a solid fuel.

Charcoal A material formed from the incomplete combustion or destructive distillation (carbonization) of organic material in a kiln or retort and having a high-energy density, being nearly pure carbon. (If produced from coal, it is coke.) Used for cooking, the manufacture of gunpowder and steel (notably in Brazil), as an absorbent and decolorizing agent, and in sugar refining and solvent recovery.

Charge carrier A free and mobile conduction electron or hole in a semiconductor.

Charge controller An electronic device that regulates the electrical charge stored in batteries so that unsafe, overcharge conditions for the batteries are avoided.

Chemical energy The energy liberated in a chemical reaction, as in the combustion of fuels.

Chemical oxygen demand (COD) It is a wastewater quality index that chemically determines the amount of oxygen demand in water or wastewater in accordance with the standard methods.

Chemical vapor deposition (CVD) A method of depositing thin semiconductor films used to make certain types of solar photovoltaic devices. With this method, a substrate is exposed to one or more vaporized compounds, one or more of which contain desirable constituents. A chemical reaction is initiated, at or near the substrate surface, to produce the desired material that will condense on the substrate.

Chiller A device for removing heat from a gas or liquid stream for air conditioning/cooling.

Chimney A masonry or metal stack that creates a draft to bring air to a fire and to carry the gaseous by-products of combustion safely away.

Chimney effect The tendency of heated air or gas to rise in a duct or other vertical passage, such as in a chimney, small enclosure, or building, due to its lower density compared to the surrounding air or gas.

Chlorofluorocarbon (CFC) A family of chemicals composed primarily of carbon, hydrogen, chlorine, and fluorine whose principal applications are as refrigerants and industrial cleansers and whose principal drawback is the tendency to destroy the Earth's protective ozone layer.

Circuit A device, or system of devices, that allows electrical current to flow through it and allows voltage to occur across positive and negative terminals.

Circuit breaker A device used to interrupt or break an electrical circuit when an overload condition exists; usually installed in the positive circuit; used to protect electrical equipment.

Circuit lag As time increases from zero at the terminals of an inductor, the voltage comes to a particular value on the sine function curve ahead of the current. The voltage reaches its negative peak exactly 90° before the current reaches its negative peak; thus, the current lags behind by 90°.

Circulating fluidized bed A type of furnace or reactor in which the emission of sulfur compounds is lowered by the addition of crushed limestone in the fluidized bed, thus obviating the need for much of the expensive stack gas cleanup equipment. The particles are collected and recirculated, after passing through a conventional bed and cooled by boiler internals.

City waste collectors It refers to any person or firm employed by the city from time to time to collect and dispose of household waste, institutional waste, commercial waste, or building debris from within the confines of the city (Bettendorf, IA).

Class A, B, and C low-level wastes Waste classifications in the USA from the US Nuclear Regulatory Commission's 10 CFR Part 61 rule. Maximum concentration limits are set for specific isotopes. (1) Class A waste disposal is minimally restricted with respect to the form of the waste. (2) Class B waste must meet more rigorous requirements to ensure physical stability after disposal. (3) Greater concentration limits are set for the same isotopes in Class C waste, which also must meet physical stability requirements. Moreover, special measures must be taken at the disposal facility to protect against inadvertent intrusion.

Class I sludge management facility Publicly owned treatment works (POTWs), required to have an approved pretreatment program under 40 *CFR* 403.8(a), including any POTW located in a state that has elected to assume local pretreatment program responsibilities under 40 *CFR* 403.10(e). In addition, the regional administrator or, in the case of approved state programs, the regional administrator in conjunction with the state director has the discretion to designate any treatment works treating domestic sewage (TWTDS) as a Class I sludge management facility.

Clean power generator A company or other organizational unit that produces electricity from sources that are thought to be environmentally cleaner than traditional sources. Clean, or green, power is usually defined as power from renewable energy that comes from wind, solar, biomass energy, etc. There are various definitions of clean resources. Some definitions include power produced from waste-to-energy and wood-fired plants that may still produce significant air emissions. Some states have defined certain local resources as clean that other states would not consider clean. For example, the State of Texas has defined power from efficient natural gas-fired power plants as clean. Some northwest states include power from large hydropower projects as clean, although these projects damage fish populations. Various states have disclosure and labeling requirement for generation source and air emissions that assist customers in comparing electricity characteristics other than price. This allows customers to decide for themselves what they consider to be "clean." The federal government is also exploring this issue.

Clean water act (CWA) The US Federal Water Pollution Control Act Amendments of 1972, Public Law 92–500. It contians a number of provisions to restore and maintain the quality of the US water resources.

Cleavage of lateral epitaxial films for transfer (CLEFT) A process for making inexpensive gallium arsenide (GaAs) photovoltaic cells in which a thin film of GaAs is grown atop a thick, single-crystal GaAs (or other suitable material) substrate and then is cleaved from the substrate and incorporated into a cell, allowing the substrate to be reused to grow more thin-film GaAs.

Clerestory A window located high in a wall near the eaves that allows daylight into a building interior and may be used for ventilation and solar heat gain.

Climate The prevailing or average weather conditions of a geographic region.

Climate change A term used to describe short- and long-term effects on the Earth's climate as a result of human activities such as fossil fuel combustion and vegetation clearing and burning.

Close coupled An energy system in which the fuel production equipment is in close proximity, or connected, to the fuel using equipment.

Closed A facility or portion thereof that is currently not receiving or accepting wastes and has undergone final closure.

Closed cycle A system in which a working fluid is used over and over without introduction of new fluid, as in a hydronic heating system or mechanical refrigeration system.

Closed-loop biomass As defined by the Comprehensive National Energy Act of 1992 (or the Energy Policy Act): any organic matter from a plant which is planted for the exclusive purpose of being used to produce energy. This does not include wood or agricultural wastes or standing timber.

Closed-loop geothermal heat pump systems Closed-loop (also known as "indirect") systems circulate a solution of water and antifreeze through a series of sealed loops of piping. Once the heat has been transferred into or out of the solution, the solution is recirculated. The loops can be installed in the ground horizontally or vertically, or they can be placed in a body of water, such as a pond. See horizontal ground loop, vertical ground loop, slinky ground loop, and surface water loop for more information on the different types of closed-loop geothermal heat pump systems.

Closure period It refers to the period of time after the operation period during which facility closure is carried out and site closure and stabilization is completed.

Closure plan It refers to the plan for site closure and stabilization prepared as required by the state or federal environmental laws or regulations.

Coal One of the fuels formed in the ground from the remains of dead plants and animals. It takes millions of years to form fossil fuels. Coal is a black mineral material that burns. Other fossil fuels include oil, natural gas, etc.

Coalbed It refers to a geological layer or stratum of coal parallel to the rock stratification.

Coal-fired (Thermoelectric) power plant A power plant that produces electricity by the force of steam through a turbine that spins a generator. The steam is produced by burning the coal.

Codes Legal documents that regulate construction to protect the health, safety, and welfare of people. Codes establish minimum standards but do not guarantee efficiency or quality.

Coefficient of heat transmission (U-value) A value that describes the ability of a material to conduct heat. The number of Btu that flows through 1 square foot of material, in one hour. It is the reciprocal of the R-value (U-value = 1/R-value).

Coefficient of performance (COP) A ratio of the work or useful energy output of a system versus the amount of work or energy inputted into the system as determined by using the same energy equivalents for energy in and out. It is used as a measure of the steady-state performance or energy efficiency of heating, cooling, and refrigeration appliances. The COP is equal to the energy efficiency ratio (EER) divided by 3.412. The higher the COP, the more efficient the device.

Coefficient of utilization (CU) A term used for lighting appliances; the ratio of lumens received on a flat surface to the light output, in lumens, from a lamp; used to evaluate the effectiveness of luminaries in delivering light.

Cofiring The use of two or more different fuels (e.g., wood and coal) simultaneously in the same combustion chamber of a power plant.

Cogeneration The generation of electricity or shaft power by an energy conversion system and the concurrent use of rejected thermal energy from the conversion system as an auxiliary energy source.

Cogenerator A class of energy producer that produces both heat and electricity from a single fuel.

Coil As a component of a heating or cooling appliance, rows of tubing, or pipe with fins attached through which a heat-transfer fluid is circulated and to deliver heat or cooling energy to a building.

Coincidence factor The ratio of the coincident, maximum demand, or two or more loads to the sum of their noncoincident maximum demand for a given period; the reciprocal of the diversity factor and is always less than or equal to one.

Coincident demand The demand of a consumer of electricity at the time of a power supplier's peak system demand.

Cold night sky The low effective temperature of the sky on a clear night.

Collective dose equivalent The sum of the dose equivalents for all the individuals comprising a defined population. The per capita dose equivalent is the quotient of the collective dose equivalent divided by the population. The unit of collective dose equivalent is person-rem or person-sievert.

Collective effective dose equivalent The sum of the effective dose equivalents for the individuals comprising a defined population. Units of measurement are person-rems or person-sieverts. The per capita effective dose equivalent is obtained by dividing the collective dose equivalent by the population. Units of measurement are rems or sieverts.

Collector The component of a solar energy heating system that collects solar radiation and that contains components to absorb solar radiation and transfer the heat to a heat-transfer fluid (air or liquid).

Collector efficiency The ratio of solar radiation captured and transferred to the collector (heat transfer) fluid.

Collector fluid The fluid, liquid (water or water/antifreeze solution), or air, used to absorb solar energy and transfer it for direct use, indirect heating of interior air or domestic water, and/or to a heat storage medium.

Collector tilt The angle that a solar collector is positioned from horizontal.

Color rendition How colors appear when illuminated by a light source. Color rendition is generally considered to be a more important lighting quality than color temperature. Most objects are not a single color, but a combination of many colors. Light sources that are deficient in certain colors may change the apparent color of an object. The color rendition index (CRI) is a 1–100 scale that measures a light source's ability to render colors the same way sunlight does. The top value of the CRI scale (100) is based on illumination by a 100-W incandescent light bulb. A light source with a CRI of 80 or higher is considered acceptable for most indoor residential applications.

Color rendition (rendering) index (CRI) A measure of light quality. The maximum CRI value of 100 is given to natural daylight and incandescent lighting. The closer a lamp's CRI rating is to 100, the better is its ability to show true colors to the human eye.

Color temperature The color of the light source. By convention, yellow-red colors (like the flames of a fire) are considered warm, and blue-green colors (like light from an overcast sky) are considered cool. Color temperature is measured in Kelvin (K) temperature. Confusingly, higher Kelvin temperatures (3,600–5,500 K) are what we consider cool and lower color temperatures (2,700–3,000 K) are considered warm. Cool light is preferred for visual tasks because it produces higher contrast than warm light. Warm light is preferred for living spaces because it is more flattering to skin tones and clothing. A color temperature of 2700–3600 K is generally recommended for most indoor general and task lighting applications.

Combined-cycle power plant A power plant that uses two thermodynamic cycles to achieve higher overall system efficiency; e.g., the heat from a gas-fired combustion turbine is used to generate steam for heating or to operate a steam turbine to generate additional electricity.

Combustible rubbish It includes miscellaneous burnable materials (Washington, DC).

Combustion The process of burning; the oxidation of a material by applying heat, which unites oxygen with a material or fuel.

Combustion air Air that provides the necessary oxygen for complete, clean combustion and maximum heating value.

Combustion chamber Any wholly or partially enclosed space in which combustion takes place.

Combustion gases The gaseous by-products of the combustion of a fuel.

Combustion power plant A power plant that generates power by combusting a fuel.

Combustion turbine A turbine that generates power from the combustion of a fuel.

Comfort zone A frequently used room or area that is maintained at a more comfortable level than the rest of the house; also known as a "warm room."

Commendable practice (self-assessment) A significant strength noted during the course of a self-assessment.

Comment or concern (self-assessment) A comment is a subjective opinion of the assessment team that may be used to improve any of the specific environmental monitoring program activities, noted in *Self-Assessments for Environmental Programs,* such as sample collection, preparation, logging, storage, and shipping; instrument and equipment calibration; data receipt and data entry; training requirements and records; and compliance with discharge permit requirements. Corrective action in response to a comment or concern is at the discretion of the cognizant staff.

Commercial building A building with more than 50% of its floor space used for commercial activities, which include stores, offices, schools, churches, libraries, museums, healthcare facilities, warehouses, and government buildings except those on military bases.

Commercial establishment (1) Any hotel, motel, apartment house, rooming house, or tourist court which contains three or more service units and any other building, business, or establishment of any nature or kind whatsoever other than a residential unit; and (2) any hotel, motel, apartment house, rooming house, or tourist court which contains three or more service units, and any other building, business, or establishment of any nature or kind whatsoever other than a residential unit (Miami, FL).

Commercial facility A facility that treats, disposes, or recycles/recovers the wastes of other facilities not under the same ownership as this facility. Commercial operations are usually made available for a fee or other remuneration. Commercial waste treatment, disposal, or recycling/recovery does not have to be the primary activity at a facility for an operation or unit to be considered "commercial."

Commercial garbage It includes all garbage produced by grocery stores; produce markets; restaurants; schools, public, private, or parochial; hospitals; or any commercial or other establishment that processes, sells, or services food or food products (Wichita, KS).

Commercial refuse It includes refuse from wholesale and retail stores, including but not limited to restaurants, florists, beauty shops, barbershops, doctor offices, variety stores, hardware stores, and other enterprises of this classification in the local government's zoning regulations (Baltimore County, MD).

Commercial sector Consists of businesses that are not engaged in transportation or manufacturing or other types of industrial activities. Standard Industrial Classification (SIC) codes for commercial establishments are 50 through 87, 89, and 91 through 97.

Commercial waste It includes garbage, rubbish, mixed refuse, and cool ashes originating in and around commercial establishments, industrial establishments, hotels, restaurants, cafeterias, grocery stores, and nonpublic institutions (Bettendorf, IA).

Commissioning The process by which a power plant, apparatus, or building is approved for operation based on observed or measured operation that meets design specifications.

Committed dose equivalent A measure of internal radiation. The predicted total dose equivalent to a tissue or organ over a fifty-year period after a known intake of a radionuclide into the body. It does not include contributions from sources of external penetrating radiation. Committed dose equivalent is measured in rems or sieverts.

Committed effective dose equivalent The sum of the committed dose equivalents to various tissues in the body, each multiplied by the appropriate weighting factor. Committed effective dose equivalent is measured in rems or sieverts.

Compact fluorescent A smaller version of standard fluorescent lamps which can directly replace standard incandescent lights. These lights consist of a gas-filled tube and a magnetic or electronic ballast.

Compactor collection vehicle It refers to any enclosed vehicles provided with special mechanical devices for conveying the refuse into the main compartment of the body and compressing the loaded materials (Washington, DC).

Complete mix digester A type of anaerobic digester that has a mechanical mixing system and where temperature and volume are controlled to maximize the anaerobic digestion process for biological waste treatment, methane production, and odor control.

Compliance findings Conditions that may not satisfy applicable environmental or safety and health regulations, US DOE Orders and memoranda, enforcement actions, agreements with regulatory agencies, or permit conditions.

Composting (1) The process of degrading organic material (biomass) by microorganisms in aerobic conditions. (2) It is a process involving collecting organic waste, such as food scraps and yard trimmings, and storing it under conditions designed to help it break down naturally. This resulting compost can then be used as a natural fertilizer (US EPA). (3) It is a controlled microbial degradation of organic waste, yielding a nuisance-free product of potential value as a soil conditioner (Washington, DC). (4) It is a biological process for biological reduction, stabilization, and conversion of putrescible and other organic solid wastes to a non-putrescible manner (Boulder, CO).

Compound paraboloid collector A form of solar concentrating collector that does not track the sun.

Compressed air storage The storage of compressed air in a container for use to operate a prime mover for electricity generation.

Compressed natural gas (CNG) Natural gas (methane) that has been compressed to a higher pressure gaseous state by a compressor; used in CNG vehicles.

Compression chiller A cooling device that uses mechanical energy to produce chilled water.

Compressor A device used to compress air for mechanical or electrical power production and in air conditioners, heat pumps, and refrigerators to pressurize the refrigerant, enabling it to flow through the system.

Concentrating (solar) collector A solar collector that uses reflective surfaces to concentrate sunlight onto a small area, where it is absorbed and converted to heat or, in the case of solar photovoltaic (PV) devices, into electricity. Concentrators can increase the power flux of sunlight hundreds of times. The principal types of concentrating collectors include compound parabolic, parabolic trough, fixed reflector moving receiver, fixed receiver moving reflector, Fresnel lens, and central receiver. A PV concentrating module uses optical elements (Fresnel lens) to increase the amount of sunlight incident onto a PV cell. Concentrating PV modules/arrays must track the sun and use only the direct sunlight because the diffuse portion cannot be focused onto the PV cells. Concentrating collectors for home or small business solar water heating applications are usually parabolic troughs that concentrate the sun's energy on an absorber tube (called a receiver), which contains a heat-transfer fluid.

Concentrated animal feeding operations (CAFO) An agricultural operation that raises livestock within a restricted space, known as feedlot.

Condensate The liquid resulting when water vapor contacts a cool surface; also the liquid resulting when a vaporized working fluid (such as a refrigerant) is cooled or depressurized.

Condensation The process by which water in air changes from a vapor to a liquid due to a change in temperature or pressure; occurs when water vapor reaches its dew point (condensation point); also used to express the existence of liquid water on a surface.

Condenser The device in an air conditioner or heat pump in which the refrigerant condenses from a gas to a liquid when it is depressurized or cooled.

Condenser coil The device in an air conditioner or heat pump through which the refrigerant is circulated and releases heat to the surroundings when a fan blows outside air over the coils. This will return the hot vapor that entered the coil into a hot liquid upon exiting the coil.

Condensing furnace A type of heating appliance that extracts so much of the available heat content from a combusted fuel that the moisture in the combustion gases condenses before it leaves the furnace. Also this furnace circulates a liquid to cool the furnace's heat exchanger. The heated liquid may either circulate through a liquid-to-air heat exchanger to warm room air, or it may circulate through a coil inside a separate indirect-fired water heater.

Condensing unit The component of a central air conditioner that is designed to remove heat absorbed by the refrigerant and transfer it outside the conditioned space.

Conditioned space The interior space of a building that is heated or cooled.

Conduction The transfer of heat through a material by the transfer of kinetic energy from particle to particle; the flow of heat between two materials of different temperatures that are in direct physical contact.

Conduction band An energy band in a semiconductor in which electrons can move freely in a solid, producing a net transport of charge.

Conductivity (thermal) This is a positive constant, k, that is a property of a substance and is used in the calculation of heat-transfer rates for materials. It is the amount of heat that flows through a specified area and thickness of a material over a specified period of time when there is a temperature difference of one degree between the surfaces of the material.

Conductor The material through which electricity is transmitted, such as an electrical wire, or transmission or distribution line.

Conduit A tubular material used to encase and protect one or more electrical conductors.

Confidence coefficient or factor The chance or probability, usually expressed as a percentage, that a confidence interval includes some defined parameter of a population. The confidence coefficients usually associated with confidence intervals are 90 %, 95 %, and 99 %.

Connected load The sum of the ratings of the electricity consuming apparatus connected to a generating system.

Connection charge An amount paid by a customer for being connected to an electricity supplier's transmission and distribution system.

Conservation To reduce or avoid the consumption of a resource or commodity.

Conservation cost adjustment A means of billing electric power consumers to pay for the costs of demand side management/energy conservation measures and programs (see also benefits charge).

Consistency The condition of showing steady conformity to practices. In the environmental monitoring program, approved procedures are in place in order to ensure that data collection activities are carried out in a consistent manner so that variability is minimized.

Constant dollars The value or purchasing power of a dollar in a specified year carried forward or backward.

Constant-speed wind turbines Wind turbines that operate at a constant rotor revolutions per minute (RPM) and are optimized for energy capture at a given rotor diameter at a particular speed in the wind power curve.

Constructed wetland It is a treatment method that uses plants (most commonly water hyacinth and duckweed) in a wetland environment to degrade organic materials.

Construction and demolition wastes They include the waste building materials and rubble, resulting from construction, remodeling, repair, and demolition operation on houses, commercial buildings, pavements, and other structures (Washington, DC).

Consumption charge The part of a power provider's charge based on actual energy consumed by the customer; the product of the kilowatt-hour rate and the total kilowatt-hours consumed.

Contact resistance The resistance between metallic contacts and the semiconductor.

Contaminated groundwater Water below the land surface in the zone of saturation which has been contaminated by landfill leachate or other sources of pollution, such as the wastewater from hydraulic fracturing operation. Contaminated groundwater occurs at landfills without liners or at facilities that have released contaminants from a process system. Groundwater may also become contaminated if the water table rises to a point where it infiltrates the landfill or the leachate collection system, or the hydraulic fracturing system, or others.

Contaminated storm water Storm water which comes in direct contact with the ground, or waste or waste storage/handling and treatment areas. Storm water which does not come into contact with the wastes is not subject to the government required limitations and standards.

Continuous fermentation A steady-state fermentation process.

Contrast The difference between the brightness of an object compared to that of its immediate background.

Convection The transfer of heat by means of air currents.

Conventional fuel The fossil fuels: coal, oil, and natural gas.

Conventional heat pump This type of heat pump is known as an air-to-air system.

Conventional pollutants Constituents of wastewater as determined by Sec. 304(a) (4) of the US Clean Water Act (CWA), including pollutants classified as biochemical oxygen demand, total suspended solids, oil and grease, fecal coliform, and pH.

Conventional power Power generation from sources such as petroleum, natural gas, or coal. In some cases, large-scale hydropower and nuclear power generation are considered conventional sources.

Conversion efficiency The amount of energy produced as a percentage of the amount of energy consumed.

Converter A device for transforming the quality and quantity of electrical energy; also an inverter.

Cooling capacity The quantity of heat that a cooling appliance is capable of removing from a room in one hour.

Cooling degree day A value used to estimate interior air-cooling requirements (load) calculated as the number of degrees per day (over a specified period) that the daily average temperature is above 65°F (or some other, specified base temperature). The daily average temperature is the mean of the maximum and minimum temperatures recorded for a specific location for a 24-h period.

Cooling load That amount of cooling energy to be supplied (or heat and humidity removed) based on the sensible and latent loads.

Cooling pond A body of water used to cool the water that is circulated in an electric power plant.

Cooling tower A structure used to cool power plant water; water is pumped to the top of the tubular tower and sprayed out into the center, is cooled by evaporation as it falls, and then is either recycled within the plant or is discharged.

Coproducts The potentially useful by-products of ethanol fermentation process.

Cord (of wood) A stack of wood 4 ft by 4 ft by 8 ft (1.22 m × 1.22 m × 2.44 m).

Cosmic radiation High-energy subatomic particles from outer space that bombard the Earth's atmosphere. Cosmic radiation is part of natural background radiation.

Coulomb A unit for the quantity of electricity transported in 1 s by a current of 1 A.

Counterflow heat exchanger A heat exchanger in which two fluids flow in opposite directions for transfer heat energy from one to the other.

Counting error The variability caused by the inherent random nature of radioactive disintegration and by the detection process.

Covenants Restrictions on the use of a property.

Crawlspace The unoccupied, and usually unfinished and unconditioned, space between the floor, foundation walls, and the slab or ground of a building.

Creosote A liquid by-product of wood combustion (or distillation) that condenses on the internal surfaces of vents and chimneys, which if not removed regularly can corrode the surfaces and fuel a chimney fire.

Critical compression pressure The highest possible pressure in a fuel-air mixture before spontaneous ignition occurs.

Crop group Individual farm fields that are managed in the same manner, with the similar yield goals, are called a *crop group*

Crop management The management involves crop group identification, crop nitrogen deficit determination, crop nitrogen fertilizer rate calculation, crop yield optimization.

Crop nitrogen deficit (CND) Crop nitrogen deficit (CND) equals to anticipated crop nitrogen fertilizer rate (CNFR) minus all past PAN sources (PAN-past) and current planned non-biosolids PAN sources (PAN-plan), in the unit of lb N/acre. Previous biosolids carry-over nitrogen is included in this calculation.

Crop nitrogen fertilizer rate (CNFR) CNFR is a rate (lb N/acre) = (Yield) (UNFR), where UNFR is the unit nitrogen fertilizer rate (lb N/unit crop yield) and Yield is the crop harvested or crop yield (bu/acre or ton/acre)

Crop year The basic time management unit is often called the *crop year* or *planting season*. The *crop year* is defined as the year in which a crop receiving the biosolids/manure treatment is harvested. For example, fall applications of biosolids/manure in 2000 intended to provide nutrients for a crop to be harvested in 2001 are earmarked for *crop year* 2001. Likewise, biosolids/manure applied immediately prior to planting winter wheat in October 2000 should be identified as fertilizer intended for *crop year* 2001 because the wheat will be harvested in the summer of 2001.

Crop yield It is the crop harvested in the unit of bu/acre or ton/acre.

Crystalline silicon photovoltaic cell A type of photovoltaic cell made from a single crystal or a polycrystalline slice of silicon. Crystalline silicon cells can be joined together to form a module (or panel).

Cube law In reference to wind energy, for any given instant, the power available in the wind is proportional to the cube of the wind velocity; when wind speed doubles, the power availability increases eight times.

Cubic foot or cubic meter (of natural gas) A unit of volume equal to 1 cubic foot or 1 cubic meter at a pressure base of 14.73 lb standard per square inch absolute (101,560 N standard per square meter absolute) and a temperature base of 60°F (15.55°C).

Cumulative pollutant loading rate (CPLR) CPLR equals to the total amount of pollutant that can be applied to a site in its lifetime by all bulk biosolids applications meeting CCL. It is the maximum amount of an inorganic pollutant that can be applied to an area of land. This term applies to bulk sewage sludge that is land applied.

Curie (Ci) (1) A unit of radioactivity equal to 37 billion (3.7×10^{10}) nuclear transformations per second; (2) amount of radioactive material which disintegrates at the rate of 37 billion atoms per second.

Current (electrical) The flow of electrical energy (electricity) in a conductor, measured in amperes.

Current cost estimate It refers to the most recent of the cost estimates prepared in accordance with the state or federal environmental laws or regulations.

Current dollars (1) The value or purchasing power of a dollar that has not been reduced to a common basis of constant purchasing power, but instead reflects anticipated future inflation; when used in computations, the assumed inflation rate must be stated. (2) It refers to the dollar value in the year a cost estimate is prepared, as opposed to a historical or future dollar value.

Customer charge An amount to be paid for energy periodically by a customer without regard to demand or energy consumption.

Customer class Categories of energy consumers, as defined by consumption or demand levels, patterns, and conditions, and generally included residential, commercial, industrial, agricultural.

Cut-in-speed The lowest wind speed at which a wind turbine begins producing usable power.

Cut-out-speed The highest wind speed at which a wind turbine stops producing power.

Cycle In alternating current, the current goes from zero potential or voltage to a maximum in one direction, back to zero, and then to a maximum potential or voltage in the other direction. The number of complete cycles per second determines the current frequency; in the USA, the standard for alternating current is 60 cycles.

Cycling losses The loss of heat as the water circulates through a water heater tank and inlet and outlet pipes.

Cyclone burner A furnace/combustion chamber in which finely ground fuel is blown in spirals in the combustion chamber to maximize combustion efficiency.

Czochralski process A method of growing large size, high-quality semiconductor crystal by slowly lifting a seed crystal from a molten bath of the material under careful cooling conditions.

Dam A structure for impeding and controlling the flow of water in a water course and which increases the water elevation to create the hydraulic head. The reservoir creates, in effect, stored energy.

Damper A movable plate used to control air flow; in a wood stove or fireplace, used to control the amount and direction of air going to the fire.

Darrieus (wind) machine A type of vertical-axis wind machine that has long, thin blades in the shape of loops connected to the top and bottom of the axle; often called an "eggbeater windmill."

Daylighting The use of direct, diffuse, or reflected sunlight to provide supplemental lighting for building interiors.

Dead animals They include those that die naturally or from disease or are accidentally killed (Washington, DC).

Decay (radioactive) Disintegration of the nucleus of an unstable nuclide by spontaneous emission of charged particles and/or photons or by spontaneous fission.

Decentralized (energy) system Energy systems supply individual, or small groups, of energy loads.

Declination The angular position of the sun at solar noon with respect to the plane of the equator.

Declining block rate An electricity supplier rate structure in which the per unit price of electricity decreases as the amount of energy increases. Normally only available to very large consumers.

Decommissioning The process of removing a power plant, apparatus, equipment, building, or facility from operation.

Decomposition The process of breaking down organic material; reduction of the net energy level and change in physical and chemical composition of organic material.

De-energize(d) To disconnect a transmission and/or distribution line; a power line that is not carrying a current; to open a circuit.

Deep discharge Discharging a battery to 20% or less of its full charge capacity.

Deep well injection A process for disposal of wastewater or hydraulic fracking liquid into a deep well such that a porous, permeable formation of a larger area and thickness is available at sufficient depth to ensure continued, permanent storage.

Deficiency (self-assessment) A condition that does not meet or cannot be documented to meet applicable requirements.

Degree day A unit for measuring the extent that the outdoor daily average temperature (the mean of the maximum and minimum daily dry-bulb temperatures) falls below (in the case of heating, see Heating Degree Day) or falls above (in the case of cooling, see Cooling Degree Day) an assumed base temperature, normally taken as 65°F, unless otherwise stated. One degree day is counted for each degree below (for heating) or above (in the case of cooling) the base, for each calendar day on which the temperature goes below or above the base.

Degree hour The product of 1 h, and usually the number of degrees Fahrenheit the hourly mean temperature is above a base point (usually 65°F); used in roughly estimating or measuring the cooling load in cases where processes heat, heat from building occupants, and humidity are relatively unimportant compared to the dry-bulb temperature.

Dehumidifier A device that cools air by removing moisture from it.

Demand The rate at which electricity is delivered to or by a system, part of a system, or piece of equipment expressed in kilowatts, kilovolt-amperes, or other suitable unit, at a given instant or averaged over a specified period of time.

Demand (tankless) water heater A type of water heater that has no storage tank, thus eliminating storage tank standby losses. Cold water travels through a pipe into the unit, and either a gas burner or an electric element heats the water only when needed.

Demand charge A charge for the maximum rate at which energy is used during peak hours of a billing period. That part of a power provider service charged for on the basis of the possible demand as distinguished from the energy actually consumed.

Demand power See peak power

Demand(ed) factor The ratio of the maximum demand on an electricity-generating and distribution system to the total connected load on the system; usually expressed as a percentage.

Demand-side management (DSM) The process of managing the consumption of energy, generally to optimize available and planned generation resources.

Dendrite A slender threadlike spike of pure crystalline material, such as silicon.

Dendritic web technique A method for making sheets of polycrystalline silicon in which silicon dendrites are slowly withdrawn from a melt of silicon whereupon a web of silicon forms between the dendrites and solidifies as it rises from the melt and cools.

Denitrification It is process that converts nitrate into atmospheric nitrogen using microorganisms known as denitrifiers.

Dependable capacity The load-carrying ability of an electric power plant during a specific time interval and period when related to the characteristics of the load to be/being supplied; determined by capability, operating power factor, and the portion of the load the station is to supply.

Derating The production of energy by a system or appliance at a level less than its design or nominal capacity.

Deregulation The process of changing regulatory policies and laws to increase competition among suppliers of commodities and services. The process of deregulating the electric power industry was initiated by the Energy Policy Act of 1992 (see also restructuring).

Derived concentration guide (DCG) The concentration of a radionuclide in air and water that, under conditions of continuous exposure for one year by one exposure mode (i.e., ingestion of water, submersion in air, or inhalation), would result in an effective dose equivalent of 100 mrem (1 mSv).

Desiccant A material used to desiccate (dry) or dehumidify air.

Desiccant cooling To condition/cool air by desiccation.

Desiccation The process of removing moisture; involves evaporation.

Design cooling load The amount of conditioned air to be supplied by a cooling system; usually the maximum amount to be delivered based on a specified number of cooling degree days or design temperature.

Design heating load The amount of heated air, or heating capacity, to be supplied by a heating system; usually the maximum amount to be delivered based on a specified number of heating degree days or design outside temperature.

Design life Period of time a system or appliance (or component of) is expected to function at its nominal or design capacity without major repair.

Design temperature The temperature that a system is designed to maintain (inside) or operate against (outside) under the most extreme conditions.

Design tip speed ratio For a wind turbine, the ratio of the speed of the tip of a turbine blade for which the power coefficient is at maximum.

Design voltage The nominal voltage for which a conductor or electrical appliance is designed; the reference voltage for identification and not necessarily the precise voltage at which it operates.

Designated use Simple narrative description of water quality expectations or water quality goals. A designated use is a legally recognized description of a desired use of the water body, such as (1) support of communities of aquatic life, (2) body contact recreation, (3) fish consumption, and (4) public drinking water supply. These are uses that the state or authorized tribe wants the water body to be healthy enough to fully support. The US Clean Water Act requires that waterbodies attain or maintain the water quality needed to support designated uses.

Desuperheater An energy-saving device in a heat pump that, during the cooling cycle, recycles some of the waste heat from the house to heat domestic water.

Detection limit or level The smallest amount of a substance that can be distinguished in a sample by a given measurement procedure at a given confidence level.

Dew point The temperature to which air must be cooled, at constant pressure and water vapor content, in order for saturation or condensation to occur; the temperature at which the saturation pressure is the same as the existing vapor pressure; also called saturation point.

Difference of potential The difference in electrical pressure (voltage) between any two points in an electrical system or between any point in an electrical system and the Earth.

Differential thermostat A type of automatic thermostat (used on solar heating systems) that responds to temperature differences (between collectors and the storage components) so as to regulate the functioning of appliances (to switch transfer fluid pumps on and off).

Diffuse solar radiation Sunlight scattered by atmospheric particles and gases so that it arrives at the Earth's surface from all directions and cannot be focused.

Diffusion The movement of individual molecules through a material; permeation of water vapor through a material.

Diffusion length The mean distance a free electron or hole moves before recombining with another hole or electron.

Digester (anaerobic) A device in which organic material is biochemically decomposed (digested) by anaerobic bacteria to treat the material and/or to produce biogas.

Dimmer A light control device that allows light levels to be manually adjusted. A dimmer can save energy by reducing the amount of power delivered to the light while consuming very little themselves.

Diode An electronic device that allows current to flow in one direction only.

Dip tube A tube inside a domestic water heater that distributes the cold water from the cold water supply line into the lower area of the water heater where heating occurs.

Direct access The ability of an electric power consumer to purchase electricity from a supplier of their choice without being physically inhibited by the owner of the electric distribution and transmission system to which the consumer is connected to (see also open access).

Direct beam radiation Solar radiation that arrives in a straight line from the sun.

Direct current A type of electricity transmission and distribution by which electricity flows in one direction through the conductor; usually relatively low voltage and high current; typically abbreviated as dc.

Direct discharger A facility that discharges or may discharge treated or untreated wastewaters into the receiving waters.

Direct solar water heater These systems use water as the fluid that is circulated through the collector to the storage tank. Also known as "open-loop" systems.

Direct vent heater A type of combustion heating system in which combustion air is drawn directly from outside and the products of combustion are vented directly outside. These features are beneficial in tight, energy-efficient homes because they will not depressurize a home and cause air infiltration and backdrafting of other combustion appliances.

Direct water heater A type of water heater in which heated water is stored within the tank. Hot water is released from the top of the tank when a hot water faucet is turned. This water is replaced with cold water that flows into the tank and down to just above the bottom plate under which are the burners.

Direct gain The process by which sunlight directly enters a building through the windows and is absorbed and stored in massive floors or walls.

Discount rate The interest rate at which the Federal Reserve System stands ready to lend reserves to commercial banks. The rate is proposed by the 12 Federal Reserve banks and determined with the approval of the Board of Governors.

Discounting A method of financial and economic analysis used to determine present and future values of investments or expenses.

Dispatchability The ability to dispatch power.

Dispatching To schedule and control the generation and delivery of electric power.

Dispersion (groundwater) The process whereby solutes are spread or mixed as they are transported by groundwater as it moves through sediments.

Displacement power A source of power (electricity) that can displace power from another source so that source's power can be transmitted to more distant loads.

Disposal (1) It refers to the isolation of radioactive wastes from the biosphere inhabited by humans and containing their food chains by emplacement in land disposal facilities. (2) It refers to the processing of solid wastes by sanitary landfill, by incineration, by composting, by grinding, or any other equivalent sanitary method (Boulder, CO).

Disposal area or site (1) Any site, location, tract of land, area, building, structure, or premises used or intended to be used for partial or total solid waste disposal (Washington, DC); (2) any land used for the disposal of solid wastes, including, but not limited to, dumps, landfills, sanitary landfills, and composting plants, but does not include a landfill site which is not used by the public either directly or through a service and which is used by the owner or tenant thereof to dispose of sawdust, bark, soil, rock, building demolition material or non-putrescible industrial waste products resulting from the process of manufacturing (Marion County, OR); and (3) any site, location, or tract of land permitted by the local government to be used for refuse disposal (Sonoma County, CA).

Disposal site or site It refers to that portion of a land disposal facility which is used for disposal of waste. It consists of disposal units and a buffer zone.

Disposal unit It refers to a discrete structure of the disposal site into which waste is placed for disposal. Disposal units include, but are not limited to, vaults, concrete modules, and the cavities of underground mined repositories.

Dissolution It refers to a space or cavity in or between rocks, formed by the solution of part of the rock material.

Distressed watershed It is a watershed which has aquatic life and health that is impaired by nutrients (nitrogen and phosphorus) from agricultural land uses, such as land application. Threats to public health, drinking water supplies, recreation, and public safety are also taken into consideration if a watershed is designated as a distressed watershed.

Distributed generation A term used by the power industry to describe localized or on-site power generation.

Distribution The process of distributing electricity; usually defines that portion of a power provider's power lines between a power provider's power pole and transformer and a customer's point of connection/meter.

Distribution line One or more circuits of a distribution system on the same line or poles or supporting structures usually operating at a lower voltage relative to the transmission line.

Distribution system That portion of an electricity supply system used to deliver electricity from points on the transmission system to consumers.

Diversity factor The ratio of the sum of the noncoincidental maximum demands of two or more loads to their coincidental maximum demands for the same period.

DOE-2.1 A computer software program that simulates energy consumption of commercial buildings; used for design and auditing purposes.

Dome (geodesic) An architectural design invented by Buckminster Fuller with a regular polygonal structure based on radial symmetry.

Domestic hot water Water heated for residential washing, bathing, etc.

Domestic refuse It includes all those types which normally originate in the residential household or apartment house (Washington, DC).

Domestic septage Either a liquid or solid material removed from a septic tank, cesspool, portable toilet, Type III marine sanitation device, or similar treatment works that receive only domestic sewage. This does not include septage, resulting from treatment of wastewater with a commercial or industrial component.

Donor In a solar photovoltaic device, an n-type dopant, such as phosphorus, that puts an additional electron into an energy level very near the conduction band; this electron is easily exited into the conduction band where it increases the electrical conductivity over than of an undoped semiconductor.

Dopant A chemical element (impurity) added in small amounts to an otherwise pure semiconductor material to modify the electrical properties of the material. An n-dopant introduces more electrons. A p-dopant creates electron vacancies (holes).

Doping The addition of dopants to a semiconductor.

Dose It refers to that quantity of ionizing radiation in rems absorbed, per unit mass, by any body tissue. Reference to a dose during a specific period of time means the total quantity of radiation so absorbed during such period.

Dosimeter A portable device for measuring the total accumulated exposure to ionizing radiation.

Double wall heat exchanger A heat exchanger in a solar water heating system that has two distinct walls between the heat-transfer fluid and the domestic water, to ensure that there is no mixing of the two.

Double-pane or glazed window A type of window having two layers (panes or glazing) of glass separated by an air space. Each layer of glass and surrounding air space reradiates and traps some of the heat that passes through, thereby increasing the windows resistance to heat loss (R-value).

Downgradient The direction of water flow from a reference point to a selected point of interest.

Downwind wind turbine A horizontal axis wind turbine in which the rotor is downwind of the tower.

Draft A column of burning combustion gases that are so hot and strong that the heat is lost up the chimney before it can be transferred to the house. A draft brings air to the fire to help keep it burning.

Draft diverter A door-like device located at the mouth of a fireplace chimney flue for controlling the direction and flow of the draft in the fireplace as well as the amount of oxygen that the fire receives.

Draft hood A device built into or installed above a combustion appliance to assure the escape of combustion by-products, to prevent backdrafting of the appliance, or to neutralize the effects of the stack action of the chimney or vent on the operation of the appliance.

Drag Resistance caused by friction in the direction opposite to that of movement (i.e., motion) of components such as wind turbine blades.

Drainback (solar) systems A closed-loop solar heating system in which the heat-transfer fluid in the collector loop drains into a tank or reservoir whenever the booster pump stops to protect the collector loop from freezing.

Draindown (solar) systems An open-loop solar heating system in which the heat-transfer fluid from the collector loop and the piping drains whenever freezing conditions occur.

Drained Free Liquids Aqueous wastes drained from waste containers (e.g., drums, etc.) prior to landfilling. Landfills which accept containerized waste may generate this type of wastewater.

Dry bulb temperature The temperature of the air as measured by a standard thermometer.

Dry steam geothermal plants Conventional turbine generators are used with the dry-steam resources. The steam is used directly, eliminating the need for boilers and boiler fuel that characterizes other steam-power-generating technologies. This technology is limited because dry-steam hydrothermal resources are extremely rare. The Geysers, in California, is the nation's only dry-steam field.

DSSTox It refers to a public forum for publishing downloadable, structure-searchable, standardized chemical structure files associated with toxicity data.

Dual duct system An air-conditioning system that has two ducts, one is heated and the other is cooled, so that air of the correct temperature is provided by mixing varying amounts of air from each duct.

Dual fuel (or flex fuel) vehicle A vehicle with an engine capable of operating on two different types of fuels.

Duct fan An axial flow fan mounted in a section of duct to move conditioned air.

Duct(s) The round or rectangular tube(s), generally constructed of sheet metal, fiberglass board, or a flexible plastic-and-wire composite, located within a wall, floor, and ceiling that distributes heated or cooled air in buildings.

Duty cycle The duration and periodicity of the operation of a device.

Dynamic head The pressure equivalent of the velocity of a fluid.

Dynamo A machine for converting mechanical energy into electrical energy by magnetoelectric induction; may be used as a motor.

Dynamometer An apparatus for measuring force or power, especially the power developed by a motor.

Dyne The absolute centimeter-gram-second unit of force; force that will impart to a free mass of one gram an acceleration of one centimeter per second.

Earth berm A mound of dirt next to exterior walls to provide wind protection and insulation.

Earth cooling tube A long, underground metal or plastic pipe through which air is drawn. As air travels through the pipe, it gives up some of its heat to the soil and enters the house as cooler air.

Earth-sheltered houses Houses that have earth berms around exterior walls.

Earth-coupled ground source (geothermal) heat pump A type of heat pump that uses sealed horizontal or vertical pipes, buried in the ground, as heat exchangers through which a fluid is circulated to transfer heat.

Earth-ship A registered trademark name for houses built with tires, aluminum cans, and earth.

Easement An incorporated right, liberty, privilege, or use of another entity's property, distinct from ownership, without profit or compensation; a right-of-way.

Eccentric A device for converting continuous circular motion into reciprocating rectilinear motion.

Economizer A heat exchanger for recovering heat from flue gases for heating water or air.

Edge-defined film-fed growth (EFG) A method for making sheets of polycrystalline silicon (for solar photovoltaic devices) in which molten silicon is drawn upward by capillary action through a mold.

Effective capacity The maximum load that a device is capable of carrying.

Efficacy The amount of energy service or useful energy delivered per unit of energy input. Often used in reference to lighting systems, where the visible light output of a luminary is relative to power input; expressed in lumens per watt; the higher the efficacy value, the higher the energy efficiency.

Efficiency Under the first law of thermodynamics, efficiency is the ratio of work or energy output to work or energy input and cannot exceed 100%. Efficiency under the second law of thermodynamics is determined by the ratio of the theoretical minimum energy that is required to accomplish a task relative to the energy actually consumed to accomplish the task. Generally, the measured efficiency of a device, as defined by the first law, will be higher than that defined by the second law.

Efficiency (appliance) ratings A measure of the efficiency of an appliance's energy efficiency.

Effluent Any treated or untreated air emission or liquid discharge, including storm water runoff, at an environmental agency specified site or facility.

Effluent limitation Any restriction, including schedules of compliance, established by a state or the federal administrator on quantities, rates, and concentrations of chemical, physical, biological, and other constituents which are discharged from point sources into navigable waters, the waters of the contiguous zone, or the ocean (CWA Sections 30 1(b) and 3 04(b)).

Effluent monitoring Sampling or measuring specific liquid or gaseous effluent streams for the presence of pollutants.

Eggbeater windmill See Darrieus (wind) machine.

Elasticity of demand The ratio of the percentage change in the quantity of a good or service demanded to the percentage change in the price.

Electric circuit The path followed by electrons from a generation source, through an electrical system, and returning to the source.

Electric energy The amount of work accomplished by electrical power, usually measured in kilowatt-hours (kWh). One kWh is 1000 W and is equal to 3413 Btu.

Electric furnace An air heater in which air is blown over electric resistance heating coils.

Electric power plant A facility or piece of equipment that produces electricity.

Electric power sector Those privately or publicly owned establishments that generate, transmit, distribute, or sell electricity.

Electric power transmission The transmission of electricity through power lines.

Electric rate The unit price and quantity to which it applies as specified in a rate schedule or contract.

Electric rate schedule A statement of the electric rate(s), terms, and conditions for electricity sale or supply.

Electric resistance heating A type of heating system where heat, resulting when electric current flows through an "element" or conductor, such as Nichrome, which has a high resistance, is radiated to a room.

Electric system The physically connected generation, transmission, and distribution facilities and components operated as a unit.

Electric system loss The total amount of electric energy loss in an electric system between the generation source and points of delivery.

Electric utility A corporation, person, agency, authority, or other legal entities that own and/or operate facilities for the generation, transmission, distribution, or sale of electricity primarily for use by the public. Also known as a power provider.

Electric vehicles A battery-powered electrically driven vehicle.

Electrical charge A condition that results from an imbalance between the number of protons and the number of electrons in a substance.

Electrical energy The energy of moving electrons.

Electrical system All the conductors and electricity using devices that are connected to a source of electromotive force (or generator).

Electrical system energy losses A measure of the amount of energy lost during the generation, transmission, and distribution of electricity.

Electricity generation The process of producing electricity by transforming other forms or sources of energy into electrical energy; measured in kilowatt-hours.

Electricity grid A common term referring to an electricity transmission and distribution system.

Electricity industry restructuring The process of changing the structure of the electric power industry from one of guaranteed monopoly over service territories, as established by the Public Utility Holding Company Act of 1935, to one of open competition between power suppliers for customers in any area.

Electrochemical cell A device containing two conducting electrodes, one positive and the other negative, made of dissimilar materials (usually metals) that are immersed in a chemical solution (electrolyte) that transmits positive ions from the negative to the positive electrode and thus forms an electrical charge. One or more cells constitute a battery.

Electrode A conductor that is brought in conducting contact with a ground.

Electrodeposition Electrolytic process in which a metal is deposited at the cathode from a solution of its ions.

Electrolysis A chemical change in a substance that results from the passage of an electric current through an electrolyte. The production of commercial hydrogen by separating the elements of water, hydrogen, and oxygen, by charging the water with an electrical current.

Electrolyte A nonmetallic (liquid or solid) conductor that carries current by the movement of ions (instead of electrons) with the liberation of matter at the electrodes of an electrochemical cell.

Electromagnetic energy Energy generated from an electromagnetic field produced by an electric current flowing through a superconducting wire kept at a specific low temperature.

Electromagnetic field (EMF) The electrical and magnetic fields created by the presence or flow of electricity in an electrical conductor or electricity consuming appliance or motor.

Electromotive force The amount of energy derived from an electrical source per unit quantity of electricity passing through the source.

Electron An elementary particle of an atom with a negative electrical charge and a mass of 1/1,837 of a proton; electrons surround the positively charged nucleus of an atom and determine the chemical properties of an atom.

Electron volt The amount of kinetic energy gained by an electron when accelerated through an electric potential difference of 1 V; equivalent to 1.603×10^{-12}; a unit of energy or work; abbreviated as eV.

Electronic ballast A device that uses electronic components to regulate the voltage of fluorescent lamps.

Electrostatic precipitator A device used to remove particulate matter from the waste gases of a combustion power plant.

Ellipsoidal reflector lamp A lamp where the light beam is focused 2 in. ahead of the lamp reducing the amount of light trapped in the fixture.

Emission A substance or pollutant emitted as a result of a process.

Emission factor A measure of the average amount of a specified pollutant or material emitted for a specific type of fuel or process.

Emissivity The ratio of the radiant energy (heat) leaving (being emitted by) a surface to that of a black body at the same temperature and with the same area; expressed as a number between 0 and 1.

Enclosure The housing around a motor that supports the active parts and protects them. They come in different varieties (open, protected) depending on the degree of protection required.

End use The purpose for which useful energy or work is consumed.

Endothermic A heat-absorbing reaction or a reaction that requires heat.

Energize(d) To send electricity through an electricity transmission and distribution network; a conductor or power line that is carrying current.

Energy The capability of doing work; different forms of energy can be converted to other forms, but the total amount of energy remains the same.

Energy audit A survey that shows how much energy you use in your house or apartment. It will help you find ways to use less energy.

Energy charge That part of an electricity bill that is based on the amount of electrical energy consumed or supplied.

Energy contribution potential Recombination occurring in the emitter region of a photovoltaic cell.

Energy crops Crops grown specifically for their fuel value. These include food crops such as corn and sugarcane, and nonfood crops such as poplar trees and switchgrass. Currently, two energy crops are under development: short-rotation woody crops, which are fast-growing hardwood trees harvested in 5 to 8 years, and herbaceous energy crops, such as perennial grasses, which are harvested annually after taking 2 to 3 years to reach full productivity.

Energy density The ratio of available energy per pound; usually used to compare storage batteries.

Energy efficiency ratio (EER) The measure of the instantaneous energy efficiency of room air conditioners; the cooling capacity in Btu/hr divided by the watts of power consumed at a specific outdoor temperature (usually 95°F).

Energy-efficient mortgages A type of home mortgage that takes into account the energy savings of a home that has cost-effective energy-saving improvements that will reduce energy costs, thereby allowing the homeowner to more income to the mortgage payment. A borrower can qualify for a larger loan amount than otherwise would be possible.

Energy end-use sectors Major energy-consuming sectors of the economy. The commercial sector includes commercial buildings and private companies. The industrial sector includes manufacturers and processors. The residential sector includes private homes. The transportation sector includes automobiles, trucks, rail, ships, and aircraft.

Energy factor (EF) The measure of overall efficiency for a variety of appliances. For water heaters, the energy factor is based on three factors: (1) the recovery efficiency, or how efficiently the heat from the energy source is transferred to the water; (2) standby losses, or the percentage of heat lost per hour from the stored water compared to the content of the water, and (3) cycling losses. For dishwashers, the energy factor is defined as the number of cycles per kWh of input power. For clothes washers, the energy factor is defined as the cubic foot capacity per kWh of input power per cycle. For clothes dryers, the energy factor is defined as the number of pounds of clothes dried per kWh of power consumed.

Energy guide labels The labels placed on appliances to enable consumers to compare appliance energy efficiency and energy consumption under specified test conditions as required by the Federal Trade Commission.

Energy intensity The relative extent that energy is required for a process.

Energy Policy Act of 1992 (EPAct) A comprehensive legislative package in the USA that mandates and encourages energy efficiency standards, alternative fuel use, and the development of renewable energy technologies. Public Law 102–486, October 24, 1992. Also authorized the Federal Energy Regulatory Commission (FERC) to order the owners of electric power transmission lines to transmit or "wheel" power for power generators including electric power providers, federal power marketing authorities, and exempt wholesale generators.

Energy security act of 1980 The US Legislation authorizing a US biomass and alcohol fuel program and that authorized loan guarantees and price guarantees and purchase agreements for alcohol fuel production.

Energy service company (ESCO) A company that specializes in undertaking energy efficiency measures under a contractual arrangement, whereby the ESCO shares the value of energy savings with their customer.

Energy storage The process of storing, or converting, energy from one form to another for later use; storage devices and systems include batteries, conventional and pumped storage hydroelectric, flywheels, compressed gas, and thermal mass.

Engineered barrier It refers to a man-made structure or device that is intended to improve the land disposal facility's ability to meet the performance objectives in the state or federal environmental laws or regulations.

Enhanced work Planning A process that evaluates and improves the program by which work is identified, planned, approved, controlled, and executed. The key elements are line management ownership, a graded approach to work management based on risk and complexity, worker involvement beginning at the earliest phases of work management, organizationally diverse teams, and organized, institution-wide communication.

Enthalpy A thermodynamic property of a substance, defined as the sum of its internal energy plus the pressure of the substance times its volume, divided by the mechanical equivalent of heat. The total heat content of air; the sum of the enthalpies of dry air and water vapor, per unit weight of dry air; measured in Btu per pound (or calories per kilogram).

Entrained bed gasifier A gasifier in which the feedstock (fuel) is suspended by the movement of gas to move it through the gasifier.

Entropy A measure of the unavailable or unusable energy in a system; energy that cannot be converted to another form.

Environment All the natural and living things around us. The earth, air, weather, plants, and animals all make up our environment.

Environmental assessment An evaluation that provides sufficient evidence and analysis for determining whether to prepare an environmental impact statement or a finding of no significant impact.

Environmental impact statement A detailed statement that includes the environmental impact of the proposed action, any adverse environmental effects that cannot be avoided should the proposal be implemented, and alternatives to the proposed action.

Environmental management system The systematic application of business management practices to environmental issues, including defining the organizational structure, planning for activities, identifying responsibilities, and defining practices, procedures, processes, and resources.

Environmental monitoring The collection and analysis of samples or the direct measurement of environmental media. Environmental monitoring consists of two major activities: effluent monitoring and environmental surveillance.

Environmental surveillance The collection and analysis of samples or the direct measurement of air, water, soil, foodstuff, and biota in order to determine compliance with applicable standards and permit requirements.

Epitaxial growth In reference to solar photovoltaic devices, the growth of one crystal on the surface of another crystal. The growth of the deposited crystal is oriented by the lattice structure of the original crystal.

Equinox The two times of the year when the sun crosses the equator and night and day are of equal length; usually occurs on March 21st (spring equinox) and September 23 (fall equinox).

Erg A unit of work done by the force of one dyne acting through a distance of one centimeter.

Erg One-billionth (1E-09) of the energy released by a 100-W bulb in 1 s.

Ethanol or ethyl alcohol (C_2H_5OH) A colorless liquid that is the product of fermentation used in alcoholic beverages, industrial processes, and as a fuel additive. Also known as grain alcohol.

Ethyl tertiary butyl ether (ETBE) A chemical compound produced in a reaction between ethanol and isobutylene (a petroleum-derived by-product of the refining process). ETBE has characteristics superior to other ethers: low volatility, low water solubility, high octane value, and a large reduction in carbon monoxide and hydrocarbon emissions.

Eutectic A mixture of substances that has a melting point lower than that of any mixture of the same substances in other proportions.

Eutectic salts Salt mixtures with potential applications as solar thermal energy storage materials.

Eutrophication (1) Enrichment of an aquatic ecosystem with nutrients (nitrogen, phosphorus) that accelerate biological productivity (growth of algae and weeds) and an undesirable accumulation of algal biomass. (2) The condition of a waterbody (particularly a lake or river) where molecular oxygen levels have been depleted due to high nutrient levels and algal blooms. When eutrophication occurs, all organisms relying on molecular oxygen to survive will die.

Evacuated-tube collector A collector is the mechanism in which fluid (e.g., water or diluted antifreeze) is heated by the sun in a solar hot water system. Evacuated-tube collectors are made up of rows of parallel, transparent glass tubes. Each tube consists of a glass outer tube and an inner tube, or absorber. The absorber is covered with a selective coating that absorbs solar energy well but inhibits radiative heat loss. The air is withdrawn ("evacuated") from the space between the tubes to form a vacuum, which eliminates conductive and convective heat loss. Evacuated-tube collectors are used for active solar hot water systems.

Evaporation The conversion of a liquid to a vapor (gas), usually by means of heat.

Evaporative cooling The physical process by which a liquid or solid is transformed into the gaseous state. For this process, a mechanical device uses the outside air's heat to evaporate water that is held by pads inside the cooler. The heat is drawn out of the air through this process, and the cooled air is blown into the home by the cooler's fan.

Evaporator coil The inner coil in a heat pump that, during the cooling mode, absorbs heat from the inside air and boils the liquid refrigerant to a vapor, which cools the house.

Evapotranspiration The combined total precipitation returned to the air through direct evaporation and by transpiration of vegetation.

Exceptional quality sewage sludge Sewage sludge that meets the most stringent limits for the three sludge quality parameters. In gauging sewage sludge quality, US EPA determined that three main parameters of concern should be considered: (1) pollutant levels, (2) the relative presence or absence of pathogenic organisms, such as salmonella and E-Coli bacteria, enteric viruses, or viable helminth ova and (3) the degree of attractiveness of the sewage sludge to vectors, such as flies, rats, and mosquitoes, that could potentially come in contact with pathogenic organisms and spread disease. Given these three variables, there can be a number of possible sewage sludge qualities. The term exceptional quality (EQ), which does not appear in the Part 503 regulation, is used to describe sewage sludge that meets the highest quality for all three of these Sewage sludge quality parameters (i.e., ceiling concentrations and pollutant concentrations in 503.13 for metals, one of the Class A pathogen reduction alternatives and one of the sewage sludge processing vector attraction reduction options 1 through 8).

Excitation The power required to energize the magnetic field of a generator.

Exempt wholesale generator An unregulated subsidiary of a power provider that is allowed to generate and sell wholesale power as an independent energy producer and is exempt from the US Public Utility Holding Company Act of 1935.

Existing grade It refers to the various elevations of the surface of the land as it actually exists upon the site (Chesterfield Township, MI).

Existing source Any facility from which there is or may be a discharge of pollutants, the construction of which is commenced before the publication of the proposed regulations prescribing a standard of performance under current environmental laws or regulations.

Exothermic A reaction or process that produces heat; a combustion reaction.

Expanded polystyrene A type of insulation that is molded or expanded to produce coarse, closed cells containing air. The rigid cellular structure provides thermal and acoustical insulation, strength with low weight, and coverage with few heat-loss paths. Often used to insulate the interior of masonry basement walls.

Expansion tank A tank used in a closed-loop solar heating system that provides space for the expansion of the heat-transfer fluid in the pressurized collector loop.

Expansion valve The device that reduces the pressure of liquid refrigerant, thereby cooling it before it enters the evaporator coil in a heat pump.

Explosive material It refers to any chemical compound, mixture, or device, which produces a substantial instantaneous release of gas and heat spontaneously or by contact with sparks or flame.

ExpoCastDB It refers to a database that consolidates observational human exposure data and links with toxicity data, environmental fate data, and chemical manufacture information.

Exposure The subjection of a target (usually living tissue) to radiation.

External combustion engine An engine in which fuel is burned (or heat is applied) to the outside of a cylinder; a Stirling engine.

Externality The environmental, social, and economic impacts of producing a good or service that are not directly reflected in the market price of the good or service.

Extruded polystyrene A type of insulation material with fine, closed cells, containing a mixture of air and refrigerant gas. This insulation has a high R-value, good moisture resistance, and high structural strength compared to other rigid insulation materials.

Facility All contiguous property owned, operated, leased, or under the control of the same person or entity.

Fallout Radioactive materials mixed into the Earth's atmosphere. Fallout constantly precipitates onto the Earth.

Fan A device that moves and/or circulates air and provides ventilation for a room or a building.

Fan coil A heat exchanger coil in which a fluid such as water is circulated, and a fan blows air over the coil to distribute heat or cool air to the different rooms.

Fan velocity pressure The pressure corresponding to the outlet velocity of a fan; the kinetic energy per unit volume of flowing air.

Farad A unit of electrical capacitance; the capacitance of a capacitor between the plates of which there appears a difference of 1 V when it is charged by one coulomb of electricity.

Farm field The farm field is the basic management unit used for all farm nutrient management, as defined as "the fundamental unit used for cropping agricultural products."

Feather In a wind energy conversion system, to pitch the turbine blades so as to reduce their lift capacity as a method of shutting down the turbine during high wind speeds.

Feed crop Crops produced primarily for consumption by animals. These include, but are not limited to, corn and grass. For a crop to be considered a feed crop, it has to be produced for consumption by animals (e.g., grass grown to prevent erosion or to stabilize an area is not considered a feed crop).

Feeder A power line for supplying electricity within a specified area.

Feedstock A raw material that can be converted to one or more products.

Fenestration The arrangement, proportion, and design of windows in a building.

Fermentation The decomposition of organic material to alcohol, methane, etc., by organisms, such as yeast or bacteria, usually in the absence of oxygen.

Fiber crop Crops, such as flax and cotton, that were included in Part 503 because products from these crops (e.g., cotton seed oil) may be consumed by humans.

Fiberglass insulation A type of insulation, composed of small diameter pink, yellow, or white glass fibers, formed into blankets or used in loose-fill and blown-in applications.

Filament A coil of tungsten wire suspended in a vacuum or inert gas-filled bulb. When heated by electricity the tungsten "filament" glows.

Fill factor The ratio of a photovoltaic cell's actual power to its power if both current and voltage were at their maxima. A key characteristic in evaluating cell performance.

Filter (air) A device that removes contaminants, by mechanical filtration, from the fresh air stream before the air enters the living space. Filters can be installed as part of a heating/cooling system through which air flows for the purpose of removing particulates before or after the air enters the mechanical components.

Fin A thin sheet of material (metal) of a heat exchanger that conducts heat to a fluid.

Finding A US Department of Energy compliance term. A finding is a statement of fact concerning a condition in the environmental, safety, and health program that was investigated during an appraisal. Findings include best management practice findings, compliance findings, and noteworthy practices. A finding may be a simple statement of proficiency or a description of deficiency (i.e., a variance from procedures or criteria).

Finish Both a noun and a verb to describe the exterior surface of building elements (walls, floors, ceilings, etc.) and furniture, and the process of applying it.

Finish grade It refers to those earth elevations established and delineated on the plot plan that will result upon completion of the proposed operation for which the permit is issued (Chesterfield Township, MI).

Fire classification Classifications of fires developed by the National Fire Protection Association.

Fireplace A wood or gas burning appliance that is primarily used to provide ambiance to a room. Conventional, masonry fireplaces without energy-saving features, often take more heat from a space than they put into it.

Fireplace insert A wood or gas burning heating appliance that fits into the opening or protrudes on to the hearth of a conventional fireplace.

Fire-rating The ability of a building construction assembly (partition, wall, floor, etc.) to resist the passage of fire. The rating is expressed in hours.

Firewall A wall to prevent the spread of fire; usually made of noncombustible material.

Firing rate The amount of BTUs/hour or kWs produced by a heating system from the burning of a fuel.

First law of thermodynamics States that energy cannot be created or destroyed but only changed from one form to another. First law efficiency measures the fraction of energy supplied to a device or process that it delivers in its output. Also called the law of conservation of energy.

Fiscal year (FY) The US Government's 12-month financial year, from October to September, of the following calendar year; e.g., FY 1998 extends from October 1, 1997 to September 30, 1988.

Fission The act or process of splitting into parts. A nuclear reaction in which an atomic nucleus splits into fragments, i.e., fission products, usually fragments of comparable mass, with the evolution of approximately 100 million to several hundred million electron volts of energy.

Five-day biochemical oxygen demand (BOD$_5$) (1) It refers to a measure of the biochemical decomposition of organic matter in a water sample. It is determined by measuring the dissolved oxygen consumed by microorganisms to oxidize the organic contaminants in a water sample under standard laboratory conditions of five days and 70 °C. BOD$_5$ is not related to the oxygen requirements in chemical combustion. (2) It is a wastewater quality index that biochemically determines the amount of oxygen required for microorganisms to degrade a given substance within a five-day period.

Flame spread classification A measure of the surface burning characteristics of a material.

Flame spread rating A measure of the relative flame spread, and smoke development, from a material being tested. The flame spread rating is a single number comparing the flame spread of a material with red oak, arbitrarily given the number 100 and asbestos cement board with a flame spread of 0. Building codes require a maximum flame spread of 25 for insulation installed in exposed locations.

Flashing Metal, usually galvanized sheet metal, used to provide protection against infiltration of precipitation into a roof or exterior wall; usually placed around roof penetrations such as chimneys.

Flashpoint The minimum temperature at which sufficient vapor is released by a liquid or solid (fuel) to form a flammable vapor-air mixture at atmospheric pressure.

Flash-steam geothermal plants When the temperature of the hydrothermal liquids is over 350°F (177°C), flash-steam technology is generally employed. In these systems, most of the liquid is flashed to steam. The steam is separated from the remaining liquid and used to drive a turbine generator. While the water is returned to the geothermal reservoir, the economics of most hydrothermal flash plants are improved by using a dual-flash cycle, which separates the steam at two different pressures. The dual-flash cycle produces 20–30 % more power than a single-flash system at the same fluid flow.

Flat plate solar photovoltaic module An arrangement of photovoltaic cells or material mounted on a rigid flat surface with the cells exposed freely to incoming sunlight.

Flat plate solar thermal/heating collectors Large, flat boxes with glass covers and dark-colored metal plates inside that absorb and transfer solar energy to a heat-transfer fluid. This is the most common type of collector used in solar hot water systems for homes or small businesses.

Flat roof A slightly sloped roof, usually with a tar and gravel cover. Most commercial buildings use this kind of roof.

Flat-black paint Non-glossy paint with a relatively high absorption.

Float-zone process In reference to solar photovoltaic cell manufacture, a method of growing a large size, high-quality crystal whereby coils heat a polycrystalline ingot placed atop a single-crystal seed. As the coils are slowly raised, the molten interface beneath the coils becomes a single crystal.

Floor The upward-facing structure of a building.

Floor space The interior area of a building, calculated in square feet or meters.

Flow condition In reference to solar thermal collectors, the condition where the heat-transfer fluid is flowing through the collector loop under normal operating conditions.

Flow restrictor A water and energy conserving device that limits the amount of water that a faucet or shower head can deliver.

Flowback water It refers to (1) the water and excess proppant (propping agent) flow up through the wellbore to the surface, after the hydraulic fracturing procedure is completed and pressure is released, and the direction of fluid flow reverses, and (2) the water that returns to the surface, after the hydraulic fracturing procedure is completed and pressure is released, and the direction of fluid flow reverses, is commonly referred to as "flowback."

Flue The structure (in a residential heating appliance, industrial furnace, or power plant) into which combustion gases flow and are contained until they are emitted to the atmosphere.

Flue gas The gas resulting from the combustion of a fuel that is emitted to the flue.

Fluffing The practice of installing blow-in, loose-fill insulation at a lower density than is recommended to meet a specified R-value.

Fluid leakoff It refers to the process by which injected fracturing fluid migrates from the created fractures to other areas within the hydrocarbon-containing formation.

Fluidized bed combustion (FBC) A type of furnace or reactor in which fuel particles are combusted while suspended in a stream of hot gas.

Fluorescent light The conversion of electric power to visible light by using an electric charge to excite gaseous atoms in a glass tube. These atoms emit ultraviolet radiation that is absorbed by a phosphor coating on the walls of the lamp tube. The phosphor coating produces visible light.

Fly ash The fine particulate matter entrained in the flue gases of a combustion power plant.

Flywheel effect The damping of interior temperature fluctuations by massive construction.

Foam (insulation) A high R-value insulation product usually made from urethane that can be injected into wall cavities, or sprayed onto roofs or floors, where it expands and sets quickly.

Foam board A plastic foam insulation product, pressed or extruded into boardlike forms, used as sheathing and insulation for interior basement or crawl space walls or beneath a basement slab; can also be used for exterior applications inside or outside foundations, crawl spaces, and slab-on-grade foundation walls.

Foam core panels A type of structural, insulated product with foam insulation contained between two facings of drywall, or structural wood composition boards such as plywood, waferboard, and oriented strand board.

Food crop Crops consumed by humans. These include, but are not limited to, fruits, grains, vegetables, and tobacco.

Food waste (garbage) It includes animal and vegetable waste resulting from the storage, handling, preparation, cooking, or serving of foods (Washington, DC).

Foot candle A unit of illuminance; 1 fc = 1 lm per square foot = 10.764 lm per square meter.

Foot pound The amount of work done in raising one pound one foot. 1 ft-pound = 0.138255 kg-m = 1.35582 J.

Force The push or pull that alters the motion of a moving body or moves a stationary body; the unit of force is the dyne or poundal; force is equal to mass time velocity divided by time.

Forced air system or furnace A type of heating system in which heated air is blown by a fan through air channels or ducts to rooms.

Forced ventilation A type of building ventilation system that uses fans or blowers to provide fresh air to rooms when the forces of air pressure and gravity are not enough to circulate air through a building.

Forest land Tract of land thick with trees and underbrush.

Formaldehyde A chemical used as a preservative and in bonding agents. It is found in household products such as plywood, furniture, carpets, and some types of foam insulation. It is also a by-product of combustion and is a strong-smelling, colorless gas that is an eye irritant and can cause sneezing, coughing, and other health problems.

Formation It means a geological formation is a body of earth material with distinctive and characteristic properties and a degree of homogeneity in its physical properties.

Fossil fuels Fuels formed in the ground from the remains of dead plants and animals. It takes millions of years to form fossil fuels. Oil, natural gas, and coal are fossil fuels.

Foundation The supportive structure of a building.

Fractional horse power motor An electric motor rated at less than one horse power (hp).

Frame (window) The outer casing of a window that sits in a designated opening of a structure and holds the window panes in place.

Framing The structural materials and elements used to construct a wall.

Francis turbine A type of hydropower turbine that contains a runner that has water passages through it formed by curved vanes or blades. As the water passes through the runner and over the curved surfaces, it causes rotation of the runner. The rotational motion is transmitted by a shaft to a generator.

Freon A registered trademark for a chlorofluorocarbon (CFC) gas that is highly stable and that has been historically used as a refrigerant.

Frequency The number of cycles through which an alternating current passes per second; in the USA the standard for electricity generation is 60 cycles per second (60 Hz).

Fresnel lens An optical device for concentrating light that is made of concentric rings that are faced at different angles so that light falling on any ring is focused to the same point.

Friction head The energy lost from the movement of a fluid in a conduit (pipe) due to the disturbances created by the contact of the moving fluid with the surfaces of the conduit or the additional pressure that a pump must provide to overcome the resistance to fluid flow created by or in a conduit.

Fuel Any material that can be burned to make energy.

Fuel cell An electrochemical device that converts chemical energy directly into electricity.

Fuel efficiency The ratio of heat produced by a fuel for doing work to the available heat in the fuel.

Fuel grade alcohol Usually refers to ethanol to 160 to 200 proof.

Fuel oil Any liquid petroleum product burned for the generation of heat in a furnace or firebox or for the generation of power in an engine. Domestic (residential) heating fuels are classed as Nos. 1, 2, 3 and industrial fuels as Nos. 4, 5, and 6.

Fuel rate The amount of fuel necessary to generate one kilowatt-hour of electricity.

Full sun The amount of power density in sunlight received at the Earth's surface at noon on a clear day (about 1000 W/square meter).

Fungi Plantlike organisms with cells with distinct nuclei surrounded by nuclear membranes, incapable of photosynthesis. Fungi are decomposers of waste organisms and exist as yeast, mold, or mildew.

Furling The process of forcing, either manually or automatically, a wind turbine's blades out of the direction of the wind in order to stop the blades from turning.

Furnace (residential) A combustion heating appliance in which heat is captured from the burning of a fuel for distribution, comprised mainly of a combustion chamber and heat exchanger.

Fuse A safety device consisting of a short length of relatively fine wire, mounted in a holder or contained in a cartridge and connected as part of an electrical circuit. If the circuit source current exceeds a predetermined value, the fuse wire melts (i.e., the fuse "blows") breaking the circuit and preventing damage to the circuit protected by the fuse.

Gallium arsenide A compound used to make certain types of solar photovoltaic cells.

Gamma isotopic (also gamma scan) An analytical method by which the quantity of several gamma ray-emitting radioactive isotopes may be determined simultaneously. Typical nuclear fuel cycle isotopes determined by this method include but are not limited to Co-60, Zr-95, Ru-106, Ag-110m, Sb-125, Cs-134, Cs-137, and Eu-154. Naturally occurring isotopes for which samples also often are analyzed are Be-7, K-40, Ra-224, and Ra-226.

Gamma radiation A form of electromagnetic, high-energy radiation emitted from a nucleus. Gamma rays are essentially the same as x-rays and require heavy shielding such as lead, concrete, or steel to be stopped.

Garbage (1) A refuse accumulation of animal, fruit, or vegetable matter that attends the preparation, use, cooking, dealing in, and storage of edibles, and any other matter, of any nature whatsoever, which is subject to decay, putrefaction, and generation of noxious or offensive gases or odors or which, during or after decay, may serve as breeding or feeding material for flies or other germ-carrying insects (Dade County, FL); (2) all organic waste or residue of animal, fruit, vegetable, or food material from kitchens and dining rooms or from the preparation or dealing in or storage of meats, fowl, fruits, grains, or vegetables (Wichita, KS); (3) animal and vegetable waste resulting from the handling, preparation, cooking, and consumption of foods, exclusive of recognized industrial by-products and human or animal feces (Baltimore County, MD); (4) animal and vegetable waste resulting from the handling, preparation, cooking, and consumption of foods, exclusive of recognized industrial by-products from canneries and other food processing industries, and human or animal feces (Prince George's County, MD); (5) putrescible animal and vegetable waste resulting from the storage, handling, preparation, cooking, and consumption of food (Los Alamos County, NM); (6) all putrescible waste, except sewage and body wastes, including waste that attends the preparation; use; cooking; dealing in or storage of meat, fish, fowl, fruit, and vegetables; and all cans, boxes, cartons, papers, or other objects which have food or other organic materials of any nature in or adhering thereto; and shall include all such wastes or accumulation of vegetable matter of residences, restaurants, hotels, and places where food is prepared for human consumption. The term "garbage" does not include recognized industrial by-products (Beaverton, OR); (7) all putrescible animal or vegetable waste resulting from the preparation, cooking, and serving of food or the storage and sale of produce (Boulder, CO); (8) all putrescible animal or vegetable wastes resulting from handling, preparation, cooking, and consumption of food in any private dwelling house, multiple dwelling, hotel restaurant, building, or institution (Bettendorf, IA); (9) all rejected food wastes including every waste accumulation of animal, fruit, or vegetable matter used or intended for food or that attends the preparation, use, cooking, dealing in, or storing of meat, fish, fowl, fruit, or vegetables (Bluffton, IN); (10) every refuse accumulation of animal, fruit, or vegetable matter that attends the preparation, use, cooking, and dealing in, or storage or edibles, and any other matter, of any nature whatsoever, which is subject to decay, putrefaction, and the generation of noxious or offensive gases or odors or which, during or after decay, may serve as breeding or feeding material for flies or other germ-carrying insects (Miami, FL); and (11) all putrescible wastes and all animal or vegetable refuse or residue that shall result in the preparation or care for, or treatment of, foodstuffs intended to be used as food, and includes all putrescible wastes having resulted from the preparation or handling of food for human consumption, or any decayed or unsound meat, fish, fruit, or vegetable (Sonoma County, CA).

Garbage can (1) Any plastic or galvanized metal container of the type commonly sold as a garbage can, of a capacity not less than twenty gallons (76 L) and not to exceed thirty gallons (114 L), having two handles upon the sides thereof, or a bail by which it may be lifted, and a tight-fitting metal or plastic top with handle, and so constructed as to permit the free discharge of its contents (Miami, FL); and (2) any galvanized metal or durable plastic container of the type commonly sold as a garbage can, of a capacity not less than twenty gallons and not to exceed thirty gallons, having two handles upon the sides thereof or a bail by which it may be lifted and a tight-fitting metal or plastic top with handle, and so constructed as to permit the free discharge of its contents.

Garden trash (1) All accumulations of leaves, grass, or shrubbery cuttings and other refuse attending the care of lawns, shrubbery vines, and trees (Miami, FL), and (2) all accumulations of leaves, grass or shrubbery cuttings, and other refuse attending the care of lawns, shrubbery, vines, and trees.

Gas condensate A liquid which has condensed in the landfill gas collection system or the hydraulic fracking system during the extraction of gas from within the operations. Gases such as methane and carbon dioxide are generated due to microbial activity within the operational sites and must be removed to avoid hazardous conditions.

Gas turbine A type of turbine in which combusted, pressurized gas is directed against a series of blades connected to a shaft, which forces the shaft to turn to produce mechanical energy.

Gasification The process in which a solid fuel is converted into a gas; also known as pyrolytic distillation or pyrolysis. Production of a clean fuel gas makes a wide variety of power options available.

Gasifier A device for converting a solid fuel to a gaseous fuel.

Gasket/seal A seal used to prevent the leakage of fluids and also maintain the pressure in an enclosure.

Gasohol A registered trademark of an agency of the State of Nebraska, for an automotive fuel containing a blend of 10% ethanol and 90% gasoline.

Gasoline A refined petroleum product suitable for use as a fuel in internal combustion engines.

Gauss The unit of magnetic field intensity equal to 1 dyn per unit pole.

Generator A device for converting mechanical energy to electrical energy.

Geologic unit It refers to the geologic media in which an underground mined repository is constructed.

Geopressurized brines These brines are hot (300°F to 400°F) (149° C to 204°C) pressurized waters that contain dissolved methane and lie at depths of 10,000 ft (3,048 m) to more than 20,000 ft (6,096 m) below the Earth's surface. The best-known geopressured reservoirs lie along the Texas and Louisiana Gulf Coast. At least three types of energy could be obtained: thermal energy from high-temperature fluids, hydraulic energy from the high pressure, and chemical energy from burning the dissolved methane gas.

Geothermal energy Energy produced by the internal heat of the earth; geothermal heat sources include hydrothermal convective systems, pressurized water reservoirs, hot dry rocks, manual gradients, and magma. Geothermal energy can be used directly for heating or to produce electric power.

Geothermal heat pump A type of heat pump that uses the ground, groundwater, or ponds as a heat source and heat sink rather than outside air. Ground or water temperatures are more constant and are warmer in winter and cooler in summer than air temperatures. Geothermal heat pumps operate more efficiently than "conventional" or "air-source" heat pumps.

Geothermal power station An electricity-generating facility that uses geothermal energy.

Gigawatt (GW) A unit of power equal to 1 billion watts; 1 million kilowatts or 1,000 MW.

Gin pole A pole used to assist in raising a tower.

Glare The excessive brightness from a direct light source that makes it difficult to see what one wishes to see. A bright object in front of a dark background usually will cause glare. Bright lights reflecting off a television or computer screen or even a printed page produces glare. Intense light sources—such as bright incandescent lamps—are likely to produce more direct glare than large fluorescent lamps. However, glare is primarily the result of relative placement of light sources and the objects being viewed.

Glauber's salt A salt, sodium sulfate decahydrate, which melts at 90°F; a component of eutectic salts that can be used for storing heat.

Glazing A term used for the transparent or translucent material in a window. This material (i.e., glass, plastic films, and coated glass) is used for admitting solar energy and light through windows.

Glazing Transparent or translucent material (glass or plastic) used to admit light and/or to reduce heat loss; used for building windows, skylights, or greenhouses or for covering the aperture of a solar collector.

Global insolation (or solar radiation) The total diffuse and direct insolation on a horizontal surface, averaged over a specified period of time.

Global warming A popular term used to describe the increase in average global temperatures due to the greenhouse effect.

Governor A device used to regulate motor speed or, in a wind energy conversion system, to control the rotational speed of the rotor.

Gradient Change in value of one variable with respect to another variable, especially vertical or horizontal distance.

Grain alcohol Ethanol.

Green certificates Green certificates represent the environmental attributes of power produced from renewable resources. By separating the environmental attributes from the power, clean power generators are able to sell the electricity they produce to power providers at a competitive market value. The additional revenue generated by the sale of the green certificates covers the above-market costs associated with producing power made from renewable energy sources. Also known as green tags, renewable energy certificates, or tradable renewable certificates.

Green power A popular term for energy produced from clean, renewable energy resources.

Green pricing A practice engaged in by some regulated utilities (i.e., power providers) where electricity produced from clean, renewable resources is sold at a higher cost than that produced from fossil or nuclear power plants, supposedly because some buyers are willing to pay a premium for clean power.

Greenhouse effect A popular term used to describe the heating effect due to the trapping of long wave (length) radiation by greenhouse gases produced from natural and human sources.

Greenhouse gases Those gases, such as water vapor, carbon dioxide, tropospheric ozone, methane, and low-level ozone that are transparent to solar radiation, but opaque to long wave radiation, and which contribute to the greenhouse effect.

Greenwood Freshly cut, unseasoned, wood.

Greywater Waste water from a household source other than a toilet. This water can be used for landscape irrigation depending upon the source of the greywater.

Grid A common term referring to an electricity transmission and distribution system.

Grid-connected system Independent power systems that are connected to an electricity transmission and distribution system (referred to as the electricity grid) such that the systems can draw on the grid's reserve capacity in times of need and feed electricity back into the grid during times of excess production.

Gross calorific value The heat produced by combusting a specific quantity and volume of fuel in an oxygen-bomb colorimeter under specific conditions.

Gross generation The total amount of electricity produced by a power plant.

Ground A device used to protect the user of any electrical system or appliance from shock.

Ground loop In geothermal heat pump systems, a series of fluid-filled plastic pipes buried in the shallow ground or placed in a body of water, near a building. The fluid within the pipes is used to transfer heat between the building and the shallow ground (or water) in order to heat and cool the building.

Ground reflection Solar radiation reflected from the ground onto a solar collector.

Groundwater (1) It refers to the supply of freshwater found beneath the Earth's surface, usually in aquifers, which supply wells and springs. It provides a major source of drinking water. (2) The body of water that is retained in the saturated zone which tends to move by hydraulic gradient to lower levels. (3) Subsurface water in the pore spaces of soil and geologic units.

Ground-source heat pump (See geothermal systems)

Guarantee It refers to a contractual undertaking to answer for the debt of another.

Guarantor It refers to the party providing a guarantee.

Guy wire Cable use to secure a wind turbine tower to the ground in a safe, stable manner.

Half-life The time in which half the atoms of a radionuclide disintegrate into another nuclear form. The half-life may vary from a fraction of a second to thousands of years.

Harmonic(s) A sinusoidal quantity having a frequency that is an integral multiple of the frequency of a periodic quantity to which it is related.

Hazardous and special waste It includes hazardous solid and liquid wastes, including but not limited to highly flammable materials, explosives, pathological wastes, and radioactive materials (Baltimore County, MD).

Hazardous refuse or solid waste (1) It includes any refuse which would, in handling, constitute a danger to city employees or to city property (Bettendorf, IA); (2) it includes soil waste that may be itself or in combination with other solid wastes, be infectious, explosive, poisonous, caustic or toxic, or otherwise dangerous or injurious to human, plant, or animal life (Marion County, OR); and (3) it refers to those wastes that can cause serious injury or disease during the normal storage, collection, and disposal cycle, including but not limited to explosives, pathological and infectious wastes, radioactive materials, and dangerous chemicals (Washington, DC).

Hazardous substance spill An accidental spillage of hazardous substances on land or surface water.

Hazardous waste (1) Any waste, including wastewater, sludge, or solids, defined as hazardous under the federal government's RCRA, TSCA, or any state environmental laws or regulations. (2) A waste or combination of wastes that because of quantity, concentration, or physical, chemical, or infectious characteristics may (2a) cause or significantly contribute to an increase in mortality or an increase in serious irreversible or incapacitating reversible illness or (2b) pose a substantial present or potential hazard to human health or the environment when improperly treated, stored, transported, disposed of, or otherwise managed.

Head A unit of pressure for a fluid, which is commonly used in water pumping and hydropower to express height a pump must lift water, or the distance waterfalls.

Heat A form of thermal energy resulting from combustion, chemical reaction, friction, or movement of electricity. As a thermodynamic condition, heat, at a constant pressure, is equal to internal or intrinsic energy plus pressure times volume.

Heat absorbing window glass A type of window glass that contains special tints that cause the window to absorb as much as 45 % of incoming solar energy to reduce heat gain in an interior space. Part of the absorbed heat will continue to be passed through the window by conduction and reradiation.

Heat balance Energy output from a system that equals energy input.

Heat content The amount of heat in a quantity of matter at a specific temperature and pressure.

Heat engine A device that produces mechanical energy directly from two heat reservoirs of different temperatures. A machine that converts thermal energy to mechanical energy, such as a steam engine or turbine.

Heat exchanger A device used to transfer heat from a fluid (liquid or gas) to another fluid where the two fluids are physically separated.

Heat gain The amount of heat introduced to a space from all heat-producing sources, such as building occupants, lights, and appliances, and from the environment, mainly solar energy.

Heat loss The heat that flows from the building interior, through the building envelope to the outside environment.

Heat pipe A device that transfers heat by the continuous evaporation and condensation of an internal fluid.

Heat pump An electricity-powered device that extracts available heat from one area (the heat source) and transfers it to another (the heat sink) to either heat or cool an interior space or to extract heat energy from a fluid.

Heat pump water heaters A water heater that uses electricity to move heat from one place to another instead of generating heat directly.

Heat rate The ratio of fuel energy input as heat per unit of net work output; a measure of a power plant thermal efficiency, generally expressed as Btu per net kilowatt-hour.

Heat recovery ventilator A device that captures the heat from the exhaust air from a building and transfers it to the supply/fresh air entering the building to preheat the air and increase overall heating efficiency.

Heat register The grilled opening into a room by which the amount of warm air from a furnace can be directed or controlled; may include a damper.

Heat sink A structure or media that absorbs heat.

Heat source A structure or media from which heat can be absorbed or extracted.

Heat storage A device or media that absorbs heat for storage for later use.

Heat storage capacity The amount of heat that a material can absorb and store.

Heat transfer The flow of heat from one area to another by conduction, convection, and/or radiation. Heat flows naturally from a warmer to a cooler material or space.

Heat transfer fluid A gas or liquid used to move heat energy from one place to another; a refrigerant.

Heat transmission coefficient Any coefficient used to calculate heat transmission by conduction, convection, or radiation through materials or structures.

Heating capacity (also specific heat) The quantity of heat necessary to raise the temperature of a specific mass of a substance by one degree.

Heating degree day(s) (HDD) The number of degrees per day that the daily average temperature (the mean of the maximum and minimum recorded temperatures) is below a base temperature, usually 65°F, unless otherwise specified; used to determine indoor space-heating requirements and heating system sizing. Total HDD is the cumulative total for the year/heating season. The higher the HDD for a location, the colder the daily average temperature(s).

Heating fuel units Standardized weights or volumes for heating fuels.

Heating fuels Any gaseous, liquid, or solid fuel used for indoor space heating.

Heating load The rate of heat flow required to maintain a specific indoor temperature; usually measured in Btu per hour.

Heating season The coldest months of the year; months where average daily temperatures fall below 65°F (18.33°C), creating demand for indoor space heating.

Heating seasonal performance factor (HSPF) The measure of seasonal or annual efficiency of a heat pump operating in the heating mode. It takes into account the variations in temperature that can occur within a season and is the average number of Btu of heat delivered for every watt-hour of electricity used by the heat pump over a heating season.

Heating value The amount of heat produced from the complete combustion of a unit of fuel. The higher (or gross) heating value is that when all products of combustion are cooled to the precombustion temperature, water vapor formed during combustion is condensed, and necessary corrections have been made. Lower (or net) heating value is obtained by subtracting from the gross heating value the latent heat of vaporization of the water vapor formed by the combustion of the hydrogen in the fuel.

Heating, ventilation, and air-conditioning (HVAC) system All the components of the appliance used to condition interior air of a building.

Heavy metals Trace elements are found in low concentrations in biosolids. The trace elements of interest in biosolids are those commonly referred to as "heavy metals." Some of these trace elements (e.g., copper, molybdenum, and zinc) are nutrients needed for plant growth in low concentrations, but all of these elements can be toxic to humans, animals, or plants at high concentrations. Possible hazards associated with a buildup of trace elements in the soil include their potential to cause phytotoxicity (i.e., injury to plants) or to increase the concentration of potentially hazardous substances in the food chain. Federal and state regulations have established standards for the following nine trace elements: arsenic (As), cadmium (Cd), copper (Cu), lead (Pb), mercury (Hg), molybdenum (Mo), nickel (Ni), selenium (Se), and zinc (Zn).

Heliochemical process The utilization of solar energy through photosynthesis.

Heliodon A device used to simulate the angle of the sun for assessing shading potentials of building structures or landscape features.

Heliostat A device that tracks the movement of the sun; used to orient solar concentrating systems.

Heliothermal Any process that uses solar radiation to produce useful heat.

Heliothermic Site planning that accounts for natural solar heating and cooling processes and their relationship to building shape, orientation, and siting.

Heliothermometer An instrument for measuring solar radiation.

Heliotropic Any device (or plant) that follows the sun's apparent movement across the sky.

Hemispherical bowl technology A solar energy concentrating technology that uses a linear receiver that tracks the focal area of a reflector or array of reflectors.

HERO It refers to a database that includes more than 300,000 scientific articles from the peer-reviewed literature used by USEPA to develop its Integrated Science Assessments (ISA) that feed into the NAAQS review. It also includes references and data from the Integrated Risk Information System (IRIS), a database that supports critical agency policymaking for chemical regulation! Risk assessment—a study which characterizes the nature and magnitude of health risks to humans and the ecosystem from pollutants and chemicals in the environment.

Hertz A measure of the number of cycles or wavelengths of electrical energy per second; US electricity supply has a standard frequency of 60 Hz.

Heterojunction A region of electrical contact between two different materials.

Higher heating value (HHV) The maximum heating value of a fuel sample, which includes the calorific value of the fuel (bone dry) and the latent heat of vaporization of the water in the fuel (see moisture content and net (lower) heating value below).

High-intensity discharge lamp A lamp that consists of a sealed arc tube inside a glass envelope or outer jacket. The inner arc tube is filled with elements that emit light when ionized by electric current. A ballast is required to provide the proper starting voltage and to regulate current during operation.

High-level waste (HLW) The highly radioactive waste material that results from the reprocessing of spent nuclear fuel, including liquid waste produced directly in reprocessing and any solid waste derived from the liquid, which contains a combination of transuranic waste and fission products in concentrations sufficient to require permanent isolation.

High-pressure sodium lamp A type of high-intensity discharge (HID) lamp that uses sodium under high pressure as the primary light-producing element. These high-efficiency lights produce a golden white color and are used for interior industrial applications, such as in warehouses and manufacturing, and for security, street, and area lighting.

Hole The vacancy where an electron would normally exist in a solid; behaves like a positively charged particle.

Home energy rating systems (HERS) A nationally recognized energy rating program that gives builders, mortgage lenders, secondary lending markets, homeowners, sellers, and buyers a precise evaluation of energy losing deficiencies in homes. Builders can use this system to gauge the energy quality in their home and also to have a star rating on their home to compare to other similarly built homes.

Homojunction The region between an n-layer and a p-layer in a single material, photovoltaic cell.

Horizontal drilling It is drilling a portion of a well horizontally to expose more of the formation surface area to the wellbore.

Horizontal ground loop In this type of closed-loop geothermal heat pump installation, the fluid-filled plastic heat exchanger pipes are laid out in a plane parallel to the ground surface. The most common layouts either use two pipes, one buried at six feet and the other at four feet or two pipes placed side-by-side at five feet in the ground in a two-foot wide trench. The trenches must be at least four feet deep. Horizontal ground loops are generally most cost-effective for residential installations, particularly for new construction where sufficient land is available. Also see closed-loop geothermal heat pump systems.

Horizontal-axis wind turbines Turbines in which the axis of the rotor's rotation is parallel to the wind stream and the ground.

Horsepower (hp) A unit of rate of operation. Electrical hp, a measure of time rate of mechanical energy output; usually applied to electric motors as the maximum output; 1 electrical hp is equal to 0.746 kW or 2545 Btu per hour. Shaft hp, a measure of the actual mechanical energy per unit time delivered to a turning shaft; 1 shaft Hp is equal to 1 electrical Hp or 550 ft pounds per second. Boiler Hp, a measure to the maximum rate to heat output of a steam generator; 1 boiler Hp is equal to 33,480 Btu per hour steam output. 1 hp $= 0.7457$ kW.

Horsepower hour (hph) One horsepower provided over one hour; equal to 0.745 kW-hour or 2545 Btu. 1 hph $= 0.7457$ kW-hour.

Hot air furnace A heating unit where heat is distributed by means of convection or fans.

Hot dry rock A geothermal energy resource that consists of high-temperature rocks above 300°F (150°C) that may be fractured and have little or no water. To extract the heat, the rock must first be fractured, and then water is injected into the rock and pumped out to extract the heat. In the western United States, as much as 95,000 square miles (246,050 square km) have hot dry rock potential.

Hot water heating systems (See hydronic)

Household waste It includes garbage, rubbish, mixed refuse, and cool ashes originating in and around private dwellings, multiple dwellings, living quarters, or dining facilities located in schools, colleges, or universities (Bettendorf, IA).

HPVIS It refers to a database that provides access to health and environmental effects information obtained through the high production volume (HPV) challenge.

Hub height The height above the ground that a horizontal axis wind turbine's hub is located.

Humidifier A device used to maintain a specified humidity in a conditioned space.

Humidity A measure of the moisture content of air; may be expressed as absolute, mixing ratio, saturation deficit, relative, or specific.

Hybrid system A renewable energy system that includes two different types of technologies that produce the same type of energy; e.g., a wind turbine and a solar photovoltaic array combined to meet a power demand.

Hydraulic conductivity The ratio of flow velocity to driving force for viscous flow under saturated conditions of a specified liquid in a porous medium; the ratio describing the rate at which water can move through a permeable medium.

Hydraulic fracturing It refers to a process which uses high pressure to pump fluid and often carries proppants into subsurface rock formations in order to improve flow into a wellbore.

Hydraulic fracturing water lifecycle It refers to the lifecycle of water in the hydraulic fracturing process, encompassing the acquisition of water, chemical mixing of the fracturing fluid, injection of the fluid into the formation, the production and management of flowback and produced water, and the ultimate treatment and disposal of hydraulic fracturing wastewaters.

Hydroelectric power plant A power plant that produces electricity by the force of water falling through a hydroturbine that spins a generator.

Hydrogen A chemical element that can be used as a fuel since it has a very high energy content.

Hydrogenated amorphous silicon Amorphous silicon with a small amount of incorporated hydrogen. The hydrogen neutralizes dangling bonds in the amorphous silicon, allowing charge carriers to flow more freely.

Hydrogeologic unit It refers to any soil or rock unit or zone which, by virtue of its porosity and permeability, or lack thereof, has a distinct influence on the storage or movement of groundwater.

Hydronic heating systems A type of heating system where water is heated in a boiler and either moves by natural convection or is pumped to heat exchangers or radiators in rooms; radiant floor systems have a grid of tubing laid out in the floor for distributing heat. The temperature in each room is controlled by regulating the flow of hot water through the radiators or tubing.

Hydrothermal fluids These fluids can be either water or steam trapped in fractured or porous rocks; they are found from several hundred feet to several miles below the Earth's surface. The temperatures vary from about 90°F to 680°F (32°C to 360°C) but roughly 2/3 range in temperature from 150°F to 250°F (65.5°C to 121.1°C). The latter are the easiest to access and, therefore, the only forms being used commercially.

Ignite To heat a gaseous mixture to the temperature at which combustion takes place.

Ignition point The minimum temperature at which combustion of a solid or fluid can occur.

Illuminance A measure of the amount of light incident on a surface; measured in foot-candles or Lux.

Illumination The distribution of light on a horizontal surface. The purpose of all lighting is to produce illumination.

Impaired water body A water body that does not meet the criteria that support its designated use.

Impoundment A body of water confined by a dam, dike, floodgate, or other artificial barrier.

Impulse turbine A turbine that is driven by high velocity jets of water or steam from a nozzle directed to vanes or buckets attached to a wheel. (A Pelton wheel is an impulse hydroturbine).

Inactive A facility or portion thereof that is currently not treating, disposing, or recycling/recovering wastes or process liquids.

Inadvertent intruder It refers to a person who might occupy the disposal site after closure and engage in normal activities, such as agriculture, dwelling construction, or other pursuits in which the person might unknowingly be exposed to radiation from the low-level radioactive waste.

Inadvertent intrusion It refers to the act of occupying the disposal site after closure and engaging in normal activities, such as agriculture, dwelling construction, or other pursuits in which a person might unknowingly be exposed to radiation from the low-level radioactive waste.

Incandescent These lights use an electrically heated filament to produce light in a vacuum or inert gas-filled bulb.

Incident solar radiation The amount of solar radiation striking a surface per unit of time and area.

Incineration (1) A process involving destruction by combustion, in which heat, not less than twelve hundred degrees Fahrenheit (649°C), is applied to all classes of waste material, within a properly designed plant (Baltimore County, MD); (2) a process involving destruction by combustion, in which heat, not less than 1,200°F (649°C), is applied to all classes of waste material, within a properly designed plant approved by the local government (Prince George's County, MD); and (3) a physicochemical process involving the controlled burning of combustible solid wastes with or without accompanying noncombustible wastes with the use of auxiliary fuel when necessary (Boulder, CO).

Incinerator (1) Any equipment, device, or contrivance and all appurtenances thereof used for the destruction by burning of solid, semisolid liquid, or gaseous combustible wastes (Washington, DC). (2) A combustion device specifically designed for the reduction by burning of solid, semisolid, or liquid combustible wastes (Marion County, OR).

Incinerator residue It includes solid materials remaining after reduction in an incinerator (Washington, DC).

Independent power producer A company or individual that is not directly regulated as a power provider. These entities produce power for their own use and/or sell it to regulated power providers.

Indicator organism An indicator organism (e.g., fecal coliform) is a nonpathogenic organism whose presence implies the presence of pathogenic organisms. Indicator organisms are selected to be conservative estimates of the potential for pathogenicity.

Indirect discharger A facility that discharges or may discharge wastewaters into a publicly owned treatment works (POTW).

Indirect solar gain system A passive solar heating system in which the sun warms a heat storage element and the heat is distributed to the interior space by convection, conduction, and radiation.

Indirect solar water heater These systems circulate fluids other than water (such as diluted antifreeze) through the collector. The collected heat is transferred to the household water supply using a heat exchanger. Also known as "closed-loop" systems.

Individual field unit An *area of cropland* that has been *subdivided* into several strips is not a single field. Rather, each strip represents an *individual field unit*.

Induction The production of an electric current in a conductor by the variation of a magnetic field in its vicinity.

Induction generator A device that converts the mechanical energy of rotation into electricity based on electromagnetic induction. An electric voltage (electromotive force) is induced in a conducting loop (or coil) when there is a change in the number of magnetic field lines (or magnetic flux) passing through the loop. When the loop is closed by connecting the ends through an external load, the induced voltage will cause an electric current to flow through the loop and load. Thus, rotational energy is converted into electrical energy.

Induction motor A motor in which a three-phase (or any multiphase) alternating current (i.e., the working current) is supplied to iron-cored coils (or windings) within the stator. As a result, a rotating magnetic field is set up, which induces a magnetizing current in the rotor coils (or windings). Interaction of the magnetic field produced in this manner with the rotating field causes rotational motion to occur.

Industrial (solid) waste or refuse (1) Any and all residue resulting directly from industrial or manufacturing operations; it shall not include waste originating from commercial operations of an industrial establishment, nor shall it include waste resulting from the commercial operations of persons, firms, or corporations engaged in the construction of buildings, the repairs of streets or buildings, demolition, or excavation. Residue or waste resulting from tree or landscaping services shall also be excluded (Bettendorf, IA). (2) Any solid wastes which result from industrial processes and manufacturing operations such as factories, processing plants, repair and cleaning establishments, refineries, and rendering plants (Washington, DC). (3) The waste products of canneries; slaughterhouses or packing plants; condemned food products; agricultural waste products; wastes and debris from brick, concrete block, roofing shingle or tile plants; and debris and wastes accumulated from land clearing, excavating, building, rebuilding, and altering of buildings, structures, roads, streets, sidewalks, or parkways; and any waste materials which, because of their volume or nature, do not lend themselves to collection and incineration co-mingled with ordinary garbage and trash, or which, because of their nature or surrounding circumstances, should be, for reasons of safety or health, disposed of oftener than the city collection service schedule provided for this ordinance (Miami, FL). (4) Any solid waste materials from factories processing plants and other manufacturing enterprises, including but not limited to putrescible garbage from food processing plants and slaughterhouses, condemned foods, building rubbish, and cinders from power plants and manufacturing refuse (Baltimore County, MD). (5) Any waste products of canneries; slaughterhouses or packing plants; condemned food products; agricultural waste products; wastes and debris from brick, concrete block, roofing shingle or tile plants; and debris and wastes accumulated from land clearing, excavating, building, rebuilding, and altering of buildings, structures, roads, streets, sidewalks or parkways, and any waste materials which, because of their volume or nature, do not lend themselves to collection and incineration co-mingled with ordinary garbage and trash or which, because of their nature or surrounding circumstances, should be, for reasons of safety or health, disposed of more often than the county collection service schedule provided for (Dade County, FL).

Industrial process heat The thermal energy used in an industrial process.

Inert gas A gas that does not react with other substances; e.g., argon or krypton; sealed between two sheets of glazing to decrease the U-value (increase the R-value) of windows.

Infrared radiation Electromagnetic radiation whose wavelengths lie in the range from 0.75 to 1,000 μm; invisible long wavelength radiation (heat) capable of producing a thermal or photovoltaic effect, though less effective than visible light.

Insolation The solar power density incident on a surface of stated area and orientation, usually expressed as watts per square meter or Btu per square foot per hour.

Installed capacity The total capacity of electrical generation devices in a power station or system.

Instantaneous efficiency (of a solar collector) The amount of energy absorbed (or converted) by a solar collector (or photovoltaic cell or module) over a 15-min period.

Institutional control period (1) It refers to a period of time after closure of the land disposal facility during which the state maintains control of access to the site and carries out a program including environmental monitoring, periodic surveillance, and minor custodial care. (2) It refers to a period of time after the post-closure period during which the state maintains control of access to the site and carries out a program including environmental monitoring, periodic surveillance, and minor custodial care.

Institutional control plan It refers to the plan for institutional control prepared as required by the state or federal environmental laws or regulations.

Institutional waste It includes garbage, rubbish, mixed refuse, and cool ashes originating in and around tax-exempt hospitals and public, charitable, philanthropic, or religious institutions conducted for the benefit of the public or a recognized section of the public. Institutions not covered by the foregoing definition shall be considered commercial establishments (Bettendorf, IA).

Insulation Materials that prevent or slow down the movement of heat.

Insulation blanket A precut layer of insulation applied around a water heater storage tank to reduce standby heat loss from the tank.

Insulator A device or material with a high resistance to electricity flow.

Integral collector storage system This simple passive solar hot water system consists of one or more storage tanks placed in an insulated box that has a glazed side facing the sun. An integral collector storage system is mounted on the ground or on the roof (make sure your roof structure is strong enough to support it). Some systems use "selective" surfaces on the tank(s). These surfaces absorb sun well but inhibit radiative loss. Also known as bread box systems or batch heaters.

Integrated heating systems A type of heating appliance that performs more than one function, for example, space and water heating.

Integrated resource plan (IRP) A plan developed by an electric power provider, sometimes as required by a public regulatory commission or agency, that defines the short- and long-term capacity additions (supply side) and demand side management programs that it will undertake to meet projected energy demands.

Integrated safety management system (ISMS). The Integrated Safety Management System (ISMS) describes the programs, policies, and procedures used by the facility and the US DOE to ensure that the facility establishes a safe workplace for the employees, the public, and the environment. The guiding principles of ISMS are line management responsibility for safety, clear roles and responsibilities, competence commensurate with responsibilities, balanced priorities, identification of safety standards and requirements, hazard controls, and operations authorization.

Interconnection A connection or link between power systems that enables them to draw on each other's reserve capacity in time of need.

Interim status (1) The status of any currently existing facility that becomes subject to the requirement to have a RCRA permit because of a new statutory or regulatory amendment to RCRA. (2) In the USA, the status of any currently existing facility that becomes subject to the requirement to have a RCRA permit because of a new statutory or regulatory amendment to RCRA.

Intermittent generators Power plants, whose output depends on a factor(s) that cannot be controlled by the power generator because they utilize intermittent resources such as solar energy or the wind.

Internal combustion electric power plant The generation of electric power by a heat engine which converts part of the heat generated by combustion of the fuel into mechanical motion to operate an electric generator.

Internal gain The heat produced by sources of heat in a building (occupants, appliances, lighting, etc.).

Internal mass Materials with high thermal energy storage capacity contained in or part of a building's walls, floors, or freestanding elements.

Internal radiation Radiation originating from a source within the body as a result of the inhalation, ingestion, or implantation of natural or man-made radionuclides in body tissues.

Internal rate of return A widely used rate of return for performing economic analysis. This method solves for the interest rate that equates the equivalent worth of an alternative's cash receipts or savings to the equivalent worth of cash expenditures, including investments. The resultant interest rate is termed the internal rate of return (IRR).

Interruptible load Energy loads that can be shut off or disconnected at the supplier's discretion or as determined by a contractual agreement between the supplier and the customer.

Interstitial The (annular) space between the inner and outer tank walls in a double-walled storage tank.

Intracompany A facility that treats, disposes, or recycles/recovers wastes generated by off-site facilities under the same corporate ownership. The facility may also treat on-site generated wastes. If any waste from other facilities not under the same corporate ownership is accepted for a fee, the facility is considered commercial.

Intrinsic layer A layer of semiconductor material (as used in a solar photovoltaic device) whose properties are essentially those of the pure, undoped, material.

Intruder barrier It refers to an engineered structure or a sufficient depth of cover over the low-level radioactive waste or disposal units that inhibit contact with the waste and help to ensure that radiation exposures to an inadvertent intruder will meet the performance objectives of the state of federal environmental laws or regulations.

Inverter A device that converts direct current electricity (from, e.g., a solar photovoltaic module or array) to alternating current for use directly to operate appliances or to supply power to an electricity grid.

Investment tax credit A tax credit granted for specific types of investments.

Investor owned utility (IOU) A power provider owned by stockholders or other investors; sometimes referred to as a private power provider, in contrast to a public power provider that is owned by a government agency or cooperative.

Ion (1) An electrically charged atom or group of atoms that has lost or gained electrons; a loss makes the resulting particle positively charged; a gain makes the particle negatively charged. (2) An atom or group of atoms with an electric charge.

Ion exchange The reversible exchange of ions contained in solution with other ions that are part of the ion-exchange material.

Ionizer A device that removes airborne particles from breathable air. Negative ions are produced and give up their negative charge to the particles. These new negative particles are then attracted to the positive particles surrounding them. This accumulation process continues until the particles become heavy enough to fall to the ground.

IRIS It refers to a human health assessment program that evaluates risk information on effects that may result from exposure to environmental contaminants.

Irradiance The direct, diffuse, and reflected solar radiation that strikes a surface.

Isolated solar gain system A type of passive solar heating system where heat is collected in one area for use in another.

Isotope Different forms of the same chemical element that are distinguished by having the same number of protons but a different number of neutrons in the nucleus. An element can have many isotopes. For example, the three isotopes of hydrogen are protium, deuterium, and tritium, with one, two, and three neutrons in the nucleus, respectively.

I-Type semiconductor A semiconductor material that is left intrinsic or undoped so that the concentration of charge carriers is characteristic of the material itself rather than of added impurities.

I-V Curve A graphical plot or representation of the current and voltage output of a solar photovoltaic cell or module as a load on the device is increased from short-circuit (no load) condition to the open-circuit condition; used to characterize cell/module performance.

Jacket The enclosure on a water heater, furnace, or boiler.

Joist A structural, load-carrying building member with an open web system that supports floors and roofs utilizing wood or specific steels and is designed as a simple span member.

Joule A metric unit of energy or work; the energy produced by a force of one Newton operating through a distance of one meter; 1 J per second equals 1 W or 0.737 ft pounds; 1 Btu equals 1055 J.

Joule's law The rate of heat production by a steady current in any part of an electrical circuit that is proportional to the resistance and to the square of the current, or the internal energy of an ideal gas depends only on its temperature.

Junction A region of transition between semiconductor layers, such as a p/n junction, which goes from a region that has a high concentration of acceptors (p-type) to one that has a high concentration of donors (n-type).

Kame delta A conical hill or short irregular ridge of gravel or sand deposited in contact with glacier ice.

Kaplan turbine A type of turbine that has two blades whose pitch is adjustable. The turbine may have gates to control the angle of the fluid flow into the blades.

Kerosene A type of heating fuel derived by refining crude oil that has a boiling range at atmospheric pressure from 400°F to 550°F.

Key finding (self-assessment) A direct and significant violation of a Department of Energy regulatory or other applicable guidance or procedural requirement or a recurring pattern of observed deficiencies that could result in such a violation. A finding is a deficiency that requires corrective action.

Kilovolt-ampere (kVa) A unit of apparent power, equal to 1000 V-amperes; the mathematical product of the volts and amperes in an electrical circuit.

Kilowatt (kW) A standard unit of electrical power equal to one thousand watts or to the energy consumption at a rate of 1000 J per second.

Kilowatt-hour A unit or measure of electricity supply or consumption of 1000 W over the period of one hour; equivalent to 3412 Btu.

Kinetic energy Energy available as a result of motion that varies directly in proportion to an object's mass and the square of its velocity.

Kneewall A wall usually about 3–4 ft high located that is placed in the attic of a home, anchored with plates between the attic floor joists and the roof joist. Sheathing can be attached to these walls to enclose an attic space.

Lacustrine sediments A sedimentary deposit consisting of material pertaining to, produced by, or formed in a lake or lakes.

Lagoon In wastewater treatment or livestock facilities, a shallow pond used to store wastewater where sunlight and biological activity decompose the waste. There are three major types of lagoons: anaerobic, aerobic, and facultative.

Lamp A light source composed of a metal base, a glass tube filled with an inert gas or a vapor, and base pins to attach to a fixture.

Land application Land application is defined as the spreading, spraying, injection, or incorporation of liquid or semiliquid organic substances, such as sewage sludge, biosolids, livestock manure, compost, septage, legumes, and other types of liquid organic waste, onto or below the surface of the land to take advantage of the soil-enhancing qualities of the organic substances. These organic substances are land applied to improve the structure of the soil. It is also applied as a fertilizer to supply nutrients to crops and other vegetation grown in the soil. The liquid or semiliquid organic substances are commonly applied to agricultural land (including pasture and range land), forests, reclamation sites, public contact sites (e.g., parks, turf farms, highway median strips, golf courses), lawns, and home gardens.

Land application site An area of land on which sewage sludge is applied to condition the soil or to fertilize crops or vegetation grown in the soil.

Land disposal facility (1) It includes, but is not limited to, underground mined repositories and the land, buildings, and equipment which are intended to be used for aboveground disposal of low-level radioactive waste: (1a) it is not classified as high-level radioactive waste, spent nuclear fuel, or by-product material as defined in the state or federal environmental laws or regulations. By-product material as defined in the US Atomic Energy Act is uranium or thorium tailings and waste; (1b) it is classified as low-level radioactive waste consistent with federal/state law and in accordance with the US Nuclear Regulatory Commission (NRC) or the International Atomic Energy Agency (IAEA); and (2) it refers to the land, buildings, and equipment which are intended to be used for the disposal of low-level radioactive waste into the subsurface of the land.

Land disposal restrictions (LDR) Regulations promulgated by the US EPA (and by the state environmental agency) governing the land disposal of hazardous wastes. The wastes must be treated using the best demonstrated available technology or must meet certain treatment standards before being disposed.

Land reclamation An action to remove pollutants from a contaminated land in order to reclaim the land for useful domestic, commercial, industrial, recreational, or agricultural applications.

Land restoration An action to remove pollutants from a contaminated land in order to restore the land's original unpolluted conditions.

Land treatment A wastewater treatment process involving the sprinkling of wastewater to vegetated soils that are slow to moderate in permeability (clay loams to sandy loams) and treatment of wastewater as it travels through the soil matrix by filtration, adsorption, ion exchange, precipitation, microbial action, and by plant uptake. An optional underdrainage system consisting of a network of drainage pipe buried below the surface may serve to recover the effluent, to control groundwater, or to minimize trespass of leachate onto adjoining property by horizontal subsurface flow.

Landfill (1) An area of land or an excavation in which wastes are placed for permanent disposal, which is not a land application or land treatment unit, surface impoundment, underground injection well, waste pile, salt dome formation, a salt bed formation, an underground mine, or a cave and (2) A disposal site operated by means of compacting and covering solid wastes at specific designated intervals, but not each operating day (Marion County, OR).

Landfill-generated wastewaters Wastewater generated by landfill activities and collected for treatment, discharge, or reuse include leachate, contaminated groundwater, storm water runoff, landfill gas condensate, truck/equipment washwater, drained-free liquids, floor washings, and recovering pumping wells.

Landfill: subtitle C See subtitle C landfill

Landfill: subtitle D See subtitle D landfill

Landscaping Features and vegetation on the outside of or surrounding a building for aesthetics and energy conservation.

Langley A unit or measure of solar radiation; 1 cal per square centimeter or 3.69 Btu per square foot.

Large volume receptacle It includes refuse receptacles of 4 cubic yards (3.06 cubic meters) of greater capacity furnished and maintained by the county which shall be subject to special collection fees as provided in the county government's ordinance (Los Alamos County, NM).

Latent cooling load The load created by moisture in the air, including from outside air infiltration and that from indoor sources such as occupants, plants, cooking, showering, etc.

Latent heat The change in heat content that occurs with a change in phase and without change in temperature.

Latent heat of vaporization The quantity of heat produced to change a unit weight of a liquid to vapor with no change in temperature.

Lattice The regular periodic arrangement of atoms or molecules in a crystal of semiconductor material.

Law(s) of thermodynamics The first law states that energy cannot be created or destroyed; the second law states that when a free exchange of heat occurs between two materials, the heat always moves from the warmer to the cooler material.

Leachate Leachate is a liquid that has passed through or emerged from solid waste and contains soluble, suspended, or miscible materials removed from such waste. Leachate is typically collected from a liner system above which waste is placed for disposal. Leachate may also be collected through the use of slurry walls, trenches, or other containment systems.

Leachate collection system The purpose of a leachate collection system is to collect leachate for treatment or alternative disposal and to reduce the depths of leachate buildup or level of saturation over the low permeability liner.

Lead acid battery An electrochemical battery that uses lead and lead oxide for electrodes and sulfuric acid for the electrolyte.

Leading edge In reference to a wind energy conversion system, the area of a turbine blade surface that first comes into contact with the wind.

Leaking electricity Related to standby power, leaking electricity is the power needed for electrical equipment to remain ready for use while in a dormant mode or operation. Electricity is still used by many electrical devices, such as TVs, stereos, and computers, even when you think they are turned "off."

Lethe A measure of air purity that is equal to one complete air change (in an interior space).

Letter of credit It is a written instrument whereby the issuer will honor drafts or other demands for payment upon compliance with the conditions specified in the letter.

Levelized life-cycle cost A total life-cycle cost divided into equal amounts.

License It refers to a license to operate a low-level radioactive waste land disposal facility or other environmental facility issued pursuant to the state of federal environmental laws or regulations.

Licensed waste collector It refers to any person, firm, or corporation who has obtained a license as specified herein from the city to collect or transport, for a consideration, bulk refuse, household waste, commercial waste, industrial waste, building waste, or building debris, regardless of the place of origin, over the streets of the city (Bettendorf, IA).

Life-cycle cost The sum of all the costs both recurring and nonrecurring, related to a product, structure, system, or service during its life span or specified time period.

Lift The force that pulls a wind turbine blade, as opposed to drag.

Light quality A description of how well people in a lighted space can see to do visual tasks and how visually comfortable they feel in that space.

Light trapping The trapping of light inside a semiconductor material by refracting and reflecting the light at critical angles; trapped light will travel further in the material, greatly increasing the probability of absorption and hence of producing charge carriers.

Light-induced defects Defects, such as dangling bonds, induced in an amorphous silicon semiconductor upon initial exposure to light.

Limit of liability It is the total amount an insurer is obligated to pay, under an insurance policy, for remediation of failures and/or for personal injury or property damage to third parties caused by the operation of the land disposal facility.

Line loss (or drop) Electrical energy lost due to inherent inefficiencies in an electrical transmission and distribution system under specific conditions.

Liner (landfill) The liner is a low permeability material or combination of materials placed at the base of a landfill to reduce the discharge to the underlying or surrounding hydrogeologic environment. The liner is designed as a barrier to intercept leachate and to direct it to a leachate collection.

Liquid manure A manure from animals that contains dry matter less than 5 %.

Liquid-based solar heating system A solar heating system that uses a liquid as the heat-transfer fluid.

Liquid-to-air heat exchanger A heat exchanger that transfers the heat contained in a liquid heat-transfer fluid to air.

Liquid-to-liquid heat exchanger A heat exchanger that transfers heat contained in a liquid heat-transfer fluid to another liquid.

Lithium-sulfur battery A battery that uses lithium in the negative electrode and a metal sulfide in the positive electrode, and the electrolyte is molten salt; can store large amounts of energy per unit weight.

Live steam Steam available directly from a boiler under full pressure.

Load The power required to run a defined circuit or system, such as a refrigerator, building, or an entire electricity distribution system.

Load analysis Assessing and quantifying the discrete components that comprise a load. This analysis often includes time of day or season as a variable.

Load duration curve A curve that displays load values on the horizontal axis in descending order of magnitude against the percent of time (on the vertical axis) that the load values are exceeded.

Load factor The ratio of average energy demand (load) to maximum demand (peak load) during a specific period.

Load forecast An estimate of power demand at some future period.

Load leveling The deferment of certain loads to limit electrical power demand or the production of energy during off-peak periods for storage and use during peak demand periods.

Load management To influence the demand on a power source.

Load profile or shape A curve on a chart showing power (kW) supplied (on the horizontal axis) plotted against time of occurrence (on the vertical axis) to illustrate the variance in a load in a specified time period.

Load shedding Turning off or disconnecting loads to limit peak demand.

Load shifting A load management objective that moves loads from on-peak periods to off-peak periods.

Local solar time A system of astronomical time in which the sun crosses the true north-south meridian at 12 noon and which differs from local time according to longitude, time zone, and equation of time.

Log law In reference to a wind energy conversion system, the wind speed profile in which wind speeds increase with the logarithmic of the height of the wind turbine above the ground.

Long ton A unit that equals 20 long hundredweight or 2,240 lb. Used mainly in England. 1 long ton = 1,016 kg.

Long-term average (LTA) For purposes of the effluent guidelines, average pollutant levels achieved over a period of time by a facility, subcategory, or technology option. LTAs were used in developing the limitations and standards in the proposed landfill regulation.

Long-wave radiation Infrared or radiant heat.

Loose fill insulation Insulation made from rockwool fibers, fiberglass, cellulose fiber, vermiculite, or perlite minerals and composed of loose fibers or granules can be applied by pouring directly from the bag or with a blower.

Loss of load probability (LOLP) A measure of the probability that a system demand will exceed capacity during a given period; often expressed as the estimated number of days over a long period, frequently 10 years or the life of the system.

Losses (energy) A general term applied to the energy that is converted to a form that cannot be effectively used (lost) during the operation of an energy-producing, energy-conducting, or energy-consuming system.

Low Btu gas A fuel gas with a heating value between 90 and 200 Btu per cubic foot.

Low flush toilet A toilet that uses less water than a standard one during flushing, for the purpose of conserving water resources.

Low-E coatings and (window) films A coating applied to the surface of the glazing of a window to reduce heat transfer through the window.

Low-emissivity windows and (window) films Energy-efficient windows that have a coating or film applied to the surface of the glass to reduce heat transfer through the window.

Lower (net) heating value The lower or net heat of combustion for a fuel that assumes that all products of combustion are in a gaseous state (see net heating value below).

Lower limit of detection (LLD) The lowest limit of a given parameter an instrument is capable of detecting. A measurement of analytical sensitivity.

Low-flow solar water heating systems The flow rate in these systems is 1/8 to 1/5, the rate of most solar water heating systems. The low-flow systems take advantage of stratification in the storage tank and theoretically allows for the use of smaller diameter piping to and from the collector and a smaller pump.

Low-level radioactive waste It refers to those low-level radioactive wastes that are acceptable for disposal in a land disposal facility pursuant to the provisions of the appropriate state or federal environmental laws or regulations.

Low-level waste (LLW) Radioactive waste not classified as high-level waste, transuranic waste, spent fuel, or uranium mill tailings (see Classes A, B, and C low-level waste).

Low-pressure sodium lamp A type of lamp that produces light from sodium gas contained in a bulb operating at a partial pressure of 0.13 to 1.3 Pa. The yellow light and large size make them applicable to lighting streets and parking lots.

Lumen An empirical measure of the quantity of light. It is based upon the spectral sensitivity of the photosensors in the human eye under high (daytime) light levels. Photometrically it is the luminous flux emitted with a solid angle (1 sr) by a point source having a uniform luminous intensity of 1 cd. As reference, a 100-W incandescent lamp emits about 1600 lm.

Lumens/watt (lpw) A measure of the efficacy (efficiency) of lamps. It indicates the amount of light (lumens) emitted by the lamp for each unit of electrical power (watts) used.

Luminaire A complete lighting unit consisting of a lamp(s), housing, and connection to the power circuit.

Luminance The physical measure of the subjective sensation of brightness; measured in lumens.

Lux The unit of illuminance equivalent to 1 l m per square meter.

Magma Molten or partially molten rock at temperatures ranging from 1260°F to 2880°F (700°C to 1600°C). Some magma bodies are believed to exist at drillable depths within the Earth's crust, although practical technologies for harnessing magma energy have not been developed. If ever utilized, magma represents a potentially enormous resource.

Magnetic ballast A type of florescent light ballast that uses a magnetic core to regulate the voltage of a florescent lamp.

Major natural phenomena It refers to rarely occurring natural events such as tornadoes, hurricanes, floods, wildfires, volcanism, and earthquakes.

Makeup air Air brought into a building from outside to replace exhaust air.

Manual J The standard method for calculating residential cooling loads developed by the Air-Conditioning and Refrigeration Institute (ARI) and the Air Conditioning Contractors of America (ACCA) based largely on the American Society of Heating, Refrigeration, and Air-Conditioning Engineer's (ASHRAE) "Handbook of Fundamentals."

Manure Any wastes discharged from livestock and other animals.

Marginal cost The cost of producing one additional unit of a product.

Masonry Material such as brick, rock, or stone.

Masonry stove A type of heating appliance similar to a fireplace, but much more efficient and clean burning. They are made of masonry and have long channels through which combustion gases give up their heat to the heavy mass of the stove, which releases the heat slowly into a room. Often called Russian or Finnish fireplaces.

Mass burn facility A type of municipal solid waste (MSW) incineration facility in which MSW is burned with only minor presorting to remove oversize, hazardous, or explosive materials. Mass burn facilities can be large, with capacities of 3,000 t (2.7 million kg) of MSW per day or more. They can be scaled down to handle the waste from smaller communities, and modular plants with capacities as low as 25 t (22.7 thousand kg) per day have been built. Mass burn technologies represent over 75 % of all the MSW-to-energy facilities constructed in the United States to date. The major components of a mass burn facility include refuse receiving and handling, combustion and steam generation, flue gas cleaning, power generation (optional), condenser cooling water, residue ash hauling, and landfilling.

Mass wasting It refers to the movement of rock or soil material under the influence of gravity either as the movement of the products of weathering down a slope or as a mass movement of rock or soil along joint planes or bedding planes. Mass wasting includes but is not limited to creep, mud flows, earth flow, soil flow, rock avalanche, landslide, land-slip, and slumping.

Maximally exposed individual A hypothetical person who remains in an uncontrolled area who would, when all potential routes of exposure from a facility's operations are considered, receive the greatest possible dose equivalent.

MCF An abbreviation for one thousand cubic feet of natural gas with a heat content of 1,000,000 Btus or 10 therms.

Mean The average value of a series of measurements.

Mean grade It refers to the arithmetic average of elevations of points on the plot plan that will result upon completion of the proposed operation for which the permit is issued (Chesterfield Township, MI).

Mean power output (of a wind turbine) The average power output of a wind energy conversion system at a given mean wind speed based on a Raleigh frequency distribution.

Mean wind speed The arithmetic wind speed over a specified time period and height above the ground (the majority of US National Weather Service anemometers are at 20 ft (6.1 m)).

Mechanical integrity It means an injection well has mechanical integrity if (1) there is no significant leak in the casing, tubing, or packer (internal mechanical integrity) and (2) there is no significant fluid movement into an underground source of drinking water through vertical channels adjacent to the injection wellbore (external mechanical integrity).

Mechanical systems Those elements of building used to control the interior climate.

Median wind speed The wind speed with 50% probability of occurring.

Medium Btu gas Fuel gas with a heating value of between 200 and 300 Btu per cubic foot.

Medium pressure For valves and fittings, it implies that they are suitable for working pressures between 125 and 175 lb per square inch (861,845–1,206,583 N per square meter).

Megawatt One thousand kilowatts or 1 million watts; standard measure of electric power plant generating capacity.

Megawatt-hour One thousand kilowatt-hours or 1 million watt-hours.

Mercury vapor lamp A high-intensity discharge lamp that uses mercury as the primary light-producing element. Includes clear, phosphor-coated, and self-ballasted lamps.

Mesophilic It is a state in an anaerobic reactor, such anaerobic digester or composting unit, where/when the temperature remains between 35–40°C.

Mesotrophic The term describes reservoirs and lakes that contain moderate quantities of nutrients and are moderately productive in terms of aquatic animal and plant life.

Met An approximate unit of heat produced by a resting person, equal to about 18.5 Btu per square foot per hour.

Metal halide lamp A high-intensity discharge lamp type that uses mercury and several halide additives as light-producing elements. These lights have the best color rendition index (CRI) of the high-intensity discharge lamps. They can be used for commercial interior lighting or for stadium lights.

Methane A colorless, odorless, tasteless gas composed of one molecule of carbon and four of hydrogen, which is highly flammable. It is the main constituent of "natural gas" that is formed naturally by methanogenic, anaerobic bacteria or can be manufactured, and which is used as a fuel and for manufacturing chemicals.

Methanol (CH_3OH; methyl alcohol or wood alcohol) A clear, colorless, very mobile liquid that is flammable and poisonous; used as a fuel and fuel additive and to produce chemicals.

Methyl tertiary butyl ether (MTBE) An ether compound used as a gasoline blending component to raise the oxygen content of gasoline. MTBE is made by combining isobutylene (from various refining and chemical processes) and methanol (usually made from natural gas).

Metric ton (Tonne) A unit of mass equal to 1,000 kg or 2,204.6 lb.

Microclimate The local climate of specific place or habitat, as influenced by landscape features.

Microcurie It refers to one one-millionth (0.000001) of a curie.

Microgroove A small groove scribed into the surface of a solar photovoltaic cell which is filled with metal for contacts.

Micrometer One millionth of a meter (10^{-6} m).

Mill A common monetary measure equal to one-thousandth of a dollar or a tenth of a cent.

Millicurie It refers to one one-thousandth (0.001) of a curie.

Millirem (mrem) (1) It refers to one one-thousandth (0.001) of a rem. (2) It is a unit of radiation dose equivalent that is equal to one one-thousandth of a rem. An individual member of the public can receive up to 500 millirems per year according to the US DOE standards. This limit does not include radiation received for medical treatment or the 100 to 360 mrem that people receive annually from background radiation.

Mineralization Most nitrogen exists in biosolids/manure as organic-N, principally contained in proteins, nucleic acids, amines, and other cellular material. These complex molecules must be broken apart through biological degradation for nitrogen to become available to crops. The conversion of organic-N to inorganic-N forms is called *mineralization*

Minimize to the extent reasonably achievable It means to reduce to the least quantity or degree which can reasonably be attained, taking into account (1) the state of technology, (2) the benefits to be gained from any possible further reduction, and (3) the impacts of the measures or efforts required to achieve any possible further reduction.

Minimum detectable concentration (MDC) Depending on the sample medium, the smallest amount or concentration of a radioactive or nonradioactive analyte that can be reliably detected using a specific analytical method. Calculations of the minimum detectable concentrations are based on the lower limit of detection.

Minority carrier A current carrier, either an electron or a hole, which is in the minority in a specific layer of a semiconductor material; the diffusion of minority carriers under the action of the cell junction voltage is the current in a photovoltaic device.

Minority carrier lifetime The average time a minority carrier exists before recombination.

Mixed low-level radioactive and hazardous waste It refers to waste that satisfies the definition of low-level radioactive waste in the appropriate state or federal environmental laws or regulations that either (1) is listed as a hazardous waste in the environmental laws or (2) causes the low-level radioactive waste to exhibit any of the hazardous waste characteristics identified in the environmental laws or regulations.

Mixed waste A waste that is both radioactive and hazardous. Also referred to as radioactive mixed waste (RMW).

Mixing valve A valve operated by a thermostat that can be installed in solar water heating systems to mix cold water with water from the collector loop to maintain a safe water temperature.

Mobile container It includes any metal 14 gauge steel garbage and waste container with the following minimum specifications: 77–3/8 in. long, 30 in. wide, and 46 in. high (197 cm L × 76 cm W × 117 cm H) at the back edge, tapering down to 41 in. (104 cm) at loading edge, and shall be mounted on four rubber wheels with roller bearings and/or metal slides. Said mobile container shall be capable of fitting hydraulic attachments for unloading. Said container shall be flyproof, ratproof, and leakproof and shall be fitted with 14 gauge steel-constructed covers (Miami, FL).

Model It refers to a conceptual description and the associated mathematical, graphical, and/or analogous representation of a system, subsystem, component, or condition that is used to predict changes from a baseline state as a function of internal and/or external stimuli and as a function of time and space.

Modified degree-day method A method used to estimate building heating loads by assuming that heat loss and gain is proportional to the equivalent heat-loss coefficient for the building envelope.

Module The smallest self-contained, environmentally protected structure housing interconnected photovoltaic cells and providing a single dc electrical output; also called a panel.

Moisture content The water content of a substance (a solid fuel) as measured under specified conditions being the dry basis, which equals the weight of the wet sample minus the weight of a (bone) dry sample divided by the weight of the dry sample times 100 (to get percent), and the wet basis, which is equal to the weight of the wet sample minus the weight of the dry sample divided by the weight of the wet sample times 100.

Moisture control The process of controlling indoor moisture levels and condensation.

Monitoring It means observing and making measurements to provide data to evaluate the performance and characteristics of the disposal site.

Monoculture The planting, cultivation, and harvesting of a single species of crop in a specified area.

Monolithic Fabricated as a single structure.

Motor A machine supplied with external energy that is converted into force and/or motion.

Motor speed The number of revolutions that the motor turns in a given time period (i.e., revolutions per minute, rpm).

Movable insulation A device that reduces heat loss at night and during cloudy periods and heat gain during the day in warm weather. A movable insulator could be an insulative shade, shutter panel, or curtain.

MTBE See methyl tertiary butyl ether (MTBE).

Multijunction device A high-efficiency photovoltaic device containing two or more cell junctions, each of which is optimized for a particular part of the solar spectrum.

Multi-zone system A building heating, ventilation, and/or air-conditioning system that distributes conditioned air to individual zones or rooms.

Municipal solid waste (MSW) (1) Waste material from households and businesses in a community that is not regulated as hazardous. (2) It is more commonly known as trash or garbage, which consists of the everyday items we use and then throw away, such as product packaging, grass clippings, furniture, clothing, bottles, food scraps, newspapers, appliances, paint, and batteries. This comes from our homes, schools, hospitals, and businesses (US EPA) (see garbage, trash).

Municipal waste As defined in the Energy Security Act (P.L. 96-294; 1980) as "any organic matter, including sewage, sewage sludge, and industrial or commercial waste, and mixtures of such matter and inorganic refuse from any publicly or privately operated municipal waste collection or similar disposal system, or from similar waste flows (other than such flows which constitute agricultural wastes or residues, or wood wastes or residues from wood harvesting activities or production of forest products)."

Municipal waste to energy project (or plant) A facility that produces fuel or energy from municipal solid waste.

Nacelle The cover for the gear box, drive train, generator, and other components of a wind turbine.

Name plate A metal tag attached to a machine or appliance that contains information such as brand name, serial number, voltage, power ratings under specified conditions, and other manufacturer supplied data.

Nanocurie It refers to one one-billionth (0.000000001) of a curie.

Narrative criteria Nonnumeric descriptions of desirable or undesirable water quality conditions.

National electrical code (NEC) The NEC is a set of regulations that have contributed to making the electrical systems in the United States one of the safest in the world. The intent of the NEC is to ensure safe electrical systems are designed and installed. The National Fire Protection Association has sponsored the NEC since 1911. The NEC changes as technology evolves and component sophistication increases. The NEC is updated every three years. Following the NEC is required in most locations.

National pollution discharge elimination system (NDPES) A US federal environmental law that regulates the quantity of waste entering navigable waters and point sources. It was first introduced by the US EPA in the Clean Water Act of 1977. Most of the industrial effluents and some agricultural effluents (such as livestock waste operation effluents) are required to have NPDES which permits to discharge to a receiving water. State legislation defines the specific operations that require NPDES permit.

National rural electric cooperative association (NRECA) This is a national organization dedicated to representing the interests of cooperative electric power providers and the consumers they serve. Members come from the 46 states that have an electric distribution cooperative.

Natural characteristics It refers to the elements which comprise the physical and biological environment, including the ecology, geochemistry, geology, hydrology, meteorology and climate, and seismology. These elements also include natural resources with agricultural, cultural, economic, recreational, and scenic values.

Natural cooling Space cooling achieved by shading, natural (unassisted, as opposed to forced) ventilation, conduction control, radiation, and evaporation; also called passive cooling.

Natural draft Draft that is caused by temperature differences in the air.

Natural gas A hydrocarbon gas obtained from underground sources, often in association with petroleum and coal deposits. It generally contains a high percentage of methane, varying amounts of ethane, and inert gases; used as a heating fuel.

Natural gas-fired (thermoelectric) power plant A power plant that produces electricity by the force of steam through a turbine that spins a generator. The steam is produced by burning the natural gas.

Natural gas or gas It refers to a naturally occurring mixture of hydrocarbon and non-hydrocarbon gases in porous formations beneath the Earth's surface, often in association with petroleum. The principal constituent is methane.

Natural gas steam reforming production A two-step process where in the first step, natural gas is exposed to a high-temperature steam to produce hydrogen, carbon monoxide, and carbon dioxide. The second step is to convert the carbon monoxide with steam to produce additional hydrogen and carbon dioxide.

Natural ventilation Ventilation that is created by the differences in the distribution of air pressures around a building. Air moves from areas of high pressure to areas of low pressure with gravity and wind pressure affecting the airflow. The placement and control of doors and windows alters natural ventilation patterns.

Naturally occurring radioactive materials (NORM) It refers to all radioactive elements found in the environment, including long-lived radioactive elements such as uranium, thorium, and potassium and any of their decay products, such as radium and radon.

N-Dodecane/tributyl phosphate An organic solution composed of 30 % tributyl phosphate (TBP) dissolved in n-dodecane used to first separate the uranium and plutonium from the fission products in the dissolved fuel and then to separate the uranium from the plutonium.

Net (lower) heating value (NHV) The potential energy available in a fuel as received, taking into account the energy loss in evaporating and superheating the water in the fuel. Equal to the higher heating value minus 1,050 W where W is the weight of the water formed from the hydrogen in the fuel and 1,050 is the latent heat of vaporization of water, in Btu, at 77°F.

Net energy production (or balance) The amount of useful energy produced by a system less the amount of energy required to produce the fuel.

Net generation Equal to gross generation less electricity consumption of a power plant.

Net metering The practice of using a single meter to measure consumption and generation of electricity by a small generation facility (such as a house with a wind or solar photovoltaic system). The net energy produced or consumed is purchased from or sold to the power provider, respectively.

Net present value The value of a personal portfolio, product, or investment after depreciation and interest on debt capital are subtracted from operating income. It can also be thought of as the equivalent worth of all cash flows relative to a base point called the present.

Neutron An electrically neutral subatomic particle in the baryon family with a mass 1839 times that of an electron, stable when bound in an atomic nucleus, and having a mean lifetime of approximately 16. 6 min as a free particle.

New source As defined in the US 40 CFR 122.2, 122.29, and 403.3 (k), a new source is any building, structure, facility, or installation from which there is or may be a discharge of pollutants, the construction of which commenced (1) for purposes of compliance with New Source Performance Standards (NSPS), after the promulgation of such standards being proposed today under Clean Water Act (CWA) section 306, or (2) for the purposes of compliance with pretreatment standards for new sources (PSNS), after the publication of proposed standards under CWA section 307 (c), if such standards are thereafter promulgated in accordance with that section.

New source performance standards See NSPS.

Nitrification It is a process of converting ammonium nitrogen (NH_4^+) into nitrate (NO_3^{2-}) with an intermediate step of producing nitrite (NO_2^-). Nitrification can be accomplished biologically by nitrogen-fixing bacteria (nitrifiers) or accomplished chemically by oxidizing chemicals.

Nitrogen dioxide (NO_2) This compound of nitrogen and oxygen is formed by the oxidation of nitric oxide (NO) which is produced by the combustion of solid fuels.

Nitrogen oxides (NOx) The products of all combustion processes formed by the combination of nitrogen and oxygen.

Nocturnal cooling The effect of cooling by the radiation of heat from a building to the night sky.

Nominal capacity The approximate energy-producing capacity of a power plant, under specified conditions, usually during periods of highest load.

Nominal price The price paid for goods or services at the time of a transaction; a price that has not been adjusted to account for inflation.

Noncombustible refuse (1) Miscellaneous refuse materials that are unburnable at ordinary incinerator temperatures (1,300–2,000°F or 704–1,093°C) (Washington, DC). (2) Refuse materials that are unburnable at ordinary incinerator temperature (800–1,800°F or 427–982°C) such as metals, mineral matter, large quantities of glass or crockery, metal furniture auto bodies or parts, and other similar material or refuse not usual to housekeeping or to operation of stores or offices (Dade County, FL). (3) Refuse materials that are unburnable at ordinary incinerator temperatures (800–1,800°F; 427–982°C) such as metals, mineral matter, large quantities of glass or crockery, metal furniture, auto bodies or parts, and other similar material or refuse not usual to housekeeping or to operation of stores or offices (Miami, FL).

Nonconventional pollutants Pollutants that are neither conventional pollutants nor priority pollutants listed at the US 40 CFR Part 401.

Nonpoint source Diffuse pollution source; a source without a single point of origin or not introduced into a receiving stream from a specific outlet. The pollutants are generally carried off the land by storm water. Common nonpoint sources are agriculture, forestry, urban areas, mining, construction, dams, channels, land disposal, saltwater intrusion, and city streets.

Non-putrescible materials It includes waste discards, dry rubbish, cardboard, wood, lumber, paper products, brick, concrete, steel shavings, metal, plastics, leather, manufactured materials not including oil, petroleum products, paint, liquid chemicals, or paint sludges (Chesterfield Township, MI).

Nonrenewable fuels Fuels that cannot be easily made or "renewed," such as oil, natural gas, and coal.

Nonutility generator/power producer A class of power generator that is not a regulated power provider and that has generating plants for the purpose of supplying electric power required in the conduct of their industrial and commercial operations.

Non-water quality environmental impact Deleterious aspects of control and treatment technologies applicable to point source category wastes, including, but not limited to, air pollution, noise, radiation, sludge and solid waste generation, and energy usage.

Normal recovery capacity A characteristic applied to domestic water heaters that is the amount of gallons raised $100\,^{\circ}F$ per hour (or minute) under a specified thermal efficiency.

Notice of violation A letter of notice from a regional water engineer in response to an instance of significant noncompliance with a NPDES/SPDES permit. Generally, an official notification from a regulatory agency of noncompliance with permit requirements.

NSPS New Source Performance Standards, applicable to new sources of direct dischargers whose construction is begun after the publication of the proposed effluent regulations under Clean Water Act (CWA) section.

N-type semiconductor A semiconductor produced by doping an intrinsic semiconductor with an electron-donor impurity (e.g., phosphorous in silicon).

Nuclear energy Energy that comes from splitting atoms of radioactive materials, such as uranium, and which produces radioactive wastes.

Nuclear power plant A power plant that produces electricity by the force of nuclear power-generated steam through a turbine system that spins a generator. Specifically water is heated through the controlled splitting of uranium atoms in the reactor core and turns to steam. Pumps force the water through the reactor at top speed, maximizing steam production. Steam drives the turbines that turn the generator that makes electricity. Cooling water from the river condenses the steam back into the water. The river water is either discharged directly back to the river or cooled in the cooling towers and reused in the plant.

Nucleus The positively charged central region of an atom, made up of protons and neutrons and containing almost all of the mass of the atom.

Numeric criteria Numeric descriptions of desirable or undesirable water quality conditions.

Nutrients Nutrients are elements required for plant growth that provide biosolids with most of their economic value. These include nitrogen (N), phosphorus (P), potassium (K), calcium (Ca), magnesium (Mg), sodium (Na), sulfur (S), boron (B), copper (Cu), iron (Fe), manganese (Mn), molybdenum (Mo), and zinc (Zn).

Observation (self-assessment) A weakness that, if not corrected, could result in a deficiency. An observation may result if an explicit procedural nonconformance is noted, but the nonconformance is an isolated incident or of minor significance. An observation requires corrective action.

Occupancy sensor An optical, ultrasonic, or infrared sensor that turns room lights on when they detect a person's presence and off after the space is vacated.

Occupied space The space within a building or structure that is normally occupied by people and that may be conditioned (heated, cooled, and/or ventilated).

Ocean energy systems Energy conversion technologies that harness the energy in tides, waves, and thermal gradients in the oceans.

Ocean thermal energy conversion (OTEC) The process or technologies for producing energy by harnessing the temperature differences (thermal gradients) between ocean surface waters and that of ocean depths. Warm surface water is pumped through an evaporator containing a working fluid in a closed Rankine-cycle system. The vaporized fluid drives a turbine/generator. Cold water from deep below the surface is used to condense the working fluid. Open-cycle OTEC technologies use ocean water itself as the working fluid. Closed-cycle OTEC systems circulate a working fluid in a closed loop. A working 10 kW, closed-cycle prototype was developed by the Pacific International Center for High Technology Research in Hawaii with US Department of Energy funding, but was not commercialized.

Offal It includes waste animal matter from butcher and slaughter or packing houses (Baltimore County, MD).

Off-peak The period of low energy demand, as opposed to maximum, or peak, demand.

Off-site Outside the boundaries of a facility.

Ohms A measure of the electrical resistance of a material equal to the resistance of a circuit in which the potential difference of 1 V produces a current of 1 A.

Ohm's law In a given electrical circuit, the amount of current in amperes (i) is equal to the pressure in volts (V) divided by the resistance, in ohms (R).

Oil (fuel) A product of crude oil that is used for space heating, diesel engines, and electrical generation.

Oil spill An accidental spillage of oil on land or surface water.

One sun The maximum value of natural solar insolation.

One-axis tracking A system capable of rotating about one axis.

On-peak energy Energy supplied during periods of relatively high system demands as specified by the supplier.

On-site The same or geographically contiguous property, which may be divided by a public or private right-of-way, provided the entrance and exit between the properties is at a crossroads intersection, and access is by crossing as opposed to going along the right-of-way. Noncontiguous properties owned by the same company or locality but connected by a right-of-way, which it controls, and to which the public does not have access, are also considered on-site properties.

On-site generation Generation of energy at the location where all or most of it will be used.

Open access The ability to send or wheel electric power to a customer over a transmission and distribution system that is not owned by the power generator (seller).

Open dump (1) Any land public or privately owned, other than a sanitary landfill, on which there is deposit and accumulation, either temporary or permanent, of any kind of organic or inorganic refuse, including but not limited to waste materials, waste products, waste paper, garbage, empty cans, broken glass, rags, and all other kinds of organic or inorganic refuse, but excluding scrap for use in manufacturing processes on the premises, or non-putrescible waste materials resulting from such processes, or resulting from the construction or elimination of facilities for such processes (Baltimore County, MD). (2) Any land publicly or privately owned, other than a sanitary landfill, on which there is deposit and accumulation, either temporary or permanent, of any kind of organic or inorganic refuse (Prince George's County, MD). (3) An area on which there is an accumulation of solid waste from one or more sources without proper cover materials (Washington, DC).

Open-circuit voltage The maximum possible voltage across a photovoltaic cell; the voltage across the cell in sunlight when no current is flowing.

Open-ended It refers to an instrument or anything which must be automatically renewed at the end of its given term.

Open-loop geothermal heat pump system Open-loop (also known as "direct") systems circulate water drawn from a ground or surface water source. Once the heat has been transferred into or out of the water, the water is returned to a well or surface discharge (instead of being recirculated through the system). This option is practical where there is an adequate supply of relatively clean water, and all local codes and regulations regarding groundwater discharge are met.

Operating cycle The processes that a work input/output system undergoes and in which the initial and final states are identical.

Operation period It refers to the period of time from the initial receipt of waste at the land disposal facility until the closure period begins.

Organic chemicals, plastics, and synthetic fibers (OCPSF) It refers to the organic chemicals, plastics, and synthetic fibers manufacturing point source category under the US 40 CFR Part 414.

Orientation The alignment of a building along a given axis to face a specific geographical direction. The alignment of a solar collector, in number of degrees east or west of true south.

Outage A discontinuance of electric power supply.

Outfall The end of a drain or pipe that carries wastewater or other effluents into a ditch, pond, or river.

Outgassing The process by which materials expel or release gases.

Outside air Air that is taken from the outdoors.

Outside coil The heat-transfer (exchanger) component of a heat pump, located outdoors, from which heat is collected in the heating mode or expelled in the cooling mode.

Overhang A building element that shades windows, walls, and doors from direct solar radiation and protects these elements from precipitation.

Overload To exceed the design capacity of a device.

Ovonic A device that converts heat or sunlight directly to electricity, invented by Stanford Ovshinsky, which has a unique glass composition that changes from an electrically nonconducting state to a semiconducting state.

Oxygenates Gasoline fuel additives such as ethanol, ETBE, or MTBE that add extra oxygen to gasoline to reduce carbon monoxide pollution produced by vehicles.

P/N A semiconductor (photovoltaic) device structure in which the junction is formed between a p-type layer and an n-type layer.

Packing factor The ratio of solar collector array area to actual land area.

Pane (window) The area of glass that fits in the window frame.

Panel (solar) A term generally applied to individual solar collectors and typically to solar photovoltaic collectors or modules.

Panel radiator A mainly flat surface for transmitting radiant energy.

Panemone A drag-type wind machine that can react to wind from any direction.

Parabolic aluminized reflector lamp A type of lamp having a lens of heavy durable glass that focuses the light. They have longer lifetimes with less lumen depreciation than standard incandescent lamps.

Parabolic dish A solar energy conversion device that has a bowl-shaped dish covered with a highly reflective surface that tracks the sun and concentrates sunlight on a fixed absorber, thereby achieving high temperatures, for process heating or to operate a heat (Stirling) engine to produce power or electricity.

Parabolic trough A solar energy conversion device that uses a trough covered with a highly reflective surface to focus sunlight onto a linear absorber containing a working fluid that can be used for medium temperature space or process heat or to operate a steam turbine for power or electricity generation.

Parallel A configuration of an electrical circuit in which the voltage is the same across the terminals. The positive reference direction for each resistor current is down through the resistor with the same voltage across each resistor.

Parallel connection A way of joining photovoltaic cells or modules by connecting positive leads together and negative leads together; such a configuration increases the current, but not the voltage.

Parameter Any of a set of physical properties whose values determine the characteristics or behavior of something (e.g.,, temperature, pressure, density of air). In relation to environmental monitoring, a monitoring parameter is a constituent of interest. Statistically, the term "parameter" is a calculated quantity, such as a mean or variance, which describes a statistical population.

Parent corporation It refers to a corporation which directly owns at least 50% of the voting stock of a corporation that is the land disposal facility permittee. The latter corporation is a subsidiary of the parent corporation.

Particulates (1) The fine liquid or solid particles contained in combustion gases. The quantity and size of particulates emitted by cars, power and industrial plants, wood stoves, etc., are regulated by the US Environmental Protection Agency. (2) Solid particles and liquid droplets small enough to become airborne.

Pass through A pollutant is determined to "pass through" a POTW when the average percentage removed by an efficiently operated POTW is less than the percentage removed by the industry's direct dischargers that are using the BAT technology.

Passivation A chemical reaction that eliminates the detrimental effect of electrically reactive atoms on a photovoltaic cell's surface.

Passive solar (building) design A building design that uses structural elements of a building to heat and cool a building, without the use of mechanical equipment, which requires careful consideration of the local climate and solar energy resource, building orientation, and landscape features, to name a few. The principal elements include proper building orientation, proper window sizing and placement and design of window overhangs to reduce summer heat gain and ensure winter heat gain, and proper sizing of thermal energy storage mass (e.g., a Trombe wall or masonry tiles). The heat is distributed primarily by natural convection and radiation, though fans can also be used to circulate room air or ensure proper ventilation.

Passive solar heater A solar water or space-heating system in which solar energy is collected and/or moved by natural convection without using pumps or fans. Passive systems are typically integral collector/storage (ICS or batch collectors) or thermosiphon systems. The major advantage of these systems is that they do not use controls, pumps, sensors, or other mechanical parts, so little or no maintenance is required over the lifetime of the system.

Passive solar home A house built using passive solar design techniques.

Passive/natural cooling To allow or augment the natural movement of cooler air from exterior, shaded areas of a building through or around a building.

Pasture Land on which animals feed directly on feed crops such as legumes, grasses, or grain stubble.

Pathogens Pathogens are disease-causing microorganisms that include bacteria, viruses, protozoa, and parasitic worms. Pathogens can present a public health hazard if they are transferred to food crops grown on land to which biosolids are applied, contained in runoff to surface waters from land application sites, or transported away from the site by vectors such as insects, rodents, and birds.

Payback period The amount of time required before the savings resulting from your system equal the system cost.

Pay-in amount It refers to the amount of each payment required to be paid to a trust in each consecutive payment.

Pay-in period It refers to the period of time from the establishment of a trust until it is fully funded.

Payment bond It refers to a surety bond under which the surety company is obligated to pay a sum of money upon the occurrence of a specified event.

Peak clipping/shaving The process of implementing measures to reduce peak power demands on a system.

Peak demand/load The maximum energy demand or load in a specified time period.

Peak power Power generated that operates at a very low capacity factor; generally used to meet short-lived and variable high-demand periods.

Peak shifting The process of moving existing loads to off-peak periods.

Peak sun hours The equivalent number of hours per day when solar irradiance averages 1 kW/m^2. For example, six peak sun hours means that the energy received during total daylight hours equals the energy that would have been received had the irradiance for six hours been 1 kW/m^2.

Peak watt A unit used to rate the performance of a solar photovoltaic (PV) cells, modules, or arrays; the maximum nominal output of a PV device, in watts (Wp) under standardized test conditions, usually 1000 W per square meter of sunlight with other conditions, such as temperature specified.

Peak wind speed The maximum instantaneous wind speed (or velocity) that occurs within a specific period of time or interval.

Peaking capacity Power generation equipment or system capacity to meet peak power demands.

Peaking hydropower A hydropower plant that is operated at maximum allowable capacity for part of the day and is either shut down for the remainder of the time or operated at minimal capacity level.

Pellet stove A space-heating device that burns pellets; are more efficient, clean burning, and easier to operate relative to conventional cord wood-burning appliances.

Pellets Solid fuels made from primarily wood sawdust that is compacted under high pressure to form small (about the size of rabbit feed) pellets for use in a pellet stove.

Pelton turbine A type of impulse hydropower turbine where water passes through nozzles and strikes cups arranged on the periphery of a runner, or wheel, which causes the runner to rotate, producing mechanical energy. The runner is fixed on a shaft, and the rotational motion of the turbine is transmitted by the shaft to a generator. Generally used for high head, low-flow applications.

Penal sum It refers to the face amount of a surety bond.

Penstock A component of a hydropower plant; a pipe that delivers water to the turbine.

Perfluorocarbon tracer gas technique (PFT) An air infiltration measurement technique developed by the Brookhaven National Laboratory to measure changes over time (one week to five months) when determining a building's air infiltration rate. This test cannot locate exact points of infiltration, but it does reveal long-term infiltration problems.

Performance bond It refers to a surety bond under which the surety is obligated to pay a sum of money or perform a specified activity at the occurrence of a specified event.

Performance ratings Solar collector thermal performance ratings based on collector efficiencies, usually expressed in Btu per hour for solar collectors under standard test or operating conditions for solar radiation intensity, inlet working fluid temperatures, and ambient temperatures.

Perimeter heating A term applied to warm-air heating systems that deliver heated air to rooms by means of registers or baseboards located along exterior walls.

Permeance A unit of measurement for the ability of a material to retard the diffusion of water vapor at 73.4° F (23°C). A perm, short for permeance, is the number of grains of water vapor that pass through a square foot of material per hour at a differential vapor pressure equal to one inch of mercury.

Person A legal term which means any individual; public, private, or government corporation; joint stock company; industry; partnership; copartnership; firm; association; trust; estate; public or private institution; group; government agency, department or bureau of the state, or political subdivision thereof; and any legal subsidiary, successor, representative, agent, or agency of the foregoing, or any other legal entity whatsoever.

Personal injury (1) It refers to injury to the body, sickness, or disease, including death, resulting from any of these. (2) It refers to bodily injury as that term is given meaning by applicable state or federal laws. However, this term usually does not include those liabilities which, consistent with standard industry practices, are excluded from coverage in liability policies for bodily injury.

Person-rem The sum of the individual radiation dose equivalents received by members of a certain group or population. It may be calculated by multiplying the average dose per person by the number of persons exposed. For example, a thousand people each exposed to one millirem would have a collective dose of one person-rem.

pH It is a measure of the degree of acidity or alkalinity of a substance, such as water, wastewater, sludge, biosolids, or soil. The pH of biosolids is often raised with alkaline materials to reduce pathogen content and attraction of disease-spreading organisms (vectors). High pH (greater than 11) kills virtually all pathogens and reduces the solubility, biological availability, and mobility of most metals. Lime also increases the gaseous loss (volatilization) of the ammonia form of nitrogen (ammonia-N), thus reducing the N-fertilizer value of biosolids.

Phantom load Any appliance that consumes power even when it is turned off. Examples of phantom loads include appliances with electronic clocks or timers, appliances with remote controls, and appliances with wall cubes (a small box that plugs into an AC outlet to power appliances).

Phase Alternating current is carried by conductors and a ground to residential, commercial, or industrial consumers. The waveform of the phase power appears as a single continuous sine wave at the system frequency whose amplitude is the rated voltage of the power.

Phase change The process of changing from one physical state (solid, liquid, or gas) to another, with a necessary or coincidental input or release of energy.

Phase-change material A material that can be used to store thermal energy as latent heat. Various types of materials have been and are being investigated such as inorganic salts, eutectic compounds, and paraffins, for a variety of applications, including solar energy storage (solar energy heats and melts the material during the day, and at night it releases the stored heat and reverts to a solid state).

Photobiological hydrogen production A hydrogen production process that process uses algae. Under certain conditions, the pigments in certain types of algae absorb solar energy. An enzyme in the cell acts as a catalyst to split water molecules. Some of the bacteria produces hydrogen after they grow on a substrate.

Photocurrent An electric current induced by radiant energy.

Photoelectric cell A device for measuring light intensity that works by converting light falling on, or reach it, to electricity and then measuring the current; used in photometers.

Photoelectrochemical cell A type of photovoltaic device in which the electricity induced in the cell is used immediately within the cell to produce a chemical, such as hydrogen, which can then be withdrawn for use.

Photoelectrolysis hydrogen production The production of hydrogen using a photoelectrochemical cell.

Photogalvanic processes The production of electrical current from light.

Photon A particle of light that acts as an individual unit of energy.

Photovoltaic (conversion) efficiency The ratio of the electric power produced by a photovoltaic device to the power of the sunlight incident on the device.

Photovoltaic (PV; solar) array A group of solar photovoltaic modules connected together.

Photovoltaic (solar) cell Treated semiconductor material that converts solar irradiance to electricity.

Photovoltaic (solar) module or panel A solar photovoltaic product that generally consists of groups of PV cells electrically connected together to produce a specified power output under standard test conditions, mounted on a substrate, sealed with an encapsulant, and covered with a protective glazing. Maybe further mounted on an aluminum frame. A junction box, on the back or underside of the module, is used to allow for connecting the module circuit conductors to external conductors.

Photovoltaic (solar) system A complete PV power system composed of the module (or array) and balance-of-system (BOS) components including the array supports, electrical conductors/wiring, fuses, safety disconnects, and grounds, charge controllers, inverters, battery storage, etc.

Photovoltaic device A solid-state electrical device that converts light directly into current electricity of voltage-current characteristics that are a function of the characteristics of the light source and the materials in and design of the device. Solar photovoltaic devices are made of various semiconductor materials including silicon, cadmium sulfide, cadmium telluride, and gallium arsenide and in single crystalline, multi-crystalline, or amorphous forms.

Photovoltaic peak watt See peak watt.

Photovoltaic-thermal (PV/T) systems A solar energy system that produces electricity with a PV module and collects thermal energy from the module for heating.

Physical vapor deposition A method of depositing thin semiconductor photovoltaic films. With this method, physical processes, such as thermal evaporation or bombardment of ions, are used to deposit elemental semiconductor material on a substrate.

Phytoremediation A land restoration method involving the use of plants for removal of heavy metals, toxic organics, and other pollutants from soil.

P-I-N A semiconductor (photovoltaic) device structure that layers an intrinsic semiconductor between a p-type semiconductor and an n-type semiconductor; this structure is most often used with amorphous silicon PV devices.

Pitch control A method of controlling a wind turbine's speed by varying the orientation, or pitch, of the blades and thereby altering its aerodynamics and efficiency.

Plant available nitrogen (PAN) Only a portion of the total nitrogen present in biosolids/manure is available for plant uptake. This *plant available nitrogen (PAN)* is the actual amount of N in the biosolids/manure that is available to crops during a specified period.

Planting and harvesting periods The cycle of crop *planting and harvesting periods*, not the calendar year, dictates the timing of biosolids and manure land application activities. Winter wheat and perennial forage grasses are examples of crops that may be established and harvested in different calendar years.

Planting season The basic time management unit is often called the *crop year* or *planting season*. The *crop year* is defined as the year in which a crop receiving the biosolids/manure treatment is harvested.

Play It refers to a set of oil or gas accumulations sharing similar geologic and geographic properties, such as source rock, hydrocarbon type, and migration pathways.

Plenum The space between a hanging ceiling and the floor above or roof; usually contains HVAC ducts, electrical wiring, fire suppression system piping, etc.

Plug flow digester A type of anaerobic digester that has a horizontal tank in which a constant volume of material is added and forces material in the tank to move through the tank and be digested.

Plume The distribution of a pollutant in air or water after being released from a source.

Point source (1) Any discernable, confined, and discrete conveyance from which pollutants are or may be discharged. (2) A stationary location or fixed facility from which pollutants are discharged; any single identifiable source of pollution, such as a pipe, ditch, ship, ore pit, or factory smokestack.

Point-contact cell A high-efficiency silicon photovoltaic concentrator cell that employs light trapping techniques and point-diffused contacts on the rear surface for current collection.

Poisonous materials It includes poison grain, insecticides, or insecticide containers or other similar materials (Beaverton, OR).

Pollutant A contaminant in a concentration or amount that adversely alters the physical, chemical, or biological properties of the natural environment.

Pollutant concentration limits (PCL) *Pollutant concentration limits* are the maximum concentrations of heavy metals for biosolids whose trace element pollutant additions do not require tracking (i.e., calculation of CPLR (cumulative pollutant loading rate)). PCL are the most stringent pollutant limits included in US Federal Regulation Part 503 for land application. Biosolids meeting pollutant concentration limits are subject to fewer requirements than biosolids meeting ceiling concentration limits.

Pollutants of interest Pollutants commonly found in landfill-generated wastewaters. For the purposes of a specific environmental or energy project, a pollutant of interest is a pollutant that is detected three or more times above a treatable level at an environmental facility and must be present at more than one facility.

Polycrystalline A semiconductor (photovoltaic) material composed of variously oriented, small, individual crystals.

Polyethylene A registered trademark for plastic sheeting material that can be used as a vapor retarder. This plastic is used to make grocery bags. It is a long chain of carbon atoms with 2 hydrogen atoms attached to each carbon atom.

Polystyrene (See foam insulation)

Porous media A solid that contains pores; normally, it refers to interconnected pores that can transmit the flow of fluids. (The term refers to the aquifer geology when discussing sites for CAES.)

Portfolio standard The requirement that an electric power provider generate or purchase a specified percentage of the power it supplies/sells from renewable energy resources and thereby guarantee a market for electricity generated from renewable energy resources.

Post-closure period It refers to the period of time after completion of closure and before the beginning of the institutional control period during which the permittee maintains control of the land disposal facility and carries out a program of monitoring and maintenance as required by this part to ensure that the land disposal facility is stable and ready for institutional control.

Potable water Water that is suitable for drinking, as defined by local health officials.

Potential energy Energy available due to position.

POTW (Publicly owned treatment works) See publicly owned treatment works.

Pound of steam It is one pound (0.454 kg) of water in vapor phase; It is NOT steam pressure, which is expressed as pounds per square inch (psi).

Power Energy that is capable or available for doing work; the time rate at which work is performed, measured in horsepower, watts, or Btu per hour. Electric power is the product of electric current and electromotive force.

Power (output) curve A plot of a wind energy conversion device's power output versus wind speed.

Power (solar) tower A term used to describe solar thermal, central receiver, power systems, where an array of reflectors focus sunlight onto a central receiver and absorber mounted on a tower.

Power coefficient The ratio of power produced by a wind energy conversion device to the power in a reference area of the free wind stream.

Power conditioning The process of modifying the characteristics of electrical power (e.g., inverting dc to ac).

Power density The amount of power per unit area of a free wind stream.

Power factor (PF) The ratio of actual power being used in a circuit, expressed in watts or kilowatts, to the power that is apparently being drawn from a power source, expressed in volt-amperes or kilovolt-amperes.

Power generation mix The proportion of electricity distributed by a power provider that is generated from available sources such as coal, natural gas, petroleum, nuclear, hydropower, wind, or geothermal.

Power plant A facility which is used for generation of electricity by using wind energy (wind power plant), hydraulic energy (hydroelectric power plant and tidal power plant), coal energy (coal-fired power plant), natural gas energy (natural gas-fired power plant), solar energy (solar power plant), refuse-derived fuel (RDF power plant), or nuclear energy (nuclear power plant).

Power provider A company or other organizational units that sell and distribute electrical power (e.g., private or public electrical utility), either to other distribution and wholesale businesses or to end users. Sometimes power providers also generate the power they sell.

Power towers See central receiver solar power plants.

Power transmission line An electrical conductor/cable that carries electricity from a generator to other locations for distribution.

Precision The degree of reproducibility of a measurement under a given set of conditions. Precision in a data set is assessed by evaluating results from duplicate field or analytical samples.

Pre-closure plan It refers to the period from the effective date of a permit issued by the state environmental agency for the land disposal facility until facility closure activities have begun. This period includes the construction phase and operation period.

Preheater (solar) A solar heating system that preheats water or air that is then heated more by another heating appliance.

Premises It includes a building, with any fences, walls, sheds, garages, or other accessory buildings appurtenant to such building, and the area of land surrounding the building and actually or by legal construction forming one enclosure in which such building is located.

Preparer Either the person who generates sewage sludge during the treatment of domestic sewage in a treatment works or the person who derives a material from sewage sludge.

Present value The amount of money required to secure a specified cash flow at a future date at a specified return.

Pressure drop The loss in static pressure of a fluid (liquid or gas) in a system due to friction from obstructions in pipes, from valves, fittings, regulators, burners, etc., or by a breech or rupture of the system.

Pressurization testing A technique used by energy auditors, using a blower door, to locate areas of air infiltration by exaggerating the defects in the building shell. This test only measures air infiltration at the time of the test. It does not take into account changes in atmospheric pressure, weather, wind velocity, or any activities the occupants conduct that may affect air infiltration rates over a period of time.

Pretreatment standards for existing sources of indirect discharges (PSES) See PSES.

Pretreatment standards for new sources of indirect discharges (PSNS) See PSNS.

Primary air The air that is supplied to the combustion chamber of a furnace.

Primary public water supply aquifer It refers to a highly productive water-bearing formation identified by the department consisting of unconsolidated (non-bedrock) geologic deposits, which (1) receives substantial recharge from the overlying land surface and (2) is presently utilized as a major source of water for public water supply.

Prime mover Any machine capable of producing power to do work.

Principal aquifer It means unconsolidated (non-bedrock) geologic deposits identified by the state or federal government (1) receive substantial recharge from the overlying land surface, (2) are known to be highly productive or whose geology suggests a potentially abundant source of water, and (3) are not presently used as a major source of water for public water supply.

Priority pollutant One hundred twenty-six compounds that are a subset of the 65 toxic pollutants and classes of pollutants outlined in the US Section 307 of the Clean Water Act (CWA). The priority pollutants are specified in the US Natural Resources Defense Council (NRDC) settlement agreement.

Private waste collector It refers to any person, firm, or corporation who collects bulk refuse, household waste, building waste, or building debris from premises owned or occupied by him within the corporation limits of the city and transports said wastes over the streets of the city (Bettendorf, IA).

Process heat Thermal energy that is used in agricultural and industrial operations.

Produced water It means the mixture of water containing both the returned fracturing fluid and natural formation water is produced along with the natural gas and moves back through the wellhead with the gas.

Producer gas Low or medium Btu content gas, composed mainly of carbon monoxide, nitrogen(N_2), and hydrogen(H_2) made by the gasification of wood or coal.

Products of combustion The elements and compounds that result from the combustion of a fuel.

Proglacial lake A lake occupying a basin in front of a glacier; generally in direct contact with the ice.

Programmable thermostat A type of thermostat that allows the user to program into the devices' memory a preset schedule of times (when certain temperatures occur) to turn on HVAC equipment.

Projected area The net south-facing glazing area projected on a vertical plane. Also, the solid area covered at any instant by a wind turbine's blades from the perspective of the direction of the wind stream (as opposed to the swept area).

Propane A hydrocarbon gas, C_3H_8, occurring in crude oil, natural gas, and refinery cracking gas. It is used as a fuel, a solvent, and a refrigerant. Propane liquefies under pressure and is the major component of liquefied petroleum gas (LPG).

Propeller (hydro) turbine A turbine that has a runner with attached blades similar to a propeller used to drive a ship. As water passes over the curved propeller blades, it causes rotation of the shaft.

Proppant (propping agent) It refers to a granular substance (such as sand grains, aluminum pellets, or other material) that is carried in suspension by the fracturing fluid and keeps the cracks open when fracturing fluid is withdrawn after a fracture treatment.

Prospective case study It refers to an investigation (1) which is conducted at the sites where hydraulic fracturing will occur after the research is initiated, (2) which allows sampling and characterization of the site prior to, and after, water extraction, drilling, hydraulic fracturing fluid injection, flowback, and gas production, and which collects data during prospective case studies, allowing US EPA to evaluate changes in water quality over time and to assess the fate and transport of chemical contaminants.

Proton A stable, positively charged subatomic particle in the baryon family with a mass of 1836 times that of an electron.

Proximate analysis A commonly used analysis for reporting fuel properties; may be on a dry (moisture free) basis, as "fired," or on an ash- and moisture-free basis. Fractions usually reported include volatile matter, fixed carbon, moisture, ash, and heating value (higher heating value).

PSES Pretreatment standards for existing sources of indirect discharges, under the US Sec. 307(b) of the Clean Water Act (CWA).

Pseudo-monitoring point A theoretical monitoring location rather than an actual physical location; a calculation based on analytical test results of samples obtained from other associated, tributary monitored locations. It is classified as a "pseudo" monitoring point because samples are not actually physically collected at that location.

PSNS Pretreatment standards for new sources of indirect discharges, applicable to new sources whose construction has begun after the publication of proposed standards under Clean Water Act (CWA) section 307 (c), if such standards are thereafter promulgated in accordance with that section.

Psychrometer An instrument for measuring relative humidity by means of wet and dry-bulb temperatures.

Psychrometrics The analysis of atmospheric conditions, particularly moisture in the air.

Psychrophilic It is a state in an anaerobic reactor, such as anaerobic digester or composting unit where/when the temperature remains below 20°C.

P-type semiconductor A semiconductor in which holes carry the current; produced by doping an intrinsic semiconductor with an electron acceptor impurity (e.g., boron in silicon).

Public contact site Land with a high potential for contact by the public, including public parks, ball fields, cemeteries, nurseries, turf farms, and golf courses.

Public disposal It refers to the disposal of garbage and rubbish which has been removed from premises used, owned, or leased by one or more persons, firms, corporations, or associations and transported to other premises and disposed either with or without the payment of a fee (Bluffton, IN).

Public utilities regulatory policy act (PURPA) of 1978 A US law that requires electric utilities to purchase electricity produced from qualifying power producers that use renewable energy resources or are cogenerators. Power providers are required to purchase power at a rate equal to the avoided cost of generating the power themselves (see avoided costs and qualifying facility).

Public utility holding company act (PUHCA) of 1935 A US law to protect consumers and investors. It placed geographic restrictions on mergers and limitations on diversification into nonutility lines of business and takeovers of electric and gas utilities and also established regulated monopoly markets or service territories for utilities.

Public utility or services commissions (PUC or PSC) These are state government agencies in the USA responsible for the regulation of public utilities within a state or region. A state legislature oversees the PUC by reviewing changes to power generator laws, rules, and regulations and approving the PUC's budget. The commission usually has five commissioners appointed by the governor or legislature. PUCs typically regulate electric, natural gas, water, sewer, telephone services, trucks, buses, and taxicabs within the commission's operating region. The PUC tries to balance the interests of consumers, environmentalists, utilities, and stockholders. The PUC makes sure a region's citizens are supplied with adequate, safe power provider service at reasonable rates.

Public water system It refers to a system for providing the public with water for human consumption (through pipes or other constructed conveyances) that has at least 15 service connections or regularly serves at least 25 individuals.

Publicly owned treatment works (POTW) Any device or system, owned by a state or municipality, used in the treatment (including recycling and reclamation) of municipal sewage or industrial wastes of a liquid nature that is owned by a state or municipality. This includes sewers, pipes, or other conveyances only if they convey wastewater to a POTW providing treatment.

Pulse-width modulated (PWM) wave inverter A type of power inverter that produces a high-quality (nearly sinusoidal) voltage at minimum current harmonics.

Pumped storage facility A type of power-generating facility that pumps water to a storage reservoir during off-peak periods and uses the stored water (by allowing it to fall through a hydroturbine) to generate power during peak periods. The pumping energy is typically supplied by lower cost base power capacity, and the peaking power capacity is of greater value, even though there is a net loss of power in the process.

Putrescible materials or wastes (1) It includes garbage, produce, food products, fruit, vegetables, chemicals, oil, petroleum products, paints, liquid chemicals, and paint sludge (Chesterfield Township, MI), and (2) they include wastes that are capable of being decomposed by microorganisms with sufficient rapidity as to cause nuisances from odors, gases, and similar objectionable conditions. Kitchen's wastes, offal, and dead animals are examples of putrescible components of solid waste (Washington, DC).

Pyranometer A device used to measure total incident solar radiation (direct beam, diffuse, and reflected radiation) per unit time per unit area.

Pyrheliometer A device that measures the intensity of direct beam solar radiation.

Pyrolysis The transformation on a compound or material into one or more substances by heat alone (without oxidation). Often called destructive distillation. Pyrolysis of biomass is the thermal degradation of the material in the absence of reacting gases and occurs prior to or simultaneously with gasification reactions in a gasifier. Pyrolysis products consist of gases, liquids, and char generally. The liquid fraction of pyrolisized biomass consists of an insoluble viscous tar and pyroligneous acids (acetic acid, methanol, acetone, esters, aldehydes, and furfural). The distribution of pyrolysis products varies depending on the feedstock composition, heating rate, temperature, and pressure.

Pyrophoric It refers to, when referring to a liquid, any liquid that ignites spontaneously in dry or moist air at or below 130°F (54.4°C). A pyrophoric solid is any solid material, other than one classified as an explosive, which, under normal conditions, is liable to cause fires through friction, retained heat from manufacturing or processing, or which can be ignited readily and, when ignited, burns so vigorously and persistently as to create a serious transportation, handling, or disposal hazard. Included are spontaneously combustible and water-reactive materials.

Quad One quadrillion Btu. (1,000,000,000,000,000 Btu)

Qualifying facility A category of electric power producer established under the US Public Utility Regulatory Policy Act (PURPA) of 1978, which includes small-power producers (SPP) who use renewable sources of energy such as biomass, geothermal, hydroelectricity, solar (thermal and photovoltaic), and wind or cogenerators who produce both heat and electricity using any type of fuel. PURPA requires utilities to purchase electricity from these power producers at a rate approved by a state utility regulatory agency under federal guidelines. PURPA also requires power providers to sell electricity to these producers. Some states have developed their own programs for SPPs and utilities.

Quality factor The extent of tissue damage caused by different types of radiation of the same energy. The greater the damage, the higher the quality factor. More specifically, the factor by which absorbed doses are multiplied to obtain a quantity that indicates the degree of biological damage produced by ionizing radiation (see radiation dose). The factor is dependent upon radiation type (alpha, beta, gamma, or x-ray) and exposure (internal or external).

Rad Radiation-absorbed dose. One hundred ergs of energy absorbed per gram.

Radiant barrier A thin, reflective foil sheet that exhibits low radiant energy transmission and under certain conditions can block radiant heat transfer; installed in attics to reduce heat flow through a roof assembly into the living space.

Radiant ceiling panels Ceiling panels that contain electric resistance heating elements embedded within them to provide radiant heat to a room.

Radiant energy Energy that transmits away from its source in all directions.

Radiant floor A type of radiant heating system where the building floor contains channels or tubes through which hot fluids such as air or water are circulated. The whole floor is evenly heated. Thus, the room heats from the bottom up. Radiant floor heating eliminates the draft and dust problems associated with forced air heating systems.

Radiant heating system A heating system where heat is supplied (radiated) into a room by means of heated surfaces, such as electric resistance elements, hot water (hydronic) radiators, etc.

Radiation (1) The transfer of heat through matter or space by means of electromagnetic waves. (2) The process of emitting energy in the form of rays or particles that are thrown off by disintegrating atoms. The rays or particles emitted may consist of alpha, beta, or gamma radiation.

Radiation dose The amount of energy from any kind of ionizing radiation: (1) absorbed dose, (2) collective dose equivalent, (3) collective effective dose equivalent, (4) committed dose equivalent, and (5) committed effective dose equivalent.

Radiative cooling The process of cooling by which a heat-absorbing media absorb heat from one source and radiate the heat away.

Radiator A room heat delivery (or exchanger) component of a hydronic (hot water or steam) heating system; hot water or steam is delivered to it by natural convection or by a pump from a boiler.

Radiator vent A device that releases pressure within a radiator when the pressure inside exceeds the operating limits of the vent.

Radioactive waste Materials left over from making nuclear energy. Radioactive waste can kill or harm living organisms, animals and plants if it is not stored safely.

Radioactivity A property possessed by some elements (such as uranium) whereby alpha, beta, or gamma rays are spontaneously emitted.

Radioisotope A radioactive isotope of a specified element. Carbon-14 is a radioisotope of carbon. Tritium is a radioisotope of hydrogen (see isotope).

Radionuclide A radioactive nuclide. Radionuclides are variations (isotopes) of elements. They have the same number of protons and electrons but different numbers of neutrons, resulting in different atomic masses. There are several hundred known nuclides, both man-made and naturally occurring.

Radon A naturally occurring radioactive gas found in the USA in nearly all types of soil, rock, and water. It can migrate into most buildings. Studies have linked high concentrations of radon to lung cancer.

Rafter A construction element used for ceiling support.

Rammed earth A construction material made by compressing earth in a form; used traditionally in many areas of the world and widely throughout North Africa and the Middle East.

Range land Open land with indigenous vegetation.

Rankine cycle The thermodynamic cycle that is an ideal standard for comparing performance of heat engines, steam power plants, steam turbines, and heat pump systems that use a condensable vapor as the working fluid; efficiency is measured as work done divided by sensible heat supplied.

Rate schedule A mechanism used by electric utilities to determine prices for electricity; typically defines rates according to amounts of power demanded/consumed during specific time periods.

Rated life The length of time that a product or appliance is expected to meet a certain level of performance under nominal operating conditions; in a luminaire, the period after which the lumen depreciation and lamp failure is at 70 % of its initial value.

Rated power The power output of a device under specific or nominal operating conditions.

Rayleigh frequency distribution A mathematical representation of the frequency or ratio that specific wind speeds occur within a specified time interval.

RCRA The Resource Conservation and Recovery Act of 1976 (RCRA) (42 U.S.C. Section 6901) which regulates the generation, treatment, storage, disposal, or recycling of solid and hazardous wastes.

Reactive power The electrical power that oscillates between the magnetic field of an inductor and the electrical field of a capacitor. Reactive power is never converted to nonelectrical power. Calculated as the square root of the difference between the square of the kilovolt-amperes and the square of the kilowatts. Expressed as reactive volt-amperes.

Real price The unit price of a good or service estimated from some base year in order to provide a consistent means of comparison.

Receiver The component of a central receiver solar thermal system where reflected solar energy is absorbed and converted to thermal energy.

Recirculated air Air that is returned from a heated or cooled space, reconditioned and/or cleaned, and returned to the space.

Recirculation systems A type of solar heating system that circulates warm water from storage through the collectors and exposes piping whenever freezing conditions occur, obviously a not very efficient system when operating in this mode.

Reclamation site Drastically disturbed land, such as strip mines and construction sites, which is reclaimed using sewage sludge.

Recovery It means removing from a disposal unit waste that has been permanently disposed in a land disposal facility.

Rectifier An electrical device for converting alternating current to direct current. The chamber in a cooling device where water is separated from the working fluid (e.g., ammonia).

Recuperator A heat exchanger in which heat is recovered from the products of combustion.

Recurrent costs Costs that are repetitive and occur when an organization produces similar goods or services on a continuing basis.

Recycling (1) The process of converting materials that are no longer useful as designed or intended into a new product. (2) It refers to the recovery of useful materials, such as paper, glass, plastic, and metals, from the trash to be used in making new products, reducing the amount of virgin raw materials needed (US EPA).

Redox (reduction-oxidation) reaction It is a chemical reaction involving transfer of electrons from one element to another.

Reflectance The amount (percent) of light that is reflected by a surface relative to the amount that strikes it.

Reflective coatings Materials with various qualities that are applied to glass windows before installation. These coatings reduce radiant heat transfer through the window and also reflect outside heat and a portion of the incoming solar energy, thus reducing heat gain. The most common type has a sputtered coating on the inside of a window unit. The other type is a durable "hard-coat" glass with a coating, baked into the glass surface.

Reflective glass A window glass that has been coated with a reflective film and is useful in controlling solar heat gain during the summer.

Reflective insulation (see also radiant barrier) An aluminum foil-fabricated insulator with backings applied to provide a series of closed air spaces with highly reflective surfaces.

Reflective window films A material applied to window panes that controls heat gain and loss, reduces glare, minimizes fabric fading, and provides privacy. These films are retrofitted on existing windows.

Reflector lamps A type of incandescent lamp with an interior coating of aluminum that reflects light to the front of the bulb. They are designed to spread light over specific areas.

Refraction The change in direction of a ray of light when it passes through one media to another with differing optical densities.

Refrigerant The compound (working fluid) used in air conditioners, heat pumps, and refrigerators to transfer heat into or out of an interior space. This fluid boils at a very low temperature, enabling it to evaporate and absorb heat.

Refrigeration The process of the absorption of heat from one location and its transfer to another for rejection or recuperation.

Refrigeration capacity A measure of the effective cooling capacity of a refrigerator, expressed in Btu per hour or in tons, where one (1) ton of capacity is equal to the heat required to melt 2,000 lb of ice in 24 h or 12,000 Btu per hour.

Refrigeration cycle The complete cycle of stages (evaporation and condensation) of refrigeration or of the refrigerant.

Refuse container It is a metal or nonabsorbent and fire-resistant container, which shall be equipped with a tightly fitting n~tal or nonabsorbent and fire-resistant cover or lid, but shall not include incinerators other than those herein defined, or ash pits (Boulder, CO).

Refuse disposal site It refers to any location designated by the city where any approved final treatment utilization, processing, or depository of solid wastes occurs (Boulder, CO).

Refuse hauler It refers to any person engaged in the business of collecting, storing, and transporting refuse in the city and who is licensed thereof by the city (Boulder, CO).

Refuse or solid waste (1) All solid wastes (Bettendorf, IA); (2) all waste materials, combustible or noncombustible, from all public and private establishments and residences, other than sewage (Baltimore County, MD); (3) all putrescible and non-putrescible solid wastes (except body wastes) including any and all garbage, rejected or waste food, offal, swill, ashes, slop, rubbish, and waste or unwholesome material of every kind and character (Los Alamos County, NM); (4) all solid wastes (Beaverton, OR); (5) all solid wastes, garbage, and rubbish, whether combustible or noncombustible, including rubble (Boulder, CO); (6) solid waste that includes both garbage and rubbish (Sonoma County, CA); and (7) all putrescible and non-putrescible solid wastes, except body wastes, and including abandoned vehicles, food waste (garbage), rubbish, ashes, incinerator residue, street cleanings, tree debris, and solid market and industrial wastes (Washington, DC). Also see solid waste.

Refuse-derived fuel (RDF) A solid fuel produced by shredding municipal solid waste (MSW). Noncombustible materials such as glass and metals are generally removed prior to making RDF. The residual material is sold as is or compressed into pellets, bricks, or logs. RDF processing facilities are typically located near a source of MSW, while the RDF combustion facility can be located elsewhere. Existing RDF facilities process between 100 and 3,000 t per day.

Refuse-derived fuel (thermoelectric) power plant A power plant that produces electricity by the force of steam through a turbine that spins a generator. The steam is produced by burning the refuse-derived fuel (RDF).

Regenerative cooling A type of cooling system that uses a charging and discharging cycle with a thermal or latent heat storage subsystem.

Regenerative heating The process of using heat that is rejected in one part of a cycle for another function or in another part of the cycle.

Relamping The replacement of a nonfunctional or ineffective lamp with a new, more efficient lamp.

Relative humidity A measure of the percent of moisture actually in the air compared with what would be in it if it were fully saturated at that temperature. When the air is fully saturated, its relative humidity is 100%.

Reliability This is the concept of how long a device or process can operate properly without needing maintenance or replacement.

Rem (1) It refers to a unit of dose equivalent for any type of ionizing radiation absorbed by the body tissue in terms of its estimated biological effect relative to an exposure of one roentgen of x-rays or gamma rays. The dose equivalent in rems is numerically equal to the absorbed dose in rads multiplied by the quality factor, distribution factor, and any other necessary modifying factors. (2) An acronym for roentgen equivalent man. A unit of radiation exposure that indicates the potential effect of radiation on human cells.

Remediation An action, process, or method for remediating/correcting the environment (usually land and groundwater).

Remediation of failures It refers to land disposal facility repair and environmental cleanup and restoration of the land disposal facility and other affected areas including but not limited to activities undertaken to eliminate, remove, abate, control, or monitor actual or potential health and environmental hazards.

Remote-handled waste At the facility, waste that has an external surface dose rate that exceeds 100 millirem per hour or a high level of alpha and/or beta surface contamination.

Renewable energy Energy derived from resources that are regenerative or for all practical purposes cannot be depleted. Types of renewable energy resources include moving water (hydro, tidal, and wave power), thermal gradients in ocean water, biomass, geothermal energy, solar energy, and wind energy. Municipal solid waste (MSW) is also considered to be a renewable energy resource.

Residential unit It includes any structure or shelter or any part thereof used, or constructed for use, as a residence for one family.

Residential well It refers to a pumping well that serves one home or is maintained by a private owner.

Residue It includes the solid materials remaining after burning, comprising ash, metal, glass, ceramics, and unburned organic substances (Washington, DC).

Resistance The inherent characteristic of a material to inhibit the transfer of energy. In electrical conductors, electrical resistance results in the generation of heat. Electrical resistance is measured in ohms. The heat-transfer resistance properties of insulation products are quantified as the R-value.

Resistance heating A type of heating system that provides heat from the resistance of an electrical current flowing through a conductor.

Resistive voltage drop The voltage developed across a cell by the current flow through the resistance of the cell.

Resistor An electrical device that resists electric current flow.

Resource conservation and recovery act See RCRA.

Resource recovery The process of converting municipal solid waste to energy and/or recovering materials for recycling.

Restructuring The process of changing the structure of the electric power industry from one of guaranteed monopoly over service territories, as established by the Public Utility Holding Company Act of 1935, to one of open competition between power suppliers for customers in any area.

Retail wheeling A term for the process of transmitting electricity over transmission lines not owned by the supplier of the electricity to a retail customer of the supplier. With retail wheeling, an electricity consumer can secure their own supply of electricity from a broker or directly from the generating source. The power is then wheeled at a fixed rate or at a regulated "nondiscriminatory" rate set by a utility commission.

Retrieval It refers to the recovery of waste in an intact container.

Retrofit The process of modifying a building's structure.

Retrospective case study It refers to (1) a study of sites that have had active hydraulic fracturing practices, with a focus on sites with reported instances of drinking water resource contamination or other impacts in areas where hydraulic fracturing has already occurred, or (2) a study that uses existing data and possibly field sampling, modeling, and/or parallel laboratory investigations to determine whether reported impacts are due to existing, ongoing hydraulic fracturing activities.

Return air Air that is returned to a heating or cooling appliance from a heated or cooled space.

Return duct The central heating or cooling system contains a fan that gets its air supply through these ducts, which ideally should be installed in every room of the house. The air from a room will move toward the lower pressure of the return duct.

Reverse thermosiphoning When heat seeks to flow from a warm area (e.g., heated space) to a cooler area, such as a solar air collector at night without a reverse flow damper.

Reversing valve A component of a heat pump that reverses the refrigerant's direction of flow, allowing the heat pump to switch from cooling to heating or heating to cooling.

R-factor See R-value.

Ribbon (photovoltaic) cells A type of solar photovoltaic device made in a continuous process of pulling material from a molten bath of photovoltaic material, such as silicon, to form a thin sheet of material.

Rigid insulation board An insulation product made of a fibrous material or plastic foams, pressed or extruded into boardlike forms. It provides thermal and acoustical insulation strength with low weight and coverage with few heat-loss paths.

Rock bin A container that holds rock used as the thermal mass to store solar energy in a solar heating system.

Rock wool A type of insulation made from virgin basalt, an igneous rock, and spun into loose fill or a batt. It is fire resistant and helps with soundproofing.

Roof A building element that provides protection against the sun, wind, and precipitation.

Roof pond A solar energy collection device consisting of containers of water located on a roof that absorbs solar energy during the day so that the heat can be used at night or that cools a building by evaporation at night.

Roof ventilator A stationary or rotating vent used to ventilate attics or cathedral ceilings; usually made of galvanized steel or polypropylene.

Rotor An electric generator consists of an armature and a field structure. The armature carries the wire loop, coil, or other windings in which the voltage is induced, whereas the field structure produces the magnetic field. In small generators, the armature is usually the rotating component (rotor) surrounded by the stationary field structure (stator). In large generators in commercial electric power plants, the situation is reversed.

Rubbish (1) All solid waste other than garbage, offal, and ashes from homes, hotels, stores, institutions, markets, and other establishments (Prince George's County, MD). (2) All refuse accumulations of paper, excelsior, rags or wooden or paper boxes or containers, sweepings, and all other accumulations of a nature other than garbage, which are usual to housekeeping and to the operation of stores, offices, and other business places, and also any bottles, cans, or other containers which, due to their ability to retain water, may serve as breeding places for mosquitoes or other water-breeding insects; rubbish shall not include noncombustible refuse, as defined above (Dade County, FL). (3) All refuse other than garbage (tin cans, bottles, ashes, paper, pasteboard, cardboard or wooden boxes, brush, weeds, leaves and cuttings from trees, lawns, shrubs, and garden, or other waste materials provided in the normal course of living) (Beaverton, OR). (4) All cardboard, plastic, metal or glass containers, waste paper, rags, sweepings, small pieces of wood, excelsior, rubber, leather, leaves, lawn cuttings, tree trimmings and hedge trimmings not over five inches (12.7 cm) in diameter nor over four feet (1.22 m) in length, tree roots or stumps small enough to be contained in a bushel basket, articles not more than eight feet (2.44 m) in length nor more than fifty pounds (22.7 kg) in weight, and small rocks and similar waste materials that ordinarily accumulate around a home, business, or industry. It shall not include garbage, ashes, bulk refuse, dead animals, hazardous refuse, industrial waste, or building waste (Bettendorf, IA). (5) All ashes, cans, metalware, broken glass, crockery, dirt sweepings, boxes, wood, grass, weeds, or litter or any kind (Bluffton, IN). (6) All non-putrescible waste materials such as paper, cartons, rags, boxes, excelsior, rubber, leather, tree branches, tin cans, bottles, scrap automotive bodies and other metallic junk, mineral matter and street sweepings, crockery, discarded furniture, dirt, ashes, and all other refuse not included in the term garbage (Los Alamos County, NM). (7) Solid waste that includes non-putrescible wastes, including, but not limited to, unusable, unwanted, or discarded material and debris resulting from normal community or business activities or materials which by their presence may injuriously affect the health, safety, and comfort of persons or may depreciate property values in its vicinity or both (Sonoma County, CA). (8) All non-putrescible solid wastes, consisting of both combustible and noncombustible wastes, including, but not limited to, paper, ashes, cardboard, tin cans, yard clippings, wood, glass, rags, discarded clothes or wearing apparel of any kind, or any other discarded object or thing, not exceeding three feet (0.91 m) in length (Boulder, CO). (9) All solid waste, other than garbage, offal, and ashes, from homes, hotels, stores, institutions, markets, and other establishments; further classified as combustible and noncombustible (Baltimore County, MD). (10) All refuse accumulations of paper, excelsior, rags, or wooden or paper boxes or containers, sweepings, and all other accumulations of a nature other than garbage, which are usual to housekeeping and to the operation of stores, offices, and other business places, and also any bottles, cans, or other containers which, due to their ability to retain water, may serve as breeding places for mosquitoes or other water-breeding insects; rubbish shall not include noncombustible refuse (Miami, FL). (11) All non-putrescible solid wastes, including ashes, consisting of both combustible and noncombustible wastes, such as paper, cardboard, tin cans, yard rubbish, wood, glass, bedding, crockery, or litter of any kind (Washington, DC). and (12) all refuse such as paper, tin cans, bottles, glass containers, rags, ashes, lawn trimmings, tree trimmings, tree branches, limbs, tree trunks and stumps, and waste materials from premises including that produced from remodeling or construction, paper sacks, boxes, packing materials and like materials from dwellings, and business, commercial, or industrial establishments and the offices thereof, except the following: (12a) garbage; (12b) sewage; (12c) dirt, rock, and concrete or masonry materials; (12d) accumulations from mud traps and settling basins; (12e) dead animals or animal excrement; and (12f) salvage materials (Wichita, KS).

Rubble It includes large brush wood, large cardboard boxes or parts thereof, large and/or heavy yard trimmings, discarded fence posts, crates, motor vehicle tires, junk motor vehicle bodies or parts thereof, scrap metal, bed springs, water heaters, and discarded similar object or thing which cannot conveniently be cut into sizes less than three feet (0.91 m) in length (Boulder, CO).

Run-of-river hydropower A type of hydroelectric facility that uses the river flow with very little alteration and little or no impoundment of the water.

Rural electrification administration (REA) An agency of the US Dept. of Agriculture that makes loans to states and territories in the USA for rural electrification and the furnishing of electric energy to persons in rural areas who do not receive central station service. It also furnishes and improves electric and telephone service in rural areas, assists electric borrowers to implement energy conservation programs and on-grid and off-grid renewable energy systems, and studies the condition and progress of rural electrification.

R-Value A measure of the capacity of a material to resist heat transfer. The R-value is the reciprocal of the conductivity of a material (U-value). The larger the R-value of a material, the greater is its insulating properties.

Sacrificial anode A metal rod placed in a water heater tank to protect the tank from corrosion. Anodes of aluminum, magnesium, or zinc are the more frequently metals. The anode creates a galvanic cell in which magnesium or zinc will be corroded more quickly than the metal of the tank, giving the tank a negative charge and preventing corrosion.

Safety disconnect An electronic (automatic or manual) switch that disconnects one circuit from another circuit. These are used to isolate power generation or storage equipment from conditions such as voltage spikes or surges, thus avoiding potential damage to equipment.

Salt gradient solar ponds Consist of three main layers. The top layer is near ambient and has low salt content. The bottom layer is hot, typically 160 °F–212 °F (71 °C–100 °C), and is very salty. The important gradient zone separates these zones. The gradient zone acts as a transparent insulator, permitting the sunlight to be trapped in the hot bottom layer (from which useful heat is withdrawn). This is because the salt gradient, which increases the brine density with depth and counteracts the buoyancy effect of the warmer water below (which would otherwise rise to the surface and lose its heat to the air). An organic Rankine-cycle engine is used to convert the thermal energy to electricity.

Salvage materials It includes waste paper, scrap materials, building materials, or any other type of waste material that has a value to the producer, owner, or occupant of the premises upon which it is produced or stored over and above the actual cost of collection and disposal (Wichita, KS).

Salvaging It refers to the controlled removal of reusable materials (Bluffton, IN).

Sanitary landfill (1) An active portion of the landfill site where refuse is being dumped, compacted, arid covered (Boulder, CO); (2) any planned and systematic method of refuse disposal whereby the waste material is placed in the earth in layers and then compacted and covered with earth or other approved cover material at the end of each day's operation (Prince George's County, MD); (3) a planned and systematic method of refuse disposal by burying (Baltimore County, MD); (4) a method of disposing of refuse on land without creating nuisances or hazards to public health or safety, by utilizing principles of engineering to confine the refuse to the smallest practical area, to reduce it to the smallest practical volume, and to cover it with a layer of suitable cover at the conclusion of each day's operation or at more frequent intervals as necessary (Bluffton, IN); (5) any disposal site operated by means of compacting and covering solid wastes at specific designated intervals, but not each operating day (Marion County, OR).

Scavenging It is the uncontrolled picking or sorting of solid wastes either before, during, or following collection (Washington, DC).

Scribing The cutting of a grid pattern of grooves in a semiconductor material, generally for the purpose of making interconnections.

Sealed combustion heating system A heating system that uses only outside air for combustion and vents combustion gases directly to the outdoors. These systems are less likely to backdraft and to negatively affect indoor air quality.

Seasonal energy efficiency ratio (SEER) A measure of seasonal or annual efficiency of a central air conditioner or air-conditioning heat pump. It takes into account the variations in temperature that can occur within a season and is the average number of Btu of cooling delivered for every watt-hour of electricity used by the heat pump over a cooling season.

Seasonal performance factor (SPF) Ratio of useful energy output of a device to the energy input, averaged over an entire heating season.

Seasoned wood Wood, used for fuel, which has been air-dried so that it contains 15–20% moisture content (wet basis).

Second law efficiency The ratio of the minimum amount of work or energy required to perform a task to the amount actually used.

Second law of thermodynamics This law states that no device can completely and continuously transform all of the energy supplied to it into useful energy.

Seebeck effect The generation of an electric current, when two conductors of different metals are joined at their ends to form a circuit, with the two junctions kept at different temperatures.

Selectable load Any device, such as lights, televisions, and power tools, which is plugged into your central power source and used only intermittently.

Selective absorber A solar absorber surface that has high absorbance at wavelengths corresponding to that of the solar spectrum and low emittance in the infrared range.

Selective surface coating A material with high absorbance and low emittance properties applied to or on solar absorber surfaces.

Self-assessment Self-assessments are appraisals conducted by the facility to identify and correct any existing deficiencies in the environmental monitoring program. Under a facility's environmental monitoring procedure *Self-Assessments for Environmental Programs*, information obtained from an appraisal is categorized as follows: (1) key finding, (2) observation, (3) comment or concern, (4) commendable practice, and (5) deficiency.

Semiconductor Any material that has a limited capacity for conducting an electric current. Certain semiconductors, including silicon, gallium arsenide, copper indium diselenide, and cadmium telluride, are uniquely suited to the photovoltaic conversion process.

Semisolid manure A manure from animals that contains 5–10 % dry matter.

Sensible cooling effect The difference between the total cooling effect and the dehumidifying effect.

Sensible cooling load The interior heat gain due to heat conduction, convection, and radiation from the exterior into the interior and from occupants and appliances.

Sensible heat The heat absorbed or released when a substance undergoes a change in temperature.

Sensible heat storage A heat storage system that uses a heat storage medium and where the additional or removal of heat results in a change in temperature.

Septage Septage means the liquid and solid material pumped from a septic tank, cesspool, or similar domestic sewage treatment system or holding tank when the system is cleaned or maintained.

Septic tank absorption bed system It is a traditional on-site wastewater treatment system involving the use of a buried septic tank followed by a soil absorption bed. The anaerobic septic tank removes the scums, grease, oil, and settleable solids and anaerobically digests the solid pollutants. The absorption bed consists of many porous distribution pipes buried underground and is supported by graded gravels or similar synthetic aggregates for uniform distribution and leaching the septic tank effluent to the sandy soil where soil microorganisms remove organic pollutants.

Series A configuration of an electrical circuit in which the positive lead is connected to the negative lead of another energy-producing, energy-conducting, or energy-consuming device. The voltages of each device are additive, whereas the current is not.

Series connection A way of joining photovoltaic cells by connecting positive leads to negative leads; such a configuration increases the voltage.

Series resistance Parasitic resistance to current flow in a cell due to mechanisms such as resistance from the bulk of the semiconductor material, metallic contacts, and interconnections.

Service unit It includes any four sleeping rooms or a fraction thereof, where no cooking privileges are provided, located in any commercial establishment.

Setback thermostat A thermostat that can be set to automatically lower temperatures in an unoccupied house and raise them again before the occupant returns.

Sewage sludge The solid, semisolid, or liquid residue generated during the treatment of domestic sewage in a treatment works. Sewage sludge includes, but is not limited to, domestic septage, scum, and solids removed during primary, secondary, or advanced wastewater treatment processes. The definition of sewage sludge also includes a material derived from sewage sludge (i.e., sewage sludge whose quality is changed either through further treatment or through mixing with other materials).

Shading coefficient A measure of window glazing performance that is the ratio of the total solar heat gain through a specific window to the total solar heat gain through a single sheet of double-strength glass under the same set of conditions; expressed as a number between 0 and 1.

Shale It refers to a fine-grained sedimentary rock that is composed mostly of consolidated clay or mud and is the most frequently occurring sedimentary rock.

Shallow land burial It refers to emplacement of low-level radioactive waste in or within the upper 30 m of the surface of the Earth in trenches, holes, or other excavations: (1) in which only soil provides (1a) structural integrity, (1b) a barrier to migration of low-level radioactive waste from or subsurface water into such excavation, or (1c) a barrier to entry of surface water to such excavation; or (2) in a manner that fails to allow for monitoring and control of releases of radioactivity during the institutional control period.

Sheathing A construction element used to cover the exterior of wall framing and roof trusses.

Short circuit An electric current taking a shorter or different path than intended.

Short circuit current The current flowing freely through an external circuit that has no load or resistance; the maximum current possible.

Shunt load An electrical load used to safely use excess-generated power when not needed for its primary uses. A shunt load in a residential photovoltaic system might be domestic water heating, such that when power is not needed for typical building loads, such as operating lights or running HVAC system fans and pumps, it still provides value and is used in a constructive, safe manner.

Shutter An interior or exterior movable panel that operates on hinges or slides into place, used to protect windows or provide privacy.

Siding A construction element applied to the outermost surface of an exterior wall.

Sievert A unit of dose equivalent from the International System of Units (Systeme Internationale). Equal to one joule per kilogram.

Sigma heat The sum of sensible heat and latent heat in a substance above a base temperature, typically 32°F (0°C).

Silicon A chemical element, of atomic number 14, that is semimetallic and an excellent semiconductor material used in solar photovoltaic devices; commonly found in sand.

Simple CS (caulk and seal) A technique for insulating and sealing exterior walls that reduces vapor diffusion through air leakage points by installing precut blocks of rigid foam insulation over floor joists, sheet subfloor, and top plates before drywall is installed.

Sine wave The type of alternative current generated by alternating current generators, rotary inverters, and solid-state inverters.

Single glaze or pane One layer of glass in a window frame. It has very little insulating value (R-1) and provides only a thin barrier to the outside and can account for considerable heat loss and gain.

Single-crystal material In reference to solar photovoltaic devices, a material that is composed of a single crystal or a few large crystals.

Single-package system A year-round heating and air-conditioning system that has all the components completely encased in one unit outside the home. Proper matching of components can mean more energy-efficient operation compared to components purchased separately.

Single-phase A generator with a single armature coil, which may have many turns, and the alternating current output consists of a succession of cycles.

Site closure and stabilization or closure It refers to those actions that are taken upon completion of operations that prepare the disposal site for custodial care and that assure that the disposal site will remain stable and will not need ongoing active maintenance.

Site remediation An action, process, or method for remediating/correcting the land and groundwater.

Sizing The process of designing a solar system to meet a specified load given the solar resource and the nominal or rated energy output of the solar energy collection or conversion device.

Skylight A window located on the roof of a structure to provide interior building spaces with natural daylight, warmth, and ventilation.

Slab A concrete pad that sits on gravel or crushed rock, well-compacted soil either level with the ground or above the ground.

Slab on grade A slab floor that sits directly on top of the surrounding ground.

SlinkyTM ground loop In this type of closed-loop, horizontal geothermal heat pump installation, the fluid-filled plastic heat exchanger pipes are coiled like a SlinkyTM to allow more pipe in a shorter trench. This type of installation cuts down installation costs and makes horizontal installation possible in areas where it would not be with conventional horizontal applications. Also see closed-loop geothermal heat pump systems.

Slumping It refers to a landsliding characterized by movement of a generally independent mass of rock or earth along a slip surface and about an axis parallel to the slope rim which it descends and by backward tilting of the mass with respect to that slope so that the slump surface often exhibits a reversed slope facing uphill.

Smart window A term used to describe a technologically advanced window system that contains glazing that can change or switch its optical qualities when a low voltage electrical signal is applied to it or in response to changes in heat or light.

Sodium lights A type of high-intensity discharge light that has the most lumens per watt of any light source.

Soffit A panel which covers the underside of a roof overhang, cantilever, or mansard.

Soil It refers to all unconsolidated earthy material overlying bedrock.

Solar access or rights The legal issues related to protecting or ensuring access to sunlight to operate a solar energy system or use solar energy for heating and cooling.

Solar air heater A type of solar thermal system where air is heated in a collector and either transferred directly to the interior space or to a storage medium, such as a rock bin.

Solar altitude angle The angle between a line from a point on the Earth's surface to the center of the solar disk and a line extending horizontally from the point.

Solar array A group of solar collectors or solar modules connected together.

Solar azimuth The angle between the sun's apparent position in the sky and true south, as measured on a horizontal plane.

Solar cell A solar photovoltaic device with a specified area.

Solar collector A device used to collect, absorb, and transfer solar energy to a working fluid. Flat plate collectors are the most common type of collectors used for solar water or pool heating systems. In the case of a photovoltaics system, the solar collector could be crystalline silicon panels or thin-film roof shingles, for example.

Solar constant The average amount of solar radiation that reaches the Earth's upper atmosphere on a surface perpendicular to the sun's rays; equal to 1353 W per square meter or 492 Btu per square foot.

Solar cooling The use of solar thermal energy or solar electricity to power a cooling appliance. There are five basic types of solar cooling technologies: absorption cooling, which can use solar thermal energy to vaporize the refrigerant; desiccant cooling, which can use solar thermal energy to regenerate (dry) the desiccant; vapor compression cooling, which can use solar thermal energy to operate a Rankine-cycle heat engine; and evaporative coolers ("swamp" coolers) and heat pumps and air conditioners that can be powered by solar photovoltaic systems.

Solar declination The apparent angle of the sun north or south of the Earth's equatorial plane. The Earth's rotation on its axis causes a daily change in the declination.

Solar distillation The process of distilling (purifying) water using solar energy. Water can be placed in an airtight solar collector with a sloped glazing material, and as it heats and evaporates, distilled water condenses on the collector glazing and runs down where it can be collected in a tray.

Solar energy Electromagnetic energy transmitted from the sun (solar radiation). The amount that reaches the Earth is equal to one billionth of total solar energy generated or the equivalent of about 420 trillion kilowatt-hours.

Solar energy collector See solar collector.

Solar energy industries association (SEIA) A national trade association of solar energy equipment manufacturers, retailers, suppliers, installers, and consultants.

Solar energy research institute (SERI) A federally funded institute, created by the Solar Energy Research, Development, and Demonstration Act of 1974, that conducted research and development of solar energy technologies. It became the National Renewable Energy Laboratory (NREL) in 1991.

Solar film A window glazing coating, usually tinted bronze or gray, used to reduce building cooling loads, glare, and fabric fading.

Solar fraction The percentage of a building's seasonal energy requirements that can be met by a solar energy device(s) or system(s).

Solar furnace A device that achieves very high temperatures by the use of reflectors to focus and concentrate sunlight onto a small receiver.

Solar gain The amount of energy that a building absorbs due to solar energy striking its exterior and conducting to the interior or passing through windows and being absorbed by materials in the building.

Solar irradiation The amount of solar radiation, both direct and diffuse, received at any location.

Solar mass A term for materials used to absorb and store solar energy.

Solar module (panel) A solar photovoltaic device that produces a specified power output under defined test conditions, usually composed of groups of solar cells connected in series, in parallel, or in series-parallel combinations.

Solar noon The time of the day, at a specific location, when the sun reaches its highest, apparent point in the sky; equal to true or due, geographic south.

Solar one A solar thermal electric central receiver power plant ("power tower") located in Barstow, California, and completed in 1981. The Solar One had a design capacity of 10,000 peak kilowatts and was composed of a receiver located on the top of a tower surrounded by a field of reflectors. The concentrated sunlight created steam to drive a steam turbine and electric generator located on the ground.

Solar panel See photovoltaic module.

Solar pond A body of water that contains brackish (highly saline) water that forms layers of differing salinity (stratifies) that absorb and trap solar energy. Solar ponds can be used to provide heat for industrial or agricultural processes, building heating, and cooling and to generate electricity.

Solar power plant A power plant that produces electricity by the solar energy transmitted by the photovoltaic modules on solar panels.

Solar power satellite A solar power station investigated by NASA that entailed a satellite in geosynchronous orbit that would consist of a very large array of solar photovoltaic modules that would convert solar-generated electricity to microwaves and beam them to a fixed point on the Earth.

Solar radiation A general term for the visible and near-visible (ultraviolet and near-infrared) electromagnetic radiation that is emitted by the sun. It has a spectral, or wavelength, distribution that corresponds to different energy levels; short wavelength radiation has a higher energy than long wavelength radiation.

Solar simulator An apparatus that replicates the solar spectrum and is used for testing solar energy conversion devices.

Solar space heater A solar energy system designed to provide heat to individual rooms in a building.

Solar spectrum The total distribution of electromagnetic radiation emanating from the sun. The different regions of the solar spectrum are described by their wavelength range. The visible region extends from about 390 to 780 nm (a nanometer is one billionth of one meter). About 99% of solar radiation is contained in a wavelength region from 300 nm (ultraviolet) to 3000 nm (near-infrared). The combined radiation in the wavelength region from 280 nm to 4000 nm is called the broadband, or total, solar radiation.

Solar thermal electric systems Solar energy conversion technologies that convert solar energy to electricity, by heating a working fluid to power a turbine that drives a generator. Examples of these systems include central receiver systems, parabolic dish, and solar trough.

Solar thermal parabolic dishes A solar thermal technology that uses a modular mirror system that approximates a parabola and incorporates two-axis tracking to focus the sunlight onto receivers located at the focal point of each dish. The mirror system typically is made from a number of mirror facets, either glass or polymer mirror, or can consist of a single-stretched membrane using a polymer mirror. The concentrated sunlight may be used directly by a Stirling, Rankine, or Brayton cycle heat engine at the focal point of the receiver or to heat a working fluid that is piped to a central engine. The primary applications include remote electrification, water pumping, and grid-connected generation.

Solar thermal systems Solar energy systems that collect or absorb solar energy for useful purposes. Can be used to generate high-temperature heat (for electricity production and/or process heat), medium temperature heat (for process and space/water heating and electricity generation), and low-temperature heat (for water and space heating and cooling).

Solar time The period marked by successive crossing of the earth's meridian by the sun; the hour angle of the sun at a point of observance (apparent time) is corrected to true (solar) time by taking into account the variation in the earth's orbit and rate of rotation. Solar time and local standard time are usually different for any specific location.

Solar transmittance The amount of solar energy that passes through a glazing material, expressed as a percentage.

Solar trough systems (see also parabolic trough, above) A type of solar thermal system where sunlight is concentrated by a curved reflector onto a pipe containing a working fluid that can be used to process heat or to produce electricity. The world's largest solar thermal electric power plants use solar trough technology. They are located in California and have a combined electricity-generating capacity of 240,000 kW.

Solar two Solar Two is a retrofit of the Solar One project (see above). It is demonstrating the technical feasibility and power potential of a solar power tower using advanced molten-salt technology to store energy. Solar Two retains several of the main components of Solar One, including the receiver tower, turbine, generator, and the 1818 heliostats.

Solarium A glazed structure, such as greenhouse or "sunspace."

Solenoid An electromechanical device composed of a coil of wire wound around a cylinder containing a bar or plunger, that when a current is applied to the coil, the electromotive force causes the plunger to move; a series of coils or wires used to produce a magnetic field.

Solenoid valve An automatic valve that is opened or closed by an electromagnet.

Solid fuels Any fuel that is in solid form, such as wood, peat, lignite, coal, and manufactured fuels such as pulverized coal, coke, charcoal, briquettes, pellets, etc.

Solid manure A manure from animals that contains dry matter greater than 15 %.

Solid waste management unit (SWMU) Any discernible unit at which solid wastes have been placed at any time, irrespective of whether the unit was intended for the management of solid or hazardous waste. Such units include any area at a facility at which solid wastes have been routinely and systematically released.

Solid waste or refuse (1) All putrescible and non-putrescible wastes, whether in solid or liquid form, except liquid carried industrial wastes or sewage or sewage hauled as an incidental part of a septic tank or cesspool cleaning service, but including garbage, rubbish, ashes, sewage sludge, street refuse, industrial wastes, swill, demolition and construction wastes, abandoned vehicles or parts thereof, discarded home and industrial appliances, manure, vegetable or animal solid and semisolid wastes, dead animals, and other discarded solid materials (Marion County, OR); (2) all garbage, rubbish, garden trash, noncombustible refuse, and industrial wastes (Dade County, FL); (2) all putrescible and non-putrescible solid wastes, except body wastes, and including abandoned vehicles, food waste (garbage), rubbish, ashes, incinerator residue, street cleanings, tree debris, and solid market and industrial wastes (Washington, DC); (3) all waste materials, combustible or noncombustible, from all public and private establishments and residences, including trash, garbage, rubbish, offal, industrial refuse, and commercial refuse, but not the body excrements (Prince George's County, MD).

Solidity In reference to a wind energy conversion device, the ratio of rotor blade surface area to the frontal, swept area that the rotor passes through.

Solstice The two times of the year when the sun is apparently farthest north and south of the Earth's equator, usually occurring on or around June 21 (summer solstice in northern hemisphere, winter solstice for southern hemisphere) and December 21 (winter solstice in northern hemisphere, summer solstice for the southern hemisphere).

Solutioning It refers to the chemical process by which rock material passes into solution.

Source water It refers to a water from surface or groundwater sources from which operators may withdraw it themselves or may purchase it from suppliers.

Space heater A movable or fixed heater used to heat individual rooms.

Spacer (window) Strips of material used to separate multiple panes of glass within the windows.

Specific heat The amount of heat required to raise a unit mass of a substance through one degree, expressed as a ratio of the amount of heat required to raise an equal mass of water through the same range.

Specific heat capacity The quantity of heat required to change the temperature of one unit weight of a material by one degree.

Specific humidity The weight of water vapor, per unit weight of dry air.

Specific volume The volume of a unit weight of a substance at a specific temperature and pressure.

Spectral energy distribution A curve illustrating the variation or spectral irradiance with wavelength.

Spectral irradiance The monochromatic irradiance of a surface per unit bandwidth at a particular wavelength, usually expressed in watts per square meter-nanometer bandwidth.

Spectral reflectance The ratio of energy reflected from a surface in a given waveband to the energy incident in that waveband.

Spectrally selective coatings A type of window glazing films used to block the infrared (heat) portion of the solar spectrum but admit a higher portion of visible light.

Spectrum See solar spectrum above.

Spent fuel Nuclear fuel that has been used in a nuclear reactor; this fuel contains uranium, activation products, fission products, and plutonium.

Septic tank filtration system It is an alternative on-site wastewater treatment system involving the use of a buried septic tank followed by an underground sand filtration and/or a trickling filtration module. In areas where problem soil conditions preclude the use of subsurface trenches or seepage absorption beds, or mounds, then the underground sand filtration and/or trickling filtration can be installed to treat the septic tank effluent. The anaerobic septic tank removes the scums, grease, oil, and settleable solids and anaerobically digests the solid pollutants. The underground sand filtration and/or trickling filtration units further treat the septic tank effluent for subsequent surface water discharge if permitted by local sanitary code. This treatment method is useful for land restoration and lake restoration. The filtration effluent is usually disinfected before surface discharge. The use of aerobic trickling filter and anaerobic trickling filter for nitrogen removal has been demonstrated.

Septic tank mound system It is an alternative on-site wastewater treatment system involving the use of a buried septic tank followed by an aboveground soil mound. In areas where problem soil conditions preclude the use of subsurface trenches or seepage absorption beds, then mounds can be installed to raise the absorption field above ground, provide treatment, and distribute the wastewater to the underlying soil over a wide area in a uniform manner. The anaerobic septic tank removes the scums, grease, oil, and settleable solids and anaerobically digests the solid pollutants. The mound consists of many porous distribution pipes buried aboveground and is supported by graded gravels or similar synthetic aggregates for uniform distribution and leaching the septic tank effluent to the aboveground sandy soil where soil microorganisms remove organic pollutants.

Spill A spill or release is defined as "any spilling, leaking, pumping, pouring, emitting, emptying, discharging, injecting, escaping, leaching, dumping, or otherwise disposing of substances from the ordinary containers employed in the normal course of storage, transfer, processing, or use."

Spillway A passage for surplus water to flow over or around a dam.

Spinning reserve Electric power provider capacity on line and running at low power in excess of actual load.

Split spectrum photovoltaic cell A photovoltaic device where incident sunlight is split into different spectral regions, with an optical apparatus, that are directed to individual photovoltaic cells that are optimized for converting that spectrum to electricity.

Split system air conditioner An air-conditioning system that comes in two to five pieces: one piece contains the compressor, condenser, and a fan and the others have an evaporator and a fan. The condenser, installed outside the house, connects to several evaporators, one in each room to be cooled, mounted inside the house. Each evaporator is individually controlled, allowing different rooms or zones to be cooled to varying degrees.

Spray pyrolysis A deposition process whereby heat is used to break molecules into elemental sources that are then spray deposited on a substrate.

Spreader stocker A type of furnace in which fuel is spread, automatically or mechanically, across the furnace grate.

Sputtering A process used to apply photovoltaic semiconductor material to a substrate by a physical vapor deposition process where high-energy ions are used to bombard elemental sources of semiconductor material, which eject vapors of atoms that are then deposited in thin layers on a substrate.

Square wave inverter A type of inverter that produces square wave output; consists of a DC source, four switches, and the load. The switches are power semiconductors that can carry a large current and withstand a high-voltage rating. The switches are turned on and off at a correct sequence, at a certain frequency. The square wave inverter is the simplest and the least expensive to purchase, but it produces the lowest quality of power.

Squirrel cage motor This is another name for an induction motor. The motors consist of a rotor inside a stator. The rotor has laminated, thin flat steel disks, stacked with channels along the length. If the casting composed of bars and attached end rings were viewed without the laminations, the casting would appear similar to a squirrel cage.

Stability It refers to structural integrity.

Stack A smokestack or flue for exhausting the products of combustion from a combustion appliance.

Stack (heat) loss Sensible and latent heat contained in combustion gases and vapor emitted to the atmosphere.

Staebler-Wronski effect The tendency of the sunlight to electricity conversion efficiency of amorphous silicon photovoltaic devices to degrade (drop) upon initial exposure to light.

Stagnation temperature A condition that can occur in a solar collector if the working fluid does not circulate when the sun is shining on the collector.

Stakeholder Individual or organization that has a stake in the outcome of the watershed plan.

Stall In reference to a wind turbine, a condition when the rotor stops turning.

Stand-alone generator A power source/generator that operates independently of or is not connected to an electric transmission and distribution network; used to meet a load(s) physically close to the generator.

Stand-alone inverter An inverter that operates independent of or is not connected to an electric transmission and distribution network.

Stand-alone system A system that operates independent of or is not connected to an electric transmission and distribution network.

Standard air Air with a weight of 0.075 lb per cubic foot (1.2 kg/m³) with an equivalent density of dry air at a temperature of 86°F (30°C) and standard barometric pressure of 29.92 in. of mercury (760 mmHg).

Standard conditions In refrigeration, an evaporating temperature of 5°F (−15°C), a condensing temperature of 86°F (30°C), liquid temperature before expansion of 77°F (25°C), and suction temperature of 12°F (−11.1°C).

Standard container Any watertight can with a close-fitting cover, side bail handles, and having a capacity of thirty-two gallons (121 l) or less.

Standard cubic foot A column of gas at standard conditions of standard temperature (32°F or 0°C) and standard pressure (one atmosphere or 760 mmHg). 1 ft³ = 0.02832 m³.

Standard deviation An indication of the dispersion of a set of results around their average.

Standard industrial classification (SIC) code Standardized codes used to classify businesses by the type of activity they engage in.

Standby heat loses A term used to describe heat energy lost from a water heater tank.

Standby power For the consumer, this is the electricity that is used by your TVs, stereos, and other electronic devices that use remote controls. When you press "off" to turn off your device, minimal power (dormant mode) is still being used to maintain the internal electronics in a ready, quick-response mode. This way, your device can be turned on with your remote control and be immediately ready to operate.

Starting surge Power, often above an appliance's rated wattage, required to bring any appliance with a motor up to operating speed.

Starting torque The torque at the bottom of a speed (rpm) versus torque curve. The torque developed by the motor is a percentage of the full-load or rated torque. At this torque, the rotational speed of the motor as a percentage of synchronous speed is zero. This torque is what is available to initially get the load moving and begin its acceleration.

Static pressure The force per unit area acting on the surface of a solid boundary parallel to the flow.

Steam Water in vapor form; used as the working fluid in steam turbines and heating systems.

Steam boiler A type of furnace in which fuel is burned and the heat is used to produce steam.

Steam turbine A device that converts high-pressure steam, produced in a boiler, into mechanical energy that can then be used to produce electricity by forcing blades in a cylinder to rotate and turn a generator shaft.

Stirling engine A heat engine of the reciprocating (piston) where the working gas and a heat source are independent. The working gas is compressed in one region of the engine and transferred to another region where it is expanded. The expanded gas is then returned to the first region for recompression. The working gas thus moves back and forth in a closed cycle.

Stoichiometric ratio The ratio of chemical substances necessary for a reaction to occur completely.

Stoichiometry Chemical reactions, typically associated with combustion processes; the balancing of chemical reactions by providing the exact proportions of reactant compounds to ensure a complete reaction; all the reactants are used up to produce a single set of products.

Storage capacity The amount of energy an energy storage device or system can store.

Storage hydropower A hydropower facility that stores water in a reservoir during high-inflow periods to augment water during low-inflow periods. Storage projects allow the flow releases and power production to be more flexible and dependable. Many hydropower project operations use a combination of approaches.

Storage tank The tank of a water heater.

Storage water heater A water heater that releases hot water from the top of the tank when a hot water tap is opened. To replace that hot water, cold water enters the bottom of the tank to ensure a full tank.

Storm door An exterior door that protects the primary door.

Storm windows Glass, plastic panels, or plastic sheets that reduce air infiltration and some heat loss when attached to either the interior or exterior of existing windows.

Stranded investment (costs and benefits) An investment in a power plant or demand side management measures or programs that become uneconomical due to increased competition in the electric power market. For example, an electric power plant may produce power that is more costly than what the market rate for electricity is, and the power plant owner may have to close the plant, even though the capital and financing costs of building the plant have not been recovered through prior sales of electricity from the plant. This is considered a stranded cost. Stranded benefits are those power provider investments in measures or programs considered to benefit consumers by reducing energy consumption and/or providing environmental benefits that have to be curtailed due to increased competition and lower profit margins.

Street refuse It includes material picked up by manual or mechanical sweeping of alleys, streets and sidewalks, litter from public litter receptacles, and dirt removed from catch basins (Washington, DC).

Stud A popular term used for a length of wood or steel used in or for wall framing.

Subsidence It refers to a local mass movement that involves principally the gradual downward settling or sinking of the solid earth's surface with little or no horizontal motion and that does not occur along a free surface (not the result of a landslide or fracture of a slope).

Substation An electrical installation containing power conversion (and sometimes generation) equipment, such as transformers, compensators, and circuit breakers.

Substrate The physical material upon which a photovoltaic cell is applied.

Subsurface It refers to an earth material (as rock) near but not exposed at the surface of the ground.

Subtitle C landfill A landfill permitted to accept hazardous wastes under the Sections 3001

Subtitle D landfill A landfill permitted to accept only nonhazardous wastes under Sections 4001 through 4010 of the US Resource Conservation and Recovery Act (RCRA) and the regulations promulgated pursuant to these sections, including the US 40 CFR Parts 257 and 258.

Sun path diagram A circular projection of the sky vault onto a flat diagram used to determine solar positions and shading effects of landscape features on a solar energy system.

Sun tempered building A building that is elongated in the east-west direction, with the majority of the windows on the south side. The area of the windows is generally limited to about 7 % of the total floor area. A sun-tempered design has no added thermal mass beyond what is already in the framing, wall board, and so on. Insulation levels are generally high.

Sunspace A room that faces south (in the northern hemisphere) or a small structure attached to the south side of a house.

Super insulated houses A type of house that has massive amounts of insulation, airtight construction, and controlled ventilation without sacrificing comfort, health, or aesthetics.

Super solid waste management unit (SSWMU) Individual solid waste management units that have been grouped and ranked into larger units—super solid waste management units—because some individual units are contiguous or so close together as to make monitoring of separate units impractical.

Super window A popular term for highly insulating window with a heat loss so low; it performs better than an insulated wall in winter, since the sunlight that it admits is greater than its heat loss over a 24-h period.

Superconducting magnetic energy storage (SMES) SMES technology uses the superconducting characteristics of low-temperature materials to produce intense magnetic fields to store energy. SMES has been proposed as a storage option to support large-scale use of photovoltaics and wind as a means to smooth out fluctuations in power generation.

Superconductivity The abrupt and large increase in electrical conductivity exhibited by some metals as the temperature approaches absolute zero.

Supplementary heat A heat source, such as a space heater, used to provide more heat than that provided by a primary heating source.

Supply duct The duct(s) of a forced air heating/cooling system through which heated or cooled air is supplied to rooms by the action of the fan of the central heating or cooling unit.

Supply side Technologies that pertain to the generation of electricity.

Surety bond It refers to an arrangement by which a surety company assumes liability for the specified obligations of a principal, if the principal fails to pay or perform as required by the permit.

Surface facilities It refers to the auxiliary buildings and equipment located on the surface of the land above an underground mined repository.

Surface impoundment A natural topographic depression, man-made excavation, or diked area formed primarily of earthen materials (although it may be lined with man-made materials), used to temporarily or permanently treat, store, or dispose of waste, usually in the liquid form. Surface impoundments do not include areas constructed to hold containers of wastes. Other common names for surface impoundments include ponds, pits, lagoons, finishing ponds, settling ponds, surge ponds, seepage ponds, and clarification ponds.

Surface water (1) Water that is exposed to the atmospheric conditions of temperature, pressure, and chemical composition at the surface of the Earth. (2) All water forms naturally open to the atmosphere (rivers, lakes, reservoirs, ponds, streams, impoundments, seas, estuaries, etc.); (3) the water whose top surface is exposed to the atmosphere including a flowing body as well as a pond and a lake (Bluffton, IN).

Surface water loop In this type of closed-loop geothermal heat pump installation, the fluid-filled plastic heat exchanger pipes are coiled into circles and submerged at least eight feet below the surface of a body of surface water, such as a pond or lake. The coils should only be placed in a water source that meets minimum volume, depth, and quality criteria. Also see closed-loop geothermal heat pump systems.

Surveillance (1) It refers to the observation of the disposal site for purposes of visual detection of need for maintenance, custodial care, evidence of intrusion, and compliance with federal and state statutory, regulatory, license, and permit requirements; and (2) the act of monitoring or observing a process or activity to verify conformance with specified requirements.

Swamp cooler A popular term used for an evaporative cooling device.

Swept area In reference to a wind energy conversion device, the area through which the rotor blades spin, as seen when directly facing the center of the rotor blades.

Synchronous generator An electrical generator that runs at a constant speed and draws its excitation from a power source external or independent of the load or transmission network it is supplying.

Synchronous inverter An electrical inverter that inverts direct current electricity to alternating current electricity and that uses another alternating current source, such as an electric power transmission and distribution network (grid), for voltage and frequency reference to provide power in phase and at the same frequency as the external power source.

Synchronous motor A type of motor designed to operate precisely at the synchronous speed with no slip in the full-load speeds (rpm).

System mix The proportion of electricity distributed by a power provider that is generated from available sources such as coal, natural gas, petroleum, nuclear, hydropower, wind, or geothermal.

Tankless water heater A water heater that heats water before it is directly distributed for end use as required; a demand water heater.

Task lighting Any light source designed specifically to direct light a task or work performed by a person or machine.

Temperature coefficient (of a solar photovoltaic cell) The amount that the voltage, current, and/or power output of a solar cell changes due to a change in the cell temperature.

Temperature humidity index An index that combines sensible temperature and air humidity to arrive at a number that closely responds to the effective temperature; used to relate temperature and humidity to levels of comfort.

Temperature zones Individual rooms or zones in a building where temperature is controlled separately from other rooms or zones.

Temperature/pressure relief valve A component of a water heating system that opens at a designated temperature or pressure to prevent a possible tank, radiator, or delivery pipe rupture.

Tempering valve A valve used to mix heated water with cold in a heating system to provide a desired water temperature for end use.

Tennessee valley authority (TVA) A federal agency established in 1933 to develop the Tennessee River Valley region of the southeastern USA and which is now the nation's largest power producer.

Termite shield A construction element that inhibits termites from entering building foundations and walls.

Therm A unit of heat containing 100,000 Btu.

Thermal balance point The point or outdoor temperature where the heating capacity of a heat pump matches the heating requirements of a building.

Thermal capacitance The ability of a material to absorb and store heat for use later.

Thermal efficiency A measure of the efficiency of converting a fuel to energy and useful work; useful work and energy output divided by higher heating value of input fuel times 100 (for percent).

Thermal energy The energy developed through the use of heat energy.

Thermal energy storage The storage of heat energy during power provider off-peak times at night, for use during the next day without incurring daytime peak electric rates.

Thermal envelope houses An architectural design (also known as the double envelope house), sometimes called a "house-within-a-house," that employs a double envelope with a continuous airspace of at least 6–12 in. (15.24–30.48 cm) in the north wall, south wall, roof, and floor, achieved by building inner and outer walls, a crawl space or subbasement below the floor, and a shallow attic space below the weather roof. The east and west walls are single, conventional walls. A buffer zone of solar-heated, circulating air warms the inner envelope of the house. The south-facing airspace may double as a sunspace or greenhouse.

Thermal mass Materials that store heat.

Thermal resistance (R-value) This designates the resistance of a material to heat conduction. The greater the R-value, the larger the number.

Thermal storage walls (masonry or water) A thermal storage wall is a south-facing wall that is glazed on the outside. Solar heat strikes the glazing and is absorbed into the wall, which conducts the heat into the room over time. The walls are at least 8 in thick. Generally, the thicker the wall, the less the indoor temperature fluctuates.

Thermocouple A device consisting of two dissimilar conductors with their ends connected together. When the two junctions are at different temperatures, a small voltage is generated.

Thermodynamic cycle An idealized process in which a working fluid (water, air, ammonia, etc.) successively changes its state (from a liquid to a gas and back to a liquid) for the purpose of producing useful work or energy, or transferring energy.

Thermodynamics A study of the transformation of energy from one form to another and its practical application (see law(s) of thermodynamics above).

Thermoelectric conversion The conversion of heat into electricity by the use of thermocouples.

Thermography A building energy auditing technique for locating areas of low insulation in a building envelope by means of a thermographic scanner.

Thermoluminescent dosimeter (TLD) A device that luminesces upon heating after being exposed to radiation. The amount of light emitted is proportional to the amount of radiation to which the luminescent material has been exposed.

Thermophilic It is a state in an anaerobic reactor, such as anaerobic digester or composting unit, where/when the temperature remains between 51°C and 57°C.

Thermophotovoltaic cell A device where sunlight concentrated onto an absorber heats it to a high temperature, and the thermal radiation emitted by the absorber is used as the energy source for a photovoltaic cell that is designed to maximize conversion efficiency at the wavelength of the thermal radiation.

Thermopile A large number of thermocouples connected in series.

Thermosiphon The natural, convective movement of air or water due to differences in temperature. In solar passive design, a thermosiphon collector can be constructed and attached to a house to deliver heat to the home by the continuous pattern of the convective loop (or thermosiphon).

Thermosiphon System This passive solar hot water system consists of and relies on warm water rising, a phenomenon known as natural convection, to circulate water through the collectors and to the tank. In this type of installation, the tank must be above the collector. As water in the collector heats, it becomes lighter and rises naturally into the tank above. Meanwhile, cooler water in the tank flows down pipes to the bottom of the collector, causing circulation throughout the system. The storage tank is attached to the top of the collector so that thermosiphoning can occur.

Thermostat A device used to control temperatures; used to control the operation of heating and cooling devices by turning the device on or off when a specified temperature is reached.

Thin-film A layer of semiconductor material, such as copper indium diselenide or gallium arsenide, a few microns, or less in thickness, used to make solar photovoltaic cells.

Third party It refers to a party who is not the permittee, a parent corporation of the permittee, or a subsidiary of the permittee.

Threatened waterbody A waterbody that is meeting standards but exhibits a declining trend in water quality such that it will likely exceed standards.

Three-phase current Alternating current in which three separate pulses are present, identical in frequency and voltage, but separated 120° in phase.

Tidal power The power available from the rise and fall of ocean tides. A tidal power plant works on the principal of a dam or barrage that captures water in a basin at the peak of a tidal flow and then directs the water through a hydroelectric turbine as the tide ebbs.

Tidal power plant One kind of hydroelectric power plant that produces electricity by the force of water falling through a hydroturbine that spins a generator. A dam or barrage that captures water in a basin at the peak of a tidal flow and then directs the water through a hydroelectric turbine as the tide ebbs.

Tight sands It refers to a geological formation consisting of a matrix of typically impermeable, nonporous tight sands.

Tilt angle (of a solar collector or module) The angle at which a solar collector or module is set to face the sun relative to a horizontal position. The tilt angle can be set or adjusted to maximize seasonal or annual energy collection.

Time-of-use (TOU) rates The pricing of electricity based on the estimated cost of electricity during a particular time block. Time-of-use rates are usually divided into three or four time blocks per twenty-four hour period (on-peak, mid-peak, off-peak, and sometimes super off-peak) and by seasons of the year (summer and winter). Real-time pricing differs from TOU rates in that it is based on actual (as opposed to forecasted) prices which may fluctuate many times a day and are weather-sensitive, rather than varying with a fixed schedule.

Timer A device that can be set to automatically turn appliances (such as lights, water sprinklers, heaters, instruments, etc.) off and on at set times.

Tip speed ratio In reference to a wind energy conversion device's blades, the difference between the rotational speed of the tip of the blade and the actual velocity of the wind.

Ton (of air conditioning) A unit of air-cooling capacity; 12,000 Btu per hour.

Topping-cycle A means to increase the thermal efficiency of a steam electric generating system by increasing temperatures and interposing a device, such as a gas turbine, between the heat source and the conventional steam-turbine generator to convert some of the additional heat energy into electricity.

Torque (motor) The turning or twisting force generated by an electrical motor in order for it to operate.

Total nitrogen It is the summation of *ammonium nitrogen* (NH_4^+-N), *nitrate nitrogen* (NO_3^--N), nitrite nitrogen (NO_2^--N), and *organic nitrogen* (organic-N). Usually nitrite nitrogen is in negligible amount. Crops directly utilize nitrogen in its inorganic forms, principally nitrate-N and ammonium-N.

Total dissolved solids (TDS) It refers to all material that passes the standard glass river filter; also called total filterable residue. Term is used to reflect salinity.

Total harmonic distortion The measure of closeness in shape between a waveform and its fundamental component.

Total heat The sum of the sensible and latent heat in a substance or fluid above a base point, usually 32°F.

Total incident radiation The total radiation incident on a specific surface area over a time interval.

Total internal reflection The trapping of light by refraction and reflection at critical angles inside a semiconductor device so that it cannot escape the device and must be eventually absorbed by the semiconductor.

Total Kjeldahl nitrogen (TKN) TKN is the summation of *ammonium nitrogen* (NH_4^+-N) and *organic nitrogen* (organic-N).

Total maximum daily load (TMDL) The amount, or load, of a specific pollutant that a waterbody can assimilate and still meet the water quality standard for its designated use. For impaired waterbodies, the TMDL reduces the overall load by allocating the load among current pollutant loads (from point and nonpoint sources), background or natural loads, a margin of safety, and sometimes an allocation for future growth.

Total solids (TS) Total solids (TS) include suspended and dissolved solids and are usually expressed as the concentration present in biosolids. TS depend on the type of wastewater process and biosolids' treatment prior to land application. Typical solid contents of various biosolids are liquid (2–12 %), dewatered (12–30 %), and dried or composted (50 %).

ToxCastDB It refers to a database (1) that links biological, metabolic, and cellular pathway data to gene and in vitro assay data for the chemicals screened in the ToxCast HTS assays; (2) that includes human disease and species homology information, which correlate with ToxCast assays that affect specific genetic loci; and (3) that is designed to make it possible to infer the types of human disease associated with exposure to these chemicals.

Toxic pollutants Pollutants declared "toxic" under the Section 307(a)(1) of the US Clean Water Act (CWA).

ToxRefDB It refers to a database that collects in vivo animal studies on chemical exposures.

Trace elements Trace elements are found in low concentrations in biosolids. The trace elements of interest in biosolids are those commonly referred to as "heavy metals".

Tracking solar array A solar energy array that follows the path of the sun to maximize the solar radiation incident on the PV surface. The two most common orientations are (1) one axis where the array tracks the sun east to west and

(2) two-axis tracking where the array points directly at the sun at all times. Tracking arrays use both the direct and diffuse sunlight. Two-axis tracking arrays capture the maximum possible daily energy.

Trailing edge The part of a wind energy conversion device blade, or airfoil, that is the last to contact the wind.

Transfer station It refers to any loading site where solid waste is transferred from one vehicle to another for transfer to a permanent refuse disposal site (Boulder, CO).

Transformer An electromagnetic device that changes the voltage of alternating current electricity. It consists of an induction coil having a primary and secondary winding and a closed iron core.

Transmission The process of sending or moving electricity from one point to another; usually defines that part of an electric power provider's electric power lines from the power plant buss to the last transformer before the customer's connection.

Transmission and distribution losses The losses that result from inherent resistance in electrical conductors and transformation inefficiencies in distribution transformers in a transmission and distribution network.

Transmission lines Transmit high-voltage electricity from the transformer to the electric distribution system.

Transuranic waste It refers to radioactive waste containing alpha-emitting radionuclides of atomic number 93 or higher (including neptunium, plutonium, americium, and curium)

Trash It includes rubbish and ashes (Prince George's County, MD).

Traveling grate A furnace grate that moves fuel through the combustion chamber.

Treatment works Federally owned, publicly owned, or privately owned device or system used to treat (including recycle or reclaim) either domestic sewage or a combination of domestic sewage and industrial waste of a liquid nature.

Treatment works treating domestic sewage A POTW or other sewage sludge or wastewater treatment system or device, regardless of ownership used in the storage, treatment, recycling, and reclamation of municipal or domestic sewage, including land dedicated for the disposal of sewage sludge.

Trellis An architectural feature used to shade exterior walls; usually made of a lattice of metal or wood; often covered by vines to provide additional summertime shading.

Trickle (solar) collector A type of solar thermal collector in which a heat-transfer fluid drips out of header pipe at the top of the collector and runs down the collector absorber and into a tray at the bottom where it drains to a storage tank.

Triple pane (window) This represents three layers of glazing in a window with an airspace between the middle glass and the exterior and interior panes.

Trombe wall A wall with high thermal mass used to store solar energy passively in a solar home. The wall absorbs solar energy and transfers it to the space behind the wall by means of radiation and by convection currents moving through spaces under, in front of, and on top of the wall.

Truck/equipment washwater Wastewater generated during either truck or equipment washes at a landfill site or a hydraulic fracking site, or others. During routine maintenance or repair operations, trucks and/or equipment used within the environmental or industrial operational site (e.g., loaders, compactors, or dump trucks) are washed, and the resultant washwaters are collected for treatment.

True power The actual power rating that is developed by a motor before losses occur.

True south The direction at any point on the Earth that is geographically in the northern hemisphere, facing toward the South Pole of the Earth. Essentially a line extending from the point on the horizon to the highest point that the sun reaches on any day (solar noon) in the sky.

Tube (fluorescent light) A fluorescent lamp that has a tubular shape.

Tube-in-plate absorber A type of solar thermal collector where the heat-transfer fluid flows through tubes formed in the absorber plate.

Tube-type collector A type of solar thermal collector that has tubes (pipes) that the heat-transfer fluid flows through that are connected to a flat absorber plate.

Tungsten halogen lamp A type of incandescent lamp that contains a halogen gas in the bulb, which reduces the filament evaporation rate increasing the lamp life. The high operating temperature and need for special fixtures limits their use to commercial applications and for use in projector lamps and spotlights.

Turbidity It refers to a cloudy condition in water due to suspended silt or organic matter.

Turbine A device for converting the flow of a fluid (air, steam, water, or hot gases) into mechanical motion.

Turn down ratio The ratio of a boiler's or gasifier's maximum output to its minimum output.

Two-axis tracking A solar array tracking system capable of rotating independently about two axes (e.g., vertical and horizontal).

Two-tank solar system A solar thermal system that has one tank for storing solar-heated water to preheat the water in a conventional water heater. U.S.C. (Section 1251), as amended by the Clean Water Act of 1977 (Pub. L. 95-217).

Ultimate analysis A procedure for determining the primary elements in a substance (carbon, hydrogen, oxygen, nitrogen, sulfur, and ash).

Ultraviolet Electromagnetic radiation in the wavelength range of 4–400 nm.

Underground home A house built into the ground or slope of a hill or which has most or all exterior surfaces covered with earth.

Underground injection well (UIC) It refers to a steel- and concrete-encased shaft into which hazardous waste, mining effluent, radioactive water, industrial effluent, municipal waste, or carbon dioxide is deposited underground by force and under pressure.

Underground mined repository It refers to a land disposal facility in which low-level radioactive waste is placed within the earth at a depth greater than 30 m below the surface of the Earth.

Underground source of drinking water (USDW) It refers to an aquifer currently being used as a source of drinking water or capable of supplying a public water system. USDWs have a TDS content of 10,000 mg per liter or less and are not exempted aquifers.

Unglazed solar collector A solar thermal collector that has an absorber that does not have a glazed covering. Solar swimming pool heater systems usually use unglazed collectors because they circulate relatively large volumes of water through the collector and capture nearly 80% of the solar energy available.

Unit nitrogen fertilizer rate (UNFR) UNFR is a rate in lb N per unit crop yield, where the unit can either bushel or ton. (Note: 1 bu (US bushel) = 1.2444 ft^3; 1 British bushel = 1.2843 ft^3; 1 t (British ton) = 2,000 lb; 1 T (metric ton) = 1,000 kg)

Unitary air conditioner An air conditioner consisting of one or more assemblies that move, clean, cool, and dehumidify air.

Unvented heater A combustion heating appliance that vents the combustion by-products directly into the heated space. The latest models have oxygen sensors that shut off the unit when the oxygen level in the room falls below a safe level.

Upgradient Referring to the flow of water or air, "upgradient" is analogous to upstream. Upgradient is a point that is "before" an area of study that is used as a baseline for comparison with downstream data. See Gradient and Downgradient.

Useful heat Heat stored above room temperature (in a solar heating system).

Utility A regulated entity which exhibits the characteristics of a natural monopoly (also referred to as a power provider). For the purposes of electric industry restructuring, "utility" refers to the regulated, vertically integrated electric company. "Transmission utility" refers to the regulated owner/operator of the transmission system only. "Distribution utility" refers to the regulated owner/ operator of the distribution system which serves retail customers.

U-Value (see coefficient of heat transmission) The reciprocal of R-value. The lower the number, the greater the heat-transfer resistance (insulating) characteristics of the material.

Vacuum evaporation The deposition of thin films of semiconductor material by the evaporation of elemental sources in a vacuum.

Vadose zone It is the zone between land surface and the water table within which (1) the moisture content is less than saturation (except in the capillary fringe); (2) pressure is less than atmospheric; (3) soil pore space also typically contains air or other gases; and (4) the capillary fringe is included in the vadose zone.

Valence band The highest energy band in a semiconductor that can be filled with electrons.

Vapor retarder A material that retards the movement of water vapor through a building element (walls, ceilings) and prevents insulation and structural wood from becoming damp and metals from corroding. Often applied to insulation batts or separately in the form of treated papers, plastic sheets, and metallic foils.

Variability factor The daily variability factor is the ratio of the estimated 99th percentile of the distribution of daily values divided by the expected value, median or mean, of the distribution of the daily data. The monthly variability factor is the estimated 95th percentile of the distribution of the monthly averages of the data divided by the expected value of the monthly averages.

Variable-speed wind turbines Turbines in which the rotor speed increases and decreases with changing wind speed, producing electricity with a variable frequency.

Vector attraction Characteristics (e.g., odor) that attract birds, insects, and other animals that are capable of transmitting infectious agents.

Vectors Vectors include rodents, birds, insects that can transport pathogens away from the land application site.

Vent A component of a heating or ventilation appliance used to conduct fresh air into, or waste air or combustion gases out of, an appliance or interior space.

Vent damper A device mounted in the vent connector that closes the vent when the heating unit is not firing. This traps heat inside the heating system and house rather than letting it draft up and out the vent system.

Vent pipe A tube in which combustion gases from a combustion appliance are vented out of the appliance to the outdoors.

Vented heater A type of combustion heating appliance in which the combustion gases are vented to the outside, either with a fan (forced) or by natural convection.

Ventilation The process of moving air (changing) into and out of an interior space either by natural or mechanically induced (forced) means.

Ventilation air That portion of supply air that is drawn from outside, plus any recirculated air that has been treated to maintain a desired air quality.

Vertical ground loop In this type of closed-loop geothermal heat pump installation, the fluid-filled plastic heat exchanger pipes are laid out in a plane perpendicular to the ground surface. For a vertical system, holes (approximately four inches in diameter) are drilled about 20 ft (6.1 m) apart and 100–400 ft (30.48–121.92 m) deep. Into these holes go two pipes that are connected at the bottom with a U-bend to form a loop. The vertical loops are connected with horizontal pipe (i.e., manifold), placed in trenches, and connected to the heat pump in the building. Large commercial buildings and schools often use vertical systems because the land area required for horizontal ground loops would be prohibitive. Vertical loops are also used where the soil is too shallow for trenching, or for existing buildings, as they minimize the disturbance to landscaping. Also see closed-loop geothermal heat pump systems.

Vertical-axis wind turbine (VAWT) A type of wind turbine in which the axis of rotation is perpendicular to the wind stream and the ground.

Visible light transmittance The amount of visible light that passes through the glazing material of a window, expressed as a percentage.

Visible radiation The visible portion of the electromagnetic spectrum with wavelengths from 0.4 to 0.76 μm.

Volatile solids (VS) Volatile solids (VS) provide an estimate of the readily decomposable organic matter in biosolids and are usually expressed as a percentage of total solids. VS are an important determinant of potential odor problems at land application sites.

Volatilization It is a phase change process that converts constituents from liquid, semiliquid, or solid form into gaseous form. The most common volatilization experienced is ammonia volatilization or the conversion of ammonium nitrogen to ammonia nitrogen. This is problematic for agricultural operations because plant's nitrogen is lost for plant uptake.

Volt A unit of electrical force equal to that amount of electromotive force that will cause a steady current of one ampere to flow through a resistance of one ohm.

Voltage The amount of electromotive force, measured in volts, that exists between two points.

Volt-ampere A unit of electrical measurement equal to the product of a volt and an ampere.

Wafer A thin sheet of semiconductor (photovoltaic material) made by cutting it from a single crystal or ingot.

Wall A vertical structural element that holds up a roof, encloses part or all of a room, or stands by itself to hold back soil.

Wall orientation The geographical direction that the primary or largest exterior wall of a building faces.

Waste (1) It includes useless, unwanted, or discarded materials resulting from normal community activities and may be solids, liquids, and gases (Washington, DC), and (2) it includes garbage, rubbish, garden trash, noncombustible refuse, and industrial wastes (Miami, FL).

Water jacket A heat exchanger element enclosed in a boiler. Water is circulated with a pump through the jacket where it picks up heat from the combustion chamber after which the heated water circulates to heat distribution devices. A water jacket is also an enclosed water-filled chamber in a tankless coiled water heater. When a faucet is turned on, water flows into the water heater heat exchanger. The water in the chamber is heated and transfers heat to the cooler water in the heat exchanger and is sent through the hot water outlet to the appropriate faucet.

Water quality standards Standards that set the goals, pollution limits, and protection requirements for each waterbody. These standards are composed of designated (beneficial) uses, numeric and narrative criteria, and anti-degradation policies and procedures.

Water source heat pump A type of (geothermal) heat pump that uses well (ground) or surface water as a heat source. Water has a more stable seasonal temperature than air, thus making for a more efficient heat source.

Water table The upper surface in a body of groundwater; the surface in an unconfined aquifer or confining bed at which the pore water pressure is equal to atmospheric pressure.

Water turbine A turbine that uses water pressure to rotate its blades; the primary types are the Pelton wheel, for high heads (pressure); the Francis turbine, for low to medium heads; and the Kaplan for a wide range of heads. Primarily used to power an electric generator.

Water wall An interior wall made of water-filled containers for absorbing and storing solar energy.

Water wheel A wheel that is designed to use the weight and/or force of moving water to turn it, primarily to operate machinery or grind grain.

Watershed (1) A watershed is the area of land where all of the water that is under it or drains off of it goes into the same place; land area that drains to a common waterway, such as a stream, lake, estuary, wetland, or ultimately the ocean. (2) The area contained within a drainage divide above a specified point on a stream.

Watershed approach A flexible framework for managing water resource quality and quantity within specified drainage area or watershed. This approach includes stakeholder involvement and management actions supported by sound science and appropriate technology.

Watershed plan A document that provides assessment and management information for a geographically defined watershed, including the analyses, actions, participants, and resources related to development and implementation of the plan.

Watt The rate of energy transfer equivalent to one ampere under an electrical pressure of one volt. One watt equals 1/746 hp or one joule per second. It is the product of voltage and current (amperage).

Watt-hour A unit of electricity consumption of one watt over the period of one hour.

Wattmeter A device for measuring power consumption.

Wave form The shape of the phase power at a certain frequency and amplitude.

Wave power The concept of capturing and converting the energy available in the motion of ocean waves to energy.

Wavelength The distance between similar points on successive waves.

Weatherization Caulking and weatherstripping to reduce air infiltration and exfiltration into/out of a building.

Weatherstripping A material used to seal gaps around windows and exterior doors.

Wheeling The process of transmitting electricity over one or more separately owned electric transmission and distribution systems (see wholesale and retail wheeling).

Whole house fan A mechanical/electrical device used to pull air out of an interior space; usually located in the highest location of a building, in the ceiling, and venting to the attic or directly to the outside.

Wholesale wheeling The wheeling of electric power in amounts and at prices that generally have been negotiated in long-term contracts between the power provider and a distributor or very large power customer.

Wind energy Energy available from the movement of the wind across a landscape caused by the heating of the atmosphere, earth, and oceans by the sun.

Wind energy conversion system (WECS) or device An apparatus for converting the energy available in the wind to mechanical energy that can be used to power machinery (grain mills, water pumps) and to operate an electrical generator.

Wind generator A wind energy conversion system (WECS) designed to produce electricity.

Wind power plant A group of wind turbines interconnected to a common power provider system through a system of transformers, distribution lines, and (usually) one substation. Operation, control, and maintenance functions are often centralized through a network of computerized monitoring systems, supplemented by visual inspection. This is a term commonly used in the United States. In Europe, it is called a generating station.

Wind resource assessment The process of characterizing the wind resource, and its energy potential, for a specific site or geographical area.

Wind rose A diagram that indicates the average percentage of time that the wind blows from different directions, on a monthly or annual basis.

Wind speed The rate of flow of the wind undisturbed by obstacles.

Wind speed duration curve A graph that indicates the distribution of wind speeds as a function of the cumulative number of hours that the wind speed exceeds a given wind speed in a year.

Wind speed frequency curve A curve that indicates the number of hours per year that specific wind speeds occur.

Wind speed profile A profile of how the wind speed changes with height above the surface of the ground or water.

Wind turbine A term used for a wind energy conversion device that produces electricity; typically having one, two, or three blades.

Wind turbine-rated capacity The amount of power a wind turbine can produce at its rated wind speed, e.g., 100 kW at 20 mph (32.186 km/h). The rated wind speed generally corresponds to the point at which the conversion efficiency is near its maximum. Because of the variability of the wind, the amount of energy a wind turbine actually produces is a function of the capacity factor (e.g., a wind turbine produces 20–35 % of its rated capacity over a year).

Wind velocity The wind speed and direction in an undisturbed flow.

Windmill A wind energy conversion system (WECS) that is used to grind grain and that typically has a high-solidity rotor; commonly used to refer to all types of WECS.

Window A generic term for a glazed opening that allows daylight to enter into a building and can be opened for ventilation.

Windpower curve A graph representing the relationship between the power available from the wind and the wind speed. The power from the wind increases proportionally with the cube of the wind speed.

Windpower profile The change in the power available in the wind due to changes in the wind speed or velocity profile; the windpower profile is proportional to the cube of the wind speed profile.

Wingwall A building structural element that is built onto a building's exterior along the inner edges of all the windows and extending from the ground to the eaves. Wingwalls help ventilate rooms that have only one exterior wall which leads to poor cross ventilation. Wingwalls cause fluctuations in the natural wind direction to create moderate pressure differences across the windows. They are only effective on the windward side of the building.

Wire (electrical) A generic term for an electrical conductor.

Wood stove A wood-burning appliance for space and/or water heating and/or cooking.

Working fluid A fluid used to absorb and transfer heat energy.

Wound rotor motors A type of motor that has a rotor with electrical windings connected through slip rings to the external power circuit. An external resistance controller in the rotor circuit allows the performance of the motor to be tailored to the needs of the system and to be changed with relative ease to accommodate system changes or to vary the speed of the motor.

X-Ray Penetrating electromagnetic radiations having wave lengths shorter than those of visible light. They are usually produced by bombarding a metallic target with fast electrons in a high vacuum. In nuclear reactions, it is customary to refer to photons originating in the nucleus as gamma rays and those originating in the extranuclear part of the atom as x-rays. These rays are sometimes called roentgen rays after their discoverer, W.C. Roentgen.

Yard rubbish It includes prunings, grass clippings, weeds, leaves, and general yard and garden wastes (Washington, DC).

Yaw The rotation of a horizontal axis wind turbine around its tower or vertical axis.

Yield It is the crop harvested in the unit of bu/acre or ton/acre.

Yurt An octagonal-shaped shelter that originated in Mongolia and traditionally made from leather or canvas for easy transportation.

Zero discharge No discharge of pollutants to waters of a country or to a POTW. Also included in this definition is alternative discharge or disposal of pollutants by way of evaporation, deep-well injection, off-site transfer, and land application

Zone An area within the interior space of a building, such as an individual room(s), to be cooled, heated, or ventilated. A zone has its own thermostat to control the flow of conditioned air into the space.

Zoning The combining of rooms in a structure according to similar heating and cooling patterns. Zoning requires using more than one thermostat to control heating, cooling, and ventilation equipment.

References

1. Oil and Gas Mineral Services (2010) Oil and gas terminology. Retrieved 20 Jan 2011, from http://www.mineralweb.com/library/oil-and-gas-terms
2. US EPA (2011) Terms of environment: glossary, abbreviations and acronyms. US Environmental Protection Agency, Washington, DC.http://www.epa.gov/OCEPAterms/ aterms. html
3. Harris DC (2003) Quantitative chemical analysis, 6th edn. W. H. Freeman and Company, New York
4. Geology Dictionary (2011) Aquiclude. http://www.alcwin.org/DictionaryOf_Geology_Description-136-A.htm
5. Agnes M (1999) Webster's new world college dictionary, 4th edn. Macmillan, Cleveland, OH
6. NYS Department of Environmental Conservation (2011) Supplemental generic environmental impact statement on the oil, gas and solution mining regulatory program (revised draft). Well permit issuance for horizontal drilling and high-volume hydraulic fracturing to develop the Marcellus Shale and other low-permeability gas reservoirs. NYS Department of Environmental Conservation, Albany, NY. ftp://ftp.dec.state.ny.us/dmn/download/ OGdSG EISFu ll.pdf
7. US EPA (2011) Glossary of underground injection control terms. US Environmental Protection Agency, Washington, DC. http://www.epa.gov/rswater/uic/glossary.htm#ltds
8. USDOE (2009) Modern shale gas development in the US: a primer. US Department of Energy, Ground Water Protection Council, Washington, DC. http://www.netl.doe.gov/technologies/oil-gas/publications/EPreports/Shale_Gas_Primer_2009.Pdf
9. USDOI (2015) Bureau of ocean energy management, regulation and enforcement: offshore minerals management glossary. US Department of the Interior, Washington, DC. http://www.mms.gov/glossary/d. Htm
10. US EPA (2011) Definition of a public water system. US Environmental Protection Agency, Washington, DC. http://water.epa.gov/infrastructu re/drinkingwater/pws/pwsdef2.cfm
11. Merriam-Webster's Dictionary (2011) Subsurface. http://wwwmerriam-webster.com/dictio nary/subsurface
12. Society of Petroleum Engineers (2011) SPE E&P Glossary. http://www.spe.org/glossary/wiki/doku.php/welcome#terms_of_use
13. US EPA (2011) Expocast. US Environmental Protection Agency, Washington, DC. http://www.epa.gov/ncct/expocast/
14. US EPA (2011) The HERO database. US Environmental Protection Agency, Washington, DC. http://hero.epa.gov/
15. Judson R, Richard A, Dix D, Houck K, Elloumi F, Martin M, Cathey T, Transue TR, Spencer R, Wolf M (2008) ACTOR - Aggregated computational toxicology resource. Toxicol Appl Pharmacol 233:7–13
16. Martin MT, Judson RS, Reif DM, Kavlock RJ, Dix DJ (2009) Profiling chemicals based on chronic toxicity results from the U.S. EPA ToxRef database. Environ Health Perspect 117 (3):392–399
17. U.S. Environmental Protection Agency (2011) The HERO database. http://actor.epa.gov/actor/faces/ToxCastDB/Home.jsp
18. IGSHPA (2009) Closed-loop/geothermal heat pump systems. International Ground Source Heat Pump Association, and Oklahoma State University, Stillwater, OK
19. Hiller C (2000) Grouting for vertical geothermal heat pump systems. Electric Power Research Institute, International Ground Source Heat Pump Association, and Oklahoma State University, Stillwater, OK
20. IGSHPA (1988) Closed-loop/ground-source heat pump systems: installation guide. International Ground Source Heat Pump Association, and Oklahoma State University, Stillwater, OK
21. IGSHPA (1989) Soil and rock classification for the design o ground-coupled heat pump systems: field manual. International Ground Source Heat Pump Association, and Oklahoma State University, Stillwater, OK

22. IGSHPA (2009) Ground source heat pump residential and light commercial design and installation guide. International Ground Source Heat Pump Association, and Oklahoma State University, Stillwater, OK
23. Wang LK (1974) Environmental engineering glossary. Calspan Corporation, Buffalo, NY, p 439
24. Wang MHS, Wang LK (2014) Glossary ad conversion factors for water resources engineers. In: Wang LK, Yang CT (eds) Modern water resources engineering. Humana Press, New York, pp 759–851
25. Wang MHS, Wang LK (2015) Environmental water engineering glossary. In: Yang CT, Wang LK (eds) Advances in water resources engineering. Springer Science + Business Media, New York, pp 471–556
26. US EPA (1998) Development document for proposed effluent limitations guidelines and standards for the landfills point source category. US Environmental Protection Agency, Washington, DC
27. US EPA (2011) Plan to study the potential impacts of hydraulic fracturing on drinking water resources. US Environmental Protection Agency, Washington, DC
28. CDC (2015) Glossary of Terms & Processes. State of California Department of Conservation. Sacramento, CA. www.conservation.ca.gov/index/pages/glossary-frk.aspx
29. Wang LK, Pereira NC (1980) Solid waste processing and resource recovery. Humana Press, Totowa, NJ, p 480
30. Wang LK, Shammas NK, Evanylo GK, Wang MHS (2014) Engineering management of agricultural land application for watershed protection. In: Wang LK, Yang CT (eds) Modern water resources engineering. Humana Press, New York, pp 571–642
31. Allaby MA (1977) Dictionary of the environment. Van Nostrand Reinhold Co., New York

Index

A

Anaerobic digestion
  biogas, AgSTAR, 19–21
  carbon price *vs.* operation size, 20, 21
  characteristics, 21, 22
  cost of, 23
  covered lagoon, 21–22
  digesters per head *vs.* number of head, 20
  fixed-film digester, 23
  IBR, 23
  ISPAD, 24
  plug-flow digester, 23
  short fatty acids, 19
  SMAD, 24
  temperature, moisture, and solid content, 19
  TPAD, 23–24
Animal waste management software tool
      (AWM), 63
Aquaculture-wetland system
  applications, 80
  aquatic vegetation, 78
  constructed wetlands, 78–80
  design, 81
  limitations, 80
  natural wetlands, 78
  performance, 81–82
Aquifers
  casing and well bore, 158
  casing inspection tests, 157
  confining strata, 158–160
  drinking water, 165
  failure, 139
  injection zone, 160–163
  protected portion, 163–165

Archaeological and natural analogues
      studies, 413

B

Biochemical oxygen demand (BOD$_5$), 11
Biocovers, 57

C

Calcining process, 354–355
Carbonation, 415
Carcinogenic effects, 243
Cardiovascular effects, 243
Cekmece Waste Processing and Storage
      Facility (CWPSF), 433
Cementitious materials
  features and limitation, 404, 405
  function, 404, 407
  hydration
    alite, 412
    ASTM specifications, 412
    controlling final hydrated cement
        properties, 411
    experimental assessment
    archaeological and natural analogues, 413
    barrier, 413, 414
    field/in-situ tests, 413
    standardized laboratory tests, 413
    hardened cement paste, 412
    kinetics, 410–411
    mechanisms, 409–410
    portland cements, 412
    Unhydrated and hydrated phases, 408

© Springer International Publishing Switzerland 2016
L.K. Wang, M.-H.S. Wang, Y.-T. Hung and N.K. Shammas (eds.),
Natural Resources and Control Processes, Handbook of Environmental
Engineering, Volume 17, DOI 10.1007/978-3-319-26800-2

Cementitious materials (*cont.*)
   International and National Assessment
      Programs, 423–424
   predictive assessment (*see* Predictive
      models)
   radioactive waste disposal (*see* Radioactive
      waste disposal)
Cement solidification, 353–354
Centralised Waste Management Facility
   (CWMF), 431
Chemical extraction, 353
Chemical oxygen demand (COD), 11
Chemical precipitation method, 431
Chemical stabilization, 351–353
Chisel-type knives, 42
Chlorides removal
   chloride behavior, 362
   chloride reduction characteristics, 358
   chloride speciation, 359–360
   in fly ash, 357
   treatment
      cement solidification, 353–354
      chemical extraction, 353
      chemical stabilization, 351–353
      melting, 351
      recycling, 355–357
      sintering/calcining process, 354–355
Consumer Product Safety Commission
   (CPSC), 273
Cooperative research program (CRP), 423
Cooperative research projects, 424
Corrosive wastes, 147
Cumulative fractional release (CFR), 427

D
Decision support system (DSSs), 63
Deep-injection well system
   aquifers
      casing and well bore, 158
      casing inspection tests, 157
      confining strata, 158–160
      drinking water, 165
      failure, 139
      injection zone, 160–163
      protected portion, 163–165
   backflushing, 166
   Belle Glade, FL, 170–172
   casing program, 126
   cement, 132
   confinement conditions, 133–134
   confining layer failure, 139–140
   economic evaluation, 142
   fiberglass casing, 141

   fluid movements, 138
   hazardous wastes management
      chemical treater, 153
      clarifier, 152
      clear-waste tank, 153
      construction, 153–154
      filter, 152
      geographic distribution, 151
      identification, 147–148
      injection pump, 153
      oil separator, 152
      radioactive contaminants, 154–156
      sources, amounts, and composition,
         148–149
      sump tank, 152
      USA, 146
   high back pressure, 132
   history, 122
   human error, 141
   injection plus monitoring, 128
   leaked fluids, 141
   monitoring wells, 166
   nondegradation, 133
   Pensacola, FL
      chemical process, 169–170
      injection facility, 167–168
      injection zone, 168
   receptor zone, 134–135
   regulations
      aquifers, 123
      mining wells, 124, 125
      oil and gas injection wells, 124
      radioactive injection wells, 125
      Resource Conservation and
         Recovery Act, 124
      SDWA, 123
      shallow injection wells, 125, 126
      UIC program, 126
      US EPA, 123
   self-inspection, 128
   subsurface hydrodynamics, 136–138
   surface equipment, 132
   underground injection, 121
   US yearly average cost index, 180
   wastewater management
      engineering feasibility, 143
      industrial wastewaters, 145, 146
      municipal wastewaters, 144–145
      treatment, 144
   Wilmington, NC
      chemical process, 174–176
      confining zone lithology, 174
      injection facility, 173, 174
Deep-well waste disposal, 119–180

Diffusion model, 419
Disperse dyes, 191
Dissolution model, 419
Dosing rate (DR), 299
Dye wastewater
    activated carbon, 213–215
    aromatics, 189
    batch process, 193, 194
    biological treatment
        aerobic treatment, 203–204
        configurations, 207
        fungal treatment, 207–209
        membrane bioreactor, 206
        microbial electrolysis cell, 209
        sequencing batch reactor, 206
    characteristics, 196–197
    coagulation-flocculation, 218–219
    definition, 190
    discharged dyes, 189
    ECF, 217–218
    history, 197–199
    legislation
        Clean Air Act, 200
        Clean Water Act, 200
        Emergency Planning and Community
            Right-to-Know Act, 201
        Pollution Prevention Act, 202
        RCRA, 201
    membrane filtration, 215–217
    oxidation treatment
        Fenton process, 210–212
        hydrogen peroxide/pyridine/copper(II),
            212
        WAO, 210
    ozone, 219–221
    wet processing, 193

E
Electrical conductivity (EC), 464
Electrocoagulation-electroflotation (ECF),
    217–218
Emergency Planning and Community
    Right-to-Know Act, 201
Energy resources engineering, 493
Evapotranspiration (ET) system
    applications, 84
    bed system, 83
    costs, 85–86
    design, 84–85
    limitations, 84
    performance, 85
    surface/groundwater contamination, 83
Environmental protection, 73

Experimental–Industrial Test Site (EITS), 391
Extraction procedure (EP) toxicity, 147

F
Field tests, 413
First-order reaction model, 418
Floating aquatic plant systems (FAP), 27
Fly ash, 349
Free-water surface (FWS) system, 25–27,
    79–80

G
Glossary, 64, 222, 342, 493–621
Ground-granulated blast furnace slag
    (GGBS), 432

H
Hard-to-treat wastes, 193
Health effects
    carcinogenic effects, 243
    cardiovascular effects, 243
    on children
        lead poisoning, 244–245
        in soil, 244
        in surface dust, 244
    federal public health agencies, 241
    hematologic and renal effects, 242
    neurotoxic effects, 241–242
    occupational lead exposure, 240–241
    reproductive and developmental effects, 242
    on women and infants
        gestational hypertension, 246
        growth and neurodevelopment, 247
        maternal hypertension, 246
        pregnancy outcomes, 246
        sexual maturation and fertility, 246
        signs and symptoms, 246
        workers, 247
Hematologic and renal effects, 242
Hydraulic process, 418

I
Immobilization techniques, 425
Incinerator, 349
Incremental fractional release (IFR), 427
Induced blanket reactor (IBR), 23
In-Storage Psychrophilic Anaerobic Digester
    (ISPAD), 24
Integrated Swine Manure Management
    (ISMM), 63

Intermediate-level waste (ILW), 434
International Agency for Research on
        Cancer (IARC), 240
International and National Assessment
        Programs, 423–424
International Atomic Energy Agency
        (IAEA), 376, 379
Ion exchanged water (IEW), 358
Irrigation system, 41

K
Kozloduy Nuclear Power Plant (NPP), 440

L
Land application, 37
Land resources engineering, 493
Land treatment system
    overland flow system
        application, 99
        costs, 101, 102
        crop production, 98, 99
        design, 99, 100
        flow diagram, 97
        limitations, 99
        performance, 100
    rapid rate system
        algae-laden wastewater, 87
        applications, 88
        costs, 90–91
        denitrification, 87
        design, 88, 89
        limitations, 88
        native soils and climate, 86
        nitrogen control benefits, 86
        performance, 89–90
        preapplication treatment, 87
        surface wastewater applications, 86
        underdrains, 86, 87
    slow-rate system
        adverse effects, 93
        applications, 93–94
        costs, 95–97
        crop management, 92
        design, 94
        flow diagram, 92
        limitations, 94
        performance, 95
        preapplication treatment, 92, 93
Leachability index, 428
Lead hazard abatement
    methods
        encapsulation, 251

enclosure, 250
        paint removal (*see* Paint removal
            methods)
        replacement, 250
    operational procedures
        building component replacement
            procedures, 259
        encapsulation procedures, 262–263
        enclosure installation procedures, 250
        lead paint removal procedures, 260–262
        process, 258–259
    practices, 263–265
Lead-Based Paint Renovation, Repair and
        Painting (RRP) Rule, 274
Livestock waste management
    anaerobic digestion
        biogas, AgSTAR, 19–21
        carbon price *vs.* operation size, 20, 21
        characteristics, 21, 22
        cost of, 23
        covered lagoon, 21–22
        digesters per head *vs.* number
            of head, 20
        fixed-film digester, 23
        IBR, 23
        ISPAD, 24
        plug-flow digester, 23
        short fatty acids, 19
        SMAD, 24
        temperature, moisture, and solid
            content, 19
        TPAD, 23–24
    composting
        bins, 35, 36
        dead livestock, 36
        factors, 35
        materials, 35
        values, 35
        windrows/piles, 35
    computer modeling, 63
    constructed wetlands
        design, 27–28
        FAP, 27
        FWS, 25–27
        processes, 24
        quality effluent, 24
        subsurface flow wetlands, 25
        surface flow wetlands, 25
        vegetated submerged system, 27
    decision-making process, 5
    federal regulations, 6–7
    feedlot runoff control system
        clean water diversion, 48
        discharge runoff control, 48–51

sources, 48
VFS, 50–52
greenhouse gas emissions
    agricultural carbon sequestration, 60
    crop and soil management, 60
    enteric fermentation, 59, 60
    manure management, 60
    scientific community, 61
    in United States, 59, 60
infiltration, 4
lagoons
    aerobic, 32
    algal blooms, 29
    anaerobic, 30–31
    cross-sectional area, 28, 29
    definition, 28
    design, 29
    effluent storage, 28
    facultative lagoons, 33
    sizing, 29
    sludge storage, 28, 30
    soil and groundwater study, 29
    treatment, 28
    weeds and grasses, 30
land application
    classification, 37
    effects, 38, 39
    gases, 38
    liquid manure, 40–42
    nutrient content, 38
    rate of, 42–43
    semisolid manure, 40
    solid manure, 39–40
    time of, 42
    treatment processes, 38, 39
milk house and diary wastewater
    ammonia nitrogen and chlorides, 15
    characteristics, 15
    equipment and pipeline cleaning, 16
    phosphorus, 15
    solids, 15
    sources, 14–15
    treatment methods, 14–16
    water conservation, 19
nitrogen and phosphorus processes,
    11, 12
odors
    ammonia emission, 54
    anaerobic degradation, 53
    animal nutrition, 55–56
    aromatic compounds, 53, 54
    causes, 53
    dust particles, 53
    hydrogen sulfate, 54

manure treatment and handling,
    56–57
mitigation, 54, 55
operation management, 58, 59
sulfate-reducing bacteria, 54
VFAs, 53
waste treatment method, 57–58
pathogens, 61–63
physical and chemical properties, 10
state regulations, 7, 8
swine lagoon analysis, 11–14
thermochemical processes
    biofuels, 33
    direct liquefaction, 34
    gasification, 34–35
    pyrolysis, 33–34
types, 13
vermicomposting, 37
waste storage
    design, 47
    estimated cost, 44
    factors, 43
    liquid manure, 45–46
    semisolid manure, 45
    solid manure, 44
    storage time, 43–44

M
Marine
    Arctic Ocean, 380–383
    Atlantic Ocean, 384
    chronological sequence, 376
    early assessment, 377
    history, 379
    operation, 375, 376
    Pacific Ocean, 383–384
Materials Characterization Centre
    (MCC), 427
Materials Testing Reactor (MTR), 432
Mechanical integrity testing, 154
Melting, 351
Municipal solid waste, 349

N
Nagra's Grimsel Test Site (GTS), 440
National Ambient Air Quality Standards
    (NAAQs), 6
National Pollutant Discharge Elimination
    System (NDPES), 7
National Toxicology Program (NTP), 243
Natural processes, 73
Neurotoxic effects, 241–242

New York State (NYS)
  Advisory Council
    design, 276
    powers and duties, 276–277
  child care/preschool enrollees, 278
  children and pregnant women, 277
  consumer products, 281
  enforcement agencies, 280
  lead poisoning conditions, 279–280
  lead poisoning prevention program,
    274–276
  leaded paint, 279
  reporting lead exposure levels, 278–279
Nitrification trickling filters (NTFs)
  vs. BOD removal, 309–310
  design procedure
    hydraulics, 312
    media selection, 312
    medium-density cross flow media, 313
    NH_4-N concentration, 312
    NTF influent $cBOD_5$ and TSS, 311
    oxygen availability, 311
    temperature, 312

O

Occupational Safety and Health
    Administration (OSHA), 240
Odors
  ammonia emission, 54
  anaerobic degradation, 53
  animal nutrition, 55–56
  aromatic compounds, 53, 54
  causes, 53
  dust particles, 53
  hydrogen sulfate, 54
  manure treatment and handling, 56–57
  mitigation, 54, 55
  operation management, 58, 59
  sulfate-reducing bacteria, 54
  VFAs, 53
  waste treatment method, 57–58
Operation and maintenance (O&M), 23
Ordinary portland cement (OPC), 353, 405
Ozone, 219–221

P

Paint removal methods
  heat, 254–255
  mechanical
    abrasive blasting, 254
    drawbacks, 252

    sanding, 253
    scraping, 252
  off-site chemical stripping, 257
  on-site chemical stripping, 257
Plastic trickling filters, 285
Predictive models
  assessment models, 422–423
  deformed cementitious barrier, 421
  degradation process
    carbonation, 415
    chloride attack, 417
    concrete protects steel
      reinforcement, 416
    empirical models, 415
    hydraulic process, 418
    mechanistic models, 416
    sulfate and magnesium ions, 415
  radionuclide leaching
    diffusion model, 419
    dissolution model, 419
    first-order reaction model, 418
  radionuclide transport
    diffusion mechanism, 419
    two-dimensional solute advective
      dispersive tarnsport, 420
    unsaturated cementitious barriers, 420
    velocities and dispersion coefficient
      component, 421

R

Radioactive waste disposal
  alkali-activated cements, 431
  aqueous durability, 432
  backfill/buffer, 439–441
  cementitious waste form, 434, 435
  characteristics, 394, 395
  chemical precipitation method, 431
  concrete performance, 437–439
  container materials, 436–437
  CWPSF, 433
  deep-well injection
    EITS, 391–393
    hydro-fracture grouting, 391, 392
    principle, 390
  Fick's law, 433
  GGBS/super-cement, 432
  global inventory, 372
  ILW, 434
  immobilization techniques, 425, 426
  influence leaching behavior,
    mechanisms, 428
  International and National Trends, 425

leachability indices, 428, 434
leaching and mechanical performance, 426
leaching test and mechanical strength, 434
long-term field tests, 430
marine
    Arctic Ocean, 380–383
    Atlantic Ocean, 384
    chronological sequence, 376
    early assessment, 377
    history, 379
    operation, 375, 376
    Pacific Ocean, 383–384
matrix materials and waste package,
    428, 429
MCC tests, 427
mechanical and radionuclide retention
    performance tests, 426
mine disposal, 393–394
mortar and concrete testing project, 432
multi-barrier, 395
near-surface
    early assessments, 384–385
    environmental impact, 385–389
nuclear industry, 374
OPC–ILW matrices, 434
OPC–ILW–polymer-immobilized
    matrices, 434
preexisting cavities, 393–394
radionuclides, 372
sample preparation, 434
sludge, kinds of, 431
sorption experiments, 433
standard leaching tests, 428
standard tests, 427, 428
UCS, 429
wet chemistry experiments, 433
XAFS measurements, 433
Radioactive Waste Management Funding and
    Research Centre (RWMC), 440
Radionuclide leaching
    diffusion model, 419
    dissolution model, 419
    first-order reaction model, 418
Reproductive and developmental effects, 242
Resource Conservation and Recovery Act
    (RCRA), 148, 201
Root zone method, 25
Roughing filter/activated sludge (RF/AS), 316

S
Safe Drinking Water Act (SDWA), 123
Self mixed anaerobic digester (SMAD), 24

Sequencing batch reactor (SBR), 206
Shihmen Reservoir
    automated TDR measurement system, 476
    back-end module facilitates services and
        database management, 477
    effective storage volume, 471
    feasibility studies, 474
    field embedded system and data center, 476
    hydraulic facility, 471
    hydrograph, 481
    inflow sediment hydrograph, 479
    large remediation projects, 472
    location, 471
    monitoring station, 476
    sediment discharge rating curve, 479
    sediment inflow and outflow, 480
    sediment transport and sluicing operation,
        482–487
    sediment transport monitoring scheme, 475
    storage loss, 471
    suspended-sediment rating curve, 479
    upstream boundary, 475
Sintering process, 354–355
Solidification process, 426
Stabilization process, 426
Subsurface flow system (SFS), 79–80
Subsurface infiltration
    anaerobic digestion, 103
    applications, 105
    buried tank, 103
    denitrification, 103
    design, 106
    distribution pipes, 103
    dosing tank, 104
    elevated sand mound
        crested site, 108–110
        sloping site, 107, 110–113
    limitations, 105
    mound system, 104, 105, 110, 111
    performance, 106–108
    precast concrete tanks, 103
    pretreatment filter, 105
    resting/conservative loadings, 103
    trench depth, 104
Sulfur dyes, 192
Suspended-sediment concentration (SSC)
    Shihmen Reservoir
        automated TDR measurement
            system, 476
        back-end module facilitates services
            and database management, 477
        effective storage volume, 471
        feasibility studies, 474

Suspended-sediment concentration (SSC)
        (*cont.*)
    field embedded system and data
        center, 476
    hydraulic facility, 471
    hydrograph, 481
    inflow sediment hydrograph, 479
    large remediation projects, 472
    location, 471
    monitoring station, 476
    sediment discharge rating curve, 479
    sediment inflow and outflow, 480
    sediment transport and sluicing
        operation, 482–487
    sediment transport monitoring
        scheme, 475
    storage loss, 471
    suspended-sediment rating curve, 479
    upstream boundary, 475
    surrogate techniques
        acoustic/ultrasonic, 459–461
        advantages, 454
        differential pressure approach, 461
        disadvantages, 454
        limitations, 454
        operating principles, 454
        optical turbidity, 454–459
    TDR
        experimental verification, 468–470
        measurement installation, 463–465
        probe design and data reduction,
            465–468
        technique, 462

T
Temperature-phased anaerobic digestion
        (TPAD), 23–24
Theis met, 137
Time-domain reflectometry (TDR), 462
    experimental verification, 468–470
    measurement installation, 463–465
    probe design and data reduction, 465–468
    technique, 462
Toxic lead
    complications, 238
    encapsulation, 239
    enclosure, 239
    health effects (*see* Health effects)
    lead hazard abatement (*see* Lead hazard
        abatement)
    lead hazard control planning
        historic preservation considerations, 270
        long-term/short-term response, 269–270

NYS (*see* New York State (NYS))
OSHA's lead
    in construction standard, 272
    in general industry standard, 272
    regulations, 271–272
paint removal, 239
replacement, 239
risk evaluation, 239
risk management
    assessment, 265
    evaluation, 266
    hazard screen, 266
    paint inspection, 267–269
symptoms, 237
US Federal Lead Standards
    lead content standards, 273–274
    RRP rule, 274
worldwide awareness and regulations,
        270–271
Transuranic (TRU) waste disposal design, 440
Trickling filter
    vs. activated sludge processes
        RF/AS, 316
        TF/SC, 314–316
    applications
        carbon-oxidizing filters, 296
        loading rates, 295–296
        nitrification filter, 296
        roughing filters, 296
        tertiary nitrification, 297
    construction
        containment structures, 328
        media integrity, 326–328
        media protection grating, 329
        subfloor and drainage, 328
        ventilation, 328–329
    containment structures, 293
    dome, 293
    evolution of, 289–291
    filter media, 292
    flow distribution system, 292
    industrial applications, 322–326
    NTFs (*see* Nitrification trickling filters
        (NTFs))
    nutrient removal
        coarse media denitrification filter,
            319–320
        phosphorus removal, 320
        process modifications, 317–318
    operation and maintenance considerations
        cold weather operation, 331–332
        distribution rate and flushing
            operation, 330
        hydraulic wetting rates, 330

    macrofauna, 331
    odor control, 332
    plastic media replacement, 332
    recirculation, 330
    treatment processes, 331
    uniform distribution, 330
    ventilation, 330–331
  plant upgrades, 320–322
  plastic
    Albertson and Okey's procedure 339–342
    clarification, 303
    DR, 299
    media selection, 304–306
    nitrification, 336–339
    pretreatment, 299
    recirculation, 299
    temperature and pH, 298
    using Germain formula, 333–335
    ventilation, 300, 302
    wastewater characteristics and
      treatability, 298
  process description, 317–318
  single and multistage
    Germain formula, 308
    treatment coefficient, 309
    Velz formula, 307–309
  underdrain system, 293
  ventilation, 294
Trickling filter/solids contact (TF/SC), 314–316

U
Ultrafiltration, 215
Unconfined compressive strength (UCS),
  353, 429

Underground Injection Control (UIC)
  regulations, 125, 126
Upflow anaerobic sludge blanket digester
  (UASB), 23
US Federal Lead Standards, 273–274

V
Vegetated infiltration basin (VIB), 51, 52
Vegetative filter strips (VFS), 50–52
Vegetative treatment area (VTA), 51, 52
Volatile fatty acids (VFAs), 53
Volatile solid (VS) loading rate, 31

W
Water hyacinth system
  applications, 75–76
  batch treatment, 75
  design, 76–77
  flow-through systems, 75
  limitations, 76
  performance, 77
  wastewater treatment, 75
Water resources protection, 1, 119
Wet air oxidation (WAO), 210
Wet chemistry experiments, 433
Wisconsin National Resource Conservation
  Service (NRCS), 15

X
X-ray absorption fine structure (XAFS)
  measurements, 433